Lecture Notes in Computer Science 14147

Founding Editors

Gerhard Goos
Juris Hartmanis

The series Lecture Notes in Computer Science (LNCS), including its subseries Lecture Notes in Artificial Intelligence (LNAI) and Lecture Notes in Bioinformatics (LNBI), has established itself as a medium for the publication of new developments in computer science and information technology research, teaching, and education.

LNCS enjoys close cooperation with the computer science R & D community, the series counts many renowned academics among its volume editors and paper authors, and collaborates with prestigious societies. Its mission is to serve this international community by providing an invaluable service, mainly focused on the publication of conference and workshop proceedings and postproceedings. LNCS commenced publication in 1973.

Christine Strauss · Toshiyuki Amagasa ·
Gabriele Kotsis · A Min Tjoa · Ismail Khalil
Editors

Database and Expert Systems Applications

34th International Conference, DEXA 2023
Penang, Malaysia, August 28–30, 2023
Proceedings, Part II

 Springer

Editors
Christine Strauss
University of Vienna
Vienna, Austria

Toshiyuki Amagasa
University of Tsukuba
Ibaraki, Japan

Gabriele Kotsis
Johannes Kepler University Linz
Linz, Austria

A Min Tjoa 🄳
Vienna University of Technology
Vienna, Austria

Ismail Khalil
Johannes Kepler University Linz
Linz, Austria

ISSN 0302-9743   ISSN 1611-3349 (electronic)
Lecture Notes in Computer Science
ISBN 978-3-031-39820-9   ISBN 978-3-031-39821-6 (eBook)
https://doi.org/10.1007/978-3-031-39821-6

This Springer imprint is published by the registered company Springer Nature Switzerland AG
The registered company address is: Gewerbestrasse 11, 6330 Cham, Switzerland

Preface

Welcome to the proceedings of the 34th International Conference on Database and Expert Systems Applications (DEXA 2023). This gathering of brilliant minds from around the world serves as a testament to the tremendous progress made in the fields of data management, intelligent systems, and advanced algorithms. It is our great pleasure to present this compilation of papers, capturing the essence of groundbreaking research and innovative ideas that were shared during the conference.

The rapid advancements in technology have ushered in an era where data has become an invaluable asset, and its effective management and analysis have become critical for organizations across various domains. The integration of expert systems, artificial intelligence, and machine learning techniques has revolutionized the way we approach data, enabling us to extract insights, make informed decisions, and create intelligent systems that can adapt and learn from the vast amounts of information available.

This conference has provided a platform for researchers, academics, industry experts, and practitioners to come together and exchange their knowledge, experiences, and ideas. The proceedings reflect the diverse range of topics covered during the event, including but not limited to data modeling, database design, query optimization, knowledge representation, rule-based systems, natural language processing, deep learning, and neural networks.

The papers included in this volume represent the collective efforts of the authors who have dedicated their time and expertise to advancing the frontiers of database systems, expert systems, artificial intelligence, and machine learning. Each paper has undergone a rigorous review process by a panel of experts in the respective fields, ensuring the highest standards of quality and relevance.

As you delve into the pages of these proceedings, you will witness the fascinating discoveries, novel methodologies, and practical applications that are shaping the future of data-driven decision making. From the development of efficient algorithms for data processing to the creation of intelligent systems capable of autonomous reasoning and decision making, the papers within this volume illuminate the vast potential of interdisciplinary research.

We are proud to report that authors from more than 37 different countries submitted papers to DEXA this year. Our program committees have conducted close to five hundred single-blind reviews, with each submission receiving three reviews, on average. From 155 submitted papers the program committee decided to accept 49 full papers, 35 short papers with an acceptance rate of 31%, a rate lower than previous DEXA conferences.

We would like to express our heartfelt gratitude to the authors for their contributions and the dedication they have shown in presenting their work. We also extend our sincere appreciation to the members of the program committee, whose rigorous evaluation and insightful feedback have played a crucial role in shaping this collection.

Finally, we would like to thank the conference organizers, keynote speakers, and attendees for their invaluable support in making this event a resounding success. Their commitment to advancing the frontiers of knowledge and fostering collaboration has been instrumental in creating an environment conducive to intellectual growth and innovation.

It is our hope that these proceedings serve as a source of inspiration and knowledge for researchers, students, and professionals alike, as they embark on their own journeys to unravel the mysteries of data, expert systems, artificial intelligence, and machine learning.

August 2023

Christine Strauss
Toshiyuki Amagasa
Gabriele Kotsis
A Min Tjoa
Ismail Khalil

Organization

Program Committee Chairs

Christine Strauss   University of Vienna, Austria
Toshiyuki Amagasa   University of Tsukuba, Japan

Steering Committee

Gabriele Kotsis   Johannes Kepler University Linz, Austria
A Min Tjoa   Vienna University of Technology, Austria
Robert Wille   Software Competence Center Hagenberg, Austria
Bernhard Moser   Software Competence Center Hagenberg, Austria
Ismail Khalil   Johannes Kepler University Linz, Austria

Program Committee Members

Seth Adjei   Northern Kentucky University, USA
Riccardo Albertoni   CNR-IMATI, Italy
Sabri Allani   University of Tunis El Manar, Tunisia
Idir Amine Amarouche   USTHB, Algeria
Nidhi Arora   Amazon, India
Mustafa Atay   Winston-Salem State University, USA
Radim Bača   VSB - Technical University of Ostrava, Czech Republic
Ladjel Bellatreche   LIAS/ENSMA, France
Nadia Bennani   INSA de Lyon, France
Djamal Benslimane   Lyon 1 University, France
Karim Benouaret   Université Claude Bernard Lyon 1, France
Vasudha Bhatnagar   University of Delhi, India
Athman Bouguettaya   University of Sydney, Australia
Omar Boussaid   Université Lumière Lyon 2, France
Stephane Bressan   National University of Singapore, Singapore
Barbara Catania   University of Genoa, Italy
Brice Chardin   ENSMA, France
Asma Cherif   King Abdulaziz University, Saudi Arabia
Ruzanna Chitchyan   University of Bristol, UK
Soon Chun   City University of New York, USA

Camelia Constantin	Sorbonne University, France
Deborah Dahl	Conversational Technologies, USA
Matthew Damigos	Ionian University, Greece
Jérôme Darmont	Université Lyon 2, France
Soumyava Das	Teradata Labs, USA
José Gaviria de la Puerta	Universidad de Deusto, Spain
Vincenzo Deufemia	University of Salerno, Italy
Sabrina De Capitani Di Vimercati	Università degli Studi di Milano, Italy
Ivanna Dronyuk	Lviv Polytechnic National University, Ukraine
Cedric du Mouza	CNAM, France
Joyce El Haddad	Université Paris Dauphine - PSL, France
Markus Endres	University of Applied Sciences Munich, Germany
Nora Faci	Université Lyon 1, France
Bettina Fazzinga	University of Calabria, Italy
Jan Fell	Vienna University of Economics and Business, Austria
Flavio Ferrarotti	Software Competence Centre Hagenberg, Austria
Lukas Fischer	Software Competence Center Hagenberg, Austria
Flavius Frasincar	Erasmus University Rotterdam, The Netherlands
Bernhard Freudenthaler	Software Competence Center Hagenberg, Austria
Bouchra Frikh	Sidi Mohamed Ben Abdellah University, Morocco
Steven Furnell	University of Nottingham, UK
Pablo García Bringas	University of Deusto, Spain
Zoltan Geller	University of Novi Sad, Serbia
Manolis Gergatsoulis	Ionian University, Greece
Joseph Giovanelli	University of Bologna, Italy
Anna Gorawska	Silesian University of Technology, Poland
Sven Groppe	University of Lübeck, Germany
Wilfried Grossmann	University of Vienna, Austria
Giovanna Guerrini	University of Genoa, Italy
Allel Hadjali	ENSMA, France
Sana Hamdi	University of Tunis El Manar, Tunisia
Abdelkader Hameurlain	Paul Sabatier University, France
Hieu Hanh Le	Ochanomizu University, Japan
Sven Hartmann	Clausthal University of Technology, Germany
Manfred Hauswirth	TU Berlin, Germany
Ionut Iacob	Georgia Southern University, USA
Hamidah Ibrahim	Universiti Putra Malaysia, Malaysia
Sergio Ilarri	University of Zaragoza, Spain
Abdessamad Imine	Loria, France
Ivan Izonin	Lviv Polytechnic National University, Ukraine
Stéphane Jean	ISAE-ENSMA and University of Poitiers, France

Peiquan Jin   University of Science and Technology of China,
                                       China
Eleftherios Kalogeros   Ionian University, Greece
Anne Kayem   Hasso Plattner Institute, Germany
Carsten Kleiner   Hochschule Hannover - University of Applied
                                       Science and Arts, Germany
Michal Kratky   VSB-Technical University of Ostrava, Czech
                                       Republic
Petr Křemen   Czech Technical University in Prague and
                                       Cognizone, Czech Republic
Josef Küng   Johannes Kepler University Linz, Austria
Lynda Said Lhadj   Ecole nationale Supérieure d'Informatique,
                                       Algeria
Lenka Lhotska   Czech Technical University in Prague, Czech
                                       Republic
Jorge Lloret   University of Zaragoza, Spain
Qiang Ma   Kyoto University, Japan
Hui Ma   Victoria University of Wellington, New Zealand
Elio Masciari   Federico II University, Italy
Massimo Mecella   Sapienza Università di Roma, Italy
Sajib Mistry   Curtin University, Australia
Jun Miyazaki   Tokyo Institute of Technology, Japan
Lars Moench   University of Hagen, Germany
Riad Mokadem   Paul Sabatier University, France
Yang-Sae Moon   Kangwon National University, South Korea
Franck Morvan   IRIT and Université Paul Sabatier, France
Amira Mouakher   University of Perpignan, France
Philippe Mulhem   LIG-CNRS, France
Emir Muñoz   Genesys Telecommunications, Ireland
Francesc D. Muñoz-Escoí   Universitat Politècnica de València, Spain
Ismael Navas-Delgado   University of Málaga, Spain
Javier Nieves   Azterlan, Spain
Makoto Onizuka   Osaka University, Japan
Brahim Ouhbi   ENSAM, Morocco
Marcin Paprzycki   Systems Research Institute, Polish Academy of
                                       Sciences, Poland
Chihyun Park   Kangwon National University, South Korea
Louise Parkin   LIAS, France
Dhaval Patel   IBM, USA
Nikolai Podlesny   Hasso Plattner Institute and University of
                                       Potsdam, Germany
Simone Raponi   NATO STO CMRE, Italy
Tarmo Robal   Tallinn University of Technology, Estonia

External Reviewers

Gaetano Cimino	University of Salerno, Italy
Kaushik Das Sharma	India
Chaitali Diwan	International Institute of Information Technology, Bangalore, India
Daniel Dorfmeister	Software Competence Center Hagenberg GmbH, Austria
Myeong-Seon Gil	Kangwon National University, South Korea
Ramón Hermoso	University of Zaragoza, Spain
A. K. M. Tauhidul Islam	Informatica, USA
Andrii Kashliev	Eastern Michigan University, USA
Khalid Kattan	University of Michigan - Dearborn, USA
Yuntao Kong	JAIST, Japan
Mohit Kumar	Software Competence Center Hagenberg GmbH, Austria
Yudi Li	Southwest Jiaotong University, China
Junjie Liu	Southwest Jiaotong University, China
Junhao Luo	Southwest Jiaotong University, China
Jorge Martinez-Gil	Software Competence Center Hagenberg GmbH, Austria
Moulay Driss Mechaoui	University of Abdelhamid-Ibn-Badis (UMAB), Algeria
Sankita Patel	Sardar Vallabhbhai National Institute of Technology, India
Maxime Prieur	CNAM & Airbus Defense and Space, France
Gang Qian	University of Central Oklahoma, USA
María del Carmen Rodríguez-Hernández	Technological Institute of Aragon, Spain
Hannes Sochor	Software Competence Center Hagenberg GmbH, Austria
Tung Son Tran	University of Applied Sciences Munich, Germany
Óscar Urra	Technological Institute of Aragon, Spain
Alexander Völz	University of Vienna, Austria
Haihan Wang	Southwest Jiaotong University, China
Yi-Hung Wu	Chung Yuan Christian University, Taiwan (R.O.C.)
Qin Yang	Southwest Jiaotong University, China
Chengyang Ye	Kyoto University, Japan
Kun Yi	Kyoto University, Japan
Chih-Chang Yu	Chung Yuan Christian University, Taiwan (R.O.C.)
Foutse Yuehgoh	Pôle Universitaire Léonard de Vinci, France
Shilong Zhu	Southwest Jiaotong University, China

Organizers

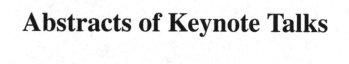

Abstracts of Keynote Talks

Physics-Informed Machine Learning

Stéphane Bressan

National University of Singapore, Singapore

Abstract. In 1687, Isaac Newton published his groundbreaking work, "Philosophiæ Naturalis Principia Mathematica". Newton's remarkable discoveries unveiled the laws of motion and the law of universal gravitation, propelling humanity's understanding of the physical world to new heights. In a letter to Robert Hooke in 1675, in response to an invitation to collaborate, Newton humbly remarked, "If I have seen further, it is by standing on the shoulders of giants." This metaphor swiftly became a powerful symbol of intellectual and scientific progress, signifying the idea that knowledge is built upon foundations laid by brilliant minds that came before us.

Fast-forwarding to the present, we find ourselves amidst a triumphant statistical machine learning revolution. In 2016, Google's AlphaGo, a deep reinforcement learning algorithm, astounded the world by outperforming a professional Go player. The following year, CheXNet, a deep convolutional neural network developed at Stanford University, surpassed radiologists in accurately detecting pneumonia from chest X-ray images. And in 2020, AlphaFold, a neural network model created by DeepMind, revolutionised protein structure prediction, surpassing other existing methods.

These advancements stand on the shoulders of giants. They owe their existence to the work of logicians, mathematicians, physicists, neurobiologists, computer scientists, and cyberneticists who have paved the way for the birth of modern machine learning models and algorithms. They also owe their existence to the work of material, electrical, electronics and other engineers, whose ingenuity has birthed the computer hardware and technology enabling such performance.

However, the remarkable ascent of machine learning is not solely reliant on these contributions. It thrives on the vast amounts of data permeating the global information infrastructure, enabling the construction of accurate representations of the world. What about knowledge?

In this context, we propose exploring and discussing how machine learning can both leverage and contribute to scientific knowledge. We explore how the training of a machine learning model can be informed by the fundamental principles of the very systems it seeks to comprehend and how it can create symbolic scientific knowledge. We explore applications in classical mechanics, fluid mechanics, quantum many-body systems, macroeconomics, chemistry, and astronomy. Along this journey, we cross the paths of such great minds as William Rowan Hamilton, Ernst Ising, Richard Feynman, and Johannes Kepler.

Data Integration Revitalized: from Data Warehouse through Data Lake to Data Mesh

Robert Wrembel

Faculty of Computing and Telecommunications,
Poznan University of Technology Poland

Abstract. For years, data integration (DI) architectures evolved from those supporting virtual integration (mediated, federated), through physical integration (data warehouse), to those supporting both virtual and physical integration (data lake, lakehouse, polystore, data mesh/fabric). Regardless of its type, all of the developed DI architectures include an integration layer. This layer is implemented by a sophisticated software, which runs the so-called DI processes. The integration layer is responsible for ingesting data from various sources (typically heterogeneous and distributed) and for homogenizing data into formats suitable for future processing and analysis. Nowadays, in all business domains, large volumes of highly heterogeneous data are produced, e.g., medical systems, smart cities, precision/smart agriculture, which require further advancements in the data integration technologies. In this paper, I present my subjective view on still-to-be developed data integration techniques, namely: (1) novel agile/flexible integration techniques, (2) cost-based and ML-based execution optimization of DI processes, and (3) quality assurance techniques in complex multi-modal data systems.

Contents – Part II

Natural Language Processing

Deep Learning

Neural Networks

Contents – Part I

Database Design

Knowledge Representation

Rule-Based Systems

RD-Classifier: Reduced Dimensionality Classifier for Alzheimer's Diagnosis Support System

Soualihou Ngnamsie Njimbouom[1], Gelany Aly Abdelkader[1], Candra Zonyfar[1], Hyun Lee[1,2], and Jeong-Dong Kim[1,2,3(✉)]

[1] Department of Computer Science and Electronic Engineering, Sun Moon University, Asan 31460, Republic of Korea
{salehrico,gelany740,candrazonyfar,mahyun91,kjd4u}@sunmoon.ac.kr
[2] Division of Computer Science and Engineering, Sun Moon University, Asan 31460, Republic of Korea
[3] Genome-Based BioIT Convergence Institute, Sun Moon University, Asan 31460, Republic of Korea

Abstract. Neurodegenerative diseases are a group of conditions resulting from the progressive degeneration of neurons and their functions in the nervous system. Alzheimer's disease, currently incurable, is one of the most common neurodegenerative diseases. Early detection is crucial for delaying disease progression. Current methods for early detection of Alzheimer's rely on handwriting tasks through which a considerable amount of data is generated. The resulting data is characterized by high dimensionality, obscuring the importance of relevant attributes. The various methods used in detecting or classifying Alzheimer's disease could be impaired by the Curse of dimensionality (sparsity present in data due to the large number of features given the lower number of data samples, hindering model performance in identifying a pattern in the data). In this work, we propose a method for enhancing the performance of early Alzheimer's detection by identifying essential features while avoiding the Curse of dimensionality. Our proposed method outperformed the current state-of-the-art benchmarks and demonstrated the effectiveness of our approach. The reported findings suggest that our method can significantly improve the performance of early detection of Alzheimer's disease and provide a foundation for early interventions and better patient outcomes.

Keywords: RD-Classifier · Alzheimer's diagnosis · Curse of dimensionality · Decision Support System

1 Introduction

Alzheimer's Disease (AD) is a progressive neurodegenerative illness that results in the loss of cognitive function due to the degradation of synapses and neurons in the cerebral cortex, as defined by the National Institute of Neurological Disorders and Stroke (NINDS) [1]. The global prevalence of AD and related dementias is estimated to be around 50 million people, with projections indicating a tripling of this number to approximately 150 million individuals by 2050 [2].

Despite the significant strides in medical research, an absolute cure for AD remains elusive. Present treatment can only slow the disease progression [2]. Alarmingly, despite the warning symptoms such as memory loss, language difficulties, and personality changes, the diagnosis of Alzheimer's remains challenging, with a failure rate of over 20% [3]. This high failure rate poses substantial obstacles to developing effective therapies, emphasizing the need for early detection [4].

Detection at an early stage offers the best chance to intervene therapeutically, with estimations suggesting that it can occur up to 9 years prior to the manifestation of diagnosable dementia symptoms [5]. AD's known influence on motor and cognitive functions, which correlate closely with handwriting movements, has inspired innovative diagnostic methodologies [6, 7]. Researchers have explored the dynamics of handwriting and drawing tasks, generating a wide array of features that have been employed in AD and Parkingson's disease diagnosis using machine learning (ML) and deep learning (DL) techniques [8, 9].

However, the high dimensionality of the dataset generated in previous studies, characterized by an excessive number of features, can hamper the performance of ML and DL methods. [9–11]. This high dimensionality could lead to various problems referred to as the "curse of dimensionality", which has yet to be comprehensively addressed in previous research.

In this context, the present study proposes a novel method that reduces dimensionality and selectively retains the most significant features. Our approach is designed to enhance the learning of feature maps, mitigating overfitting, and decreasing the computational complexity of the ML models. By generating a concise feature subset, our proposed approach seeks to improve the performance of ML models for AD detection, enabling neurologists to focus on specific features during patients' diagnosis. This approach can potentially facilitate earlier detection and treatment, ultimately improving patient outcomes.

The remainder of the research paper is structured as follows: The related work section provides a comprehensive literature review of relevant studies. In the proposed model section, we describe our proposed approach, which includes the data acquisition, the proposed RD-Classifier algorithm, and evaluation metrics. In the Experimentation section, we provide a detailed account of the experimental environment, including the hardware and software used in our study. We also provide statistical information about our dataset's composition. Furthermore, we present the results of our experiments using various metrics and simultaneously discuss the results. Finally, in the Conclusion section, we summarize our findings and discuss their implications for AD diagnosis using ML. We also acknowledge the limitations of our study and suggest potential future research directions for the scientific community.

2 Related Work

In the past few years, numerous studies have investigated early diagnosis of neurode-generative diseases. Awasthi et al. [13] proposed a novel approach to AD diagnosis that combines MRI, immunoassay, and biomarker analysis through a multimodal approach to enhance treatment development. Diogo V.S et al. [14] developed a multi-diagnostic app-roach to diagnosing mild cognitive impairment and AD, exploring the impact of graph technology on classification performance and interpretability of predicted results. Sev-eral studies [15–20] have also explored early-stage AD identification through handwrit-ing tasks, utilizing machine learning methods due to their cost-effectiveness, time effi-ciency, and simplicity. These studies suggest that handwriting tasks can serve as a marker for neurodegenerative diseases due to their reliance on visual-spatial abilities and eye-hand coordination, which can be compromised in individuals with neurodegenerative diseases [21–23].

Diagnosis of AD using handwriting tasks is an area of active research, where various datasets with different sets of features and dimensionality are commonly used. ML models can leverage these features to detect AD more effectively. However, in ML, the size of the dataset used to train the model is an important factor that affects the performance of the model [24–26]. Although a larger dataset leads to better performance, an increase in the number of dimensions can cause the data to become noisy, which negatively impacts the model's performance. This is known as the curse of dimensionality [12, 27], which can lead to overfitting [28, 29]. To address this issue, it is essential to find an even distribution of data points across the dataset's dimensions. An appropriate feature map can be applied to the data to achieve this, and then ML models can be experimented with to perform classification on less noisy data. This approach has not been considered by previous studies on Alzheimer's detection [2].

3 Proposed Model

This section presents our proposed approach for detecting Alzheimer's disease (AD) in patients using handwriting task-driven features. Figure 1 outlines three key stages of the approach: the Data Preparation stage, RD-Classifier stage, and Evaluation stage. In the Data Preparation stage, relevant data is collected, and valuable features are extracted. The RD-Classifier stage employs dimensionality reduction methods such as PCA, Forward-Feature-Selection/Backward-Feature-Selection, and several ML models, including K-Nearest Neighbors (K-NN), Random Forest (RF), Decision Tree (DT), Logistic Regres-sion (LR), Gaussian Naive Bayes (GNB), and Support Vector Machine (SVM), to develop an effective classification model. Finally, the Evaluation stage involves assessing the performance of the proposed model in accurately identifying AD.

3.1 Data Preparation

Our study utilized a novel dataset from the benchmarking study [2], which was generated using an acquisition protocol detailed in [30]. The Data Module encompassed both data collection and feature extraction. During the data collection stage, participants completed

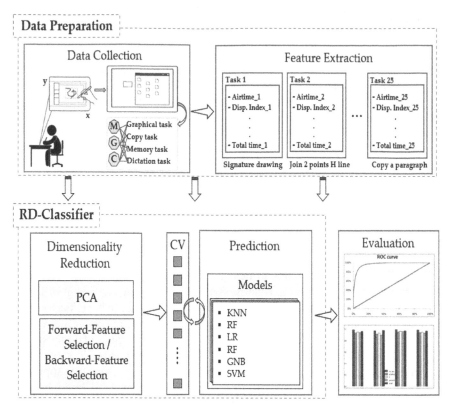

Fig. 1. Overall framework of the proposed RD-Classifier. M (Memory and dictation tasks), G (Graphic tasks), and C (Copy task).

25 handwriting tasks using a Wacom's Bambo tablet equipped with a pen and an A4 size paper connected to a Personal Computer.

The resulting data were processed to acquire the necessary features. The feature extraction stage utilized Wacom's Bambo tablet coordinates, the pressure exerted on the tablet, the timestamp of the task performed, and the break time while the pen was removed from the device to extract a total of 18 features for each task. For a more comprehensive understanding of the data collection and feature extraction stages, please refer to [2].

Our study benefited from this unique dataset and the careful execution of the acquisition protocol, which enabled the precise collection and extraction of relevant features.

3.2 RD-Classifier

This section presents our proposed RD-Classifier (Reduced Dimensionality Classifier), which comprises three primary components: Dimensionality Reduction, Training through Cross Validation, and Prediction Models, as illustrated in the RD-Classifier Module (Fig. 1).

The dimensionality reduction step involves the application of three methods, namely Principal Component Analysis (PCA), forward feature selection, and backward feature selection, to sequentially select the features that maximize the accuracy of the models in detecting the disease. PCA is a widely used data reduction technique that transforms high-dimensional data into a set of orthogonal variables called principal components (PCs). PCs are linear combinations of original variables representing different data space directions. The first PC captures the most variance, and each subsequent one captures the remaining variance in order of importance. This enables a more efficient representation of data with fewer variables, identifying important features or patterns [31]. Forward and backward feature selections are popular techniques in ML for dimensionality reduction. Forward selection starts with an empty feature set and adds one feature at a time based on performance. In contrast, backward selection begins with all features and eliminates the least important ones iteratively. Both methods aim to identify the most relevant features, improving model performance and reducing computational complexity [32]. We opted for the forward feature selection technique to reduce the dimensionality and constructed the appropriate dataset with a specified number of features for further steps.

In the second step of the RD-classifier, we performed cross-validation to ensure our model generalizes well to new, unseen data and avoids overfitting or underfitting. This involved dividing our available data into several subsets, training the model on the training set with different combinations of hyperparameters, and evaluating its performance on a held-out validation set. By doing so, we obtained a more accurate estimate of the model's performance and selected the best hyperparameters for our problem.

The reduced set of features was used in the third step of RD-Classifier to experiment with different classifiers (K-NN, RF, DT, LR, GBN, and SVM) and evaluate their performance using various metrics such as accuracy, precision, f1-score, and recall. The proposed RD-Classifier is presented with pseudocode in Algorithm 1.

Algorithm 1: *RD-Classifier*

Input

- Input data: $f_i = (x_i, y_i)$ where $f \in F$, $t \subseteq T$ and $i = (0,1,2,3,\ldots,n)$. f is the subset of features in task t. F and T the total number features and list of tasks respectively.
- Parameter: $k_{neighbors}$ and $k_{features}$ (the k neighbors of the K-NN model and feature numbers to be selected, respectively)

Output: Most performing classifier C_{max} and the subset $S_d \in F$ of the most important features.

1. **Begin algorithm:**
2. **foreach** i^{th} task $t \subseteq T$ **do:**
3.   **foreach** f_i in F **do:**
4. $x_\varphi := argMax(KNN, k_{neighbors}, param(S_d + f_i))$
5. $S_d.append(x_\varphi)$
6.   **if** $len(S_d) == k_{features}$:
7.   **break**
8.   **endfor**
9. **endfor**
10. **Perform** *Alzheimer_detection*
11.   Classifiers = [RF, K-NN, DT, SVM, GBN, LR]
12.   **foreach** clf in Classifiers **do:**
13.   clf = **GridSearchCV**$(param\_grid, cv\_iteration)$
14. $clf_{metrics} = eval(clf, S_d, metrics)$
15.   **endfor**
16. $C_{max} = Max\_metrics(clf_{metrics})$
17. **Return** C_{max}, S_d
18. **END algorithm**

3.3 Evaluation

To evaluate the performance of the proposed RD-Classifier in predicting AD, we employed a range of commonly used performance metrics, including accuracy, precision, f1-score, and recall. We used k-fold cross-validation (k = 10) to present the model generalization (see Sect. 3) on the data while fine-tuning. Additionally, we compared the AUC-ROC to demonstrate how different methods performed in correctly classifying patients with AD. The following formulas represent the measurement indicators used to assess the different methods.

$$\text{Accuracy (Acc.)} = \frac{TP + TN}{TP + FP + TN + FN} \times 100 \qquad (1)$$

$$\text{Precision (Pre.)} = \frac{TP}{TP + FP} \times 100 \qquad (2)$$

$$\text{Recall (sensitivity)} = \frac{TP}{TP + FN} \times 100 \qquad (3)$$

$$\text{F1 - score} = \frac{2TP}{2TP + FP + FN} \qquad (4)$$

$$\text{AUC - ROC} = \frac{1 + TPR - FPR}{2} \tag{5}$$

where TP, FP, TN, FN, TPR, and PR describe true positive, false positive, true negative, false negative, true positive rate, and false positive rate.

4 Experimentation

This section outlines our experimental setup, dataset composition, results, and discussion. We first report the attended outcomes before comparing our findings to those in the relevant literature review. By providing a detailed account of our methodology, experimental setup, and results, we ensure transparency and reproducibility of our study, allowing others to reproduce our results and further advance the field of research.

4.1 Experimental Environment Setup

In this study, we conducted all experiments using Jupyter notebook with TensorFlow 2.2.0 and Python 3.7.13. Our experiments were run on an Intel Core i7-6850k PC with a 3.60 GHz CPU, GeForce GTX 1080 Ti GPU, and 32GB RAM. These specifications are summarized in the following Table 1.

Table 1. Environmental setup of the development environment

Components	Details
CPU	Core i7-6850k CPU @ 3.60GZ x 12
GPU	GeForce GTX 1080 Ti
RAM	32GB
Web-based tools	Jupyter notebook IPython 7.34.0
TensorFlow	2.2.0
Python	3.7.13

4.2 Dataset

In this study, we used the DARWIN (Diagnosis AlzheimeR With handwriting) dataset provided by [2], which consisted of 174 samples from patients assessed for AD. The dataset was divided into two classes: label "0" for healthy patients and label "1" for patients with AD, with 85 and 89 samples, respectively. A total of 452 features were extracted from the 25 handwriting tasks, categorized into memory and dictation, graphic, and copy tasks (see Table 2). Because the dataset was already balanced between the two classes, no additional balancing mechanism was required. This dataset provides a valuable resource for developing accurate and reliable algorithms for diagnosing AD.

Table 2. Statistics of the DARWIN dataset [2].

# Patients	# Tasks	Type and number of tasks	# Feature
85 (Healthy)	25	5 (Memory & dictation task)	452 (18 features from each task + label and ID)
		6 (Graphic task)	
89 (Unhealthy)			
		14 (Copy task)	

4.3 Results and Discussion

We evaluated the proposed RD-Classifier after experimenting with various ML models and dimensionality reduction techniques, namely PCA and FFS. Our results in Table 3 indicate that the RD-Classifier with RF and FFS achieved the best performance, with an accuracy score of 87.75%, precision of 90.03%, recall of 86.58%, and f1-score of 87.30%. Conversely, when employing PCA, LR offered the highest efficiency of 79.89%, 78.07%, 80.50%, and 78.72% for accuracy, precision, recall, and f1-score, respectively.

Table 3. Mean accuracy of the RD-Classifier for the different ML models.

Model	PCA				FSS			
	Acc.	f1-score	Pre.	recall	Acc.	f1-score	Pre.	recall
RF	79.84 (±6.69)	77.45	85.92	72.41	87.75 (±5.93)	87.30	90.03	86.58
K-NN	75.71 (±8.50)	73.42	75.69	73.42	79.23 (±7.90)	79.81	73.07	88.75
DT	64.07 (±9.12)	64.29	67.51	64.67	79.18 (±7.93)	79.81	79.01	83.67
SVM	78.41 (±7.50)	77.91	77.11	81.00	81.92 (±4.84)	81.51	82.02	83.75
GNB	51.70 (±7.89)	21.86	49.33	15.33	51.70 (±6.10)	18.91	48.33	12.00
LR	79.89 (±8.19)	78.72	78.07	80.50	84.84 (±8.48)	84.25	83.42	87.08

Our experiments showed that using FFS in dimensionality reduction significantly improved the model performance, as demonstrated in Fig. 2. The proposed RD-Classifier (RF + FFS) showed the highest ability to correctly classify AD when the comparison was performed using the AUC-ROC scores.

The superior performance of the RD-Classifier demonstrated on the AD dataset used in this study, compared to FFS or PCA methods paired with KNN, DT, LR, or SVM, can be primarily attributed to the unique characteristics offered the combination of components of our proposed classifier (RF and FFS). Notably, the salient attributes of the RD-Classifier comprise Feature importance, as both RF and FFS are equipped with the ability to evaluate the significance of each attribute within the dataset in relation to the target variable; And robustness, primarily attributed to the ensemble nature of the RF, which is composed of multiple DTs, hence making it less susceptible to overfitting. In addition, FFS further enhances the RF's robustness by selecting the most significant features to reduce the dataset's dimensionality, as opposed to PCA, which transforms the original dataset into new components with a different spatial distribution than the original data.

We compared our proposed approach to the benchmarking method from a previous study [2]. As shown in Table 4, the RD-Classifier reduced computational complexity, as evidenced by reduced computing times. We also conducted hyperparameter optimization via cross-validation grid search, and Table 5 presents the performance comparison results. In most cases, the mean metric values obtained using RD-Classifier outperformed or were similar to the ML models used in the benchmarking approach.

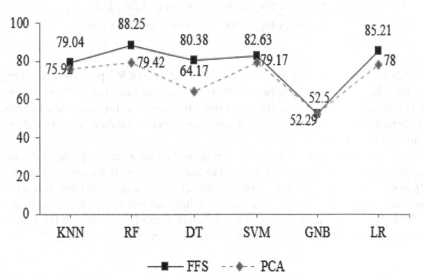

Fig. 2. AUC-ROC comparison results of the proposed approach based on the dimensionality reduction techniques.

Table 4. Comparison of computational time (in second).

	RD-Classifier		[2]	
	Training time	Test time	Training time	Test Time
RF	0.0433	0.0214	0.3601	0.0288
K-NN	0.0004	0.034	0.0035	0.0103
DT	0.0087	0.0001	0.0192	0.0037
SVM	0.156	0.0004	0.0122	0.0047
GBN	0.0012	0.0005	0.0048	0.0044
LR	0.0055	0.0001	0.0219	0.0043

Regarding disease classification for each task performed, our proposed approach demonstrated the highest efficiency with the highest accuracy and recall (specificity) for

Table 5. Mean metrics comparison of proposed RD-Classifier and benchmarking model [2].

	RD-Classifier		[2]	
	Acc.	Sensitivity	Acc.	Sensitivity
RF	87.75 (±5.93)	86.58	85.29 (±6.03)	88.06
K-NN	79.23 (±7.90)	88.75	77.29 (±7.15)	68.00
DT	79.18 (±7.93)	83.67	78.57 (±7.21)	82.50
SVM	81.92 (±4.84)	83.75	81.86 (±4.57)	80.56
GNB	51.70 (±6.10)	12.00	85.14 (±5.53)	83.61
LR	84.84 (±8.48)	87.08	83.86 (±4.57)	83.00

the K-NN algorithm on 24 of the 25 tasks and, in the case of RF, performed similarly or better than the benchmarking approach in some tasks, as shown in Table 6. Moreover, the RD-Classifier, through dimensionality reduction and selecting the essential features from the best task, showed better results when compared to the method proposed by [2], as demonstrated in Table 5.

Our proposed RD-Classifier algorithm achieved superior accuracy when performing dimensionality reduction on the entire set of features compared to the approach presented in [2]. Specifically, our approach outperformed [2]'s results obtained using task-specific classifiers by 5.72% with the K-NN algorithm while yielding similar results with the RF and SVM algorithms. However, our proposed algorithm's LR, GNB, and DT classifiers were outperformed by [2]'s approach. These findings are illustrated in Fig. 3.

Table 6. Mean accuracy and specificity comparison between RD-Classifier and benchmarking model [2] on each task.

No	RD-Classifier (Our approach)				Cilia, Nicole. D			
	K-NN		RF		K-NN		RF	
	Acc.	recall	Acc.	recall	Acc.	recall	Acc.	recall
1	65.44 (±6.92)	71.38	62.58 (±7.57)	77.89	49.47 (±9.81)	40.56	65.86 (±8.22)	63.89
2	68.96 (±8.37)	66.15	66.15 (±8.81)	78.98	55.29 (±7.43)	61.39	67.14 (±7.78)	65.28
3	61.81 (±9.81)	71.25	64.73 (±7.57)	74.98	51.86 (±9.19)	61.67	66.57 (±8.90	61.67
4	61.04 (±5.37)	69.71	69.78 (±6.13)	74.04	64.71 (±5.58)	61.11	71.29 (±6.78)	66.67
5	67.64 (±7.90)	74.24	72.03 (±6.81)	76.04	65.86 (±5.90)	60.00	72.14 (±6.35)	72.22
6	74.12 (±7.82)	83.49	75.49 (±7.00)	77.19	65.57 (±10.22)	54.72	72.43 (±7.66)	66.94
7	76.32 (±6.89)	77.48	74.12 (±7.81)	83.74	72.00 (±5.76)	59.17	78.00 (±7.54)	70.83
8	71.15 (±10.04)	80.28	69.84 (±6.92)	80.54	57.00 (±8.57)	51.94	64.86 (±6.36)	67.50
9	83.41 (±7.06)	84.39	79.78 (±9.77)	91.08	74.86 (±5.83)	63.06	77.43 (±7.69)	70.83
10	63.35(±7.21)	63.86	66.15 (±5.59)	80.33	58.14 (±8.95)	44.17	69.29 (±6.80	69.44
11	69.84(±8.87)	71.46	70.55 (±8.93)	63.56	53.71 (±6.12)	41.67	64.86 (±6.23	66.94
12	61.76 (±9.25)	58.99	59.78 (±7.59)	64.44	55.29 (±7.43)	52.50	67.14 (±7.78)	56.67
13	71.21 (±7.21)	76.96	71.15 (±10.25)	76.98	51.86 (±9.19)	45.00	66.57 (±8.90)	61.94
14	72.69 (±6.46)	76.99	66.21 (±8.93)	74.04	64.71 (±5.58)	57.22	71.29 (±6.78)	70.00
15	72.69 (±7.22)	76.54	74.78 (±7.88)	83.32	65.86 (±5.90)	51.94	72.14 (±6.35)	69.44
16	67.64 (±9.90)	73.15	71.87 (±6.19)	84.51	65.57 (±10.22)	63.61	72.43 (±7.66)	67.78
17	79.07 (±7.67)	81.97	73.96 (±6.14)	75.46	72.00 (±5.76)	70.28	78.00 (±7.54)	76.67

(continued)

Table 6. (*continued*)

| No | RD-Classifier (Our approach) | | | | Cilia, Nicole. D | | | |
| | K-NN | | RF | | K-NN | | RF | |
	Acc.	recall	Acc.	recall	Acc.	recall	Acc.	recall
18	66.10 (±9.65)	68.03	68.30 (±8.03)	70.82	57.00 (±8.57)	55.83	64.86 (±6.36)	65.00
19	75.55 (±6.03)	75.44	72.58 (±5.03)	75.74	74.86 (±5.83)	70.28	77.43 (±7.69)	74.17
20	78.30 (±8.95)	85.82	75.38 (±10.19)	79.24	60.29 (±7.47)	45.56	71.43 (±8.03)	68.89
21	70.33 (±8.15)	67.79	72.47 (±6.93)	67.48	64.43 (±7.21)	69.17	72.29 (±6.56)	76.39
22	72.64 (±7.38)	79.24	77.03 (±8.14)	85.13	68.43 (±6.04)	67.22	75.00 (±7.80)	75.28
23	76.92 (±7.70)	74.49	77.64 (±4.81)	75.25	68.71 (±7.21)	69.17	80.00 (±5.48)	82.78
24	64.01 (±6.83)	72.38	68.46 (±8.56)	77.34	61.29 (±6.25)	56.67	72.14 (±4.98)	67.78
25	71.10 (±5.90)	72.44	72.53 (±7.42)	72.82	68.43 (±6.78)	63.89	73.71 (±7.79)	74.17

Our findings demonstrate the effectiveness of the proposed approach for developing and refining ML algorithms in detecting AD through handwriting tasks. Although the present RD-Classifier has advantages such as avoiding the curse of dimensionality and reducing computational complexity, there are still gaps that can be investigated for future research.

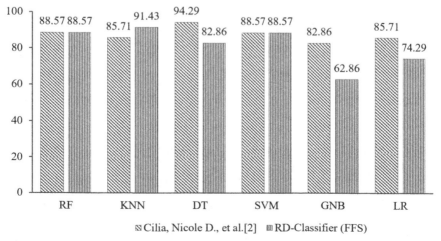

Fig. 3. RD-Classifier (dimensionality reduction on the entire set of features) Vs benchmarking model [2] (combining task-specific classifiers).

5 Conclusion

AD is a growing public health concern, and early detection is crucial for timely intervention and better patient outcomes. In this study, we proposed a novel approach to detecting AD using high-dimensional handwriting data from the non-invasive approach introduced by [2]. We employed three dimensionality reduction methods together with six ML classifiers and identified a subset of the most important features for accurate diagnosis support. Our results showed that using the most important features led to similar or better performance compared to state-of-the-art models. Moreover, our method significantly improved the accuracy of detecting patients with AD, outperforming current benchmarks [2]. The importance of each feature was evaluated, and the most important ones were used to enhance the performance of the ML model and reduce computational complexity. Overall, our proposed method offers a significant advance in detecting AD and provides a foundation for developing more accurate and efficient diagnosis support systems.

While our approach demonstrated promising results, further improvements are possible. For example, an effective hyperparameter tuning method could optimize the performance of the different ML models used. Additionally, novel approaches could be developed and compared to the current study to achieve higher prediction accuracy.

References

1. NIH Alzheimer's Disease. https://www.ninds.nih.gov/health-information/disorders/alzheimers-disease?search-term=alzheimer. Accessed 20 Feb 2023
2. Cilia, N.D., De Gregorio, G., De Stefano, C., Fontanella, F., Marcelli, A., Parziale, A.: Diagnosing Alzheimer's disease from on-line handwriting: a novel dataset and performance benchmarking. Eng. Appl. Artif. Intell. **111**, 104822 (2022)
3. Archer, M.C., Hall, P.H., Morgan, J.C.: [P2–430]: Accuracy of clinical diagnosis of Alzheimer's disease in Alzheimer's disease centers (ADCS). Alzheimer's Dement. **13**(7S_Part_16), P800–P801 (2017)
4. Lock, M.M.: The Alzheimer Conundrum: Entanglements of Dementia and Aging. Princeton University Press, Princeton (2013).http://site.ebrary.com/id/10773745. Accessed 20 Feb 2023
5. Swaddiwudhipong, N., Whiteside, D.J., Hezemans, F.H., Street, D., Rowe, J.B., Rittman, T.: Pre-diagnostic cognitive and functional impairment in multiple sporadic neurodegenerative diseases. Alzheimer's Dement. **19**(5), 1752–1763 (2023)
6. Myszczynska, M.A., et al.: Applications of machine learning to diagnosis and treatment of neurodegenerative diseases. Nat. Rev. Neurol. **16**(8), 440–456 (2020)
7. Albu, A., Precup, R.E., Teban, T.A.: Results and challenges of artificial neural networks used for decision-making and control in medical applications. Facta Univ. Ser.: Mech. Eng. **17**(3), 285–308 (2019)
8. Tanveer, M., et al.: Machine learning techniques for the diagnosis of Alzheimer's disease: a review. ACM Trans. Multimed. Comput. Commun. Appl. (TOMM) **16**(1s), 1–35 (2020)
9. Mei, J., Desrosiers, C., Frasnelli, J.: Machine learning for the diagnosis of Parkinson's disease: a review of literature. Front. Aging Neurosci. **13**, 633752 (2021)
10. Abedini, M., Kirley, M.: CoXCS: a coevolutionary learning classifier based on feature space partitioning. In: Nicholson, A., Li, X. (eds.) AI 2009. LNCS (LNAI), vol. 5866, pp. 360–369. Springer, Heidelberg (2009). https://doi.org/10.1007/978-3-642-10439-8_37

11. Berlanga, F.J., del Jesus, M.J., Herrera, F.: A novel genetic cooperative-competitive fuzzy rule-based learning method using genetic programming for high dimensional problems. In: 2008 3rd International Workshop on Genetic and Evolving Systems, pp. 101–106. IEEE (2008)

12. Debie, E., Shafi, K.: Implications of the curse of dimensionality for supervised learning classifier systems: theoretical and empirical analyses. Pattern Anal. Appl. **22**(2), 519–536 (2019). https://doi.org/10.1007/s10044-017-0649-0

13. Weiner, M.W., et al.: Alzheimer's disease neuroimaging initiative the Alzheimer's disease neuroimaging initiative 3: continued innovation for clinical trial improvement. Alzheimers Dement. **13**(5), 561–571 (2017)

14. Diogo, V.S., Ferreira, H.A., Prata, D.: Early diagnosis of Alzheimer's disease using machine learning: a multi-diagnostic, generalizable approach. Alzheimer's Res. Therapy **14**(1), 107 (2022)

15. Impedovo, D., Pirlo, G., Vessio, G.: Dynamic handwriting analysis for supporting earlier Parkinson's disease diagnosis. Information **9**(10), 247 (2018)

16. Moetesum, M., Siddiqi, I., Vincent, N., Cloppet, F.: Assessing visual attributes of handwriting for prediction of neurological disorders—A case study on Parkinson's disease. Pattern Recogn. Lett. **121**, 19–27 (2019)

17. Vessio, G.: Dynamic handwriting analysis for neurodegenerative disease assessment: a literary review. Appl. Sci. **9**(21), 4666 (2019)

18. Pozna, C., Precup, R.E.: Applications of signatures to expert systems modelling. Acta Polytech. Hung. **11**(2), 21–39 (2014)

19. De Gregorio, G., Desiato, D., Marcelli, A., Polese, G.: A multi classifier approach for supporting alzheimer's diagnosis based on handwriting analysis. In: Del Bimbo, A., et al. (eds.) ICPR 2021. LNCS, vol. 12661, pp. 559–574. Springer, Cham (2021). https://doi.org/10.1007/978-3-030-68763-2_43

20. Parziale, A., Della Cioppa, A., Marcelli, A.: Mimicking the immune system to diagnose Parkinson's disease from handwriting. In: 2022 26th International Conference on Pattern Recognition (ICPR), pp. 2496–2502. IEEE (2022)

21. Senatore, R., Marcelli, A.: A paradigm for emulating the early learning stage of handwriting: performance comparison between healthy controls and Parkinson's disease patients in drawing loop shapes. Hum. Mov. Sci. **65**, 89–101 (2019)

22. Parziale, A., Marcelli, A.: Should we look at curvature or velocity to extract a motor program? In: Carmona-Duarte, C., Diaz, M., Ferrer, M.A., Morales, A. (eds.) Intertwining Graphonomics with Human Movements, IGS 2022, vol. 13424, pp. 203–216. Springer, Cham (2022). https://doi.org/10.1007/978-3-031-19745-1_15

23. Meulemans, C., Leijten, M., Van Waes, L., Engelborghs, S., De Maeyer, S.: Cognitive writing process characteristics in Alzheimer's disease. Front. Psychol. **13**, 872280 (2022)

24. Meng, Z., Yang, D., Huo, J., Zhuo, P., Bao, Y.: Development and performance evaluation of an integrated disc cutter system for TBMs. Appl. Sci. **11**(2), 644 (2021)

25. Barbedo, J.G.A.: Impact of dataset size and variety on the effectiveness of deep learning and transfer learning for plant disease classification. Comput. Electron. Agric. **153**, 46–53 (2018)

26. Dawson, H.L., Dubrule, O., John, C.M.: Impact of dataset size and convolutional neural network architecture on transfer learning for carbonate rock classification. Comput. Geosci. **171**, 105284 (2023)

27. Sammut, C., Webb, G.I.: Encyclopedia of Machine Learning and Data Mining. Springer, Cham (2017). https://doi.org/10.1007/978-1-4899-7687-1

28. Pham, H.T., Awange, J., Kuhn, M.: Evaluation of three feature dimension reduction techniques for machine learning-based crop yield prediction models. Sensors **22**(17), 6609 (2022)

29. Jia, W., Sun, M., Lian, J., Hou, S.: Feature dimensionality reduction: a review. Complex Intell. Syst. **8**(3), 2663–2693 (2022). https://doi.org/10.1007/s40747-021-00637-x

30. Cilia, N.D., De Stefano, C., Fontanella, F., Di Freca, A.S.: An experimental protocol to support cognitive impairment diagnosis by using handwriting analysis. Procedia Comput. Sci. **141**, 466–471 (2018)

31. Jolliffe, I.T., Cadima, J.: Principal component analysis: a review and recent developments. Philos. Trans. Roy. Soc. A: Math. Phys. Eng. Sci. **374**(2065), 20150202 (2016)

32. Jović, A., Brkić, K., Bogunović, N.: A review of feature selection methods with applications. In: 2015 38th International Convention on Information and Communication Technology, Electronics and Microelectronics (MIPRO), pp. 1200–1205. IEEE (2015)

User Interaction-Aware Knowledge Graphs for Recommender Systems

Ru Wang[1,2], Bingbing Dong[1,2], Tianyang Li[3], Meng Wu[1,2],
Chenyang Bu[1,2], and Xindong Wu[4,1]($\boxtimes$)

[1] Key Laboratory of Knowledge Engineering with Big Data
(The Ministry of Education of China), Hefei University of Technology, Hefei, China
{wangru,blingdong,wu}@mail.hfut.edu.cn
[2] School of Computer Science and Information Engineering,
Hefei University of Technology, Hefei, China
{chenyangbu,xwu}@hfut.edu.cn
[3] School of Software, Hefei University of Technology, Hefei, China
litianyang@mail.hfut.edu.cn
[4] Research Center for Knowledge Engineering, Zhejiang Lab, Hangzhou, China

Abstract. The performance of recommender systems can be improved
effectively by using knowledge graphs as auxiliary information. However,
most of the knowledge graph-based recommendations focus on learning item representations in knowledge graphs, capture the collaborative
signals between user interactions inadequately. The user-item bipartite
graph contains explicit preference information of users, and the collaborative signals of user-item interactions help to enhance representations of
users. A user interaction-aware knowledge graph recommendation model
(UIKR) is proposed, which enhances user representation and introduces
the higher-order collaborative signals in user interactions into the representation learning of items in knowledge graphs. Specifically, the high-
order collaborative signals hidden in the user-item bipartite graph are
captured to strengthen user representations. Then, the enhanced user
representation is applied to the representation learning of items in knowledge graphs. A hybrid attention function is proposed to aggregate neighbor representation of items, which augments the propagation of user
preferences in knowledge graphs and helps to learn personalized item
representations. Finally, the user interaction-aware item representations
and the enhanced user representations are used for recommendations.
Extensive experiments are conducted on two standard datasets and the
results show that proposed UIKR model outperforms current state-of-
the-art baselines.

Keywords: Recommender system · Knowledge graph · Graph neural
network

1 Introduction

Recommender system is a powerful information filtering tool, which can solve
information profusion effectively and provide users with personalized services

C. Strauss et al. (Eds.): DEXA 2023, LNCS 14147, pp. 18–32, 2023.
https://doi.org/10.1007/978-3-031-39821-6_2

[14]. Collaborative filtering (CF) [3] is a widely used recommendation technique, which uses historical behavioral preference data to build models. Although CF-based methods are effective, they still have the problems of sparse user-item interactions and the cold start. To solve the above limitations, researchers try to combine CF with other auxiliary information (e.g., user-item attributes, user social network information, etc.), which can help alleviate the sparsity problem and improve the performance of recommendation [10].

Knowledge graph (KG) is a structured network that stores knowledge entities and relationships between entities and contains abundant information of items, which can effectively alleviate the problem of data sparsity and provides an effective tool for the study of the interpretability and accuracy of the recommender systems [21]. KG contains many types of entities and relations, it is a challenge to learn the feature representations of users and items based on complex graph structures. Graph Neural Networks (GNN) [19,20,25] can be used to model knowledge graphs, which can learn high-order relationships between users and items, capture the user's interest and the characteristics of items, and improve the recommendation performance effectively. In recent years, the KG-aware approach has achieved significant success, but there are still two problems [6]. (1) This approach focuses on extracting information from KG, while not combining higher-order signals in user interactions with representation learning in KG. For example, KGCN [13] and KGNN-LS [12] fail to model user preferences and neglect the collaborative signals in historical interactions of users that can help enhance user representations. In addition, some models (such as KGAT [16], MKGAT [8], and CKAN [17]) construct a unified graph with user interactions and KG, but they still only perform higher-order aggregation propagation in KG and do not explicitly model historical interactions of users. (2) Most of the existing efforts apply relational-aware attention to aggregate neighbor representations of items, which does not spread user preferences adequately and learn personalized item representations in KG with few types of relations [18]. It is a challenge to design a reasonable and effective attention function to aggregate the neighbor representations of items.

Based on these limitations, a user interaction-aware knowledge graph recommendation model (UIKR) is proposed. The motivation of the UIKR model is shown in Fig. 1. The items that users have interacted with can reveal the preferences of users, and the higher-order collaborative signals hidden in the user-item bipartite graph are actually the user's high-order interactions, which are advantageous for augmenting user representations. First, we enhance user representations based on high-order collaborative signals in user interactions, and apply them to the representation learning of items in KG. Then, a hybrid attention function is proposed to aggregate neighbor representations of items. The neighbor information is weighted according to the user's preferences for relations and neighbor entities when updating item representations in KG, which combines the high-order collaborative signals in user-item interactions with the representation learning of items in KG effectively and assists to learn personalized item representations. The higher-order collaborative signals in the user-item bipartite graph and a hybrid attention function is introduced into the representation learning of items in KG, which enhances the propagation of user preferences in KG, and can effectively improve the performance of personalized recommendations.

Our contributions can be summarized as follows:

- A user interaction-aware knowledge graph recommendation model (UIKR) is proposed, which explicitly models the user representations based on the historical interactions of users, combines the high-order collaborative signals in user-item interactions with the representation learning of items in KG, and improves the performance of recommendations effectively.
- A hybrid attention function is proposed for assigning weights to neighborhoods. The weights are assigned according to the user's preferences for relations and linked entities, which enhance the propagation of user preferences in KG.
- Extensive experiments are conducted on two public datasets to demonstrate that our model is superior to the current state-of-the-art baselines.

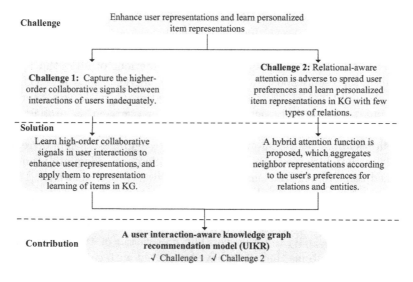

Fig. 1. Motivation of the proposed model.

2   Related Work

This section discusses existing work on KG-based recommendation and GNN techniques, which are closely related to our work.

2.1   KG-Based Recommendation

KG contains abundant entities and relations, which is effective auxiliary information in recommender systems. Generally, the existing KG-based recommendation methods can be divided into three types: embedding-based methods, path-based methods, and GNN-based methods [9].

Embedding-based methods usually use knowledge graph embedding (KGE) [15] algorithms to represent entities and relations in KG, which is conducive to expanding the semantic information of original items and users. The embedded method is very flexible in KG-based recommender systems, but it is more suitable for the graph tasks in knowledge graph, such as link prediction and knowledge graph completion, because the commonly used knowledge graph embedding algorithms focus on modeling strict semantic relevance.

Path-based methods provide auxiliary information for the recommender systems by mining various association paths among items in KG. These paths visually present the propagation of user preferences. FMG [24] and PER [22] treated KG as a heterogeneous information network, and the multiple relations between user and item were obtained by extracting potential features based on meta-path/meta-graph. However, this method largely relies on manually designed meta-paths/meta-graphs, so it is difficult to optimize in practice.

In recent years, GNN has shown great potential in the graph task, and more and more researchers apply GNN to the KG-based recommendations. Given the graph data, the main idea of GNN is to iteratively aggregate information from neighbors during propagation and update current node representation by the aggregated information. KGCN applies graph convolutional networks (GCN) [4] to item KGs and uses an attention mechanism to capture users' personalized preference on relations.

2.2   Graph Neural Network Techniques

Recently, systems based on variants of GNN have shown breakthrough performance in many tasks related to graph data. In this part, we will briefly introduce the existing GNN frameworks, and summarize the work of applying GNN in KG-based recommendations.

In general, there are five typical GNN frameworks that are used in recommender systems [18]. GCN approximates the first-order eigendecomposition of the graph Laplacian to iteratively aggregate information from neighbors. Graph-SAGE [2] samples a fixed-size neighborhood for each node, proposes mean/ sum/ max-pooling aggregator, and adopts concatenation operation for the update of entities. GAT utilizes the attention mechanism to distinguish the contribution of neighbors and updates the vector of each node according to the weight of neighbors. GGNN [5] uses a gated recurrent unit (GRU) [5] in the update step. HGNN [1] is a hypergraph neural network, which encodes high-order data correlation in a hypergraph structure.

At present, the application of GNN in KG-based recommendations mainly focuses on the construction of graphs and relation-aware aggregation [18]. In the construction of graphs, one way is to explicitly incorporate user nodes into the knowledge graph, and another way is to use user nodes implicitly to distinguish the importance of different relationships. There are also two approaches to relation-aware aggregation. For the collaborative graph that merges user nodes and KG, user preferences will be propagated to the entities in KG. As for the works that do not merge user nodes and KG, usually characterize users' interests in relations according to the score function between users and relations.

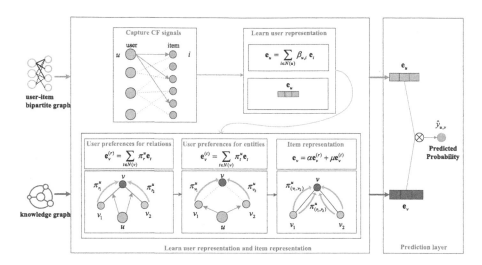

Fig. 2. Illustration of the proposed UIKR framework, which consists of three modules: learning user representations based on the user-item bipartite graph, learning item representations based on KG and predicting the probability of a user interacting with a candidate item.

3   Problem Formulation

In this section, the formulation of the problem we studied is introduced. The input of the model includes historical interaction records of users to items and KG. We use $U = \{u_1, u_2, ...u_n\}$ to denote the sets of users, and $V = \{v_1, v_2, ...v_m\}$ to denote the sets of items. The user-item interaction matrix $Y = \{(u, v)|u \in U, v \in V\}$ is defined according to the interaction records of users. The value of y_{uv} is 0 or 1. If user u has engaged with item v, such as clicking, browsing, or purchasing, then $y_{uv} = 1$; otherwise, $y_{uv} = 0$. In addition, we use $G = \{(h, r, t)|(h, t) \in \mathcal{E}, r \in \mathcal{R}\}$ to represent the knowledge graph, which is constructed in the form of entity-relation-entity triples (h, r, t). The $\mathcal{E}$ and $\mathcal{R}$ denote the sets of entities and relations in KG, respectively.

Given the user-item interaction matrix Y and the knowledge graph G, the goal of our model is to learn the prediction function $\hat{y}_{uv} = \mathcal{F}(u, v|Y, G, \theta)$, where $\hat{y}_{uv}$ denotes the probability that the user u interact with the item v, with a value between 0 and 1, and θ is the model parameters.

4 Methodology

In this section, we detail the proposed UIKR model, the model framework is shown in Fig. 2, which mainly consists of three parts: (1) Learn user representation based on the user-item bipartite graph, the user representation can be augmented by capturing the high-order cooperative signals hidden in it. (2) Learn item representation based on KG. GCN framework is used to model KG, iteratively aggregate information from neighbors during propagation, and update current entity representation according to the hybrid attention function. (3) Predict the probability of interaction based on user representation and item representation.

4.1 Learn User Representation

The higher-order collaborative signals hidden in the user-item bipartite graph can effectively help recommendation, so we try to explicitly model user preferences based on it. We inherit the method of learning the higher-order collaborative signals from the user-item bipartite graph in UltraGCN [7] to enhance the user representation. In this approach, the message passing is removed in the representation learning of users and the constraint loss is constructed by setting the convergence mode to approximate the convergence state after the infinite-layer message passing. More details in the original paper [7].

The expected user representation is shown in the following formula [7]:

$$\mathbf{e}_u = \sum_{i \in \mathcal{N}(u)} \beta_{u,i}\, \mathbf{e}_i, \quad \beta_{u,i} = \frac{1}{d_u} \sqrt{\frac{d_u+1}{d_i+1}} \tag{1}$$

where $\mathbf{e}_u$ and $\mathbf{e}_i$ are the representations of user u and item i, $\mathcal{N}(u)$ is a set of items for user u interacted, d_u denotes the number of items that user u has interacted with, and d_i denotes the number of times that item i has been interacted. When each user node satisfies the following formula, the model approximates the convergence state of message passing. The constraint loss is constructed by maximizing the cosine similarity between e_u and e_i.

$$\max \sum_{i \in \mathcal{N}(u)} \beta_{u,i} \mathbf{e}_u^\top \mathbf{e}_i, \quad \forall u \in \mathcal{U} \tag{2}$$

For the convenience of optimization, the sigmoid activation function and negative log likelihood are introduced. In addition, we choose to perform negative sampling during training to alleviate the over-smoothing problem. After performing negative sampling, the ultimate constraint loss is as follows [7]:

$$\mathcal{L}_C = -\sum_{(u,i)\in\mathcal{N}^+} \beta_{u,i} \log\left(\sigma(\mathbf{e}_u^\top \mathbf{e}_i)\right) - \sum_{(u,j)\in\mathcal{N}^-} \beta_{u,j} \log\left(\sigma(-\mathbf{e}_u^\top \mathbf{e}_j)\right) \tag{3}$$

where the σ is the sigmoid function, $\mathcal{N}^+$ and $\mathcal{N}^-$ represent the sets of positive pairs and randomly sampled negative pairs, and $\beta_{u,i}$ is the constraint coefficients.

In addition to the constraint loss, the BCE(binary cross-entropy) loss is used as the main optimization objective in the user-item bipartite graph [7].

$$\mathcal{L}_O = -\sum_{(u,i)\in\mathcal{N}^+} \log\left(\sigma(\mathbf{e}_u^\top \mathbf{e}_i)\right) - \sum_{(u,i)\in\mathcal{N}^-} \log\left(\sigma(-\mathbf{e}_u^\top \mathbf{e}_j)\right) \tag{4}$$

For simplicity, $\mathcal{L}_O$ and $\mathcal{L}_C$ use the same sets of sample pairs. Hence, the total loss on the user-item bipartite graph is:

$$\mathcal{L}_U = \mathcal{L}_O + \mathcal{L}_C \tag{5}$$

Instead of modeling user preference by explicit multi-hop information propagation in the user-item bipartite graph, the enhanced user representation is obtained by constructing loss which approximates the convergence state of multi-hop message passing. This approach not only learns high-order collaborative signals from interactions to enhance user representations, but also avoids the over-smooth problem that easily occurs in multi-layer information propagation.

4.2   Learn Item Representation

To obtain the representations of items, the GCN framework is used to aggregate and propagate messages in KG. The neighborhood information is weighted according to the personalized interests of users when calculating the representation of a given entity in KG.

KGCN [13] is a typical KG-aware recommendation model, which can efficiently aggregate the attributes of items in KG and obtain satisfactory results in multiple datasets. Users have different interests in different relationships. For example, one user may be more interested in the "director" of a movie while another user is more concerned about the "actor" of the movie. Therefore, the KGCN model aggregates neighbor information based on user preferences for relations. The attention function is as follows:

$$\pi_r^u = g(\ \mathbf{e}_u,\ \mathbf{e}_r) \tag{6}$$

where $\mathbf{e}_u$ and $\mathbf{e}_r$ are the representations of user u and relation r, respectively, π_r^u shows the importance of relation r to user u.

To learn personalized item representations and enhance the propagation of user preferences in KG, user preferences for item attributes also need to be considered. Assuming that there are two triples in KG: (The Shawshank Redemption, actor, Tim Robbins) and (The Shawshank Redemption, director, Frank Darabont). Two types of relations connect two different neighbors with the same central node. Suppose that user A is very interested in "Tim Robbins" and not very interested in "Frank Darabont", when recommending movies to user A, "Tim Robbins" plays a more important role than "Frank Darabont" in calculating the representation of "The Shawshank Redemption". In other words, "Tim Robbins" needs to make a greater contribution when aggregating neighbor information. Considering the different preferences of users for item attributes, a hybrid attention function is Proposed as follows:

$$\pi_t^u = g(\ \mathbf{e}_u,\ \mathbf{e}_t)$$
$$\pi_r^u = \frac{\exp(\pi_r^u)}{\sum_{e \in \mathcal{N}(v)} \exp(\pi_r^u)}, \pi_t^u = \frac{\exp(\pi_t^u)}{\sum_{e \in \mathcal{N}(v)} \exp(\pi_t^u)} \tag{7}$$

$$\pi = \alpha\pi_r^u + \mu\pi_t^u \tag{8}$$

where π_r^u and π_t^u shows the importance of relation r and attribute node t to user u, respectively, α and μ are the weights of attention scores π_r^u and π_t^u, t is the attribute node of candidate item v, and $N(v)$ represents the neighborhood set of item v. For ease of calculation, the number of entities in neighborhood set $N(v)$ is fixed, $|N(v)| = K$, where K is a configurable constant. Figure 2 describes an example of aggregating neighbor information using the hybrid attention function, which considers not only the user's preferences for relations, but also the user's preferences for entities. With regard to the weight of two attention scores, ablation experiments are performed in the experimental part, which proved the validity of our proposed hybrid attention weights.

After receiving the attention, the neighborhood information of item i is aggregated according to the attention score. As shown below:

$$\mathbf{e}_{N(v)}^{(1)} = \sum_{t \in N(v)} \pi \mathbf{e}_t \tag{9}$$

where $\mathbf{e}_{N(v)}^{(1)}$ represents the first-layer neighborhood representation of item v, $\mathbf{e}_t$ is the initial representations of entity t. Similar to the KGCN model, a fixed-size neighborhood set is sampled for each entity in KG for ease of calculation. The neighborhood representation of an item is weighted by the user's preferences for relations and neighborhood entities.

The representation of item v is updated by itself and its neighborhood representation. It has been proved by LightGCN that feature transformation and nonlinear activation function have no substantial effect on GCN model. Inspired by this, feature transformation and nonlinear activation function are discarded when aggregating neighbor information. The ultimate item representation is shown in the following formulas:

$$\mathbf{e}_v^{(h)} = \mathbf{e}_v^{(h-1)} + \mathbf{e}_{N_{(v)}}^{(h)} \tag{10}$$

$$\mathbf{e}_{N_{(v)}}^{(h)} = \sum_{t \in N(v)} \pi \mathbf{e}_t^{(h-1)} \tag{11}$$

where $\mathbf{e}_v^{(h)}$ represents the h-layer representation of item v, and $\mathbf{e}_{N_v}^{(h)}$ represents the h-layer neighborhood representation of item v.

After obtaining the ultimate h-layer item representation, using the inner product to predict the probability of user u interacting with item v.

$$\hat{y}_{uv} = \mathbf{e}_u^T \mathbf{e}_v^{(h)}. \tag{12}$$

where $\mathbf{e}_u$ is the user representation learned on the user-item bipartite graph.

The BCE loss and negative sampling strategy are also used for optimization when learning item representation in KG. The total loss on KG is:

$$\mathcal{L}_{KG} = -\sum_{(u,v) \in \mathcal{N}^+} \mathcal{I}(\hat{y}_{uv}, y_{uv}) - \sum_{(u,j) \in \mathcal{N}^-} \mathcal{I}(\hat{y}_{uj}, y_{uj}) \tag{13}$$

where $\mathcal{I}$ is BCE loss, $\hat{y}_{uv}$ is a prediction, and y_{uv} is a true label.

The final total loss function of the UIKR model is as follows:

$$\mathcal{L} = \mathcal{L}_U + \mathcal{L}_{KG} + \gamma \|\theta\|_2^2 \tag{14}$$

where γ is the parameter for $L2$ regularization.

5 Experiments

In this section, extensive experiments are conducted on two realistic scenarios to determine the validity of our proposed model. Firstly, the datasets and evaluation metrics used in the experiment are introduced. Then, the experiment settings are discussed. Finally, the experimental results are presented and analyzed.

5.1 Evaluation Datasets and Metrics

Two typical datasets were used in our experiment: Last.FM and Book-Crossing. All datasets are publicly accessible and vary in size and sparsity. KGs of these datasets are provided by KGCN [13] which are constructed using Satori, a commercial KG built by Microsoft. The statistics of the two datasets are recorded in Table 1.

- **Last.FM.** This is a music dataset from Last.FM online music system, which contains music listening events created by Last.FM users.
- **Book-Crossing.** It is a frequently used book dataset, which includes user rating data (ratings) and book metadata (description, category information, price, and brand).

Table 1. Statistics of the datasets.

	Last.FM	Book-Crossing
Users	1,872	17,860
Items	3,846	14,967
Interactions	42,346	139,746
Entities	9366	77,903
Relations	60	25
KG triples	15,518	151,500

We evaluate the model in the experimental scenario of click-through rate (CTR) prediction, which uses the trained model to predict each interaction in the test set. The metrics of AUC and F1 to evaluate the results of CTR prediction. The AUC is equivalent to the probability of positive samples are ranked higher than negative samples. F1 is the harmonic mean of accuracy and recall of the model, which can reflect the robustness of the model.

5.2 Baselines

To demonstrate the effectiveness of our proposed model, UIKR is compared with recent state-of-the-art KG-based recommendation models, including CKE [23], PER [22], RippleNet [11], KGAT [16], KGCN [13], and CKAN [17]. The details are as follows:

- **CKE** [23]. CKE introduces structure information, text data, image data, and other knowledge base information into the recommendation system to improve the performance of the recommender systems.
- **PER** [22]. PER treats KG as a heterogeneous information network, and the multiple relations between user and item were obtained by extracting potential features based on meta-path/meta-graph.
- **RippleNet** [11]. RippleNet is an end-to-end framework for KG-based recommendation, which automatically discovers the potential interests of users by means of preference propagation in KGs.
- **KGAT** [16]. KGAT trains recommendation and KGE alternately, which assigns the weight according to the distance between the linked entities in space of the relation.
- **KGCN** [13]. KGCN applies GCN to discover the high-order structural and semantic information from item KGs, which aggregates information based on user preferences for relation.
- **CKAN** [17]. CKAN explicitly encodes the collaborative signals that are latent in user-item interactions and naturally combines them with knowledge associations in an end-to-end.

5.3   Experimental Setup

For each dataset, we randomly select 60% of interaction history of each user to constitute the training set, and treat the remaining as the test set. For comparisons, the batch size of data for all of the models was fixed as 256, the embedding parameters were initialized by the Xavier method, and our model was optimized by the Adam optimizer. The optimal parameter settings were determined by grid search, the learning rate is explored between $\{0.001, 0.002, 0.0015, 0.01\}$, the $L2$ normalization coefficients are set $\{0.0001, 0.00002, 0.0002, 0.00015\}$, and the embedding size is searched in $8, 16, 32, 64, 128$. For some baselines, we report the results from their papers to keep consistency. They are also comparable since we use the exactly same datasets and experimental settings provided by them. For other baselines, we mainly use their official open-source code and carefully tune the parameters to achieve the best performance for fair comparisons.

5.4   Results

Table 2 shows the experimental results of all models in CTR prediction. We have the following observations from the results of the experiment:

- The UIKR has significantly improved over the state-of-the-art baselines in both datasets. The values of AUC and F1 were increased by 2.6% and 1.2% in the Last.FM and by 5.2% and 7.9% in the Book-Crossing, respectively, which shows the effectiveness of our model.
- The performance improvement of the UIKR in the Book-Crossing was higher than in the Last.FM, which may be caused by the fewer relationship types in Book-Crossing than in Last.FM. On average, there are 6060 triples per

relationship in Book-Crossing (compared with 258 triples per relationship in
Last.FM). In this case, aggregation of neighborhood representations based on
user's preference for relations easy to learn similar item representations.
– The performance of GNN-based model (KGAT, KGCN, and CKAN) is better than that of non-GNN-based model (CKE, PER, and RippleNet), which
  shows that superior item representations can be learned by using the GNN
  framework to model the KG.

Table 2. Overall performance comparison. The best results are highlighted in bold.

	Last.FM		Book-Crossing	
Models	AUC	F1	AUC	F1
CKE	0.747	0.674	0.676	0.623
PER	0.641	0.603	0.605	0.572
RippleNet	0.776	0.702	0.721	0.647
KGCN	0.829	0.725	0.813	0.728
KGAT	0.829	0.742	0.731	0.654
CKAN	0.842	0.769	0.753	0.673
UIKR(Ours)	**0.864**	**0.778**	**0.856**	**0.786**
%imp	2.60%	1.20%	5.20%	7.90%

5.5   Ablation Study of UIKR

In order to verify the effectiveness of the proposed model, we performed the
ablation experiment from the following two aspects.

First, we investigate the effect of the perception of user interactions and
the proposed hybrid attention function on the performance of the model. Other
parameters are consistent, UIKR-w indicates that the model uses the hybrid
attention function, but the perception of user interactions are ignored. UIKR-u
indicates that the model captures the perception of user interactions, but the
attention function is determined only by the user's preference for the relations.

Then, the effect of different weights in attention function on the performance
of the model is explored. Set $\alpha + \mu = 1$, we changed the values of α and μ to
explore the performance of UIKR after using the attention function with different
coefficients.

Impact of Different Improved Modules. The ablation experiment results of
different improved modules on the performance of the model are shown in Fig. 3.
We found that in both datasets, UIKR performed best, followed by UIKR-u, and
UIKR-w performed worst, which indicates that both of our proposed improvements can help to improve recommendation performance, and the perception

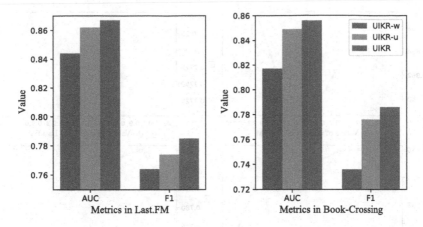

Fig. 3. Performance comparison w.r.t different improved modules. UIKR-w indicates that the model introduces the hybrid attention function, UIKR-u indicates that the model captures the perception of user interactions, and UIKR indicates that both improved modules are introduced.

of user interactions is more effective than the hybrid attention function. The experimental results show that the cooperative signals on the user-item bipartite graph can effectively enhance representations of users and the use of hybrid attention functions can also help spread the user's personalized preferences.

Impact of Weights in Attention Function. The experimental results of the model using hybrid attention functions with different coefficients are shown in Fig. 4. We can observe that the model of $\alpha = 0.9$ and $\mu = 0.1$ performs best in the Last.FM, and the model of $\alpha = 0.1$ and $\mu = 0.9$ performs best in the Book-Crossing, which shows the effectiveness of our proposed hybrid attention function. In the Last.FM, the more even the proportion of α and μ, the lower the performance of the model, which may be due to the rich types of relations (60 types) in the Last.FM. The aggregated representations of items may tend to be similar when considering the user's preference for relationship and the user's preference for neighbor nodes on average, which is not conducive to improving the performance of model. Furthermore, aggregation of neighborhoods based on user's preferences for relations is more efficient in the Last.FM, so the model performs best when $\alpha = 0.9$. In the Book-Crossing, the performance of UIKR decreases with increasing values of α, which indicates that it is more efficient for Book-Crossing to aggregate neighbor information based on user's preferences for neighbor entities. Therefore, the UIKR performs best when $\alpha = 0.1$.

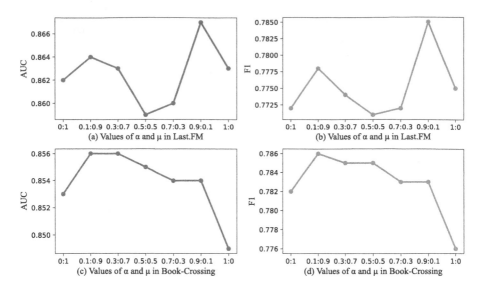

Fig. 4. Performance comparison w.r.t different attention function.

6   Conclusion and Future Work

In this paper, a new model named UIKR is proposed, which enhances user representations and introduces higher-order collaborative signals in user interactions into the representation learning of items in knowledge graphs. First, user representations are enhanced by capturing high-order collaborative signals in user interactions. Then, the augmented user representations are applied to representation learning of items in KG. Finally, a hybrid attention function is proposed to aggregate neighbor information according to user's preferences for relations and entities when learning item representations. The experiments on two public datasets have demonstrated that UIKR outperforms the current state-of-the-art baselines.

It is an important and hot research topic to apply the CNN framework to model graph data for improving recommendations. We consider two directions for future work. (1) To make the time complexity of the UIKR model predictable, we have used a random strategy for selecting a fixed number of neighbors. We can possibly explore neighborhood samples in more effective ways than random sampling. (2) With the development of multimodal KGs, the data in a KG includes not only text, but also pictures and videos. It will be worth our while to use multimodal KGs for enhancing recommendations.

Acknowledgements. This work is supported in part by the National Natural Science Foundation of China under Grants [62120106008, 61806065], and the Fundamental Research Funds for the Central Universities [JZ2022HGTB0239].

References

1. Feng, Y., You, H., Zhang, Z., Ji, R., Gao, Y.: Hypergraph neural networks. In: Proceedings of the AAAI Conference on Artificial Intelligence, vol. 33, pp. 3558–3565 (2019)
2. Hamilton, W., Ying, Z., Leskovec, J.: Inductive representation learning on large graphs. In: Advances in Neural Information Processing Systems, vol. 30 (2017)
3. He, X., Liao, L., Zhang, H., Nie, L., Hu, X., Chua, T.S.: Neural collaborative filtering. In: Proceedings of the 26th International Conference on World Wide Web, pp. 173–182 (2017)
4. Kipf, T.N., Welling, M.: Semi-supervised classification with graph convolutional networks. arXiv preprint arXiv:1609.02907 (2016)
5. Li, Y., Tarlow, D., Brockschmidt, M., Zemel, R.: Gated graph sequence neural networks. arXiv preprint arXiv:1511.05493 (2015)
6. Ma, Y., et al.: Enhancing recommendations with contrastive learning from collaborative knowledge graph. Neurocomputing **523**, 103–115 (2023)
7. Mao, K., Zhu, J., Xiao, X., Lu, B., Wang, Z., He, X.: UltraGCN: ultra simplification of graph convolutional networks for recommendation. In: Proceedings of the 30th ACM International Conference on Information & Knowledge Management, pp. 1253–1262 (2021)
8. Sun, R., et al.: Multi-modal knowledge graphs for recommender systems. In: Proceedings of the 29th ACM International Conference on Information & Knowledge Management, pp. 1405–1414 (2020)
9. Wang, F., Li, Y., Zhang, Y., Wei, D.: KLGCN: knowledge graph-aware light graph convolutional network for recommender systems. Expert Syst. Appl. **195**, 116513 (2022)
10. Wang, H., Zhang, F., Hou, M., Xie, X., Guo, M., Liu, Q.: Shine: signed heterogeneous information network embedding for sentiment link prediction. In: Proceedings of the Eleventh ACM International Conference on Web Search and Data Mining, pp. 592–600 (2018)
11. Wang, H., et al.: RippleNet: propagating user preferences on the knowledge graph for recommender systems. In: Proceedings of the 27th ACM International Conference on Information and Knowledge Management, pp. 417–426 (2018)
12. Wang, H., et al.: Knowledge-aware graph neural networks with label smoothness regularization for recommender systems. In: Proceedings of the 25th ACM SIGKDD International Conference on Knowledge Discovery & Data Mining, pp. 968–977 (2019)
13. Wang, H., Zhao, M., Xie, X., Li, W., Guo, M.: Knowledge graph convolutional networks for recommender systems. In: The World Wide Web Conference, WWW 2019, pp. 3307–3313. Association for Computing Machinery, New York (2019)
14. Wang, J., Huang, P., Zhao, H., Zhang, Z., Zhao, B., Lee, D.L.: Billion-scale commodity embedding for e-commerce recommendation in Alibaba. In: Proceedings of the 24th ACM SIGKDD International Conference on Knowledge Discovery & Data Mining, pp. 839–848 (2018)
15. Wang, Q., Mao, Z., Wang, B., Guo, L.: Knowledge graph embedding: a survey of approaches and applications. IEEE Trans. Knowl. Data Eng. **29**(12), 2724–2743 (2017). https://doi.org/10.1109/TKDE.2017.2754499
16. Wang, X., He, X., Cao, Y., Liu, M., Chua, T.S.: KGAT: knowledge graph attention network for recommendation. In: Proceedings of the 25th ACM SIGKDD International Conference on Knowledge Discovery & Data Mining, pp. 950–958 (2019)

17. Wang, Z., Lin, G., Tan, H., Chen, Q., Liu, X.: CKAN: collaborative knowledge-aware attentive network for recommender systems. In: Proceedings of the 43rd International ACM SIGIR Conference on Research and Development in Information Retrieval, pp. 219–228 (2020)
18. Wu, S., Sun, F., Zhang, W., Xie, X., Cui, B.: Graph neural networks in recommender systems: a survey. ACM Comput. Surv. **55**(5), 1–37 (2022)
19. Wu, Z., Pan, S., Chen, F., Long, G., Zhang, C., Yu, P.S.: A comprehensive survey on graph neural networks. IEEE Trans. Neural Netw. Learn. Syst. **32**(1), 4–24 (2021). https://doi.org/10.1109/TNNLS.2020.2978386
20. Xia, F., et al.: Graph learning: a survey. IEEE Trans. Artif. Intell. **2**(2), 109–127 (2021). https://doi.org/10.1109/TAI.2021.3076021
21. Yang, Z., Dong, S.: HAGERec: hierarchical attention graph convolutional network incorporating knowledge graph for explainable recommendation. Knowl.-Based Syst. **204**, 106194 (2020)
22. Yu, X., et al.: Personalized entity recommendation: a heterogeneous information network approach. In: Proceedings of the 7th ACM International Conference on Web Search and Data Mining, pp. 283–292 (2014)
23. Zhang, F., Yuan, N.J., Lian, D., Xie, X., Ma, W.Y.: Collaborative knowledge base embedding for recommender systems. In: Proceedings of the 22nd ACM SIGKDD International Conference on Knowledge Discovery and Data Mining, pp. 353–362 (2016)
24. Zhao, H., Yao, Q., Li, J., Song, Y., Lee, D.L.: Meta-graph based recommendation fusion over heterogeneous information networks. In: Proceedings of the 23rd ACM SIGKDD International Conference on Knowledge Discovery and Data Mining, pp. 635–644 (2017)
25. Zhou, J., et al.: Graph neural networks: a review of methods and applications. AI Open **1**, 57–81 (2020)

How Does the System Perceive Me? — A Transparent and Tunable Recommender System

Mingman Xu, Qiong Chang(ID), and Jun Miyazaki(✉)(ID)

School of Computing, Tokyo Institute of Technology,
2-12-1 Oookayama, Meguro City 152-8550, Tokyo, Japan
xu@lsc.c.titech.ac.jp, {q.chang,miyazaki}@c.titech.ac.jp

Abstract. We present a transparent and tunable recommender system using neural networks in which the user's preference for each tag calculated from his or her rating history is extracted as a user feature, and latent knowledge about the relationship between an item and a tag is extracted as an item feature. To improve user satisfaction with recommender systems, researchers have been focusing not only on a system's recommendation accuracy but also on its transparency, novelty, and serendipity as evaluation indices. Furthermore, the degree of user involvement in the recommendation process has been shown to substantially affect user satisfaction. Therefore, we propose showing a tag cloud as the user's profile as captured by the system from the user's interaction history and providing the user with a way to tune the recommendation results so that user satisfaction can be improved.

Keywords: recommender system · user profile · transparent recommendation · tunable recommendation

1 Introduction

With the advancement of information technology and the rapid spread of the Internet, the cost of storing and processing data has decreased dramatically, and an enormous amount of information is spread over the web. While a wide variety of information has already been made available to everyone, it is becoming increasingly difficult to identify useful information from the vast amount of data. Hence, recommender systems that support user decision-making are attracting attention and being utilized in many fields.

Many conventional recommender systems are based on the assumption that recommending items that match the user's preferences leads to higher user satisfaction. However, recent research has shown that user satisfaction cannot be improved by focusing only on recommendation accuracy and that new evaluation indices such as diversity, novelty, and serendipity are required [7].

© The Author(s), under exclusive license to Springer Nature Switzerland AG 2023
C. Strauss et al. (Eds.): DEXA 2023, LNCS 14147, pp. 33–48, 2023.
https://doi.org/10.1007/978-3-031-39821-6_3

Research on satisfaction with recommender systems suggests that the transparency of recommendations, i.e., the ability of users to understand the recommendation process and reasons for the recommendations, contributes to the acceptability of recommended items, trust in the system, and satisfaction [6,15]. While many systems now provide users with reasons for recommending items, such as "this item is recommended because ...", they do not tell users how their preferences were captured. This is because it is hard to directly show it to users, in particular, when neural networks are used. To the best of our knowledge, there is no recommender system that gives the user the entire picture of user preferences.

Furthermore, one of the most important factors in determining user satisfaction is the degree of user involvement in the recommender system. Although conventional recommender systems are often studied from the viewpoint of how to reduce the user's load, several experiments have demonstrated that user involvement in the recommendation process improves satisfaction even if the user's load increases [8].

Many web services currently use the user's personal information and interaction history to make recommendations, which leaves the user with very limited means to influence the recommendations based on his or her preferences. For example, TikTok, a video-sharing application, provides users with a means to update its recommendation algorithm in response to user feedback such as "likes" or "not interested" on recommended videos. However, it is very difficult to change all past relevant ratings if the user's interests change substantially. In addition, if a user has an interest that is not in his or her interaction history, it will not be reflected in the recommendation results unless the user searches for and evaluates items related to that interest. Under these circumstances, users are more likely to unknowingly fall into a *filter bubble*, which reduces their exposure to information outside their interests and to new information [11].

Therefore, we aim to improve user satisfaction by providing users with a higher level of involvement while increasing the transparency of the recommender system. Specifically, by making use of tags which are keywords labeled on items, we have devised a system that presents a tag cloud of an individual user's tag preferences calculated on the basis of the user's rating history in order to provide an overall picture of the user's preferences and to enable users to adjust their tag profile to match their preferences. A user can then adjust the tag profile in accordance with his or her preferences to obtain a new recommendation result.

However, in the real world, user preferences often change over the course of using a system due to a variety of factors. For the system to recommend items that match the user's current preferences, it is necessary to consider the relationship between the user's previous and subsequent interaction histories. Therefore, the recommendation algorithm of the proposed system extracts the relationship between items and tags as item features and time-series data for the user's preference and rating history for each tag as user features. Both features are used for making recommendations using neural networks.

This paper is organized as follows. Section 2 describes related work. Section 3 introduces our method. We show experimental results and discuss them in Sect. 4. Section 5 summarizes the key points.

2 Related Work

2.1 Explainable Recommendation

Explainable recommendation refers to a recommendation algorithm that recommends items and at the same time provides the reason the items are recommended to the user. For instance, an explainable recommender system provides explanations like "this item has been bought by people similar to you" or "these items are similar to the item you are currently browsing" on Amazon and other online shopping service sites. Recommendation explanations in recommender systems are expected to have various effects, such as increasing trust and satisfaction with the system and assisting the user's decision-making.

The category of explanation is divided into three components: (1) input, or user models, (2) process, or algorithms, and (3) output, or recommended items. An explainable recommender system that focuses on process aims to understand how the algorithm works. Various studies have focused on this point. Given an arbitrary trained recommendation model, Ribeiro et al. [13] proposed a method for learning input-output pairs for the model by using linear regression and for interpreting the model with the weight of each feature as the feature's importance.

Recommender systems that focus on explaining outputs regard the recommendation process as a black box and try to justify the recommendation results. Liu et al. [10] provided a visual and knowledge-level explanation of the interactions between users and items with a knowledge graph constructed by users' reviews. Wang et al. [17] fused a knowledge graph and sentiment extracted from review aspect information to generate recommendation reasons.

Recommender systems that focus on explanations for inputs have become increasingly important in recent years. This is because users are concerned about the collection and use of their personal data in the system, and there is a growing interest in transparency of the user model as input. An example of this type of recommender system is System U developed by Badenes et al. [1]. System U uses psycho-linguistic analysis to automatically derive personality traits from social media information.

2.2 Interactive Recommendation

Recommender systems in web services such as Amazon and Netflix aim to present recommended items that fit well with the user's overall preferences in order to reduce their decision-making efforts.

However, users may feel too dominated by the system, which reduces their exposure to more diverse and novel information. In particular, in cold-start situations, where it is difficult to make recommendations for targets that do not yet have sufficient interaction history, considerable effort is required on the part of the user before an appropriate recommendation can be obtained. In light of these limitations, research on interactive recommendation has recently focused on increasing the degree of user involvement in the recommendation process and improving the transparency of the system.

In the Movie tuner interface proposed by Vig et al. [16], the user adjusts his or her preferences for tags to obtain recommendations for movies that differ from the currently recommended ones in terms of certain features while they are similar overall.

Many other studies have focused on the degree of involvement. The TEA (Tunable and Explainable Attributes) system [5] devised by Farooq Butt et al. enables users to adjust the importance of demographic features such as age, gender, and occupation. Chen et al. developed a technique called *critiquing* that enables users to manipulate recommendation results by rating the attributes of recommended items [4]. Taste Weights [3], a model devised by Bostandjiev et al., makes recommendations on the basis of different sources of information available on social networks and based on user feedback. Denis et al. [12] developed a hybrid recommendation system that enables the user to control how the algorithms are combined. These approaches aim to increase the transparency of the recommender system and provide users with more satisfactory recommendations.

2.3   Tag-Based Recommendation

The use of tags, or keywords, on web sites such as Delicious, Flickr, and Amazon to search and post contents has become widespread. This is because tags are easily manageable and searchable by users. Tags can describe what a particular item is, what it is about, and what characteristics it has, and they are useful in explaining recommendations. Shirky [14] asserted that since tags are created by the user, they represent concepts that are meaningful to the user. Tags serve as a bridge for users to understand unknown relationships between items and themselves. The user's attitude toward tags can also be seen as a reflection of their preferences.

There are many studies on tag-based recommendation systems, and they have shown that tags can help users discover and understand items and make decisions [2,9]. However, none of them applies tags to a user feedback mechanism.

3   Proposed System

Our proposed recommender system shows how it perceives a user's profile on the basis of his or her tag preferences as a tag cloud and provides a user feedback mechanism based on the tag cloud. The principle of deciding this design is to provide more intuitive and easier user interface to users.

An overview of our system is shown in Fig. 1. The left side shows the flow of the tag-based recommendation model. Specifically, the item part (pink boxes) embeds item and tag relevance information, and the user part (green boxes) embeds user interaction history. The generated item and user embeddings are each learned by a neural network and used to generate latent factors. The ratings of the items to be recommended to this user are predicted by calculating the inner product of the item feature and the user feature. On the basis of the predicted ratings, the Top-k recommended items are generated and presented to the user. At the same time, on the basis of the user's tag preferences, a tag cloud representing the user's profile is generated and presented to the user. The user can then adjust the tag preferences in accordance with his or her current preferences. The system immediately updates the recommendations in accordance with the adjusted tag preferences.

The proposed system can be divided into three components: item modeling, user modeling, and rating prediction. We explain the user feedback mechanism as well as these three components in detail in the following sections.

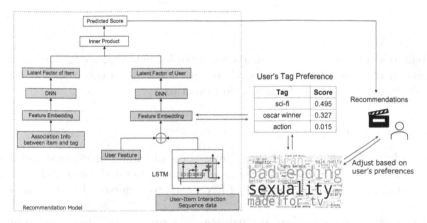

Fig. 1. Overview of proposed recommender system

3.1 Item Modeling

Item modeling uses a neural network to learn item-tag associations on the basis of the tag genome proposed by Vig et al. [16]. They are created by extracting content features such as user reviews, ratings, and tags on the basis of human judgment of tag relevance derived from user surveys conducted by the Movie-Lens group. As a biological genome contains all the genetic information of DNA, a tag genome contains information about the relationships between items and common tags in a vector space. The tag genome indicates the degree of association between each item and its tag by a sequence of values ranging from 0 to 1 (0 means no association at all, and 1 means very strong association). For example, Table 1 shows that the movie "Toy Story" has a higher association with the tag "toys" than with the tag "computer animation".

Table 1. Example of tag genome

Toy Story (1995)	
Tag	Relev.
toys	0.999
computer animation	0.998
pixar animation	0.995
kids and family	0.995
animation	0.987
animated	0.979
children	0.973
cartoon	0.947
pixar	0.941
disney	0.929

Item modeling is performed using a fully coupled neural network. The number of units in the input layer corresponds to the number of tags. Item features $\tilde{I}$ are calculated using the ReLU activation function.

3.2 User Modeling

User modeling is performed by combining the embedding of inferred user tag preference information and that of time-series data generated using LSTM (long short-term memory), which is a type of RNN used to understand changes in user interests.

RNNs are effective for learning patterns of temporal variation from continuous data. In an RNN, the output of one layer is not only used as the input of the next layer but also as an output, whereas the input of one layer in a conventional neural network is only used as input for the next layer. RNNs are not well suited to handling data with *long-term dependencies* beyond the gradient explosion and disappearance problem of convolutional neural networks. To avoid this long-term dependence problem, our user model uses LSTM to learn the long-term dependencies in a user's rating history.

Specifically, we represent each user as an $L_{ts} \times E_u$ matrix, where L_{ts} is the maximum length of the time series, and E_u is the number of dimensions of the user embedding. Potential dependencies in the rating history are learned and used as additional information for defining the user's features, as represented in Eq. (1), where w_1 and w_2 are the weights of the user embedding and time-series embedding, respectively. User modeling, like item modeling, uses a fully coupled neural network. However, the user's tag preference information is used as input.

$$U_{embedding} = w_1 \cdot U_{feature\_embedding} + w_2 \cdot U_{seq\_embedding} \qquad (1)$$

3.3   Rating Prediction

The rating value determined from the item and user features for user u_i is calculated using

$$\hat{r}_{ui} = \tilde{U} \cdot \tilde{I}, \tag{2}$$

where $\hat{r}$ is the predicted value, and $\tilde{U}$ and $\tilde{I}$ are the user and item features, respectively. Loss function $J(\theta)$ is defined as the squared error between the actual and predicted values:

$$J(\theta) = \sum_{r_{ui} \in R} (\hat{r_{ui}} - r_{ui})^2. \tag{3}$$

To further improve the accuracy of the model, we introduce a bias term representing the average rating per user and per item, while several techniques, such as a z-score, to deal with the bias exist. User bias can be considered as the inherent tendency of users to give higher or lower ratings than the average for the system as a whole. Similarly, item bias represents the tendency for a particular item to receive a higher or lower rating than the average. When the bias term is taken into account, the predicted values of the ratings are modified using

$$\hat{r}_{ui} = \tilde{U} \cdot \tilde{I} + \mu + b_i + b_u, \tag{4}$$

where μ is the mean of all non-missing values, b_i is computed by subtracting μ from the average rating of item i and is the item's rating bias term, and b_u is the user's rating bias, calculated by subtracting μ from the average rating value of user u.

3.4   User Feedback

To improve the acceptability of recommended items to a user, trust in the system, and satisfaction with its use, we introduce a mechanism that requires users to provide feedback. To help users intuitively understand how their preferences are perceived by the system, a tag cloud like that shown in Fig. 2 is generated from the user embedding like that shown in Table 2 and presented to the user as a tag cloud corresponding to the user's profile. From the tag cloud in Fig. 2, the user can understand that the system perceives that he or she is interested in movies that are highly related to the tags "bad ending", "sexuality", and "made for tv".

This tag cloud enables the user to adjust the tag size in accordance with his or her preferences. The embedding weight of a tag is updated in response to the adjustment by using Eq. (5). The terms $weight_{tag}$ and $weight_{t\hat{a}g}$ are the embedding weights of the tag before and after adjustment, respectively, and $weight_{variation}$ is the amount the embedding weight is adjusted.

$$weight_{t\hat{a}g} = \frac{weight_{tag} \sum weight_{tag}}{\sum weight_{tag} + \sum weight_{variation}} \tag{5}$$

As an example, if we adjust the embedding weights of the tags "pixar animation" and "007 (series)" to be 0.0499 and 0.0500, respectively, the tag "pixar

Table 2. Example of user embedding

tag	embedding weight
bad ending	0.049998
sexuality	0.049998
made for tv	0.049923
iran	0.049886
better than expected	0.049879
neo-nazis	0.049811
romantic	0.049804
male nulity	0.049743
witch	0.049578
competition	0.049498

Fig. 2. Example of tag cloud

animation" is placed between the tags "made for tv" and "iran" and the tag "007 (series)" between the tags "romantic" and "male nudity". The updated user embedding profile and tag cloud are shown in Table 3 and Fig. 3, respectively.

Table 3. Example of user embedding after tag adjustment

tag	embedding weight
bad ending	0.049888
sexuality	0.049888
made for tv	0.049814
pixar animation	0.049790
iran	0.049777
better than expected	0.049770
neo-nazis	0.049702
romantic	0.049694
007 (series)	0.049691
male nudity	0.049634

Fig. 3. Example of tag cloud after tag adjustment

In accordance with this user feedback, the system updates the feature embedding, which was learned by combining the user feature shown in Fig. 1 and the embedding generated by using LSTM for time-series data. This update causes the user profile to be recalculated and the user features to be updated, which results in a new recommendation list being generated.

4   Experimental Evaluation

Our aim was to construct a transparent and tunable recommendation model. Therefore, we first needed to design an underlying highly accurate recommender system based on our proposed tag-based model. We then evaluate our transparent and tunable recommendation model which is built on top of this recommender system.

After determining the best underlying recommender system out of four design choices, we evaluate our transparent and tunable recommendation model based on user feedback through case studies. This is because being able to see the generated tag profile as a cloud enables the user to understand how the system understands his or her preferences, to efficiently adjust his or her preferences, and to obtain recommendations that match those preferences.

4.1   Datasets

We conducted experiments using the MovieLens-1M (ml-1m) and MovieLens-25M (ml-25m) datasets, which are publicly available from GroupLens, to evaluate the effectiveness of the proposed method. From each dataset, we selected only users who had rated at least 20 movies. A summary of the datasets is shown in Table 4. For the experiments, the datasets were sorted on the basis of time stamps, with 80% of the data from the oldest to the newest used as training data and the remaining 20% used as test data.

Table 4. Summary of datasets

Dataset	ml-1m	ml-25m
Users	6,019	162,541
Movies	2,618	13,816
Ratings	404,273	24,674,113
Tags	1,128	1,128
Rating Scale	$\{1, 2, 3, 4, 5\}$	$\{0.5, 1, \cdots, 4.5, 5\}$

4.2   Evaluation Metrics

We evaluated recommendation accuracy using RMSE, which is a measure of accuracy in predicting ratings and is defined by

$$RMSE = \sqrt{\frac{1}{n}\sum_{i=1}^{n}(\hat{r_i} - r_i)^2}, \tag{6}$$

where n is the total number of predictions, and r_i and $\hat{r_i}$ are the measured and predicted values, respectively.

4.3   Design Choices

To evaluate the effect of using LSTM and the bias term on the underlying rec-ommender system performance, we created four systems, each with a different design, and conducted experiments. For training, the hyperparameters were set so that the number of dimensions of the hidden layer was 512, the learning rate was 0.001, and the batch size was 1024, on the basis of the exploratory experimental results.

– Rec: This model was trained using item embedding for the association between items and tags and user embedding for the inferred user tag prefer-ences.
– Rec++: This model was trained using item embedding for the item-tag asso-ciation and user embedding for the inferred user tag preferences with a bias term.
– RecLSTM: This model was trained using item embedding for the item-tag association and user embedding plus LSTM-trained time-series data for the inferred user tag preferences. The maximum length of the time series to be trained was set to 20.
– RecLSTM++: This model was trained using item embedding for the item-tag association and user embedding plus LSTM-trained time-series data for the inferred user tag preferences with a bias term. The maximum length of the time series to be trained was set to 20.

Table 5. RMSEs of recommender systems

Design	ml-1m	ml-25m
Rec	3.785	3.737
Rec++	1.154	1.248
RecLSTM	1.094	0.986
RecLSTM++	0.971	0.925

For both datasets in Table 5, the RecLSTM++ design system showed the best recommendation accuracy. The Rec++ and RecLSTM++ design systems, which take account into account user and item evaluation biases, showed better recom-mendation accuracy than Rec and RecLSTM, which do not. On the other hand, the RecLSTM and RecLSTM++ design systems, which take into account the user's history, showed better recommendation accuracy than Rec and Rec++, which do not. These results demonstrate that it is possible to improve perfor-mance and provide more accurate recommendations by considering evaluation bias and learning time-series data.

From the above discussion, we decided to use the RecLSTM++ as an under-lying recommender system for our proposed transparent and tunable system for the following two case studies.

Table 6. User embedding of target user (Case 1)

tag	embedding weight
independent film	0.090688
action	0.070832
talky	0.070338
black comedy	0.067924
weird	0.067227
enigmatic	0.065787
off-beat comedy	0.063943
low budget	0.060274
funny	0.060227
creepy	0.060054

Fig. 4. Tag cloud of target user before adjustment (Case 1)

4.4 Case Studies

We evaluated the transparency and tunable features of our proposed system through case studies because it is difficult to evaluate such features in an automatic and systematic way.

For the first case study, we randomly selected a user from the MovieLens-25m dataset. We adjusted the tag profile of the target user on the basis of user feedback and presented new recommendations in accordance with the adjusted tag profile.

Table 6 shows the part of the user embedding that contains the target user's preferences, and Fig. 4 shows the tag cloud generated from the user embedding. From this tag cloud, we can see that the system perceives that the target user is likely interested in movies that are highly related to the tags "independent film", "action", "talky", "black comedy", and "weird".

The system presented the tag cloud to the user and, at the same time, presented a list of the recommendations generated by the recommendation algorithm, which is shown in the second column in Table 8.

To test the user feedback mechanism, we changed the size of the tags "sci-fi", "adventure", and "drama" and adjusted their weights accordingly. The adjusted user embedding profile and tag cloud are shown in Table 7 and Fig. 5, respectively.

In response to this user feedback, the user embedding was updated and the ratings were recalculated and predicted. The newly generated recommendation list is shown in the third column in Table 8.

Table 9 shows the associations between the recommended items before and after adjustment and the Top-5 tags of high interest to the target user ("independent film", "action", "talky", "black comedy", and "weird") and the three adjusted tags ("sci-fi", "adventure", and "drama"). It shows that the movies with high relevance to the tags of high user interest were ranked higher. Comparing

Table 7. User embedding after adjustment (Case 1)

tag	embedding weight
independent film	0.090290
sci-fi	0.079649
adventure	0.074671
action	0.070521
talky	0.070029
drama	0.069693
black comedy	0.067626
weird	0.066932
enigmatic	0.065499
off-beat comedy	0.063662

Fig. 5. Tag cloud after adjustment (Case 1)

Table 8. Recommendation lists for target user before and after feedback (Case 1)

Top-10	Before adjustment	After adjustment
1	Heavenly Creatures	The City of Lost Children
2	The City of Lost Children	Heavenly Creatures
3	Space Jam	Space Jam
4	That Thing You Do!	That Thing You Do!
5	Artemisia	Yes, Madam
6	Glengarry Glen Ross	Artemisia
7	Miami Rhapsody	Glengarry Glen Ross
8	The White Balloon	Miami Rhapsody
9	Loaded	What's Eating Gilbert Grape
10	Second Jungle Book	The White Balloon

the before and after adjustment recommendation lists, we see that "The City of Lost Children," which was in second place before the adjustment, is now in first place due to the user feedback. This is because "The City of Lost Children" is more relevant to the adjusted tags than "Heavenly Creatures", which was in first place in the original recommendation list. In addition, "Yes, Madam" and "What's Eating Gilbert Grape", which were not in the Top-10 recommended list before the adjustment, are now in the Top-10 as a result of the user feedback. These two movies were apparently recommended due to their very high relevance to the tags "action" and "weird", respectively, and because they have higher relevance to the three adjusted tags than the other movies.

Table 9. Relationships between recommended movies and tags (Case 1)

Movie	indepen-dent film	action	talky	black comedy	weird	sci-fi	adven-ture	drama
Heavenly Creatures	0.484	0.070	0.509	0.266	0.805	0.023	0.111	0.631
The City of Lost Children	0.273	0.105	0.542	0.289	0.984	0.602	0.312	0.330
Space Jam	0.131	0.602	0.209	0.088	0.257	0.087	0.313	0.137
That Thing You Do!	0.170	0.153	0.220	0.103	0.276	0.043	0.212	0.360
Artemisia	0.253	0.116	0.263	0.102	0.428	0.029	0.104	0.420
Glengarry Glen Ross	0.527	0.161	0.828	0.374	0.470	0.019	0.086	0.931
Miami Rhapsody	0.486	0.101	0.376	0.177	0.225	0.043	0.117	0.279
The White Balloon	0.280	0.067	0.403	0.147	0.466	0.019	0.197	0.196
Loaded	0.625	0.223	0.323	0.198	0.418	0.071	0.147	0.331
Second Jungle Book	0.101	0.184	0.085	0.065	0.357	0.045	0.313	0.150
Yes, Madam	0.074	0.959	0.229	0.090	0.179	0.058	0.200	0.192
What's Eating Gilbert Grape	0.444	0.076	0.503	0.226	0.527	0.020	0.079	0.790

Fig. 6. Tag cloud of target user before adjustment (Case 2)

Fig. 7. Tag cloud of target user after adjustment (Case 2)

Table 10. Recommendation lists before and after adjustment (Case 2)

Top-10	Before adjustment	After adjustment
1	Blackmail	Blackmail
2	Gordy	Black Beauty
3	Burnt by the Sun	Gordy
4	Black Beauty	Burnt by the Sun
5	Nemesis 2: Nebula	Nemesis 2: Nebula
6	Pretty in Pink	Mother Night
7	For Whom the Bell Tolls	Pretty in Pink
8	The Philadelphia Story	For Whom the Bell Tolls
9	Broken Arrow	The Philadelphia Story
10	House Party 3	Broken Arrow

To verify the effectiveness of the proposed transparent and tunable recommender system, another case study was performed for a different user. The tag cloud for this user before adjustment is shown in Fig. 6, and the one after adjust-

Table 11. Associations between recommended items and tags (Case 2)

Movie	money	dark hero	paranoid	007 (series)	coming-of- age	oscar winner	vampires
Blackmail	0.066	0.818	0.624	0.019	0.278	0.710	0.003
Gordy	0.100	0.608	0.480	0.020	0.189	0.510	0.003
Burnt by the Sun	0.052	0.424	0.302	0.043	0.332	0.832	0.002
Black Beauty	0.067	0.337	0.263	0.019	0.275	0.948	0.003
Nemesis 2: Nebula	0.023	0.160	0.103	0.024	0.280	0.861	0.003
Pretty in Pink	0.081	0.449	0.235	0.019	0.449	0.797	0.003
For Whom the Bell Tolls	0.049	0.394	0.242	0.034	0.301	0.750	0.003
The Philadelphia Story	0.024	0.426	0.197	0.038	0.256	0.795	0.003
Broken Arrow	0.017	0.401	0.142	0.034	0.143	0.843	0.005
House Party 3	0.632	0.150	0.342	0.015	0.124	0.454	0.003
Mother Night	0.038	0.458	0.616	0.018	0.216	0.962	0.003

ment with larger tags "oscar winner" and "vampires" is shown in Fig. 7. Table 10 shows the recommendations before and after adjustment. From Table 11, it can be seen that movies with higher associations with the tag "oscar winner" are ranked higher due to user feedback.

5   Conclusion

Our proposed recommender system is aimed at improving transparency and user involvement. The former aim is achieved by presenting the user's tag preferences corresponding to the user's profile to the user in the form of a tag cloud, and the second is achieved by enabling the user to adjust the tag profile in accordance with his or her preferences. A new recommendation list is then presented that is based on the adjusted profile.

We built and compared four neural-network-based recommender systems with different designs as an underlying system of our proposed system, in which item embedding represents the relationship between items and tags and user embedding represents the time-series data of the user's preference and rating history for each tag. The results demonstrated that considering bias and learning time-series data improved performance.

Evaluation using two case studies with user feedback demonstrated that our proposed system enables users to obtain an overall picture of their preferences and that adjusting their preferences on the basis of user feedback yields new recommendation results. This should lead to increased user satisfaction with the recommender system.

Although the proposed recommender system can provide new recommendation results to users in accordance with their adjusted preferences, it may not be able to do so if the amount of user adjustment is small. Moreover, since the system does not consider relationships among tags, when user embedding is updated in response to user feedback, tags that are highly similar to the tags to be adjusted may be updated in the direction opposite to the user preferences. This may cause the system to learn different user preferences for the same preference, which may cause the system to operate improperly. We intend to investigate these limitations and improve the system to obtain more satisfactory recommendation results. Future work also includes assessing the effectiveness of the proposed system in the real world through user satisfaction surveys, and dealing with scalability issues of the proposed system.

Acknowledgment. This work was partially supported by JST, CREST Grant Number JPMJCR22M2, and Japan Society for the Promotion of Science (JSPS) KAKENHI Grant Numbers 23H03401 and 23H03694, Japan.

References

1. Badenes, H., et al.: System U: automatically deriving personality traits from social media for people recommendation. In: Proceedings of the 8th ACM Conference on Recommender Systems, pp. 373–374. ACM (2014)
2. Bogers, T.: Tag-based recommendation. In: Brusilovsky, P., He, D. (eds.) Social Information Access. LNCS, vol. 10100, pp. 441–479. Springer, Cham (2018). https://doi.org/10.1007/978-3-319-90092-6_12
3. Bostandjiev, S., O'Donovan, J., Höllerer, T.: TasteWeights: a visual interactive hybrid recommender system. In: Proceedings of the Sixth ACM Conference on Recommender Systems, pp. 35–42. ACM (2012)
4. Chen, L., Pu, P.: Critiquing-based recommenders: survey and emerging trends. User Model. User-Adapted Interact. **22**(1), 125–150 (2012)
5. Farooq Butt, M.H., Li, J.P., Saboor, T.: A tunable and explainable attributes (TEA) for recommendation system. In: 2020 17th International Computer Conference on Wavelet Active Media Technology and Information Processing (ICCWAMTIP), pp. 39–43 (2020)
6. Gedikli, F., Jannach, D., Ge, M.: How should I explain? A comparison of different explanation types for recommender systems. Int. J. Hum. Comput. Stud. **72**(4), 367–382 (2014)
7. Herlocker, J.L., Konstan, J.A., Terveen, L.G., Riedl, J.T.: Evaluating collaborative filtering recommender systems. ACM Trans. Inf. Syst. **22**(1), 5–53 (2004)
8. Hijikata, Y., Kai, Y., Nishida, S.: The relation between user intervention and user satisfaction for information recommendation. In: Proceedings of the 27th Annual ACM Symposium on Applied Computing, pp. 2002–2007. ACM (2012)
9. Kuan, J.: How to Build a Simple Movie Recommender System with Tags, December 2019. https://towardsdatascience.com/how-to-build-a-simple-movie-recommender-system-with-tags-b9ab5cb3b616. Accessed 30 May 2023
10. Liu, Y., Miyazaki, J.: Knowledge-aware attentional neural network for review-based movie recommendation with explanations. Neural Comput. Appl. **35**(3), 2717–2735 (2022)
11. Pariser, E.: The Filter Bubble: What The Internet Is Hiding From You. Penguin, London (2012)
12. Parra, D., Brusilovsky, P., Trattner, C.: See what you want to see: visual user-driven approach for hybrid recommendation. In: Proceedings of the 19th International Conference on Intelligent User Interfaces, pp. 235–240. ACM (2014)

13. Ribeiro, J., Carmona, J., Mısır, M., Sebag, M.: A recommender system for process discovery. In: Sadiq, S., Soffer, P., Völzer, H. (eds.) BPM 2014. LNCS, vol. 8659, pp. 67–83. Springer, Cham (2014). https://doi.org/10.1007/978-3-319-10172-9_5

14. Shirky, C.: Ontology is Overrated: Categories, Links, and Tags, April 2005. https://gwern.net/doc/technology/2005-04-shirky-ontologyisoverratedcategoriesli nksandtags.html. Accessed 2 Apr 2023

15. Sinha, R., Swearingen, K.: The role of transparency in recommender systems. In: CHI '02 Extended Abstracts on Human Factors in Computing Systems, pp. 830–831. ACM (2002)

16. Vig, J., Sen, S., Riedl, J.: Navigating the tag genome. In: Proceedings of the 16th International Conference on Intelligent User Interfaces, pp. 93–102. ACM (2011)

17. Wang, Z., Li, Y., Fang, L., Chen, P.: Joint knowledge graph and user preference for explainable recommendation. In: 2019 IEEE 5th International Conference on Computer and Communications (ICCC), pp. 1338–1342 (2019)

MERIHARI-Area Tour Planning by Considering Regional Characteristics

Sotaro Moritake[1]([✉]), Hidekazu Kasahara[2], and Qiang Ma[3][ID]

[1] Kyoto University, Kyoto, Japan
moritake.sotaro.b18@kyoto-u.jp
[2] Osaka Seikei University, Osaka, Japan
kasahara.hidekazu.k13@kyoto-u.jp
[3] Kyoto Institute of Technology University, Kyoto, Japan
qiang@kit.ac.jp

Abstract. There are various styles of sightseeing, such as sightseeing tours and city walk, and support is required to meet the various needs derived from these styles. Most existing methods specialize in a single sightseeing style and optimize time and distance. We proposes a "MERIHARI Area Tourism" style in which sightseeing spots, POI (Point Of Interest) are clustered into multiple areas and visiting style is varied for each area, so that visitors can enjoy the "representative" of the sightseeing city. The proposed method divides a tourist city into multiple areas (clusters) by clustering tourist POIs. Next, features obtained from trajectory data and image data are analyzed to calculate representative spots that symbolize area characteristics, and multiple area visitation courses (plans) that include these places are generated. The system then optimizes these courses based on the user's time and other conditions and generates a sightseeing plan, contrasting the number of visiting spots in each area.

Keywords: Smart Tourism · Tour Planning · POI Mining

1 Introduction

There are various styles of sightseeing. For example, there is the "Sightseeing tours" style, which aims to visit as many famous and well-known tourist spots as possible tourists. On the other hand, the "City walking" style, in which tourists aim to experience the atmosphere of the city by walking around the area. These styles differ regarding the selection of POIs (Point Of Interest), duration of stay, and area transition. However, tourists may prefer to a combination of multiple tourist styles, except in the case of group tours where the plan is decided by travel agents. Therefore, by combining multiple tourism styles, it will be possible to support tourists with high satisfaction.

This work was partly supported by JSPS KAKENHI (23H03404, 21K12140) and MIC SCOPE (201607008).

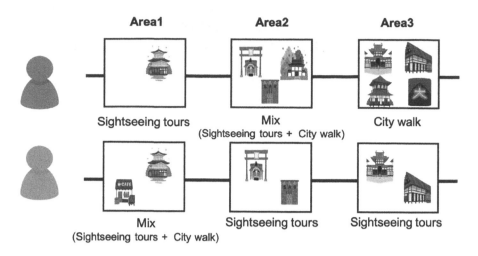

Fig. 1. Overview of the MERIHARI Area Tourism

Existing research has mainly focused on a single style of sightseeing. For example, some research tries planning a tour by clustering POIs and optimizing the plan in terms of time and distance [2]. In contrast, others analyze POIs suitable for city walking by evaluating the number of nearby POIs to walk around major tourist POIs [3]. In addition, the characteristics (appearance, name recognition, etc.) of individual POIs are not considered in detail when compared and analyzed with those of other POIs. The comparison and analysis of the characteristics of POIs will enable tourism planning that allows visitors to enjoy the attractions of a city or region and visit a wide range of tourist POIs with different characteristics.

In this study, we propose and support "MERIHARI Area Tourism style," in which we divide tourist spots into multiple areas and vary the granularity of the selection of tourist spots in each area to allow visitors to enjoy the "representative" of the tourist city. The word "MERIHARI" is derived from the Japanese word meaning "to make things strong or weak" or "not monotonous". An overview of the MERIHARI Area Tourism is shown in Fig. 1. We divide a tourist city into several areas, and each area determines its visitation style of sightseeing. MERIHARI area planning can choose different sightseeing styles for each area, and even for the same area, different sightseeing styles can be selected for each user. In the proposed method, We construct plans by performing four major processes: clustering, extracting representative spots, course generation, and planning for a set of POIs the user has selected as candidates to visit among tourist cities.

The contributions of this study are as follows:

– We advocate a new sightseeing style, "MERIHARI Area Tourism". It proposes a mechanism of combining multiple sightseeing styles. We also propose a method that considers the value of POIs in tourist attractions by introducing the concept of representative spots, generating multiple visit plans to correspond to each tourist style, and generating a plan optimized according to the user's conditions (Sect. 3).

- We proposed a method for extracting representative spots in tourist cities and areas. We extract representative spots by evaluating the degree of recognition and visual characteristics obtained from the Normalized Google Distance and Vision Transformer (Sect. 3.2).

2   Related Work

D. Gavalas et al. [2] performed route generation by clustering POIs and then sequentially selecting POIs from multiple clusters according to time budget to optimize the user's time budget and efficiency of travel for planning. However, this study regarded POIs as data points, and individual preferences and regional characteristics are not considered. In our study, POIs selection considers regional characteristics, and courses are set for each area, thus enabling the weighting of stay time for areas and POIs. Furthermore, the novelty of our method is that it allows visitors to visit more places with different characteristics by considering differences in tourist destinations (areas and POIs).

Hara et al. [3] calculated the degree of suitability for city walking based on the total number of nearby walking spots for a predefined set of major sightseeing spots. However, switching between multiple sightseeing styles is limited to supporting the user's decision-making and does not consider the sightseeing spots' regional and urban characteristics. The novelty of our proposed method is that it switches sightseeing styles within the plan by selecting spots that consider the characteristics of regions and cities, called "representative spots," and by optimizing multiple courses generated based on them according to conditions.

Y. Shen and M. Ge et al. [8] estimated the value of tourism in terms of natural and cultural values and analyzed geosocial images. J. Sun et al. [9] extracted POIs from trajectory data collected from tourists and applied a weighted HITS-based algorithm [5] to classify the nature of POIs. J. Sun et al. [10] analyzed tourists' travel patterns from trajectory data and proposed a travel route recommendation method that combines knowledge of locations and transitions and rewards both.

3   MERIHARI Area Tour Planning

Figure 2 shows an overview of the proposed method. The main flow of the method consists of trajectory data, image data, the user's time budget, and a set of candidate POIs as inputs, and a sightseeing plan (MERIHARI area plan) as an output. First, we construct a POI database from trajectory data and image data. From this database, the user selects a set of candidate POIs to visit. We perform clustering on the candidate POI sets based on the actual travel distance to generate areas (clusters). We then extract a "representative spot" for each area, symbolizing the tourist city and area, and consider the regional

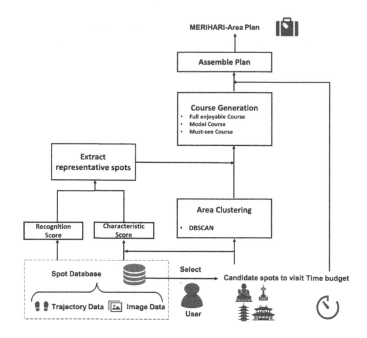

Fig. 2. Overview of the MERIHARI Area Tourism Framework

characteristics. Based on the representative spots for each area, we generate multiple courses (visit plans) to express the sightseeing style. Then, we combine and optimize these courses according to the user's time and budget to generate a MERIHARI area plan.

3.1 Area Clustering

We use the well-known density clustering method DBSCAN to generate areas from a set of candidate POIs, using the walking time calculated by the route search function of the Open Route Service (ORS)[1]. ORS allows us to consider the cost of taking a passable route. In this case, the radius of the clusters is set to a travel range of 8 to 9 min on foot by ORS, and the existence of a cluster (area) with a single data point (POI) is allowed.

3.2 Extraction of Representative Spots

We describe a method for extracting representative spots in the areas generated in Sect. 3.1. The definitions of representative spots are as follows.

Representative Spots
POIs in the generated areas adequately represent the characteristics and visibility of the city or area in particular.

---

[1] https://openrouteservice.org.

Representative spots aim to extract POIs that have features that are universal in the same area but not unique in other areas or that have specific features that can only be seen in that spot throughout the city. In this study, we calculate the property of "representative" by recognition and uniqueness.

- **Recognition**
  - The cycle of many tourists visiting the city and the reviews spreading through various media by those who have visited the city.
  - For example, the degree to which Kyoto is widely recognized, such as "Kyoto is known as XX".
- **Uniqueness**
  - Focus on the "character" of the city or area, which is the motivation for tourism.
  - Indicates "unique" characteristics that cannot be found in other cities or areas.

Calculation of Recognition. Normalized Google Distance (NGD) [4] is a measure of the degree of co-occurrence between two words and is calculated by substituting the total number of search results for the two words in a Google search into the Eq. 1, where N is the total number of web pages searched by Google, $f(x)$ and $f(y)$ are the number of hits for each search term x and y, and $f(x, y)$ is the number of hits for both x and y simultaneously. In this method, we use the query "(Name of city to visit) sightseeing" and the name of each POI (e.g., "kyoto sightseeing" gion), and calculate the results to obtain the degree of recognition of a city or area and its POIs. To reduce the bias in the number of results when searching in each language, we created queries in both Japanese and English, calculated the NGD, and used the harmonic mean as the score as in the Eq. 2.

$$NGD(x, y) = \frac{\max\{\log f(x), \log f(y)\} - \log f(x, y)}{\log N - \min\{\log f(x), \log f(y)\}} \tag{1}$$

$$Score(x, y) = \frac{2(NGD_{Japanese}(x, y) \times NGD_{English}(x, y))}{NGD_{Japanese}(x, y) + NGD_{English}(x, y)} \tag{2}$$

Calculation of Uniqueness. We use the Vision Transformer (ViT) [1] to extract the features of an area or POI from an image as a feature vector and compare the similarity of the vectors inside and among the area. The similarity within an area compares the feature vector of the corresponding POI with the feature vector of each POI belonging to the same area, and the similarity among areas compares the feature vector of the corresponding POI with the average feature vector of the POIs belonging to that area for each area.

The results of these comparisons can be categorized as shown in Table 1. POIs with specific features that have low out-of-area (similarity among areas) similarity and low in-area similarity (similarity within area) and POIs with area-unique features that have low out-of-area similarity and high in-area similarity are judged to be high uniqueness.

Table 1. Comparison of feature vectors in and among area

| | | Similarity within area | |
		High	Low
Similarity among areas	High	POIs with universal characteristics in that city	POIs where similar features can be seen outside of the relevant area
	Low	POIs indicating the characteristics and visibility of the area	POIs that are specific to the city as a whole and symbolic of the city or area

Determining the Representative Spots. The features of each POI obtained from recognition and uniqueness are expressed as a representative score. We define the representative score by the following equation.

$$RepresentativeScore = 0.6(1-NGD)+0.3(1-Sim_{Areaout})+0.1(1-Sim_{Areain}) \tag{3}$$

where NGD is the Normalized Google Distance score, $Sim_{Areaout}$ is the similarity among areas, and Sim_{Areain} is the similarity within area, respectively. The coefficients of the scores are positioned as hyper-parameters so that POIs that are well-known, famous, and have features found only in that city or area will have higher scores.

Then, We determine the representative spots for each area according to the flow shown in Fig. 3.

3.3 Courses of Area Visiting

We generate three types of area visitation routes (called courses) from each area generated in Sect. 3.1, with varying degrees of granularity in the selection of tourist spots. Each course differs in terms of the number of places to visit and time spent in the area, and is defined as follows.

Full-enjoyable Course. Course to visit all POIs belonging to the area with an average length of stay.

Model Course. Course to visit the most popular POIs and representative spots in the area with an average length of stay.

Must-see Course. Course to visit visits only representative spots belonging to an area with a minimum length of stay (or an average).

To generate the courses, we use the extracted information on representative spots and length of stay. We calculate the time spent at each POI using the [9]'s method, which refers to the time spent associated with a stay point when the stay point is calculated from the trajectory data. The output of each course is information on the set of POIs belonging to the course, the time spent on the course (travel time and visiting time), and the satisfaction level of each course.

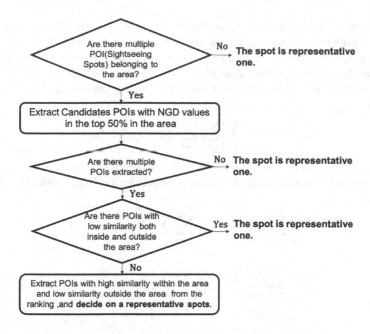

Fig. 3. Flow of determining the representative spots

We define satisfaction as the linear sum of the representative scores defined in Sect. 3.2 clause at the POIs belonging to the course, as in Eq. 4.

$$C_i^x = \sum_{k=1}^{K} RepScorex_{ik} \tag{4}$$

where C_i^x refers to the satisfaction score of course i in area x and the number of POIs belonging to course i in area x is K.

Many existing methods focus on area generation through clustering [2]. When outputting a sightseeing plan, they output a sightseeing plan that assumes that all POIs belonging to a cluster will be visited. In the proposed method, since the three types of courses are determined for each area, it is possible to create a sightseeing plan with a contrast selection of sightseeing spots according to the user's purpose and preferences.

3.4   MERIHARI Area Plan

The course generated in Sect. 3.3 is used in the context of integer programming in tourism [6,7] to define the Eq. 5 and generate a MERIHARI area plan $I = (Area_1 Course_i, \ldots, Area_N Course_i)$ that maximizes this within the user's time budget $Timebudget_u$.

We define each area and course score as C_i^x by the Eq. 4. This equation enables the generation of a flexible plan by optimizing the "area's uniqueness" and "efficiency of movement" according to the user's time budget. Therefore, instead of selecting courses from all areas one by one, a plan is generated by selecting and combining areas and courses with higher scores within the user's conditions. In addition, different sightseeing styles can be selected for each area, and even within the same area, different sightseeing styles can be selected for each user.

$$\text{maximize} \sum_{x=1}^{N} \sum_{y=1}^{N} \sum_{i=1}^{3} \sum_{j=1}^{3} (A_{x_i y_j} \times C_i^x) \tag{5}$$

where $A_{x_i y_j}$ is a binary variable set to 1 if a user visits course i in area x and then successively visits courses j in area y. Solving this Eq. 5 with the following constraints.

$$\sum_{i=1}^{3} \sum_{y=1}^{N-1} \sum_{j=1}^{3} A_{0_i y_j} = \sum_{x=1}^{N-1} \sum_{i=1}^{3} \sum_{j=1}^{3} A_{x_i N_j} = 1 \tag{6}$$

$$\sum_{x=1}^{N} \sum_{i=1}^{3} A_{x_i w_k} = \sum_{y=1}^{N} \sum_{j=1}^{3} A_{w_k y_j} \leq 1 \tag{7}$$

$$\forall w = 2 \ldots N - 1, \forall k = 1, 2, 3$$

$$\sum_{x=1}^{N-1} \sum_{y=1}^{N} \sum_{i=1}^{3} \sum_{j=1}^{3} Time_{x_i y_j} \times A_{x_i y_j} \leq Timebudget_u \tag{8}$$

$$u_i - u_j + 1 \leq (N - 1) \times (1 - A_{x_i y_j})$$
$$2 \leq u_i, u_j \leq N, \forall i, j = 2 \ldots N \tag{9}$$

Existing methods such as [6, 7] define equations to express movement between spots or areas, but our method generates multiple courses in each area and selects and combines them. Therefore, in our study, constraints on courses (the part of the sum calculation) are newly devised and added in Eqs. 6, 7, and 8, respectively. Equation 6 expresses that the plan starts in area 1 and ends in area N. Equation 7 expresses the constraint that the plan routes are connected, that the same area is not visited more than once, and that from each area at most one course out of three is selected. The constraints are expressed in the following way Eq. 8 expresses the constraint that all time spent on the plan does not exceed user u's time budget $Timebudget_u$, and Eq. 9 expresses the exclusion of sub-tours.

4   Experiments

4.1   Dataset

We used a dataset containing 91 spots and their trajectories in Kyoto Prefecture [10]. The dataset contains the name of the POIs, their location information in latitude and longitude, and the time spent at the POIs. The trajectory data include data estimated from photographs taken by tourists. We compared the average time spent in each spot obtained from the trajectory data with the average time spent in each spot shown on Google Maps and the official website of the POI, and if there was a discrepancy of two times or more, the latter was used. For the POI image data, we obtained POI images via Google Custom Search API and used approximately 10 to 30 images per spot.

4.2 Result of Area Clustering

The results of DBSCAN are shown in Fig. 4. The parameter $\epsilon = 550$ (within a 9-min walking distance) was set as the radius of the search from the data points, and $minPts = 1$ was set as the minimum number of data points required to form a cluster (area).

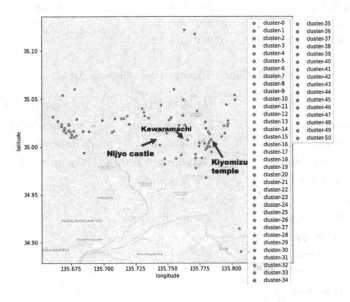

Fig. 4. Result of Area clustering

4.3 Result of Extraction of Representative Spots and Generate Courses

We were able to extract representative spots according to the decision flow in Sect. 3.2. Table 2 shows some of the results. The POIs with all values in bold in each row are the extracted representative spots. We were able to generate three courses. Table 3 shows some of the results. The POIs in boldface refer to the representative spots in the area obtained. The average time spent in the POIs of the "must-see" courses was set as the average time spent in the course.

Table 2. Example of representative spots extraction(Higashiyama Area)

POI	NGD	Similarity among Area	Similarity in Area
Sannenzaka	**0.2811**	**0.4114**	**0.4310**
Ninenzaka	**0.3426**	**0.2980**	**0.3514**
Kiyomizu Temple	**0.3552**	**0.2400**	**0.2656**
Yasaka Shrine	**0.3567**	0.6982	0.6138
Kodaiji Temple	**0.4248**	0.7339	0.6260
Kenninji Temple	**0.4464**	0.7282	0.6342
Gion	0.4801	0.5927	0.5504
Chionin	0.5219	0.7359	0.6410
Shorenin Temple	0.5302	0.5099	0.4983
Ryozen Kannon	0.5689	0.4580	0.4327
Maruyama Park	0.5692	0.3491	0.3808
Kyoto Ryozen Gokoku Shrine	0.5951	0.6992	0.6177

Table 3. Course Example(Higashiyama Area)

Full enjoyable (7.94 h)	Model (3.33 h)	Must-see (2.11 h)
Sannnenzaka	**Sannnenzaka**	**Sannnenzaka**
Ninenzaka	**Ninenzaka**	**Ninenzaka**
Kiyomizu Temple	**Kiyomizu Temple**	**Kiyomizu Temple**
Yasaka Shrine	Yasaka Shrine	
Gion	Gion	
Kenninji Temple	Kenninji Temple	
Kodaiji Temple		
Chionin Temple		
Shorenin Temple		
Ryozen Kannon		
Maruyama Park		
Kyoto Ryozen Gokoku Shrine		

4.4   Result of Generate MERIHARI Area Plan

From the areas and associated courses obtained in the previous section, an optimized plan was generated for the user's time budget. Table 4 shows the plan generated by the proposed method when the starting spot is "Kyoto Tower" (near Kyoto Station), the destination spot is "Kawaramachi" and the time budget is 8 h. Note that the satisfaction level of the course refers to the value obtained from Eq. 4 in Sect. 3.3. The results show that multiple tourist styles can be combined within a condition.

Table 4. MERIHARI Area plan Example (Start: Kyoto Tower, Goal: Kawaramachi, Time Budget: 8 h)

Order	Course	Spots to visit	Required time (hours)	Course Satisfaction Score
1	Must see	Kyoto Tower	0.32	0.793
2	Model	Kiyomizu temple, Ninenzaka Sannenzaka, Gion, Yasaka-shrine	4.10	3.51
3	Full enjoyable	Heian Shrine, Okazaki Park	1.76	1.10
4	Model	Kawaramachi	1.77	0.634

Table 5. Query used for evaluation

Start Point	End Point	Timebudget (hours)
Kyoto Tower	Kawaramachi	4, 6, 8
Kyoto Tower	Arashiyama Station	4, 6, 8
Kyoto Tower	Kinkaku Temle	4, 6, 8

4.5   Evaluation Methods

We conducted an ablation study of the index for extracting representative spots, and evaluated them using nDCG. We also conducted a user study. We prepared three queries and generate tour plans. Here, a query consisting of a start point, end point, and time budget was set, and the course plan generated from the query was evaluated.

Ablation Study. We evaluate our method of extracting representative spots (combining scores of recognition and uniqueness). In all areas consisting of multiple spots, the validity of representative spots was evaluated by comparing the combined scores and their Ablation Study rankings with the order of validity of representative spots by the author using $nDCG@k$ (k = 1, 3, 5, 10).

User Study. We conducted a user questionnaire to evaluate the proposed method. The participants were three people in their 20 s, all of whom had experience sightseeing in Kyoto Prefecture. A query (Eq. 10) consisting of a starting point, an end point, and a time budget was set, and 9 different plans (3 time budgets × 3 plans) generated from each query were evaluated. The queries used in the evaluation are shown in Table 5.

$$Query = (Startpoint, Endpoint, Timebudget) \tag{10}$$

Table 6. Results for Representative Spots Decision Score

Score Combinations	nDCG@1	nDCG@3	nDCG@5	nDCG@10	nDCG
Proposed Method	**0.9028**	**0.9699**	**0.8700**	0.9032	**0.9869**
NGD	0.8958	0.8950	0.8211	0.8599	0.9807
Similarity in Area	0.7014	0.8470	0.7196	0.8123	0.9779
Similarity among Area	0.7153	0.8141	0.8686	0.8804	0.9679
NGD+Similarity in Area	0.9028	**0.9699**	**0.8700**	0.9001	0.9864
NGD+Similarity among Area	0.9028	0.9589	0.8249	0.8889	0.9810
Similarity among Area+Similarity in Area	0.7569	0.8937	0.8686	0.9062	0.9788
NGD+Similarity among Area+Similarity in Area	0.8056	0.9588	0.8308	**0.9204**	0.9816

Table 7. Areas to be evaluated and their representative locations used in user study

Area number	Spot
1	**Kyoto Tower, Higashi Honganji Temple**
2	**Heian Shrine, Okazaki Park**
3	**Kiyomizu Temple, Sannenzaka, Ninenzaka, Gion, Yasaka Shrine,** Chionin Temple, Kenninji Temple, Kodaiji Temple, Maruyama Park Shorenin Temple, Ryozen Kannon, Kyoto Ryozen Gokoku Shrine
4	**The bamboo forest path, Togetsukyo Bridge, Arashiyama Station,** Tenryuji Temple, Arashiyama Park, Arashiyama Trokko Station Okochi Sanso Garden, Jojakkoji Temple, Seiryoji Temple Giouji Temple, Nisoin Temple, Hokyoin Temple
5	**Kinkakuji Temple, Kitano-Tenmangu Shrine,** Hirano Shrine, Wara-tenjingu Shrine

We compared three methods as follows.

Proposed Method. Both areas and courses are considered. Optimization is performed for area and course.

Area Method. Only area is considered in planning. Optimization is performed for the area.

Spot Method. Area and courses are not considered. The tour is planned by using spots, which is well studied in conventional methods.

In the user questionnaire, we asked users to review the generated plans and evaluate the adequacy of the representative spots, the adequacy of the plans

Table 8. Evaluation results for representative spots in user study

	Area 1	Area2	Area3	Area4	Area5	Average
nDCG	1.000	1.000	0.761	1.000	1.000	0.952

Table 9. Area Planning Example (Start: Kyoto Tower, End: Kawaramachi, Time Budget: 8 h)

Order	Spots to visit	Required Time (hours)
1	Kyoto Tower, Higashi Honganji Temple	0.94
2	Kyoto aquarium	1.20
3	Toji Temple	0.67
4	Heian jingu Shrine, Okazaki Park	1.81
5	Shogoin	0.73
6	Kyoto University	0.60
7	Kawaramachi	1.80

(i.e., whether visits can be completed in a timely manner with little inefficient travel), and the degree to which the plans are balanced between "sightseeing tours" and "city walking."

4.6   Experimental Results

Extraction of Representative Locations. First, we discuss the results of the evaluation of the method itself. The ablation Study for the index used to determine the representative spots, Eq. 3, is shown in Table 6. In this evaluation, all areas consisting of multiple spots were evaluated. From this result, it can be said that the proposed method is the best for determining the representative spots.

Next, we discuss the evaluation results of the user questionnaire. We evaluate the representative spots indicated by the proposed method and the degree to which they are appropriate for the representative spots given by the subjects, using nDCGs. Table 7 shows the areas to be evaluated and the names of the spots belonging to them, and Table 8 shows the nDCG for each area. For Table 7, the spot names in boldface refer to the representative spots extracted by the proposed method. From this result, it can be said that the representative spots generated by the proposed method have high validity.

The results of the user study show that in areas where the number of their member spots is small and where a certain spot is significantly better known and more accessible than others, the subjects tend to perceive the same spots as representative spots. However, when the number of spots belonging to an area is large, there is a difference between the results of the proposed method and those of the users. Specifically, in the Higashiyama area (area 3) in Table 7, Gion

Table 10. Spot Planning Example (Start: Kyoto Tower, End: Kawaramachi, Time Budget: 8 h)

Order	Spots to visit	Required Time (hours)
1	Kyoto Tower	0.31
2	Higashi Honganji Temple	0.46
3	Heian jingu Shrine	1.12
4	Okazaki Park	0.58
5	Shorenin Temple	0.60
6	Yasaka Shrine	0.49
7	Kodaiji Temple	0.58
8	Ninenzaka	0.54
9	Sannenzaka	0.71
10	Kenninji Temple	0.64
11	Gion	0.54
12	Kawaramachi	1.23

and Yasaka Shrine were ranked highly by the subjects as representative spots, although the proposed method did not extract them. As for Gion, since the recognition was calculated around October 2022, more articles about autumn leaves were detected than usual, so the recognition of other spots famous for autumn leaves was calculated higher. This indicated that recognition highly depends on the number of search results. For Yasaka Shrine, the vermilion-lacquered gate was used in the image used for feature vector generation. The gate of Kiyomizu-dera Temple and the gate of Shimogamo-jinja Shrine, both of which belong to the same area, and the vermilion gate of Shimogamo-jinja Shrine, both of which belong to a different area, are located inside the facility, and ViT judged that the color and shape are similar, so both the similarity within an area and outside an area was calculated to be high.

In addition to recognition and uniqueness, other possible indicators for selecting representative spots include the genre of the spot and the number of visitors. By considering these factors from multiple perspectives, it is thought that spots with higher representativeness can be selected.

User Study Results. For the same query as Table 5, Table 9 shows an example result of the area method, and Table 10 shows an example result of the spot method used for comparison.

We evaluated the adequacy of the courses generated by each method in terms of whether the visits could be completed within the time budget provided and whether there was no wasteful transition. We also evaluated whether the selection of spots was balanced or not on a 5-point scale. Plans for the query "Kyoto Tower - Arashiyama Station" generated by the Area Method were not evaluated

Table 11. Questionnaire results regarding the adequacy and Balance of each plan

Query	Timebudget (h)	Adequacy			Balance		
		Proposed	Area	Spot	Proposed	Area	Spot
Kyoto Tower - Kawaramachi	4	3.67	3.67	4.67	3.00	3.33	2.00
	6	4.00	3.00	4.67	3.67	4.33	3.33
	8	3.00	3.00	4.33	4.33	4.33	2.67
Kyoto Tower - Arashiyama Station	4	4.67	–	4.67	3.67	–	4.33
	6	4.00	–	4.33	3.67	–	3.67
	8	3.33	3.00	4.00	4.33	4.33	2.67
Kyoto Tower - Kinkakuji Temple	4	4.33	5.00	5.00	3.67	3.33	4.33
	6	3.67	4.33	4.00	4.00	4.33	4.00
	8	2.67	2.67	4.33	4.00	4.33	3.00

for time budgets of 4 and 6 h because the minimum time budget to satisfy the query was 8 h.

The results of the user study are shown in Table 11. The results of the user questionnaire showed that the comparative method, the spot method (which optimizes the movement between individual POIs), was the most appropriate in terms of validity. Regarding the degree of balance, it was generally clear that the proposed method was able to generate more balanced plans with larger time budgets. The reason for the higher validity of the spot method is that the spot method performs optimization for each set of POIs, so that the travel distance between POIs is calculated to be shorter compared to the proposed method and the area method. In particular, the number of POIs around Kawaramachi, Higashiyama, and Arashiyama Stations is abundant, and a plan with a short travel distance to visit them in order was generated, so it is highly likely that the subjects judged that the plan was highly appropriate (i.e., the travel was not unreasonable). In addition, the proposed method showed a higher degree of balance as the time budget became larger. In fact, the number of times the proposed method switches courses tends to increase as the time budget increases. Therefore, the proposed method and the "MERIHARI-Area Tourism" may be suitable for long-term and wide-area sightseeing.

In the future, it is expected that personalized scoring and course generation will become possible by adjusting the weights when scoring recognition and uniqueness.

5   Conclusion

In this paper, we propose a method to generate a sightseeing plan that combines multiple sightseeing styles to taste the "representative" of a sightseeing city, and to support this method, we propose a method to generate a sightseeing plan with a contrast selection of spots according to user conditions such as time and budget. Further experiments with data in other cities are planned. We also plan to apply our method to longer and more extensive sightseeing tours.

References

1. Dosovitskiy, A., et al.: An image is worth 16×16 words: transformers for image recognition at scale. In: International Conference on Learning Representations (2021)
2. Gavalas, D., Kasapakis, V., Konstantopoulos, C., Pantziou, G., Vathis, N., Zaroliagis, C.: The eCOMPASS multimodal tourist tour planner. Expert Syst. Appl. **42–21**, 7303–7316 (2015)
3. Hara, T., Ho, B.Q., Miyamoto, R., Aoike, T., Ota, J., Kurata, Y.: Support of tour planning incorporating town walk by presenting information on walking around major tourist attractions. Tour. Inf. J. Tour. Inf. Soc. Jpn. **16**(1), 33–48 (2020). (in Japanese)
4. Karve, S., Shende, V., Hople, S.: Semantic relatedness measurement from Wikipedia and WordNet using modified normalized Google distance. In: Nagabhushan, P., Guru, D.S., Shekar, B.H., Kumar, Y.H.S. (eds.) Data Analytics and Learning. LNNS, vol. 43, pp. 143–154. Springer, Singapore (2019). https://doi.org/10.1007/978-981-13-2514-4_13
5. Li, L., Shang, Y., Zhang, W.: Improvement of hits-based algorithms on web documents. In: Proceedings of the 11th International Conference on World Wide Web, pp. 527–535. Association for Computing Machinery (2002)
6. Lim, K.H., Chan, J., Leckie, C., Karunasekera, S.: Personalized trip recommendation for tourists based on user interests, points of interest visit durations and visit recency. Knowl. Inf. Syst. **54**(2), 375–406 (2018). https://doi.org/10.1007/s10115-017-1056-y
7. Miller, C.E., Tucker, A.W., Zemlin, R.A.: Integer programming formulation of traveling salesman problems. J. ACM (JACM) **7**(4), 326–329 (1960)
8. Shen, Y., Ge, M., Zhuang, C., Ma, Q.: Sightseeing value estimation by analysing geosocial images. Int. J. Big Data Intell. **5**(1/2), 31–48 (2018)
9. Sun, J., Matsushima, Y., Ma, Q.: Property analysis of stay points for POI recommendation. In: Strauss, C., Kotsis, G., Tjoa, A.M., Khalil, I. (eds.) DEXA 2021. LNCS, vol. 12923, pp. 374–379. Springer, Cham (2021). https://doi.org/10.1007/978-3-030-86472-9_34
10. Sun, J., Zhuang, C., Ma, Q.: User transition pattern analysis for travel route recommendation. IEICE Trans. Inf. Syst. **102-D**(12), 2472–2484 (2019)

Financial Argument Quality Assessment in Earnings Conference Calls

Alaa Alhamzeh[(⊠)]

University of Passau, Passau, Germany
`Alaa.Alhamzeh@uni-passau.de`

Abstract. Arguments are ubiquitous. Yet, the definition of what is a good argument depends on the goal, and the settings. Although various business communication studies confirm the crucial role of argumentation, no work has shaped financial argument quality in a way that is concise enough for a practical application.

In this paper, we aim to close this research gap by modeling the quality of managers' arguments, during the Q&A sessions of earnings conference calls. To this end, we propose various quality dimensions at both levels of argument and argument units. Our quality model establishes a well-considered link between, on the one hand, insights as they are expressed in financial text analysis literature, and, on the other hand, insights derived from empirical quality descriptions as provided by argumentation discourse linguistics and computational models.

We further conducted the related annotation study and produced *FinArgQuality*, the first financial dataset annotated with argument quality. This corpus composes of 14,146 sentences in a total of 80 earnings calls transcripts.

Our proposed quality assessment dimensions, and final annotated corpus are publicly available, and can serve as strong baselines for future work in both FinNLP and computational argumentation disciplines. We further discuss some potential optimization goals and financial applications of this data, and highlight future directions.

Keywords: Argument quality · FinNLP · Earnings Conference Calls · FinArgQuality dataset

1 Motivation

The rise of data and the development of machine learning have arrived at the foundation of the financial technology (FinTech) domain. This interdisciplinary field aims at supporting financial services with digital innovations and technology-enabled business models [1]. However, given that about 80% of today's data is unstructured information which is composed mainly of textual data, some argue that Natural Language Processing (NLP) is the most important research nowadays. That is because we need to automatically leverage this data and to process it in different frameworks, which, in financial aspect, is called Financial Natural Language Processing (FinNLP).

C. Strauss et al. (Eds.): DEXA 2023, LNCS 14147, pp. 65–81, 2023.
https://doi.org/10.1007/978-3-031-39821-6_5

Among various financial textual sources, Earnings Conference Calls (ECCs) are one of the most considerable information players in the stock market [2].

An earnings call is a quarterly organized event where public traded companies report their last quarter performance and give guidelines about the next one. The company management often discuss and detail key points, such as growth, risks, buybacks, and dividends. Explicitly, an earnings call consists of two sections: a presentation held by the company, followed by a Questions & Answers (Q&A) session where company representatives[1] answer the questions of professional analysts and other market participants. The analysts are later expected to announce their opinions about this stock in a sort of recommendation (in a scale 1–5), expected price target and sometimes a detailed report. In fact, many studies show that analyst's discussions during the question-answering session to be the most informative and impacting part on the market [3–5]. Therefore, we focus in our study on this particular section of the call.

During this session, the management team may provide additional context and information on the company's financial results and future outlook, which can help analysts better understand the company's performance and make more informed recommendations. For example, if a company reports weaker-than-expected earnings, but the management team explains that the results were affected by one-time events that are not expected to reoccur, analysts may be more likely to maintain or even upgrade their recommendations. Contrarily, if a company reports weaker-than-expected earnings, and the management team provides no clear explanation for the results, analysts may be more likely to downgrade their recommendations towards this company.

In fact, the automatic analysis of earnings calls is valuable for different financial services and applications (e.g., financial risk prediction [6,7], modeling of analysts' decision-making [2]). However, these calls are still an under-resourced text genre in computational argumentation, despite the fact that various business communication studies proved the important role of argumentation in ECCs (e.g., [8,9]).

In our previous work [10], we covered the argument structure in the managers' speech during the Q&A sessions of four top tech companies *(Apple, Facebook, Amazon, and Microsoft)* for the period of five years 2015–2019, resulting in 80 transcripts.

In this work, we extend on it, to study further the quality of managers' arguments during these public calls. In other words, we want to answer the following research question: How to handle the quality of company executives' arguments, while establishing a well-considered link between, on the one hand, insights as they are expressed in financial text analysis literature, and, on the other hand, insights derived from empirical quality descriptions as provided by argumentation discourse linguistics and computational models?

We tackle this research gap by conducting a comprehensive synthesis on earnings calls and Computational Argument Quality (CAQ) state of the art. Investigating on the same *FinArg* corpus, we have introduced in [10], our contributions

---

[1] Mainly chief executive officer and chief financial officer.

in this paper are two-fold. First, we develop a scoring model for argument quality in ECCs, that covers both argument units and overall argument. Second, we conduct the related annotation study, and contribute to the research community with *FinArgQuality*: the first financial corpus annotated with argument quality scores.

This paper is organized as follows: Sect. 2 covers a conceptual background of argument quality, financial text analysis, aligned with most related works to our research. In Sect. 3, we define and illustrate our argument quality dimensions with detailed examples and explore our rating rubrics. We further report our annotation study, inter-annotator agreement, size, and statistics of our final corpus - *FinArgQuality* in Sect. 4. We discuss our findings, the potentials of this data, and conclude future directions in Sect. 5.

2   Related Work

Our work is closely related to the following two lines of research:

2.1   Computational Argument Quality Assessment

An argument is defined as the justification made to reach a conclusion on a controversial topic. Thus, the simplest argument composes of one claim and one premise supporting it. Argument Quality is the assessment of its attributes, strength, and persuasiveness. Delving into the rich realm of argumentation theories, various quality proposals have been introduced. To the best of our knowledge, the computational argumentation literature reported only one study that comprehensively survey the argument quality assessment theories and proposals by Waschsmuth et al. [11]. By that, they derived a taxonomy of 15 dimensions covering the logical (e.g., level of support), rhetorical (e.g., persuasiveness) and dialectical (e.g., relevance) aspects of an argument.

Despite the philosophical background of argumentation and argument quality, researchers in computational argumentation looked for practical, yet considerable definitions of argument quality. They further faced this problem with different methodologies of assessment. An overview of the literature approaches entails the following categories of treatment: First, *Point-wise versus Pair-wise Rating*, meaning, either an absolute rating of the argument (e.g., ranking the strength of a student essay [12]), or a relative rating of it in comparison with another argument (e.g., which argument is more convincing by [13]). Second, with respect to the *level of granularity*, we can distinguish methods that estimate the quality of an overall argument (e.g., [14]) versus, the quality of its particular components (e.g., [15]). Furthermore, some scholars explored the interaction dynamics into debate context. For instance, [16] tried to define the wining argument on the Reddit platform using the interaction patterns. Third, regarding *the method of assessment*, the literature reported mostly direct classification (regression) models (e.g., [17,18]), with some indirect attempts. For instance, [19] investigated on a set of linguistic features that reflect the argument quality instead of considering the original text. Similarly, Gurcke et al. [20] aimed at assessing the sufficiency of arguments through conclusion generation. However, not surprisingly, direct methods outperform their peers. This discussion should give you a bird's-eye view on the diversity of computational argument quality field.

Furthermore, while many studies treat the argument in a holistic manner, Walton [21] argues, *"if the concept of an argument is defined in terms of the premises in it (providing grounds or reasons for accepting the conclusion), then we have to ask what "grounds" or "reasons" are, other than being good or reasonable arguments"*. We also follow this vision in our argument quality dimensions. Thus, we distinguish further the types of argumentative units (i.e., premises and claim). We provide further discussions all across our quality dimensions.

2.2   Text Quality in Finance and Business Communication

The analysis of available textual data has always been a topic of interest for many researchers in the financial domain. However, the end target could be widely different. For example, while [22,23] evaluated the forecasting skills of investors, [24] analyzed the managers' speech with the goal of predicting the financial risk, and [25] aimed at making a future price prediction out of detected events on news and social media. We present in the following some related work that is directly linked to our proposed quality metrics:

Zong et al. [23] used the Linguistic Inquiry and Word Count (LIWC) lexicon [26] to detect the temporal orientation of a forecaster's justifications. They found that good forecasters tend to focus more on past rather than future events. Therefore, we build on that, and we extend to more fine-grained assessment of the past level in our *temporal_history* attribute.

Besides, as we have aforesaid, various business communication studies proved the important role of argumentation in earnings conference calls. Among others, Rocci et al. [27] differentiate evidential type presented in different sections of an earnings calls to be: "common knowledge, direct, epistemic possibility, generic indirect, inference, report, and subjective". In their empirical study, they found that the subjective type to be the most frequent in the answers of company executives. Hence, we consider studying the *subjectivity* of an argument as one of our quality metrics, since we want to highlight the objective arguments.

Notably, different financial studies focus on the statement specificity as a major factor of its quality. Text "uncertainty" [23] and "hedging" [2] are only indicators of "the lack of commitment to the content of the speech" [28]. This is logical, since the qualitative analysis of a financial text cannot be separated from its quantitative property. Therefore, we also concentrate on the argument *specificity*, but further from two angles: the specificity of the answer in relation to the asked question, and the specificity of the premises and claims through identifying their particular types.

3   Argument Quality Dimensions

Given that the criteria of what is a good argument depends on the goal orientation [19,29], we define our quality attributes in collaboration with experts from the Chair of *Financial Data Analytics* at Faculty of Business, Economics, and Information Systems - University of Passau[2].

The CAQ literature reported different guidelines in terms of the annotation scale. For instance, Stab et al. [17] reported 681 (66.2%) sufficient to 348 (33.8%) insufficient arguments in their student essays corpus. Likewise, in the corpus of Persing and Ng [12] annotated with the strength attribute of 1000 student essays, they used a scale of 1.0 to 4.0 with 0.5 increments, giving a total of seven values. Among all essays, 372 are categorized as class 3.0, whereas only 2 are categorized as class 1.0; 21 with class 1.5 and merely 15 belong to class 4.0. Therefore, to avoid such a high data imbalance and to make more fair fine-grained judgment rather than binary decision, we suggest our annotation guidelines with respect to a 3-point scale of assessment, except for *objectivity* which remains binary, given its nature, and the *temporal-history* of an argument, to provide more gradual indicators. In addition, we suggest the argument quality on the unit level (claim and premise types) to be categorical rather than numerical.

In summary, our rating follows the point-wise approach, and looks at each argument from two levels:

3.1   At the Level of Argument

A holistic assessment of an argument quality is the most used approach in the literature. We present in the following the quality metrics we define at the granularity of the overall argument. In other words, considering the argument claim and premises as well as the relations between them.

- **Strong** Persing et al. [12] labeled the strength of a student essay using a scale 1.0 to 4.0. On the other hand, [30] inspected the strength of only the premise component. They defined it by "how well a single statement is contributing to persuasiveness" on a scale 1–6. Inspired by these studies, we define the strength of an argument by two factors: *how many and what type of premises are backing its claim?* For example, an argument with a statistical premise is supposed to be stronger than an argument with a hypothetical premise. Furthermore, Table 1 represents the rubrics for rating the argument strength.
- **Specific** Carlile et al. [30] studied the specificity of every single argumentative statement in a student essay (i.e., premise, claim, major-claim). They score it on a scale of 1 to 5 based on how detailed the statement is. The main source of tolerant and inexact language is using hedging expressions. Prokofieva et al. [28] defined some general guidelines for recognizing hedge expressions in English. Hedges can appear in forms like: "I think", "it is sort of", "probably", etc. In our particular case, we study the arguments presented by company managers to answer analysts' questions. Therefore, it was important for us to

---

[2] https://www.wiwi.uni-passau.de/en/financial-data-analytics.

declare the specificity in a relation to the question itself. Hence, we rate the argument specificity on a 0–2 Likert scale, as illustrated in Table 1.

- **Persuasive** The persuasiveness is the most subjective attribute to judge. Yet, it is still taken into account by many other studies. This could be due to the fact that, we have a more holistic feedback from the annotator about all argument elements, and their coordination. In addition, we can use these annotations to analyze the relations with other argument attributes (i.e., what makes a persuasive argument). Table 1 displays also our hints to label persuasiveness across arguments.

Table 1. Quality dimensions at the argument level.

Attribute	Definition	Score
Strong	How well the statement contributes to persuasiveness, considering the count and types of supporting premises?	• Strong-0: A poor, not supported argument (e.g., the claim is supported by only one premise that is doubtful) • Strong-1: A decent, fairly clear argument. The argument has at least two premises that authorize its standpoint • Strong-2: A clear and well-defended argument, supported by concrete and powerful premises
Specific	How well the statement is precise and answers directly the question?	• Specific-0: The argument is not related to the question (e.g., blaming the market, mentioning competitors) • Specific-1: The statement partially answers the question, but still implies some hedging • Specific-2: The argument is concrete and directly related to the question
Persuasive	From the annotator view, to what extent is the argument convincing?	• Persuasive-0: The argument is not easily understandable, the speaker may state some description, incident, value but does not explain why it's important. It may then persuade only listeners who are already inclined to agree with it • Persuasive-1: The argument provides acceptable reasoning, may still contain some defects that decrease its ability of convincing. Hence, it would persuade some listeners • Persuasive-2: A clear, well-structured argument that would persuade most listeners. The speaker stated precise and sound premises that remove doubts of the listener
Objective	Is the argument based on facts rather than feelings or opinions?	• Objective-0: A subjective or biased argument based on particular views and opinions • Objective-1: A logical argument supported by verifiable evidences
Temporal-history	Does the argument include any time indicator? In case of many, choose the most recent one	• Temporal-3: during this quarter • Temporal-2: up to two quarters • Temporal-1: half to one year • Temporal-0: more than one year • Temporal-1: not mentioned (if there is no explicit time indicator, choose this value, even if you think that it could be concluded from the context)

- **Objective** Being objective, is very essential from the market perspective. Arguing by opinions and particular views has less impact on investors than arguing with objective information and reached earnings. Hence, we binary classify the argument to objective or subjective based on the question: is the argument based on facts rather than feelings or opinions?.
- **Temporal-history** The temporal information assessment, composes a special phenomenon in financial opinions. Studying the time associated with given information, and estimating its impact period, are important research questions to the stock market [22,31,32]. On the other hand, in a business communication study, Crawford et al. [33] analyzed the persuasion language in economic "Crisis Corpus" in comparison to economic "Recovery Corpus". They found that executives tend to emphasize progress and future expectations in the crisis corpus, while they report achievements in their recovery time period. This is similar to the findings of [23] we have aforementioned, that providing past information reflects better forecasts. Hence, we ignore future expressions and rather weight the temporal spans of text that represents a real value for finance, by recognizing five degrees of temporal-history as shown in Table 1.

3.2 At the Level of Argument Unit

Most argument models include one type of premise. However, we can easily distinguish different types of premises in everyday discourse [34]. For example, a premise may provide empirical evidence, a fact, or a justification why the reasoning of an argument is correct. Similarly, this applies to claims.

Despite the fact, that knowing the types of the argument claim or premise(s) can give us a clear estimation about its quality, the literature reports very rare attempts towards this research direction. Moreover, the annotation of those types could be more objective and less biased itself than scoring the whole argument towards one attribute (e.g., strength, clarity, etc.). Hence, we elaborated part of the data with one of our annotators and suggest the following pragmatic types of premises and claims, as shown in Fig. 1.

Types of Claims
Clarile et al. [30] distinguish three types of claims: Fact, Value (something is good or bad), and Policy. Their study shows that fact claims seem to be the most frequent in their corpus of student essays. We distinguish the following types of a claim:

- **Fact** The earning conference call, is the event where a company shares private information with the public. Therefore, some managers' claims tend to be facts, that still need to be accepted by supporting evidences.

  > Example: *"..When it comes to our Commercial Licensing and our servers, it's the same trend, which is the big shift that's happening is our enterprise and datacenter products, being Windows Server, Systems Centers, SQL Server, are more competitive..."*

- **Value** Considering our kind of data (earnings calls), when claiming some information that reflects quantities and reports measures, the claim is classified as a numerical value.

  > Example: " *Secondly to provide a bit more color, sales of the Watch did exceed our expectations and they did so despite supply still trailing demand at the end of the quarter.*"

- **Opinion** We identify this type of claims, for all statements that reflect the company vision and its executives' standpoints. Few terms introducing an opinion are like: *we're very happy, I think.* In fact, this type of claim is very common, especially while expressing the company future hopes [33].

  > Example: "*...And so we are incredibly optimistic about what we've seen so far.*"

- **Policy** This kind of claim is used to express a plan of action, or existent rules.

  > Example: "*And so as you know, we don't make long term forecasts on here.*"

- **Reformulated** During our pilot annotation, we observed a common pattern of repeating the same claim with some reformulation, mainly at the end of the answer. Hence, we define the Reformulated claim type, which could be justified by the oral argumentation nature of our data.
  According to [35], reformulation or restatement is a rephrase of the evaluative expression without adding any significant information, where the goal is to make certain that the evaluation is clear and unambiguous. Some indicators to reformulations are: *in other words, that is to say, rather.* In our data, the reformulated claim is mostly the shorter one of the two claims. We ask the annotators not to link this claim to any premises (i.e., not to consider it as a new argument).

  > Example: "*...And I think when you take those two things, along with what Satya said, being able to balance disciplined focus and execution for us, I think we feel very good about the progress we've made.*"

- **Other**
  This label is selected when no particular claim type is recognized.

Types of Premises. Similarly to claims forms, and motivated by the works of [30,36], we set the following premise types:

- **Fact** This unit provides evidence by stating a known truth, a testimony, or reporting something that happened.

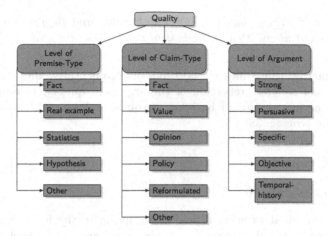

Fig. 1. Our quality dimensions at the levels of argument and argument units

> Example: *"And then at the same time, we're bringing more and more advertisers into the system and that's giving us a better selection of the ads that we can serve to the people using Facebook, and that, again, improves the quality and the relevance."*

- **Real Example** Sharing out a comparable experience, a specific event, or similar, is a common strategy in spoken language and in argumentation in general [37].

> Example: *"I also look at the first time iPhone buyers and we're still seeing very, very large numbers in the countries that you would want to see those in, like China and Russia and Brazil and so forth."*

- **Statistics** This type of premise is decisive in any argumentative discussion. Definitely, it is very common and powerful in earnings calls.

> Example: *"So we ended the year last year with 109 fulfillment centers around the world and 19 U.S. sort centers..."*

This example also implies that the automatic understanding of numerical data is more complicated in this genre of text [31].
- **Hypothesis** Besides probative deductions, hypothetical, and assumption evidences can be used. However, this type of text seems not to be frequent in our data.

> Example: *And if this works as planned, it can be big.*

- **Other** This unit is supporting the final conclusion, but none of the previous evidence characteristics applies to it.

> Example: *"...These numbers are unbelievable and they're done in an environment where it's not the best of conditions..."*

We assume that those fine-grained types of argumentative units, should provide a clear and concrete reflection of the argument quality. In addition, recognizing the argument ground basis of reasoning is inline with analyzing the argumentation scheme [38].

4   Data Creation

4.1   Annotation Study

We downloaded our data using a paid subscription to the Financial Modeling Prep API[3]. We used Label Studio[4] as a visualized annotation tool. Our annotated data concerns the quarterly earnings calls of four companies: Amazon, Apple, Microsoft, and Facebook for the period of 2015–2019. For each transcript, we determine, using a Python script, the list of all the speakers (Analyst, Representative, or an Operator). We then split each transcript into different documents. Each *document* contains one or two questions asked by a *single analyst*, along with the corresponding response(s) by the company's managers.

We elaborate with one of our annotators to define the annotation guidelines as a first step. Later on, three other annotators have joined the annotation process. All of them have a significant level of language. One of them is an international economics master student, whereas the others are all computer-science students. Therefore, our choice of companies was more tech-oriented, where industry jargon and different products are known to all. Moreover, to let annotators gain insight into the company's performance over the years, we assign one company for each of our four annotators to do all its quarters' annotations. We started by training sessions and discussions with the annotators. Based on their feedback, we were able to refine the guidelines and clarify ambiguous situations. Thereafter, a division of the data was done to 20% to be double annotated for inter-annotator agreement calculations, and 80% individual annotations. We call the output corpus: *FinArgQuality* and it is publicly available to foster future research[5].

4.2   Inter Annotator Agreement

To calculate the inter annotator agreement, we define about 20% of our data to be twice annotated. Each of the annotators had to label 4 transcripts at this stage, one from each company. At the end, they meet and discuss disagreements to proceed the final version (gold annotation) on this part. The value of this

---

[3] https://site.financialmodelingprep.com/developer.

[4] https://labelstud.io/.

[5] https://github.com/Alaa-Ah/The-FinArgQuality-dataset-Quality-of-managers-arguments-in-Eearnings-Conference-Calls.

strategy is that it guarantees direct discussions between every pair of annotators, which help at the end to unify their mindset towards the annotations.

We report in the following Cohen's kappa inter-annotator agreement [39] on our *FinArgQuality* final corpus. For all data, we measure the agreement separately for each pair of annotators, and report the average. Table 2 shows that we obtained fair to substantial agreements [40].

Table 2. Inter-annotator agreement of the overall argument quality and unit types

Company	Specific	Persuasive	Strong	Objective	Temporal-history	Claim (All types)	Premise (All types)
MSFT	0.63	0.64	0.72	0.65	0.79	0.61	0.59
FB	0.33	0.13	0.21	0.36	0.66	0.56	0.57
AAPL	0.31	0.31	0.35	0.27	0.55	0.66	0.69
AMZN	0.11	0.21	0.24	0.36	0.26	0.37	0.51
All	0.345	0.322	0.38	0.41	0.565	0.55	0.59

We compare our results to a similar study by Wachsmuth et al. [41], that introduced the *Dagstuhl15512 ArgQuality Corpus* for ranking argumentation quality based on their developed taxonomy of 15 dimensions. They also adopted a 3-point scale (low, medium, high) for rating. Therefore, we consider this data as the most relevant to compare with. They reported Krippendorf's α of all annotators ranging from 0.174 to 0.447 only.

By analyzing disagreements, we found that the main source of disagreement is the missing of unit boundaries (Speech-To-Text nature of the transcripts' data), and the multiple possible interpretations of argument structure [42–44]. This, definitely, applies to rating argument quality, which is even more inherently subjective [11]. In addition, a high proportion of disagreement is associated with arguments that include modal verbs, and uncertainty quantification (e.g., "many", "some") which may hastily perceived with low degrees of specificity, strength, and persuasiveness. Thus, extending guidelines with those cases would improve further annotations.

4.3 *FinArgQuality* Data Statistics

The overall quality dimensions are described in Table 3 in total, and per company. The percentages are based on the total number of arguments. Overall, the score 1 is always the most associated with specific, persuasive and strong quality dimensions. Argument objectivity is validated mostly when mentioning unbiased indicators, such as numerical values or time references. We also notice that label 0 (low) is the least frequent. In addition, only 0.4% of the arguments are considered bad, i.e., all four dimensions (specific, persuasive, strong and objective) are rated by zero. This small percentage reflects the overall good quality

Table 3. Statistics of overall argument quality dimensions over *FinArgQuality*

Dimension	Company									
	FB		AMZN		MSFT		AAPL		Total	
	Count	[%]	Count	[%]	Count	[%]	Count	[%]	Count	[%]
Specific 0	29.0	1.33	13.0	0.60	34.0	1.56	7.0	0.32	83.0	3.80
Specific 1	281.0	12.87	202.0	9.25	466.0	21.34	147.0	6.73	1096.0	50.18
Specific 2	180.0	8.24	220.0	10.07	309.0	14.15	296.0	13.55	1005.0	46.02
Strong 0	39.0	1.79	31.0	1.42	49.0	2.24	19.0	0.87	138.0	6.32
Strong 1	317.0	14.51	274.0	12.55	557.0	25.50	285.0	13.05	1433.0	65.61
Strong 2	134.0	6.14	130.0	5.95	203.0	9.29	146.0	6.68	613.0	28.07
Persuasive 0	70.0	3.21	20.0	0.92	37.0	1.69	11.0	0.50	138.0	6.32
Persuasive 1	254.0	11.63	209.0	9.57	370.0	16.94	221.0	10.12	1054.0	48.26
Persuasive 2	166.0	7.60	206.0	9.43	402.0	18.41	218.0	9.98	992.0	45.42
Objective 0	102.0	4.67	76.0	3.48	304.0	13.92	149.0	6.82	631.0	28.89
Objective 1	388.0	17.77	359.0	16.44	505.0	23.12	301.0	13.78	1553.0	71.11
Temp.history -1	338.0	15.48	288.0	13.19	733.0	33.56	408.0	18.68	1767.0	80.91
Temp.history 0	26.0	1.19	18.0	0.82	12.0	0.55	4.0	0.18	60.0	2.75
Temp.history 1	54.0	2.47	43.0	1.97	7.0	0.32	10.0	0.46	114.0	5.22
Temp.history 2	24.0	1.10	41.0	1.88	17.0	0.78	11.0	0.50	93.0	4.26
Temp.history 3	48.0	2.20	45.0	2.06	40.0	1.83	17.0	0.78	150.0	6.87

of arguments, and the persuasion strategies which managers often use during the earnings calls and public speech, as highlighted by Crawford [33]. The time reference itself, is defined in our guidelines only in the past, as the temporal-history dimension. To standardize the annotations, we asked the annotators not to assume their interpretations of time references if it is not explicitly mentioned. Therefore, we got a majority class of -1, while all expressed time indicators compose about 20% of our arguments.

Furthermore, Table 4 exhibits the detailed corpus size and sentence/tokens distributions, as well as statistics with respect to the claim and premise types and argument relation. We can see that 43% of claims are factual, while 36% are based on opinions. The remaining claim types (Reformulated, Policy, Value, and Other) represent approximately 21%. This is reasonable since managers mainly report facts, or explain their views and future prospects.

Moreover, the distribution of premise types confirms the financial nature of the collected data, since it mostly covers facts (71%) and statistics (13%). Nevertheless, some background information seems to be annotated as facts by our annotators, given that it is still true (happened) information that could be tricky not to consider as a fact. Similarly, Carlile et al. [30] found that 493 of their premises received a "common_knoweledge" label, out of 707 premises with 8 potential premises types defined in the annotation guidelines. In a related analysis, Villalba and Saint-Dizier [35] show how "a number of evaluative expressions with a'heavy' semantic load receive an argumentative interpretation".

Table 4. Size and statistics of argument components types and argument relation over *FinArgQuality*. The average is presented along with its standard deviation

	Attribute	Count	[%]	Avg. per doc	Avg. per company
Sentences	In-argument	9693	68.53	12 ± 6	2423 ± 423
	Out-of-argument	4453	31.47	6 ± 4	1113 ± 158
Tokens	In-argument	244253	78.84	297 ± 155	61063 ±13437
	Out-of-argument	65537	21.16	82 ± 78	16384 ± 4796
Arg. components	Premises	5078	52.40	6 ± 4	1270 ± 271
	Claims	4613	47.60	6 ± 3	1153 ± 158
Claims	Fact	2001	43.38	3 ± 2	500 ± 93
	Opinion(view)	1672	36.25	2 ± 2	418 ± 64
	Reformulated	850	18.43	2 ± 1	212 ± 56
	Policy	45	0.99	1 ± 0	11 ± 5
	Value	28	0.60	1 ± 0	7 ± 3
	Other	17	0.37	1 ± 0	4 ± 3
Premises	Fact	3624	71.37	5 ± 3	906 ± 303
	Statistic	691	13.60	2 ± 1	173 ± 92
	Real Example	496	9.77	2 ± 1	124 ± 53
	Hypothesis	46	0.91	1 ± 0	12 ± 5
	Other	221	4.35	2 ± 1	55 ± 24
Relation types	Support	4823	98.41	6 ± 4	1206 ± 271
	Attack	78	1.59	1 ± 1	20 ± 10

5   Discussion and Conclusions

Recently, financial argumentation gained momentum in different languages (e.g., [45,46]). Given that both financial NLP and computational argumentation communities suffer from the lack of labeled data, we believe that our proposed assessment model and publicly available dataset, can serve as strong baselines for future work. We expect our carefully developed corpus to prompt various directions.

On the one hand, six potential argument mining tasks could be investigated using our data: argument identification, argument unit classification, argument relation classification, premise type multi-class classification, claim type multi-class classification, and overall argument quality assessment. In addition, a future direction could be to use the argumentative unit types in order to mine the argumentation strategies [47].

On the other hand, the automatic detection and qualification of arguments in financial domain is important for several goals, including but not limited to:

- Efficiency: It allows for the analysis of large amounts of financial data and reports quickly and accurately, reducing the time and resources required to manually identify and analyze arguments.

- Objectivity: It eliminates the subjective biases that can occur when humans are manually reviewing financial data, resulting in a more objective analysis.
- Enhanced market transparency: Arguments can provide more visibility into the reasoning behind investment decisions or analysts' recommendations, improving market transparency and trust.
- Improved decision-making: Providing a comprehensive analysis of financial arguments, identifying worthiness, and detection of verified claims, can help inform and improve decision-making in finance. For example, by using argument mining to validate managers' claims (e.g., [9]), we can assess the objectivity, completeness, and credibility of these arguments, providing investors with more informed and reliable insights for making investment decisions.

Last but not least, we would like to note that a first emerging output of this work is established by the FinArg-1 Shared Task[6], in cooperation with AIST, Japan[7], as well as different other partners. FinArg–1 covers argument unit and relation classification. The data reported in this paper is planned to be used in the next editions of this task. We plan to cover different tasks, including argument quality assessment. Our ultimate goal is to improve the automatic understanding of financial text. Therefore, we hope that the work presented in this paper fuels and inspires more research in computational argumentation, stock market and their interplay.

Acknowledgement. I would like to thank, for most, Prof. Ralf Kellner and Mr. Lukas Marx, for their open discussions and insightful feedback, which were essential to refine those argument quality dimensions from a financial point of view. In addition, I would like to thank all our annotators, including M. Kürsad Lacin.

References

1. Philippon, T.: The fintech opportunity. Technical report, National Bureau of Economic Research (2016)
2. Keith, K.A., Stent, A.: Modeling financial analysts' decision making via the pragmatics and semantics of earnings calls. arXiv preprint arXiv:1906.02868 (2019)
3. Matsumoto, D., Pronk, M., Roelofsen, E.: What makes conference calls useful? the information content of managers' presentations and analysts' discussion sessions. Account. Rev. **86**(4), 1383–1414 (2011)
4. Price, S.M., Doran, J.S., Peterson, D.R., Bliss, B.A.: Earnings conference calls and stock returns: the incremental informativeness of textual tone. J. Bank. Financ. **36**(4), 992–1011 (2012)
5. Ma, Z., Bang, G., Wang, C., Liu, X.: Towards earnings call and stock price movement. arXiv preprint arXiv:2009.01317 (2020)
6. Li, J., Yang, L., Smyth, B., Dong, R.: MAEC: a multimodal aligned earnings conference call dataset for financial risk prediction. In: Proceedings of the 29th ACM International Conference on Information & Knowledge Management, pp. 3063–3070 (2020)

---

[6] http://finarg.nlpfin.com/.
[7] https://www.aist.go.jp/index_en.html.

7. Ye, Z., Qin, Y., Xu, W.: Financial risk prediction with multi-round Q&A attention network. In: IJCAI, pp. 4576–4582 (2020)
8. Palmieri, R.: The role of argumentation in financial communication and investor relations. In: Handbook of Financial Communication and Investor Relations, pp. 45–60 (2017)
9. Stenvall, J.: Management earnings forecasts: could an investor reliably detect an unduly positive bias on the basis of the strength of the argumentation? J. Bus. Commun. 48(4), 393–408 (2011)
10. Alhamzeh, A., Fonck, R., Versmée, E., Egyed-Zsigmond, E., Kosch, H., Brunie, L.: It's time to reason: annotating argumentation structures in financial earnings calls: the FinArg dataset. In: Proceedings of the Fourth Workshop on Financial Technology and Natural Language Processing (FinNLP), Abu Dhabi, United Arab Emirates (Hybrid), pp. 163–169. Association for Computational Linguistics (2022)
11. Wachsmuth, H., et al.: Argumentation quality assessment: theory vs. practice. In: Proceedings of the 55th Annual Meeting of the Association for Computational Linguistics (Volume 2: Short Papers), Vancouver, Canada, pp. 250–255. Association for Computational Linguistics (2017)
12. Persing, I., Ng, V.: Modeling argument strength in student essays. In: Proceedings of the 53rd Annual Meeting of the Association for Computational Linguistics and the 7th International Joint Conference on Natural Language Processing (Volume 1: Long Papers), pp. 543–552 (2015)
13. Habernal, I., Gurevych, I.: What makes a convincing argument? Empirical analysis and detecting attributes of convincingness in web argumentation. In: Proceedings of the 2016 Conference on Empirical Methods in Natural Language Processing, pp. 1214–1223 (2016)
14. Farra, N., Somasundaran, S., Burstein, J.: Scoring persuasive essays using opinions and their targets. In: Proceedings of the Tenth Workshop on Innovative Use of NLP for Building Educational Applications, pp. 64–74 (2015)
15. Rinott, R., Dankin, L., Perez, C.A., Khapra, M.M., Aharoni, E., Slonim, N.: Show me your evidence - an automatic method for context dependent evidence detection. In: Proceedings of the 2015 Conference on Empirical Methods in Natural Language Processing, Lisbon, Portugal, pp. 440–450. Association for Computational Linguistics (2015)
16. Tan, C., Niculae, V., Danescu-Niculescu-Mizil, C., Lee, L.: Winning arguments: interaction dynamics and persuasion strategies in good-faith online discussions. In: Proceedings of the 25th International Conference on World Wide Web, pp. 613–624 (2016)
17. Stab, C., Gurevych, I.: Recognizing insufficiently supported arguments in argumentative essays. In: Proceedings of the 15th Conference of the European Chapter of the Association for Computational Linguistics: Volume 1, Long Papers, pp. 980–990 (2017)
18. Lauscher, A., Ng, L., Napoles, C., Tetreault, J.: Rhetoric, logic, and dialectic: advancing theory-based argument quality assessment in natural language processing. In: Proceedings of the 28th International Conference on Computational Linguistics, Barcelona, Spain (Online), pp. 4563–4574. International Committee on Computational Linguistics (2020)
19. Wachsmuth, H., Werner, T.: Intrinsic quality assessment of arguments. arXiv preprint arXiv:2010.12473 (2020)
20. Gurcke, T., Alshomary, M., Wachsmuth, H.: Assessing the sufficiency of arguments through conclusion generation. arXiv preprint arXiv:2110.13495 (2021)

21. Walton, D.N.: Argument structure: a pragmatic theory. University of Toronto Press, Toronto (1996)
22. Chen, C.-C., Huang, H.-H., Chen, H.-H.: Evaluating the rationales of amateur investors. In: 2021 Proceedings of the Web Conference, pp. 3987–3998 (2021)
23. Zong, S., Ritter, A., Hovy, E.: Measuring forecasting skill from text. arXiv preprint arXiv:2006.07425 (2020)
24. Qin, Y., Yang, Y.: What you say and how you say it matters: predicting stock volatility using verbal and vocal cues. In: Proceedings of the 57th Annual Meeting of the Association for Computational Linguistics, pp. 390–401 (2019)
25. Alhamzeh, A., et al.: A hybrid approach for stock market prediction using financial news and stocktwits. In: Candan, K.S., et al. (eds.) CLEF 2021. LNCS, vol. 12880, pp. 15–26. Springer, Cham (2021). https://doi.org/10.1007/978-3-030-85251-1_2
26. Tausczik, Y.R., Pennebaker, J.W.: The psychological meaning of words: liwc and computerized text analysis methods. J. Lang. Soc. Psychol. **29**(1), 24–54 (2010)
27. Rocci, A., Raimondo, C., Puccinelli, D.: Evidentiality and disagreement in earnings conference calls: preliminary empirical findings, pp. 100–104 (2019)
28. Prokofieva, A., Hirschberg, J.: Hedging and speaker commitment. In: 5th International Workshop on Emotion, Social Signals, Sentiment & Linked Open Data, Reykjavik, Iceland (2014)
29. Johnson, R.H., Blair, J.A.: Logical self-defense. In: International Debate Education Association (2006)
30. Carlile, W., Gurrapadi, N., Ke, Z., Ng, V.: Give me more feedback: annotating argument persuasiveness and related attributes in student essays. In: Proceedings of the 56th Annual Meeting of the Association for Computational Linguistics (Volume 1: Long Papers), Melbourne, Australia, pp. 621–631. Association for Computational Linguistics (2018)
31. Chen, C.-C., Huang, H.-H., Shiue, T.-T., Chen, H.-H.: Numeral understanding in financial tweets for fine-grained crowd-based forecasting. In: 2018 IEEE/WIC/ACM International Conference on Web Intelligence (WI), pp. 136–143. IEEE (2018)
32. UzZaman, N., Llorens, H., Derczynski, L., Allen, J., Verhagen, M., Pustejovsky, J.: SemEval-2013 task 1: tempeval-3: evaluating time expressions, events, and temporal relations. In: Second Joint Conference on Lexical and Computational Semantics (* SEM), Volume 2: Proceedings of the Seventh International Workshop on Semantic Evaluation (SemEval 2013), pp. 1–9 (2013)
33. Camiciottoli, B.C.: Persuasion in earnings calls: a diachronic pragmalinguistic analysis. Int. J. Bus. Commun. **55**(3), 275–292 (2018)
34. Bentahar, J., Moulin, B., Bélanger, M.: A taxonomy of argumentation models used for knowledge representation. Artif. Intell. Rev. **33**(3), 211–259 (2010)
35. Villalba, M.P.G., Saint-Dizier, P.: Some facets of argument mining for opinion analysis. COMMA **245**, 23–34 (2012)
36. Khatib, K.A., Wachsmuth, H., Kiesel, J., Hagen, M., Stein, B.: A news editorial corpus for mining argumentation strategies. In: Proceedings of COLING 2016, the 26th International Conference on Computational Linguistics: Technical Papers, pp. 3433–3443 (2016)
37. Al-Khatib, K.: Computational analysis of argumentation strategies. Dissertation, Bauhaus-Universität Weimar (2019)
38. Walton, D., Reed, C., Macagno, F.: Argumentation Schemes. Cambridge University Press, Cambridge (2008)
39. Cohen, J.: A coefficient of agreement for nominal scales. Educ. Psychol. Measur. **20**(1), 37–46 (1960)

40. Landis, J.R., Koch, G.G.: The measurement of observer agreement for categorical data. Biometrics 159–174 (1977)
41. Wachsmuth, H., et al.: Computational argumentation quality assessment in natural language. In: Proceedings of the 15th Conference of the European Chapter of the Association for Computational Linguistics: Volume 1, Long Papers, pp. 176–187 (2017)
42. Walton, D., Reed, C.: Diagramming, argumentation schemes and critical questions. In: Van Eemeren, F.H., Blair, J.A., Willard, C.A., Snoeck Henkemans, A.F. (eds.) Anyone Who Has a View. Argumentation Library, vol. 8, pp. 195–211. Springer, Dordrecht (2003). https://doi.org/10.1007/978-94-007-1078-8_16
43. Stab, C., Gurevych, I.: Annotating argument components and relations in persuasive essays. In: Proceedings of COLING 2014, the 25th International Conference on Computational Linguistics: Technical Papers, pp. 1501–1510 (2014)
44. Henkemans, A.: State-of-the-art: the structure of argumentation. Argumentation **14**(4), 447–473 (2000)
45. Chen, C.-C., Huang, H.-H., Chen, H.-H.: From Opinion Mining to Financial Argument Mining. Springer, Heidelberg (2021). https://doi.org/10.1007/978-981-16-2881-8
46. Fishcheva, I., Osadchiy, D., Bochenina, K., Kotelnikov, E.: Argumentative text generation in economic domain. arXiv preprint arXiv:2206.09251 (2022)
47. Al-Khatib, K., Wachsmuth, H., Kiesel, J., Hagen, M., Stein, B., Göring, S.: Webis-editorials-16 (2016)

Explaining Decisions of Black-Box Models Using BARBE

Mohammad Motallebi[1,2], Md Tanvir Alam Anik[1,2],
and Osmar R. Zaïane[1,2(✉)] 🆔

[1] University of Alberta, Edmonton, AB, Canada
zaiane@cs.ualberta.ca
[2] Alberta Machine Intelligence Institute, Edmonton, Canada

Abstract. Machine learning models are ubiquitous today in most application domains and are often taken for granted. While integrated into many systems, oftentimes even unnoticed by the user, these powerful models frequently remain as black-boxes. They are black-boxes because while they are powerful predictive models, it is commonly the case that one cannot understand the decision-making process behind their predictions. Even if we understand the inner workings of a learning algorithm building a predictive model, the mechanism during inference is more often than not obscure. How can we trust that a certain prediction from a model is correct? How can we trust that the model is making reasonable predictions in general? Debugging a predictive model is unworkable in the absence of explanations.

We propose herein a new framework, called BARBE, a model-independent explainer, that learns a surrogate rule-based model on data labeled by the black-box. BARBE makes use of an interpretable associative classifier to create a rule-based model that provides various explanations, including salient features, associations between features, and rule-based representations. Our experimental analysis illustrates the effectiveness of BARBE in generating rule-based explanations for both numerical and text data, when compared to state-of-the-art explainers. Our study demonstrates the faithfulness of BARBE to black-box models. The text-based explanations generated by BARBE are more meaningful to show the fidelity and trustworthiness of the explanation.

Keywords: Machine Learning · Explainable AI · Associative Classification · Model Independent Explanation

1 Introduction

Explainable Artificial Intelligence (XAI) has attracted the attention of many researchers in recent years. This surge in interest is prompted by the need to obtain explainability in different AI domains. Providing an explanation is indeed a requirement in many jurisdictions when an AI system is used to make critical decisions for humans [1,23]. The objective of augmenting systems with explainability is to provide supplementary information on top of the main output created

© The Author(s), under exclusive license to Springer Nature Switzerland AG 2023
C. Strauss et al. (Eds.): DEXA 2023, LNCS 14147, pp. 82–97, 2023.
https://doi.org/10.1007/978-3-031-39821-6_6

by them (such as the class label in a classifier). This new information allows or empowers the user to know why the AI system provided the aforementioned output. One example is a model trained to detect colon cancer in people based on their medical records. Researchers later noticed the data included the name of medical clinics in which patients were admitted later. The presence of this feature was erroneously picked up by the trained model and had caused a significant fictitious boost in the performance of the system [26]. In all such cases, some sort of explainability providing transparency could have helped them avoid the consequences, consequences that are often far-reaching like contributing to the distrust of machine learning.

To tackle this issue, various XAI methods have been developed in recent years that attempt to provide explainability in one way or another. Most effort has been on attempting to justify or elucidate neural networks since the spotlight is currently on deep learning. However, there is also an effort for more generic approaches. For the classification task, in particular, methods have been developed in which the explanation framework is either independent of the classifier or is integrated into it. The former approach is called model-independent explanation (or sometimes called model-agnostic), while the latter is called model-dependent. One significant advantage of model-independent approach frameworks is that these frameworks can be added to different classifiers. This addition allows machine learning enthusiasts to inject some sort of explainability into any existing classifier and leverage them readily.

One problem with some of the current model-independent approaches is that they do not provide highly precise explanations. In other words, regardless of what the "real" explanation is, the user can ask for a fixed number of important features (e.g., give me top k features relevant to the decision), and the system then provides k important features accordingly. Additionally, another notable aspect is to take into account the correlations among input features. Some approaches, as we discuss in the next section, pay no attention to this aspect. In our view, this is a critical part in which the end-user should be able to depend on to better understand the underlying "reasoning" of the black-box model.

In this work, we introduce BARBE, for **B**lack-box **A**ssociation **R**ule-**B**ased **E**xplanation, a model-independent method that explains the decisions of any black-box classifier for tabular and text datasets with high precision. Moreover, the black-box classifier is not required to provide any probability score to take advantage of BARBE. Furthermore, BARBE presents explanations in three alternative forms: 1) the importance score for salient features, which many methods also benefit from; 2) significant associations between pertinent features; and 3) the construction of classification rules, which distinguishes BARBE from other methods. BARBE exploits association rules, a particular kind of rules that take into account the associations between features, helping users grasp different underlying potential causes of a decision.

The rest of this paper is organized as follows. We discuss a few of the main XAI approaches in the next section. Section 3 contains some preliminaries on associative classification. This section is needed as we take advantage of an associative classifier as the core of our work. We introduce BARBE, the main contribution of this paper, in Sect. 4. Later, we report the experiments we conducted in Sect. 5 and Sect. 6 to show correctness and fidelity to the black-box predictions. In this section, we show how BARBE performs and compare it against other methods. We conclude this work in Sect. 7 and provide some thoughts about future work.

2   Related Work

Attempts to explain classification decisions are not new. ExplainD is a tool introduced by Poulin et al. [19] that visualizes the decisions of well-known classifiers, which helps users understand their behavior during inference. Most researchers focus on explaining Deep Neural Networks (DNNs), particularly for image classification, by exposing the internals of the model using methods such as computing gradients and propagating them back to input to capture important pixels, which can be presented as the explanation [7,25] (e.g., Grad CAM [24]).

LIME [21] is a popular model-agnostic method that uses perturbed samples to train a linear regression model for explaining black-box models. It relies on the input and output of the model to generate explanations, without knowledge of the internal structure of the black-box model, and its explanation is a ranked list of important features for the prediction of each data point. Anchor [22] is an approach for explaining black-box models that provides a set of salient features in the form of a single "if-then" rule to overcome the limitation of the linear model associated with LIME. A weakness of this approach is that it does not reveal the associations among features. Also, in contrast to LIME, it cannot provide any relative feature importance scores anymore.

Guidotti et al. introduce LORE [6] that takes advantage of rules for providing explanations. In their method, they create a neighbourhood around the instance using a Genetic Algorithm. Moreover, they enforce the data point selection algorithm to choose at most half of the data points from the class of the original data point. Note that while data points are created by the genetic algorithm, the class labels are obtained by querying the black-box model. With the labeled synthetic data points they train a decision tree. They take advantage of the decision tree to produce two types of rules; a single decision rule, and a set of counter-factual rules.

Pattern Aided Local Explanation (PALEX) is another method proposed to provide explanations for black-box models. In their method, Jia et al. [10] suggest a set of patterns as the explanation using FP-Growth algorithm [9]. Their method, however, requires defining a few hyper-parameters such as minimum support, and minimum growth ratio thresholds as well as the probability score provided by the black-box model. Alternatively, LACE [17] directly learns an associative classifier by exploiting the nearest data points in training data. Their method, however, requires the training data to be available, and this may not always be realistic. Additionally, the sparsity of the training data in that neighbourhood, can have a substantial impact on the performance of their system. CoSP (Co-Selection Pick) is a recent framework proposed by Meddahi et al. [15], which aims to provide global explainability for black-box machine learning models. It does not explain a specific prediction but piggybacks on an existing explainer to co-select the most important test instances and features of the model as a whole. The framework selects individual explanations based on a similarity preserving approach, achieving a co-selection of instances and features. Unlike submodular optimization methods, CoSP considers the problem as a co-selection task and can be applied in both supervised and unsupervised scenarios with supposedly any local explainer. In their paper they used LIME.

3   Associative Classification

Rule-based classifiers, such as Ripper [4] or SigDirect [11], generate easily interpretable models by learning classification rules of the form "**If** *condition* **Then** *class*". Being known as transparent classifiers, they use attribute-value pairs as the antecedent and a class label as the consequent. During inference, applicable rules are selected based on whether their antecedent matches the instance's features, and a heuristic is used to assign the consequent as the prediction.

Associative classifiers learn their classification rules by applying association rule mining, a canonical task in data mining on the data after modeling the training data into transactions, each transaction being a set of attribute-value pairs and the class label. The rules are conjunctions of feature-values implying a class label: f_1, and f_2, and f_3, and f_4, and ..., $f_n \rightarrow class1$.

Associative classifiers have, for the most part, after the rule generation, a rule pruning phase to weed out redundant and noisy rules, and this is where the various approaches differ, in addition to the heuristics used to select rules at inference time. The most recent associative classifier approach that outperforms all preceding algorithms is SigDirect [11], which we take advantage of in our framework BARBE. The authors of SigDirect showed that not only did their algorithm outperform the other associative classifier contenders in accuracy on various datasets, it also generates a classification model with significantly less rules. Having fewer and more accurate rules is particularly pertinent for providing explanations in BARBE. Another advantage is the lack of cumbersome parameters. With other associative classifiers like CBA [13], CMAR [12] or ARC [3] hyper-parameter tuning is required. They heavily rely on support and confidence thresholds which are notoriously difficult to assess. SigDirect uses instead statistical significance to appraise rules.

SigDirect uses an Apriori-like strategy to first generate the rules and then leverage a new instance-based approach for the pruning step to only keep rules with the highest quality [11]. Similar to Apriori [2], it expands one level at a time but uses the Kingfisher algorithm [8] to find globally optimum, non-redundant dependencies with a scalable branch and bound approach with supplementary pruning by means of Fisher exact test and a P-value for statistical significance.

Therefore we use SigDirect, a strong interpretable rule-based classifier that generates a minimal number of statistically significant classification rules, as the core of our model-independent explanation framework.

4   BARBE: Black-Box Association Rule-Based Explanation

4.1   Shortcoming of Other Methods

Take what LIME generates for the text "A movie where tensions build and conflicts arise" as shown in Fig. 6A. The number below each feature is simply the "importance scores" used for ranking. For example, 0.10 for *movie* and 0.06 for *conflicts* highlight that *movie* has slightly higher importance than *conflicts* in making the sentence negative. This leads us to the conclusion that only the order among features matters to the users and not the numbers generated in the explanations. Sine LIME uses a weighted loss function for its linear model that also benefits from regularisation, it is likely that the instances which are not in the very close proximity of the original instance would be misclassified by this linear model, thus providing wrong explanations to the user. Ribeiro et al. [22], the same authors of LIME, also point out the fact that features are taken independently (see example in Fig. 6B). They introduce Anchor to overcome this issue. In their new method, an explanation is a set of features that whenever they co-occur, the class label is determined with a 95% confidence. This Anchor essentially resembles a rule (with a high confidence threshold of 95%).

The authors of LORE [6] benefit from the idea of using a set of counter-factual rules as the explanation in their method as well. Despite the fact that these methods, to some degree, overcome the problem mentioned above, one issue remains: is there always only one set of correlative features (and hence one reason) behind the final outcome of the model? What if there were multiple sets of correlative features that independently derive the final conclusion of the system [16]. Therefore an explanation should not solely focus on independent features or one unique set of associated features but on possibly a set of causes. Hence the interest in an associative classifier that can provide a set of rules as an explanation.

To overcome the above shortcomings, we introduce Black-box Association Rule-Based Explanations or BARBE. Our method, unlike LIME, provides a set of rules as the explanation, where not only do rules provide users with important features (what LIME does), but also takes care of the associations among them (what LORE and Anchor do). In addition, since we provide multiple rules as an explanation, we can hint at multiple causes that have led to that decision by the system, something that the aforementioned methods are unable to provide. Note that using a decision tree (in systems like LORE [6]) the path in the tree leading to the predicted label results in a single applicable rule which constitutes only one unique cause.

4.2   Explanations by BARBE

BARBE generates a descriptive model learned on data labeled by the black-box and provides as the explanation a subset of rules from the model that apply to the instance for which the explanation is expected. From this set of rules and their individual measure of confidence and significance, BARBE can provide an ordered set of important features as an alternative way of providing explanations. This allows the users to have the choice to look at these two types and get a better understanding of the underlying causes. Moreover, as mentioned earlier, each rule in addition to the items in its antecedent and the class label, comes with added information such as its confidence, support value, and p-value. Not only does this provide an alternative means for users to comprehend black-box models, but it also opens the door for researchers to conduct comparisons with other techniques such as LIME.

Figure 1 shows an example of what BARBE outputs for an instance of the Glass dataset [5]. In this example, BARBE produces three rules in which they not only provide important features to the users but also hint at the associations among the features. For compactness, the feature number in both the table and histogram is displayed instead of the full name. The first rule could be written as "Magnesium = [0.38, 2.13], Aluminum = [1.64, 1.76], Calcium = [7.80, 8.23] → Vehicle Window". The second rule could be expressed as "Magnesium = [0.38, 2.13], Potassium = [0.00, 0.61] → Vehicle Window" and the third rule could be written as "Potassium = [0.00, 0.61] → Vehicle Window". Here, the rules are inferring class label 3.

RULES	Support	Confidence	ln pF
3="1" AND 4="3" AND 7="3" → 3	0.0900	0.833	-17.095
3="1" AND 6="1" → 3	0.2230	0.791	-33.641
6="1" → 3	0.3730	0.696	-24.174

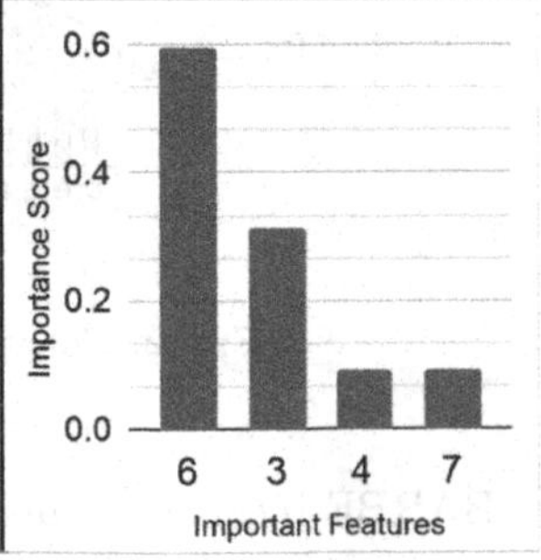

Fig. 1. The explanation provided by BARBE for an instance of the Glass dataset [5]. Here, feature tokens are shown for conciseness. The right side contains the important features ranked based on their importance. The left side contains important rules with their support, confidence, and the logarithm of the p-value reported by SigDirect.

4.3   How Does BARBE Work?

A high-level representation of BARBE's activity diagram is shown in Fig. 2. BARBE creates a neighbourhood around the instance to explain with synthetic data points produced by perturbing the features of the instance. The synthetic data points are labeled by the black-box which produces a training set for the SigDirect classifier. The outcome of the training is a set of rules. Rules from the trained model relevant to the original instance are extracted. Lastly, BARBE derives important features from these rules. BARBE needs to have access to the set of possible values for each attribute. Moreover, SigDirect, the heart of BARBE, relies on associations between discrete features. Indeed numerical values of continuous attributes need to be discretized into intervals. Associative classification rules demand ordinal features. Therefore, if buckets are not predefined, and in order to define buckets for continuous data, BARBE needs to access a sample of data from which the instance to be explained was drawn.

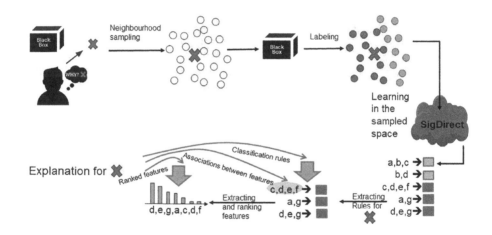

Fig. 2. Coarse representation of BARBE's framework.

BARBE receives an instance and a label to explain. If there are continuous features, and buckets for value intervals are not defined, a dataset of data points is used to discretize numerical attributes and define buckets. These buckets are then used to perturb the original instance and generate a neighbourhood of synthetic data points around the original instance. All these points are then labeled by the black-box after being converted to the input format of the black-box. For instance, before the labeling by the black-box, a reverse quantization may be required for continuous values that were mapped to the set of discrete finite buckets. This reverse transformation creates a normal distribution for each bucket of a feature and then randomly samples from the distribution. For text data, BARBE uses a simple strategy to generate a synthetic dataset around the original instance. BARBE takes advantage of random word removal from the sentence. The algorithm takes the input sentence and a number which tells the algorithm to iterate the process for n times to generate n number of synthetic text data. It selects a set of random positions within the input sentence and deletes the words from those positions. The resulting sentence after deleting random words is returned as the synthetic sentence. This process is repeated n times to generate n new sentences which form the neighbourhood dataset around the original text.

The resulting set of classification rules is relevant to the original instance since the training data is made of instances from its vicinity. The set is further reduced by selecting the most relevant rules to the instance to explain. The most important features are thereafter selected from these selected rules. We have experimented with different metrics to rank the features and found that summing the supports of all applied rules in which a feature f appears provides the best accuracy when compared with the features used by the black-box.

5   Experiments

5.1   Experiments Setup

To evaluate our method, we compare the explanation produced by BARBE against the true explanation. We replace the black-box model with a fully-transparent model and conduct experiments on this "open box". The interpretable model we leverage in our experiments is a Decision Tree (DT)[1]. Not every DT is interpretable. Its depth should be limited to a reasonable level so humans can track different paths in this data structure [7]. We limit the depth of the DT to a specific level k at train time, with k set at 5 in our experiment.

5.2   Experiments' Metrics

We use *Precision, Recall, F$_\beta$-score*, and *Rank-Biased Overlap (RBO)* as our comparative metrics in the experiments. We make use of $F_{0.5}$-*score* in our experiments to put more importance on *Precision* than *Recall*. If the explanation includes only a few of the features which are mostly tagged correctly as important (i.e., a case where *Precision* is high but *Recall* is low), then the end-user can still trust the system as this case indicates the black-box model is focusing on some of the right features. Moreover, since BARBE is presenting a ranked list of features as explanation, we take advantage of Rank-Biased Overlap (RBO) [27] to evaluate the order of important features compared to the ground truth explanation[2].

In our experiments, we compare the explanation provided by BARBE with the ground truth explanation obtained from the DT for each instance in the dataset. We then calculated the metrics for each one. These metrics are later averaged over all instances used in the experiment and finally reported in Table 1 and Fig. 3 in the next section. All results are averages for 100 experiments of explanation evaluations.

---

[1] We use scikit-learn [18] for implementing the DT.

[2] We used the implementation in https://github.com/changyaochen/rbo.

Table 1. $F_{0.5}$ and RBO for different methods and datasets with sample size at $5,000$.

dataset	$F_{0.5}$-score			RBO		
	LIME	Anchor	BARBE	LIME	Anchor	BARBE
Glass	0.736	0.534	**0.796**	0.777	0.287	**0.796**
Wine	0.554	0.656	**0.680**	0.314	**0.484**	0.416
Hungarian	0.483	0.508	**0.570**	**0.776**	0.264	0.427
Poker	**0.666**	0.331	0.637	0.353	**0.713**	0.368
Breast	0.633	0.497	**0.715**	0.300	0.246	**0.332**
Image	0.566	0.570	**0.852**	0.483	0.321	**0.500**
Magic	0.596	**0.729**	0.712	0.684	**0.835**	0.515
Vowel	0.526	0.683	**0.840**	0.671	0.444	**0.675**
Hepatitis	0.432	**0.492**	0.340	0.106	**0.218**	0.055
WPBC	0.489	0.536	**0.642**	0.543	0.594	**0.606**
WDBC	0.468	0.131	**0.494**	0.228	0.056	**0.320**
Average	0.559	0.515	**0.662**	**0.476**	0.406	0.455
Nb. Wins	1/11	2/11	**8/11**	1/11	4/11	**6/11**

5.3 Comparison with Other Explainers

We choose LIME and Anchor to compare with BARBE through different experiments with an interpretable decision tree disguised as a black-box. We exploit 11 different UCI datasets [5] (Table 1) to conduct these experiments where the results we report are averages over five runs with different random seeds. In Fig. 3, we provide *Precision*, *Recall*, and $F_{0.5}$-score for all data points that the method has predicted the class label correctly when the generated sample size around the instance increases from 1,000 to 5,000. For lack of space, we report here the results of the average for all 11 datasets.

To estimate the faithfulness of LIME, we examine its prediction score. Because the regression model is trained on values originating from a probability space, we can expect the regression model to generate a number in the same domain. Additionally, for the original point, if the interpretable model is trained properly, the predicted value should be close to 1. If the score, however, is below $\frac{1}{n}$, where n is the number of classes, that explanation is not faithful, and thus we do not include the incorrect case. Our criterion generously considers the outcome of LIME's regression model faithful, since most instances would belong to the target class and thus their probability score should be close to 1, yet, we observe BARBE providing competing results throughout our experiments. Anchor, however, does not rely on any interpretable model, and therefore, we are not able to determine its fidelity with the ground truth. Consequently, we assume in our experiments that all its explanations are faithful, and as a result, we include all instances of Anchor in the evaluation.

We also report the $F_{0.5}$ and *RBO* for the three methods and for all the datasets in Table 1 where the sample size is $5,000$.

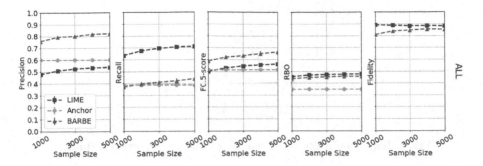

Fig. 3. Performance of LIME, Anchor, and BARBE.

In terms of $F_{0.5}$-*score*, BARBE displays the best results for all datasets but two, for which it was still a good contender. This good performance of BARBE is mainly due to the very high *Precision*. It was able to pinpoint correct features as per the ground truth. Moreover, LIME had typically higher *Recall* as LIME ranked all available features and therefore would rarely miss relevant ones. Interestingly, BARBE outperforms Anchor in not only for the $F_{0.5}$-*score*, but also *Precision* in most datasets, even though Anchor depends on a high precision rule to explain an instance. Moreover, the results demonstrate that BARBE has a better capacity to order the importance of features. In more than half the datasets, BARBE gives a better*RBO* score, followed by Anchor which beats LIME in terms of arranging the discovered salient features by importance.

5.4   Faithfulness to the Black-Box

To show the real importance of the features claimed important by BARBE on the decision of the black-box model, we make changes to the values of those features and request the black-box to do another classification. The more those changes are significant, the higher the chance that the black-box flips its decision. This is illustrated in Fig. 4. We changed the value of the most important features by 1 standard deviation up to 2 standard deviations for 100 randomly selected data points from the Pen Digits dataset of UCI repository and classified by a neural network with 2 hidden layers as a black-box. The accuracy clearly and steadily drops as we increase the extent of the change which indicates that the features highlighted by BARBE are indeed the influencers for the black-box.

6   Experiments on BARBE for Text

In this section, we discuss the settings under which we conduct experiments to evaluate BARBE for text. Our goal is to develop a framework that can be

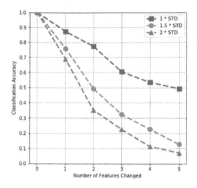

Fig. 4. Impact on Accuracy of changing values of important features.

employed not only on tabular datasets but also on text datasets. For text data, we demonstrate the efficacy of BARBE for binary classification tasks in this paper. We choose the IMDB movie review dataset [14] for the binary classification task since it is widely used in the literature for text classification.

6.1   Results

BARBE uses the data labeled by the black-box model to train a descriptive model that generates a set of rules as the explanation. Each rule has a support, confidence, and statistical significance value associated with it. The rules correspond to the features in the data which constitute the set of important features as the form of explanation. We have used support vector machine (SVM) as the underlying black-box and trained it with the IMDB movie review dataset. TF-IDF [20] has been used to convert the text into features in BARBE. When the black-box is ready, we use the neighbourhood generation process as discussed in Sect. 4.3 to create the synthetic dataset. This dataset once labeled by the black-box is used by BARBE to generate the rules in the form of explanation. Figure 5 shows the result generated by BARBE for a sentence with negative sentiment as labeled by the black-box. The sentence depicted in Fig. 5A contains words that have been highlighted with a color gradient of red for negative words and green for positive words. Figure 5B depicts the rules generated by BARBE. Here, BARBE generates a total of 5 rules for this sentence to identify the most important features. The important features are highlighted from strong red to light red for negative and strong green to light green for positive ones.

   It is evident from the figure that the words "conflicts" and "tensions" are the two negative words that are responsible for labeling the sentence as negative by the black-box model. It is noteworthy that BARBE generates a set of rules, not a single rule, and each rule contains a set of words. For the sentence of Fig. 5A, BARBE generates "conflicts arise" as the association rule that essentially has semantic context in the sentence. "conflicts arise" has more semantic significance than the word "conflicts" only. The three other rules generated by BARBE are:

"tensions", "movie", and "build" as shown in Fig. 5B. It is important to note that the word "build" has been detected as a feature having a positive label here. With 31.01% confidence, this rule has a positive label meaning this rule has little impact on making the sentence positive. Thus, BARBE not only identifies the rules that explain the negative sentiment of the sentence but also highlights the rules that may contribute slightly to the sentence being positive. Each rule has its support, confidence, and the logarithm of statistical significance values shown in the table in Fig. 5B inside brackets. BARBE represents the black-box prediction probabilities as depicted in Fig. 5C. In Fig. 5D, the vertical bar chart showcases the most important features based on their frequency within the rules, along with the corresponding weighted confidence value. It is evident from this figure that the word "conflicts" and "tensions" are the most important feature whereas "movie", and "arise" are the least important ones.

We also demonstrate the explanation obtained by BARBE for a sentence having positive sentiment. Figure 5E presents the sentence. BARBE generates 5 rules to explain why the sentence has been labeled as positive by the black-box model as shown in Fig. 5F. The words "wonderful" and "success" are enough to justify the sentence as positive. Figure 5G is the prediction probabilities of the black-box model and Fig. 5H highlights the most important features in terms of their frequency weighted by the confidence values within the rules.

6.2   Comparison with Other Explainers

We compare the explanation generated by BARBE with LIME and Anchor for the text dataset. LIME for text differs from LIME for tabular data in terms of the neighborhood data generation methodology. Starting from the original instance, new instances are created by randomly removing words from the original instance. There is a major drawback here. While generating such neighborhood instances, LIME creates a large number of empty sentences as we explore the neighborhood generation algorithm of LIME. An empty sentence does not make any sense when it is used to be labeled by the black-box model. On the other hand, Anchor deploys a perturbation-based strategy to generate local explanations for predictions of black-box model.

Figure 6 illustrates the explanation generated by LIME and Anchor for the sentence "A movie where tensions build and conflicts arise". LIME highlights the feature "movie" as the most important word for making the sentence negative. It provides fewer weights to "tensions" and "conflicts". The word "movie" cannot justify the decision of the black-box alone with such a significant weight of 0.10. Besides, there is no association within the features generated by LIME. LIME provides less significance to "tensions" and "conflicts" whereas BARBE provides higher significance to them (see Fig. 5). Moreover, BARBE discovers the rules containing a set of features e.g. a conjunction of features that makes more sense when explaining the decision of the black-box. All the rules have their support, confidence, and statistical significance values which are not available in LIME. On the other hand, Anchor generates only a single feature "conflicts" and treats this word as solely responsible for making the sentence negative. It misses the

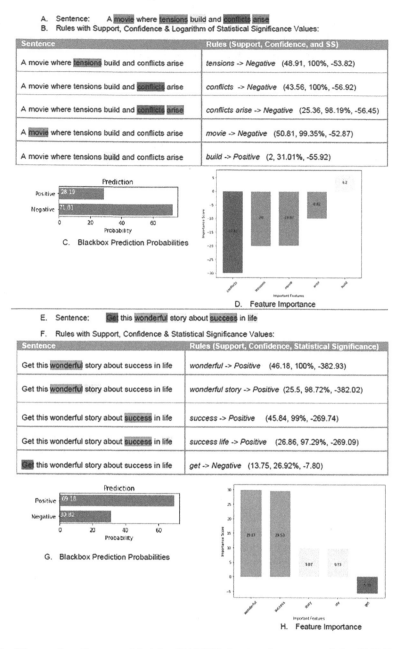

Fig. 5. The explanation provided by BARBE for two instances of the IMDB movie review dataset labeled by the black-box model. (A) shows a negative sentence with features highlighted. Red presents the negative words and green presents the positive ones. (B) presents the set of important rules with their support, confidence, and logarithm of statistical significance values Found by BARBE. (C) presents the prediction probability of the black-box, and (D) presents the histogram of important features ranked by BARBE. Figures E-F present the case for a positive sentence. (Color figure online)

other significant words present in the sentence as BARBE and LIME figure out in their explanation.

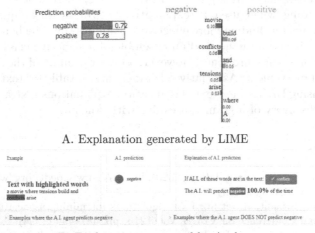

A. Explanation generated by LIME

B. Explanation generated by Anchor

Fig. 6. Comparing the explanation generated by Lime and Anchor for the sentence: "A movie where tensions build and conflicts arise". (A) presents the explanation of LIME. The probabilities on the left of Figure A are the prediction probabilities of the underlying black-box model. On the right of Figure A, the features and their corresponding importance scores generated by LIME are shown in order of their importance. (B) presents the explanation of Anchor which only depends on the single word "conflicts"

7   Conclusion and Perspectives

We have presented a model-independent explanation framework based on associative classifiers, BARBE, that provides explanations for any black-box classifier in three forms: a set of ranked salient features that are relevant in the prediction of an instance; significant associations between features; and a set of interpretable classification rules that could explain the attribution of a class label to an instance. Unlike other methods, BARBE does not require from the black-box anything more than the predicted label for a given input. For a given input and its imputed label to explain, BARBE creates a perturbed sample around the input, requests labels from the black-box, and learns from the sample classification rules using an effective associative classifier. Taking advantage of the interpretability of the model generated by the surrogate learner, BARBE can then provide useful explanations. Compared to other prevalent methods that provide salient features, we have shown that BARBE has a better *Precision*, presents a better balance between *Precision* and *Recall* via the $F_{0.5}$ score, and ranks better the features as indicated by a respectable *RBO* score. In addition, BARBE provides classification rules with associations between the features.

Demonstrating the performance of BARBE on text makes the association rule more feasible to explain. By providing a conjunction of rules e.g. conjunction of features, the semantic fidelity is preserved by BARBE.

Associative classifiers are highly accurate but can generate noisy rules, which can be misleading when used in explanations. Pruning techniques can help address this issue, but finding more effective techniques could be beneficial. High dimensionality is also an issue, and an ensemble of associative classifiers can be used to partition the feature space. However, a better pruning of the search space would be an improvement. Associative classifiers are suitable for text categorization [3], and using BARBE for text classification explanations is straightforward, allowing the discovery of n-gram causes like with Anchor.

References

1. General data protection regulation (2020). https://en.wikipedia.org/wiki/General_Data_Protection_Regulation
2. Agrawal, R., Srikant, R., et al.: Fast algorithms for mining association rules. In: International Conference on VLDB, vol. 1215, pp. 487–499 (1994)
3. Antonie, M.L., Zaiane, O.R.: Text document categorization by term association. In: IEEE International Conference on Data Mining, pp. 19–26. IEEE (2002)
4. Cohen, W.W.: Fast effective rule induction. In: Twelfth International Conference on Machine Learning (1995)
5. Dua, D., Graff, C.: UCI machine learning repository (2017). http://archive.ics.uci.edu/ml
6. Guidotti, R., Monreale, A., Ruggieri, S., Pedreschi, D., Turini, F., Giannotti, F.: Local rule-based explanations of black box decision systems. arXiv preprint arXiv:1805.10820 (2018)
7. Guidotti, R., Monreale, A., Ruggieri, S., Turini, F., Giannotti, F., Pedreschi, D.: A survey of methods for explaining black box models. ACM Comput. Surv. (CSUR) **51**(5), 93 (2018)
8. Hamalainen, W.: Efficient discovery of the top-k optimal dependency rules with fisher's exact test of significance. In: 2010 IEEE International Conference on Data Mining, pp. 196–205. IEEE (2010)
9. Han, J., Pei, J., Yin, Y.: Mining frequent patterns without candidate generation. In: ACM Sigmod Record, vol. 29, pp. 1–12 (2000)
10. Jia, Y., Bailey, J., Ramamohanarao, K., Leckie, C., Ma, X.: Exploiting patterns to explain individual predictions. Knowl. Inf. Syst. **62**, 927–950 (2019). https://doi.org/10.1007/s10115-019-01368-9
11. Li, J., Zaiane, O.R.: Exploiting statistically significant dependent rules for associative classification. Intell. Data Anal. **21**(5), 1155–1172 (2017)
12. Li, W., Han, J., Pei, J.: CMAR: accurate and efficient classification based on multiple class-association rules. In: Proceedings 2001 IEEE International Conference on Data Mining, pp. 369–376. IEEE (2001)
13. Liu, B., Hsu, W., Ma, Y., et al.: Integrating classification and association rule mining. In: KDD, vol. 98, pp. 80–86 (1998)
14. Maas, A., Daly, R.E., Pham, P.T., Huang, D., Ng, A.Y., Potts, C.: Learning word vectors for sentiment analysis. In: Proceedings of the 49th Annual Meeting of the Association for Computational Linguistics: Human Language Technologies, pp. 142–150 (2011)

15. Meddahi, K., et al.: Towards a co-selection approach for a global explainability of black box machine learning models. In: Chbeir, R., Huang, H., Silvestri, F., Manolopoulos, Y., Zhang, Y. (eds.) WISE 2022. Lecture Notes in Computer Science, vol. 13724, pp. 97–109. Springer, Cham (2022). https://doi.org/10.1007/978-3-031-20891-1_8

16. Miller, T.: Explanation in artificial intelligence: insights from the social sciences. Artif. Intell. **267**, 1–38 (2018)

17. Pastor, E., Baralis, E.: Explaining black box models by means of local rules. In: Proceedings of the 34th ACM/SIGAPP Symposium on Applied Computing, pp. 510–517 (2019)

18. Pedregosa, F., et al.: Scikit-learn: machine learning in Python. J. Mach. Learn. Res. **12**, 2825–2830 (2011)

19. Poulin, B., et al.: Visual explanation of evidence with additive classifiers. In: Proceedings of the National Conference on Artificial Intelligence, vol. 21, p. 1822. AAAI Press, MIT Press, Menlo Park, Cambridge, London (2006)

20. Ramos, J., et al.: Using TF-IDF to determine word relevance in document queries. In: Proceedings of the first instructional conference on machine learning, vol. 242, pp. 29–48. Citeseer (2003)

21. Ribeiro, M.T., Singh, S., Guestrin, C.: Why should i trust you?: explaining the predictions of any classifier. In: 22nd ACM SIGKDD International Conference on Knowledge Discovery and Data Mining, pp. 1135–1144 (2016)

22. Ribeiro, M.T., Singh, S., Guestrin, C.: Anchors: high-precision model-agnostic explanations. In: Thirty-Second AAAI Conference on Artificial Intelligence (2018)

23. Secretariat, T.B.O.C.: Directive on automated decision-making (2017). https://www.tbs-sct.gc.ca/pol/doc-eng.aspx?id=32592

24. Selvaraju, R.R., Cogswell, M., Das, A., Vedantam, R., Parikh, D., Batra, D.: Grad-CAM: Visual explanations from deep networks via gradient-based localization. In: Proceedings of the IEEE International Conference on Computer Vision, pp. 618–626 (2017)

25. Shahroudnejad, A.: A survey on understanding, visualizations, and explanation of deep neural networks. preprint arXiv:2102.01792 (2021)

26. Sukel, K.: Artificial intelligence ushers in the era of superhuman doctors (2017). https://www.newscientist.com/article/mg23531340-800-artificial-intelligence-ushers-in-the-era-of-superhuman-doctors/

27. Webber, W., Moffat, A., Zobel, J.: A similarity measure for indefinite rankings. ACM Trans. Inf. Syst. (TOIS) **28**(4), 1–38 (2010)

Efficient Video Captioning with Frame Similarity-Based Filtering

Elyas Rashno (ID) and Farhana Zulkernine(✉)(ID)

School of Computing, Queen's University, Kingston, ON, Canada
{elyas.rashno,farhana.zulkernine}@queensu.ca

Abstract. Video captioning combines computer vision and Natural Language Processing (NLP) to perform the challenging task of scene understanding. The rapid advancements in artificial intelligence have led to a growing interest in video captioning, which involves generating natural language descriptions based on the visual content of videos. In this paper, we present a novel approach to video caption generation. The proposed method first extracts frames from the video and reduces the number of frames based on their similarity. The remaining frames are then processed by a Convolution Neural Network (CNN) to extract a feature vector, which is then fed into a Long Short-Term Memory (LSTM) network to generate the captions. The results are compared with the state-of-the-art models which demonstrate that the proposed approach outperforms the existing methods on MSVD, M-VAD, and MPII-MD datasets.

Keywords: Video Caption Generation · Video frame similarity · Sequence to Sequence · Stacked LSTM

1 Introduction

The enormous amount of digital information has led researchers to explore the use of natural language processing to interpret video content. Video captioning, which involves generating written descriptions of the visual content of a video, has particularly captured the interest of researchers due to its many potential applications [1]. Some of the important applications are human-robot interaction such as chat-bot [2], describing movies for the blind [3], movie interpretation for elderly people, and video indexing [4]. Image caption generation requires handling variable lengths of text captions and videos. Holistic video representation [5], pooling over frames [6], and choosing a fixed length of frames [7] are some of the existing approaches used to handling variable lengths of input frames. We propose a novel sequence-to-sequence approach that is trained on selected video frames and has the ability to better recognize patterns from video captioning including input sequence. Our model operates by processing video frames in a sequence, reducing the number of frames, and producing words in a sequence to make it a sequence-to-sequence model.

C. Strauss et al. (Eds.): DEXA 2023, LNCS 14147, pp. 98–112, 2023.
https://doi.org/10.1007/978-3-031-39821-6_7

Open-domain videos have a diverse set of actions, objects, and attributes. Therefore, generating text descriptions in each domain offers different challenges that researchers are trying to address [8]. In addition, determining the visible and important content to interpret and explain an event properly in the detected context is difficult for open-domain videos. The proposed model is trained with video frames and their paired sentences which describe the salient actions in the context of natural language. The Long Short-Term Memory (LSTM) network [9] is a Recurrent Neural Network (RNN) that is used with time-series data, especially in sequence-to-sequence tasks such as speech processing [10], natural language processing from the Internet of Things (IOT) sensor data [11], and machine translation [12]. Since the video caption method presents a combination of the sequential nature of videos and languages, LSTM is a simple and efficient model that can handle the inherent sequential pattern in both language and video [13].

The main contribution of our work is developing a video frame selection approach that can provide a good result with existing models. It uses a standard sequence-to-sequence architecture, which enables the model to 1) accept an unpredictable number of input frames, 2) reduce the number of frames to a fixed number that enables the model to train fast, and 3) develop a language model that creates grammatical sentences naturally. Our model is trained end-to-end, taking into account both Convolution Neural Network (CNN) and LSTM networks for feature extraction and caption generation respectively. We demonstrate that our method surpasses the state-of-the-art results in three distinct datasets: MSVD [14], M-VAD [15], and MPII movie description [16].

2 Related Work

Video caption generation is a bridge between two research paradigms namely computer vision and natural language processing, which has attracted great attention recently [17]. Generating a description of the content of an image or video is a challenging topic in computer vision [18]. Kojima et al. [19] extracted body features including, head directions, and head and hand positions, which are related to detecting human postures. They generated case frames based on the syntactic components and converted these case frames into sentences based on syntactic rules. Khan et al. [20] and Hanckmann et al. [21] tried to use some strategies to enhance multimedia applications.

Later on, probabilistic graphical models were used to describe video and images [18,22,23]. For example, Farhadi et al. [22] designed three different spaces: image space, sentence space, and meaning space. For the next step, the authors combined image and sentence spaces and generated the meaning space to find meaningful connections between images and related sentences. For representing the meaning space, they used a triplet as $<object, action, scene>$. The image space was mapped into a meaning space by predicting triples from images, while the sentence space was mapped into a meaning space by extracting triples from sentences and then calculating their similarity. In addition, Rohrbach et al. proposed a method named Conditional Random Field (CRF) [23] to detect the relationship between computer vision and semantic representation. The main disadvantage of these models is their dependency on sentence templates.

Deep learning has been used by many researchers to solve the problem of video captioning [13,24–26]. For instance, Venugopalan et al. [13] used a stacked LSTM to generate effective descriptions, while Pan et al. [25] used visual-semantic embedding to learn from the semantics of entire sentences and video content. Some studies have shown that semantic attributes can greatly contribute to video captioning [27,28]. For example, Pan et al. [27] used Multiple Instance Learning (MIL) to learn semantic attributes from videos and then improve their models. Instead of using mean pooling [26,29,30], others have used attention mechanisms to tackle video captioning. Yao et al. [30] used a temporal soft attention mechanism to select the most relevant frames. Yu et al. [31] used a supervised spatial attention mechanism to guide the model to learn relevant spatial information for video captioning. In addition to our unimodal approach that utilizes only video content for caption generation, several state-of-the-art models incorporate other modalities such as audio with video frames. Yang et al. proposed Vid2Seq [32], a visual language model pretraining approach, aimed at dense video captioning. Their large-scale pretraining method effectively captures the visual and linguistic information in videos, improving the quality and accuracy of generated captions. Wang et al. introduced a powerful few-shot video-language learning approach [33] by combining language models with image descriptors. Their method demonstrates impressive performance in understanding and generating captions for videos with limited training data. Li et al. proposed Clip-event [34], a novel framework that establishes connections between text and images using event structures. By leveraging event information, their approach effectively bridges the gap between textual descriptions and visual content, enabling better understanding and interpretation of visual data.

3   Preliminaries

3.1   Tensor Dot (TD)

The TD (also known as Tensor Contraction) between two tensors A and B of shape $N_1 \times N_2 \times K$ and $K \times M_1 \times M_2$ respectively, is defined as the product of corresponding components over specific axes. This operation is denoted as per the notation in [35] as

$$C = A(i,j) \cdot B \tag{1}$$

where C is the resulting tensor of the TD operation which has a shape of $N_1 \times N_2 \times M_1 \times M_2$, and i and j represent the axes over which the sum is performed. In this example, i and j can only be 3 and 1 respectively, as they are the only axes that are of the same size (K).

3.2 Chamfer Similarity (CS)

The Chamfer Similarity (CS) is the similarity counterpart of the Chamfer Distance [36]. It involves comparing two sets of items, for example, set x with N items, and set y with M items. The similarity between each item in set x and set y is represented by a matrix S with shape $N \times M$. The CS is calculated as the average of the highest similarity between each item in set x and the items in set y, as represented by Eq. 2.

$$CS(x,y) = \frac{1}{N} \sum_{i \in N} \max_{j \in [1,M]} S(i,j) \tag{2}$$

It is important to note that the CS is not symmetric, meaning $CS(x,y)$ is not equal to $CS(y,x)$, however, a symmetric variant, $SCS(x,y)$ can be defined as the average of $CS(x,y)$ and $CS(y,x)$:

$$SCS(x,y) = (CS(x,y) + CS(y,x))/2 \tag{3}$$

3.3 LSTM for the Sequence-to-Sequence Models

We use a latent vector representation to handle varying lengths of input and output in video captioning. This representation encodes the sequence of frames in the video and then decodes it into a sentence, word by word. A popular approach to do this is to use the Long Short-Term Memory Recurrent Neural Network (LSTM-RNN). LSTM was first introduced by Hochreiter et al. [9] which uses the LSTM unit proposed in [37]. This unit computes a hidden/control state and a memory cell state at each time step, which encodes all observations made until that time as demonstrated in Eq. 4.

$$
\begin{aligned}
i_t &= \sigma \left(W_{xi} x_t + W_{hi} h_{t-1} + b_i \right) \\
f_t &= \sigma \left(W_{xf} x_t + W_{hf} h_{t-1} + b_f \right) \\
o_t &= \sigma \left(W_{xo} x_t + W_{ho} h_{t-1} + b_o \right) \\
g_t &= \phi \left(W_{xg} x_t + W_{hg} h_{t-1} + b_g \right) \\
c_t &= f_t \odot c_{t-1} + i_t \odot g_t \\
h_t &= o_t \odot \phi \left(c_t \right)
\end{aligned}
\tag{4}
$$

where a sigmoidal non-linearity σ and a hyperbolic tangent non-linearity ϕ are used to compute the hidden control state and memory cell state. These states are determined by the element-wise product of gate values, weight matrices (W_{ij}), and biases (b_j) which are learned through the model training. So, during the encoding phase, the LSTM takes in an input sequence X of length n which is composed of $x_1, x_2, ...x_n$ and computes a sequence of hidden states $h_1, h_2, ...h_n$. During the decoding phase, it defines a probability distribution over the output sequence $Y = (y_1, ...y_m)$ given the input sequence is $X = (x_1, ...x_n)$ as $p(Y|X)$.

$$p(y_1, \ldots, y_m \mid x_1, \ldots, x_n) = \prod_{t=1}^{m} p(y_t \mid h_{n+t-1}, y_{t-1}) \tag{5}$$

The distribution $p(y_t|h_{n+t})$ is determined by a SoftMax function over all the words in the vocabulary as shown in Eq. 5. It is important to note that the h_{n+t} is obtained from the previous hidden state h_{n+t-1} and the previous output y_{t-1} using the recursion in Eq. 5.

4   Proposed Method

The main contribution of this work is to develop a model that can reduce the number of video frames without missing information and can directly translate a series of frames into a sequence of words. The model is illustrated in Fig. 1. First, all the frames are extracted from the video. Then a series of consecutive frames are selected and processed through a CNN to extract the significant features of each frame. Finally, a stacked LSTM translates frame vectors into sentences. These frames are encoded one by one through a stacked LSTM. After all frames are encoded, the model starts to generate words in a sentence in the decoder stage, one at a time. The encoding and decoding of the frame and word representations are trained together. In this problem, the input is a sequence of frames and the output is a sequence of words. The sequence of inputs or video frames is denoted as x where $x = x_1, x_2, x_3, ..., x_n$ and the generated words in the output are denoted by $y = y_1, y_2, y_3, ..., y_m$. We have to estimate the probability of $p(y_1, ..., y_m|x_1, ..., x_n)$. The length of n and m may vary for each video or label.

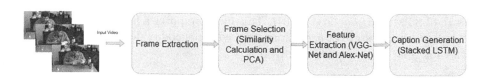

Fig. 1. Proposed Model Structure

The video caption generation problem is similar to that of machine translation in that both involve the mapping of a sequence of words from one modality to another. While the numbers of input and output tokens in machine translation are approximately the same, there can be many times more video frames than the number of words in the generated captions. In addressing this challenge, one common approach is to utilize a stacked LSTM network that contains both an encoder and a decoder. However, the imbalance in size between the number of video frames and the generated captions often leads to an imbalanced LSTM. To address this issue, it is necessary to reduce the number of video frames by removing the redundant similar frames.

This can be achieved by first creating a similarity matrix based on frame similarity as described in Sect. 4.1. Then, Principal Component Analysis (PCA) can be applied to select a subset of frames that are not similar, which results in detecting the distinctive video frames as outlined in Sect. 4.2. These methods help to reduce the number of video frames and allow for a more balanced representation in the stacked LSTM.

Auhroanme et al. [12] show the effectiveness of LSTM and RNN models in addressing sequence-to-sequence problems. We used the LSTM model to extract features from the video frame as input and interpret the frames to generate a meaningful sequence of words as the caption. In the following four subsections, we first describe data preprocessing which includes extracting frames from videos and calculating the similarity between frames. The second subsection explains the algorithms used to reduce the number of frames based on frame similarity. In the third subsection, we describe the CNN model used to extract features from selected frames. The fourth subsection illustrates the used stacked LSTM respectively.

4.1 Frame-to-Frame Similarity

To determine the similarity between two video frames, d, and b, we use Chamfer Similarity (CS) on the similarity matrix of the frames. This process is depicted in Fig. 2. First, d and b frames are decomposed into individual RGB vectors, M_d and M_b, respectively. Then, the similarity between each pair of RGB vectors is calculated to create the similarity matrix for the two frames. Finally, CS is applied to the matrix to determine the similarity between the frames. The equation for CS is given below.

$$\text{CS}_f(d,b) = \frac{1}{N^2} \sum_{i,j=1}^{N} \max_{k,l\in[1,N]} \mathbf{d}_{ij}^\top \mathbf{b}_{kl} \tag{6}$$

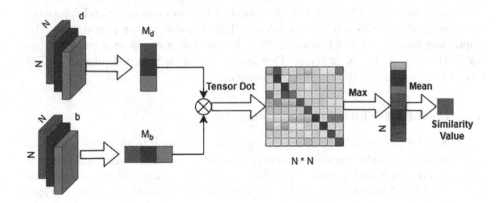

Fig. 2. Frame to frame similarity scheme

4.2   Frame Selection

The goal of this task is to select significant and distinct frames from a video. The task starts with the calculation of the similarity matrix for all frames of the video which is defined in the previous section. Then, by applying the PCA algorithm on the similarity matrix, which detects the correlations between frames, n frames with the lowest correlation are selected as the output of the task.

The principal components of the data are calculated by finding the eigenvectors of the covariance matrix. An eigenvector is a non-zero vector which, when multiplied by a matrix, returns a scalar multiple of itself. In PCA, the eigenvectors of the covariance or similarity matrix represent the directions in the data space with the most variance, and are used to construct the principal components [38]. Each principal component is a linear combination of the original variables, and the first principal component corresponds to the direction in the data space with the maximum variance. The subsequent components correspond to the directions of decreasing variance.

By transforming the similarity matrix into the space defined by the principal components, PCA helped to identify the n frames that have the lowest correlation. The frames with the lowest correlations are most different from one another, and thus represent the most significant information contained in the video.

4.3   Feature Extraction

VGGNet [39] and AlexNet [40] represent two variations of CNN architecture that are often used as feature extractors for image classification tasks. In order to generate a 4096-dimensional feature vector for each frame using VGGNet or AlexNet, the frame is first passed through the network and processed by several layers of convolutions and max pooling operations. These layers extract local and global features from the frame. The final layer of the network, usually referred to as the fully-connected layer, takes the output of the previous layer and computes a dense 4096 dimensional feature vector. This feature vector summarizes the important features of the frames and can be used as a compact representation for further processing or analysis. This feature vector is then fed into an LSTM model to train the model to generate captions.

4.4   LSTM for Video Caption Generation

A stacked LSTM model is employed to generate captions from video frames in this study. The model uses n consecutive frames as input, where each frame is represented as a 4096 dimensional feature vector. The output of the model consists of n LSTM blocks (each block generates one word) that generate captions in the form of sentences.

The encoder (green blocks) processes the input feature vectors and converts them into a compact representation that summarizes the key information from the video frames. This can be done using a series of LSTM layers where each LSTM block processes a single video frame represented by a feature vector.

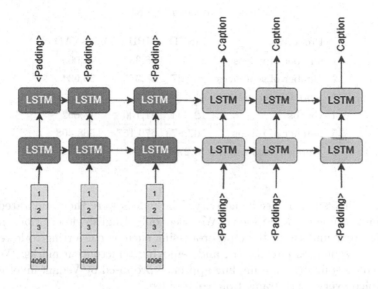

Fig. 3. LSTM structure to generate captions (Color figure online)

When all frames are fed to the model, the decoder takes the context vector as input and uses another series of LSTM layers to generate the caption. At each time step, the decoder outputs a probability distribution over the vocabulary of words that can be used to generate the caption. The stacked LSTM structure enables the model to process long-range dependencies in the data, and the use of a fixed-length context vector allows the model to handle input sequences of varying lengths (Fig. 3).

5   Results and Discussion

This section describes the datasets, evaluation metrics, and results.

5.1   Data Sets

This paper presents the evaluation of our proposed approach on three video description corpora: MSVD [14], MPII-MD [15], and M-VAD [16]. These three corpora form the largest collection of open-domain video and natural language descriptions. MSVD consists of web clips with brief human-annotated sentences, whereas MPII-MD and M-VAD include scenes from Hollywood movies and have descriptions obtained from script and audio sources. The statistics for each corpus as given in Table 1.

Microsoft Video Description Corpus (MSVD). The MSVD is a set of short videos from YouTube, collected by asking Mechanical Turk workers to

Table 1. Dataset Properties

	Properties	MSVD	MPII-MD	M-VAD
1	number of videos	1,970	68,337	46,009
2	number of sentences	80,827	68,375	56,634
3	sentence per video	41	1	1–2
4	vocabs	12,594	21,700	18,092
5	number of tokens	567,874	679,157	568,408
6	average video length	10.2 s	3.9 s	6.2 s

select clips that show a specific activity. The videos were then annotated with short, single-sentence descriptions. We used only English descriptions in this study. The text underwent basic preprocessing such as converting to lowercase, dividing the sentences into tokens, and removing punctuation marks. We used the same data splitting and sampling approach proposed by Vedantam et al. [41] and sampled every tenth frame from each video.

MPII Movie Description Dataset (MPII-MD). The MPII-MD dataset contains 68,000 video clips from 94 Hollywood movies [29]. Each video clip is accompanied by a single sentence description sourced from movie scripts and audio descriptions (AD) for the visually impaired. Although the movie clips are manually aligned with their descriptions, the data remains challenging due to the diverse visual and textual content and the fact that most clips have only one reference sentence. The study uses the training/validation/test split provided by the authors and every 5th frame is extracted (the videos are shorter than MSVD, averaging 94 frames per each video clip).

Montreal Video Annotation Dataset (M-VAD). The M-VAD movie description corpus is a collection of 49,000 short video clips from 92 movies that are similar to MPII-MD but contain only audio description data with automatic alignment. The same setup as MPII-MD is used for this dataset to extract video frames.

5.2　Evaluation Metric

The effectiveness of the models is assessed using the METEOR [42] evaluation metric, designed to measure the performance of machine translations. The METEOR score is determined by comparing a hypothetical sentence to reference sentences, taking into account exact token matches, stemmed tokens, semantic similarities, and paraphrases using WordNet synonyms. METEOR is a semantic-based metric, unlike other measures such as BLEU [43], ROUGE-L [44], or CIDEr [41]. Research has shown that METEOR performs better than BLEU and ROUGE, and outperforms CIDEr when the number of references is limited. Since MPII-MD and M-VAD only have a single reference sentence per video clip, METEOR was selected as the evaluation metric in this study. The study uses METEOR version 1.5.2 from the Microsoft COCO Evaluation Server [45]. The METEOR score consistently exhibits a low numerical value, which does not necessarily indicate poor performance. Being sensitive to Grammarly errors, sentence structure, and word choice and vocabulary, are the factors that contribute to the lower METEOR scores reported.

5.3   Results

In this work, we compare the proposed method against several state-of-the-art models in the field namely Factor Graph Model (FGM) [46], Mean-Pooled (MP) model [47], Temporal-Attention [7], and the Sequence to Sequence - Video to Text (S2VT) model [13]. The reason for selecting these models is that they are all popular baseline methods for video captioning. They have been widely used in previous studies and have been shown to produce competitive results.

FGM is presented by Thomason et al. [46], which identifies the most probable subject by using a two-step process. At first, the verb, object, and scene elements in a video are identified and then a sentence is generated based on a predefined template. We also compare our model with a different variation of the MP model [48]. The MP model uses a fixed-length vector representation of the video obtained by pooling activation from a deep neural network named AlexNet. MP model uses LSTM to generate sequence of words from the encoded vector. We compare our models with the basic MP model pre-trained on the Flickr30k [47], and an MP model pre-trained on the MSCOCO dataset [49]. Both MP models were fine-tuned on MSVD. We also compare our model with the GoogleNet [50] which is a variant of the MP model [7] is shown in row 5 Table 2. We use another variation of GoogleNet that applied temporal attention by Yao et al. [7] (row 6 Table 2). Another variant of this model uses a combination of GoogleNet and 3D-CNN which extract features over a set of video frames that is trained for activity classification (row 7 Table 2). Finally, we compare our model with variants of the S2VT [13] are shown in the row 8 and 9 of Table 2. The AlexNet or VGG are used in the two variants to extract video frame features and then feed the features into a sequence to sequence LSTM model to generate a caption.

Table 2 displays the experimental results of the above mentioned methods on the MSVD dataset. The first nine rows show results from the SOTA methods, and the rest depict variations of our method. Our method with AlexNet as the feature extractor selecting 20 frames that utilize RGB video frames (line 10 in Table 2) achieves a METEOR score of 28.1%, surpassing the basic MP model [41] (line 2, 26.9%) and the S2VT model with the same setup (line 8, 27.9%). When using the VGG-Net to extract features, our method achieves a METEOR score of 29.9%, where the S2VT method achieves only 29.2%. It is clear that using redundant frames during training can lead to over-fitting of the model with a drop in the test accuracy.

The performance of any model drops considerably when it is trained using fewer frames (lines 12 and 13 in Table 2). This low performance was likely due to missing important information when we select fewer frames. The average length

Table 2. METEOR results for MSVD dataset (higher is better)

	Method	METEOR
1	FGM [46]	23.9
	Mean Pool models	
2	AlexNet [47]	26.9
3	VGG [47]	27.7
4	AlexNet COCO pre-trained [49]	29.1
5	GoogleNet [7]	28.7
	Temporal Attention	
6	GoogleNet [7]	29.0
7	GoogleNet + 3D-CNN [7]	29.6
	Sequential Models	
8	S2VT + AlexNet [13]	27.9
9	S2VT + VGG [13]	29.2
10	Proposed Method (20 frames) + AlexNet	28.1
11	**Proposed Method (20 frames) + VGG**	**29.9**
12	Proposed Method (10 frames) + AlexNet	27.8
13	Proposed Method (10 frames) + VGG	29.0

of all videos in the MSVD dataset is 10.2 s which shows that choosing 10 frames cannot represent the whole content of the video. Therefore, in the experiment, the best performance is achieved when using VGGNet as a feature extractor with the 20 frame selected from the video for training our model.

Furthermore, we evaluate our model with the MPII-MD and M-VAD datasets and compare it with S2VT and other caption generation models. We utilize our best model from previous experiment, which is trained on RGB frames using VGG. To reduce over-fitting on the movie corpora, we implement dropout, which has proven effective on these datasets [28]. We found that the best results were achieved by applying dropout to both inputs and outputs of the LSTM layers. Optimization was done using ADAM with a momentum of 0.7. The results for MPII-MD and M-VAD datasets are shown in the second and third columns in Table 3. For the MPII-MD dataset, our approach outperforms the SMT method proposed by Rohrbach et al. [16], and achieves 7.4% in METEOR compared to 5.6% achieved by SMT. Our method achieves a 0.7% improvement compared to the MP method [48]. In addition, our result is also better than Visual-Labels [51] and S2VT in the same setup. Visual-Labels is a contemporary LSTM-based approach that lacks temporal encoding but utilizes more diverse visual features such as object detectors and activity/scene classifiers. Our results on the M-VAD dataset were not superior to those of S2VT. This is likely due to the fact that the dataset contains a smaller number of videos and vocabulary compared to the other datasets, and reducing the number of frames results in the loss of valuable information.

Table 3. METEOR results for MPII-MD and M-VAD datasets (higher is better)

	Method	MPII-MD	M-VAD
1	SMT [16]	5.6%	–
2	MP (VGG) [48]	6.7%	6.1%
3	Visual-Labels [51]	7.0%	6.3%
5	S2VT (VGG) [13]	7.1%	**6.7%**
6	Proposed Method	**7.4%**	6.6%

5.4 Training Time and Number of Parameters

Our experiments are conducted on a server with 32 CPU cores, 256 GB of RAM, and an A40 GPU (48 GB). The server is running on the ubuntu operating system and is equipped with python 3.8. All the data used for the experiments are stored on an extensible SSD storage. This hardware and software configuration provided the necessary computational resources for our experiments and allowed us to efficiently run the machine learning models. The number of parameters for the two best performing models, S2VT and the proposed method is shown in Table 4.

Table 4. Number of parameters and training time for S2VT and our proposed method

	Method	# Parameters	Training Time
1	S2VT [13]	14M	55 min
2	**Proposed Method**	9M	**37 min**

6 Conclusion

We propose a model that can reduce the number of video frames without missing information and can directly translate a series of frames into a sequence of words. Based on this study, we conclude that the proposed approach for video caption generation demonstrates superior performance compared to the existing methods on multiple datasets. By utilizing a combination of processes such as frame similarity reduction, CNN feature extraction, and LSTM caption generation, the proposed approach can generate natural language descriptions that accurately reflect the contents of the video. These results provide promising evidence for the potential of this approach to advance the field of video captioning. Future research in this area may explore further improvements to the proposed method by the application of reinforcement learning and investigating other data sampling strategies.

References

1. Donahue, J., et al.: Long-term recurrent convolutional networks for visual recognition and description. In: Proceedings of the IEEE Conference on Computer Vision and Pattern Recognition (CVPR) (2015)
2. Janssens, R., Demeester, T., Belpaeme, T.: Visual conversation starters for human-robot interaction (2022)
3. Jiang, L., Ladner, R.: Co-designing systems to support blind and low vision audio description writers. In: Proceedings of the 24th International ACM SIGACCESS Conference on Computers and Accessibility, pp. 1–3 (2022)
4. Bhooshan, R.S., Suresh, K.: A multimodal framework for video caption generation. IEEE Access **10**, 92166–92176 (2022)
5. Guadarrama, S., et al.: YouTube2Text: recognizing and describing arbitrary activities using semantic hierarchies and zero-shot recognition. In: International Conference on Computer Vision (ICCV) (2013)
6. Venugopalan, S., Xu, H., Donahue, J., Rohrbach, M., Mooney, R., Saenko, K.: Translating videos to natural language using deep recurrent neural networks. In: Conference of the North American Chapter of the Association for Computational Linguistics (NAACL) (2015)
7. Yao, L., et al.: Describing videos by exploiting temporal structure. arXiv preprint arXiv:1502.08029v4 (2015)
8. Hu, Y., Luo, C., Chen, Z.: Make it move: controllable image-to-video generation with text descriptions. In: Proceedings of the IEEE/CVF Conference on Computer Vision and Pattern Recognition, pp. 18219–18228 (2022)
9. Hochreiter, S., Schmidhuber, J.: Long short-term memory. Neural Comput. **9**(8), 1735–1780 (1997)
10. Graves, A., Jaitly, N.: Towards end-to-end speech recognition with recurrent neural networks. In: International Conference on Machine Learning (ICML) (2014)
11. Mahfuz, S., Isah, H., Zulkernine, F., Nicholls, P.: Detecting irregular patterns in IoT streaming data for fall detection. In: 2018 IEEE 9th Annual Information Technology, Electronics and Mobile Communication Conference (IEMCON), pp. 588–594. IEEE (2018)
12. Sutskever, I., Vinyals, O., Le, Q.V.: Sequence to sequence learning with neural networks. In: Conference on Neural Information Processing Systems (NIPS) (2014)
13. Venugopalan, S., Rohrbach, M., Donahue, J., Mooney, R.J., Darrell, T., Saenko, K.: Sequence to sequence-video to text. In: ICCV, pp. 4534–4542 (2015)
14. Chen, D.L., Dolan, W.B.: Collecting highly parallel data for paraphrase evaluation. In: ACL (2011)
15. Torabi, A., Pal, C., Larochelle, H., Courville, A.: Using descriptive video services to create a large data source for video annotation research. arXiv preprint arXiv:1503.01070 (2015)
16. Rohrbach, A., Rohrbach, M., Tandon, N., Schiele, B.: A dataset for movie description. In: CVPR (2015)
17. Wang, H., Zhang, Y., Yu, X., et al.: An overview of image caption generation methods. Comput. Intell. Neurosci. **2020** (2020)
18. Lee, M.W., Hakeem, A., Haering, N., Zhu, S.: SAVE: a framework for semantic annotation of visual events. In: CVPR, pp. 1–8 (2008)
19. Kojima, A., Tamura, T., Fukunaga, K.: Natural language description of human activities from video images based on concept hierarchy of actions. Int. J. Comput. Vision **50**, 171–184 (2002)

20. Khan, M.U.G., Zhang, L., Gotoh, Y.: Human focused video description. In: ICCV, pp. 1480–1487 (2011)
21. Hanckmann, P., Schutte, K., Burghouts, G.J.: Automated textual descriptions for a wide range of video events with 48 human actions. In: Fusiello, A., Murino, V., Cucchiara, R. (eds.) ECCV 2012. LNCS, vol. 7583, pp. 372–380. Springer, Heidelberg (2012). https://doi.org/10.1007/978-3-642-33863-2_37
22. Farhadi, A., Hejrati, M., Sadeghi, M.A., Young, P., Rashtchian, C., Hockenmaier, J., Forsyth, D.: Every picture tells a story: generating sentences from images. In: Daniilidis, K., Maragos, P., Paragios, N. (eds.) ECCV 2010. LNCS, vol. 6314, pp. 15–29. Springer, Heidelberg (2010). https://doi.org/10.1007/978-3-642-15561-1_2
23. Rohrbach, M., Qiu, W., Titov, I., Thater, S., Pinkal, M., Schiele, B: Translating video content to natural language descriptions. In: ICCV, pp. 433–440 (2013)
24. Guo, Z., Gao, L., Song, J., Xu, X., Shao, J., Shen, H.T.: Attention-based LSTM with semantic consistency for videos captioning. In: ACM MM, pp. 357–361 (2016)
25. Pan, Y., Mei, T., Yao, T., Li, H., Rui, Y.: Jointly modeling embedding and translation to bridge video and language. In: CVPR, pp. 4594–4602 (2016)
26. Yao, L., et al.: Describing videos by exploiting temporal structure. In: ICCV, pp. 4507–4515 (2015)
27. Pan, Y., Yao, T., Li, H., Mei, T.: Video captioning with transferred semantic attributes. In: CVPR (2017)
28. Long, X., Gan, C.-Y., de Melo, G.: Video captioning with multi-faceted attention. CoRR, abs/1612.00234 (2016)
29. Song, J., Gao, L., Guo, Z., Liu, W., Zhang, D., Shen, H.T.: Hierarchical LSTM with adjusted temporal attention for video captioning. In: Proceedings of the Twenty-Sixth International Joint Conference on Artificial Intelligence, IJCAI 2017, pp. 2737–2743 (2017)
30. Yu, Y., Choi, J.-S., Kim, Y., Yoo, K., Lee, S., Kim, G.: Supervising neural attention models for video captioning by human gaze data. In: CVPR (2017)
31. Zeiler, M.D.: AdaDelta: an adaptive learning rate method. arXiv preprint arXiv:1212.5701 (2012)
32. Yang, A., et al.: Vid2Seq: large-scale pretraining of a visual language model for dense video captioning. In: Proceedings of the IEEE/CVF Conference on Computer Vision and Pattern Recognition, pp. 10714–10726 (2023)
33. Wang, Z., et al.: Language models with image descriptors are strong few-shot video-language learners. arXiv preprint arXiv:2205.10747 (2022)
34. Li, M., et al.: Clip-event: connecting text and images with event structures. In: Proceedings of the IEEE/CVF Conference on Computer Vision and Pattern Recognition, pp. 16420–16429 (2022)
35. Yang, Y., Hospedales, T.: Deep multi-task representation learning: a tensor factorisation approach. In: International Conference on Learning Representations (2017)
36. Barrow, H.G., Tenenbaum, J.M., Bolles, R.C., Wolf, H.C.: Parametric correspondence and chamfer matching: two new techniques for image matching. Technical report, SRI AI Center (1977)
37. Zaremba, W., Sutskever, I.: Learning to execute. arXiv preprint arXiv:1410.4615 (2014)
38. Maćkiewicz, A., Ratajczak, W.: Principal components analysis (PCA). Comput. Geosci. 19(3), 303–342 (1993)
39. Koonce, B., Koonce, B.: VGG network. In: Koonce, B. (ed.) Convolutional Neural Networks with Swift for Tensorflow, pp. 35–50. Springer, Cham (2021). https://doi.org/10.1007/978-1-4842-6168-2_4

40. Krizhevsky, A.: One weird trick for parallelizing convolutional neural networks. CoRR, abs/1404.5997 (2014)
41. Vedantam, R., Zitnick, C.L., Parikh, D.: CIDEr: consensus-based image description evaluation. In: CVPR (2015)
42. Denkowski, M., Lavie, A.: Meteor universal: language specific translation evaluation for any target language. In: EACL (2014)
43. Papineni, K., Roukos, S., Ward, T., Zhu, W.-J.: BLEU: a method for automatic evaluation of machine translation. In: ACL (2002)
44. Lin, C.-Y.: ROUGE: a package for automatic evaluation of summaries. In: Text Summarization Branches Out: Proceedings of the ACL-04 Workshop, pp. 74–81 (2004)
45. Chen, X., et al.: Microsoft COCO captions: data collection and evaluation server. arXiv preprint arXiv:1504.00325 (2015)
46. Thomason, J., Venugopalan, S., Guadarrama, S., Saenko, K., Mooney, R.J.: Integrating language and vision to generate natural language descriptions of videos in the wild. In: COLING (2014)
47. Hodosh, M., Young, P., Lai, A., Hockenmaier, J.: From image descriptions to visual denotations: new similarity metrics for semantic inference over event descriptions. In: TACL (2014)
48. Venugopalan, S., Xu, H., Donahue, J., Rohrbach, M., Mooney, R., Saenko, K.: Translating videos to natural language using deep recurrent neural networks. In: NAACL (2015)
49. Lin, T.-Y., et al.: Microsoft COCO: common objects in context. In: Fleet, D., Pajdla, T., Schiele, B., Tuytelaars, T. (eds.) ECCV 2014. LNCS, vol. 8693, pp. 740–755. Springer, Cham (2014). https://doi.org/10.1007/978-3-319-10602-1_48
50. Szegedy, C., et al.: Going deeper with convolutions. In: CVPR (2015)
51. Rohrbach, A., Rohrbach, M., Schiele, B.: The long-short story of movie description. In: Gall, J., Gehler, P., Leibe, B. (eds.) GCPR 2015. LNCS, vol. 9358, pp. 209–221. Springer, Cham (2015). https://doi.org/10.1007/978-3-319-24947-6_17

Trace-Based Anomaly Detection with Contextual Sequential Invocations

Qingfeng Du[1], Liang Zhao[1(✉)], Fulong Tian[2], and Yongqi Han[1]

[1] School of Software Engineering, Tongji University, Shanghai, China
{du_cloud,1931525,2011438}@tongji.edu.cn
[2] Di-Matrix(Shanghai) Information Technology Co., Ltd., Shanghai, China

Abstract. Nowadays, microservice architecture has been widely adopted in various real systems because of the advantages such as high availability and scalability. However, microservice architecture also brings the complexity of operation and maintenance. Trace-based anomaly detection is a key step in the troubleshooting of microservice systems, which can help to understand the anomaly propagation chain and then locate the root cause. In this paper, we propose a trace-based anomaly detection approach called TICAD. Our core idea is to group the invocations according to their microservice pairs and then perform anomaly detection individually. For each distinct microservice pair, we propose a neural network based on LSTM and self-attention to automatically learn the contextual pattern in the target invocation and previous invocations. Detected invocation anomalies can be further used to infer the trace anomalies. We have verified it on a public data set and the experimental results show that our proposed method is effective compared to the existing approaches.

Keywords: Trace · Anomaly Detection · AIOps · Deep Learning

1 Introduction

Trace-based anomaly detection is a key step in the troubleshooting of microservice systems because the structure of trace can help operators to understand the anomaly propagation chain and then locate the root cause.

Existing approaches can be divided into two categories: trace-level approaches and invocation-level approaches. For those trace-level approaches such as [3,5], they need further analysis to locate the abnormal microservice. For those invocation-level approaches such as [2,4], they can directly detect the invocation anomalies but highly depend on the accuracy.

We propose a supervised trace anomaly detection method called TICAD (Trace Invocation Callee Anomaly Detection), which can effectively learn the sequential patterns in the invocations and then infer the anomalies in the traces. Firstly, TICAD reorganizes the invocations of traces. Then, each invocation's callee state will be represented as a vector using multiple metrics. TICAD will mine the inherent relationship between the previous invocations and the current one through a neural network based on LSTM and self-attention. After the invocation anomalies are detected, whether a trace is abnormal can be inferred from them.

Our main contributions are listed below: We propose the TICAD, which detects invocation anomalies and subsequently infers the anomalies of the traces. We further propose a neural network based on LSTM and self-attention to detect anomalies in the invocations, which can learn the contextual dependencies and patterns between the invocation vectors. We conduct extensive experiments on TICAD to verify its effectiveness on the public dataset.

2   Related Works

Supervised Machine Learning Approaches: MEPFL [8] is proposed to predict multiple tasks such as latent error detection in the trace log, which is collected from both normal and faulty versions of the application. And Seer [1] is presented to detect Qos violations in the massive trace data. A deep learning model, which contains CNN and LSTM layers, is trained in Seer to predict the abnormal microservices.

Unsupervised Machine Learning Approaches: Among all the unsupervised approaches, most of them are based on the normal assumption. AVEB [4] trains a variational autoencoder to learn the response time feature of normal cases for each microservice. Then the target data with significant reconstruct errors will be determined as anomalous. TraceAnomaly [3] also trains a variational autoencoder with posterior flow to model the normal pattern of trace. In [5], a multimodal LSTM model is proposed to learn the sequential pattern of the invocation type and response time.

3   TICAD Design

As shown in the Fig. 1, we first reorganize all the traces. After that, each invocation will be transformed into a vector according to the metrics. For each invocation, its vector will be fed into a neural network based on LSTM and self-attention along with those of the previous invocations. Finally, it will automatically learn the potential features associated with anomalies.

3.1   Trace Pre-processing

In this section, we process the original trace data and fine tune the data structure to better detect anomalies. We group all the invocations with the same microservice pair and reorder them by their timestamp. After that, the original dataset is divided into n_c datasets where n_c is the number of unique microservice pairs. In the following steps, invocations of different groups can be processed and

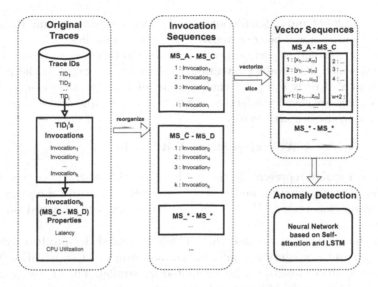

Fig. 1. The framework of TICAD

learned in parallel without affecting each other. Next, we vectorize the invocations from different perspectives. More precisely, we vectorize the callee state of the invocation, which means the label represents whether the callee is normal when the invocation occurs. To avoid the problem caused by using the latency alone, we use additional resource utilization metrics of the callee to enhance the representation of the invocation. We directly concatenate the latency and resource utilization metrics to form the vector of the invocation and standardize the values of each dimension in the vector.

3.2 Anomaly Detection

After the vectorization of the invocations, we will detect whether each invocation is abnormal. For the target invocation, in addition to its own feature vector information, we also use the extra vectors of the previous invocations to enrich the current information. Instead of relying only on the vector of the target invocation, this kind of learning method can help to decrease the false positives caused by the noise data. In practice, a reasonable window will be selected to slice the invocations. For each invocation waiting to be detected, the input is a matrix X consists of $w + 1$ vectors:

$$X_i = [v_{i-w}, v_{i-w+1}, \ldots, v_i]^\top \tag{1}$$

where v_i is the vector of current invocation and w is the window length.

Now that we have the input matrix X, TICAD demands a neural network to automatically learn the relation between the input and the fact whether the current invocation is anomalous. Therefore, we propose a neural network based on both LSTM and self-attention [6]. Briefly, The same input matrix will be learned with LSTM and self-attention neural network separately, and the output of target invocation will be concatenated to detect anomalies.

For the self-attention part, a neural network will take the input and use the multi-head scaled dot-product attention mechanism to aggregate the information. Instead of directly using the whole encoder of Transformer, we only utilize a few parts which are easy but effective. We first scale the input and add positional information to the original vector:

$$X_h = dropout(\sqrt{h}(X_i W^h) + W^{pos}) \tag{2}$$

where h is a scalar representing the hidden size. $W^h \in R^{n_f \times h}$ is the weight of the linear transformation and n_f is the number of features, i.e., the size of original invocation vector. $W^{pos} \in R^{(w+1) \times h}$ represents the learned positional embedding.

Then the X_h will be fed into the multi-head scaled dot-product attention layer, which can aggregate the information according to the attention scores. In practise, multiple heads can be calculated in parallel. The Q_i, K_i and V_i of each head, which are the indispensable elements of attention mechanism, will be transformed from the same input X_h:

$$Q_i = XW_i^Q, K_i = XW_i^K, V_i = XW_i^V \tag{3}$$

where $W^Q, W^K, W^V \in R^{h \times d_{head}}, d_{head} = h/n_{heads}$, n_{heads} is the number of heads.

For each head, scaled dot-product attention mechanism will calculate the attention scores and then get the weighted sum of values, which is shown in the following equation:

$$head_i = softmax(\frac{Q_i K_i^\top}{\sqrt{d_{head}}})V_i \tag{4}$$

All the results of the heads will be concatenated and transformed to X_h' which is shown below:

$$X_{h'} = (head_1 \oplus head_2 \oplus \cdots \oplus head_i)W^{h'} \tag{5}$$

where $W^{h'} \in R^{h \times h}$.

The final part of self-attention consists of layer normalization and residual dropout and the aggregated vector of target invocation is represented as v_s, which is shown in the following equation:

$$X_f = LayerNormalization(X_h + dropout(X_{h'})) \tag{6}$$

$$v_s = X_f[w] \tag{7}$$

For the LSTM part, we adopt a variant of LSTM called Bi-LSTM (Bidirectional Long Short Term Memory), whose detailed structure is shown in the Fig. 2. Each row of X_h will be input into the Bi-LSTM model at each time step.

As shown in the figure, $hf_w \in R^{h/2}$ and $hb_0 \in R^{h/2}$ are the hidden state vectors at the last time step, which will be concatenated to represent the result of Bi-LSTM:

$$v_l = h_w^l \oplus h_0^r \tag{8}$$

Finally, v_s and v_l will be concatenated to calculate the anomaly probability:

$$Anomaly\_Probability = \sigma((v_s \oplus v_l)^\top W^a + b^a) \tag{9}$$

where σ represents the sigmoid function and $W^a \in R^{2*h \times 1}$.

If a trace has at least one abnormal invocation, the trace will be judged as abnormal.

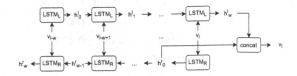

Fig. 2. The structure of LSTM based neural network

4   Evaluation

4.1   Datasets and Criteria

Datasets. To make the experiments more convincing, we use the public dataset which is proposed in TraceRCA [2] to evaluate the effectiveness of TICAD. This dataset collected traces from the Train Ticket [7] system, which is one of the largest open source microservice systems.

Baselines. To demonstrate the effectiveness of TICAD, We compare it with TraceAnomaly [3] and MEPFL-RF [8]. MEPFL-RF refers to the Random Forest version of MEPFL. Parameters of them are set best for accuracy.

Evaluation Metrics. As with previous researches, we use three evaluation metrics: *precision, recall* and *F1 score*, which are calculated as follows: Precision = TP/(TP + FP), Recall = TP/(TP + FN), F1 score = (2*Precision*Recall)/ (Precision + Recall).

4.2   Preparation Experiments

For TICAD, if we directly divides all the invocations, it's likely that there will be no complete trace in the test set. This will result in the inability to compare the effectiveness of methods because TICAD can't infer trace-level results from incomplete trace invocations. Therefore, we randomly select 5% normal trace ID and 5% abnormal trace ID as the reserved ID, which means all the invocations of these traces will be reserved for test set. For supervised methods such as TICAD and MEPFL, we directly copy the abnormal traces or invocations to solve the lack of positive samples.

4.3   Experiments on Trace-Level Anomaly Detection

In this section, we use different methods to perform trace-level anomaly detection, which can show the effectiveness of different methods on whole traces. The results are shown in the Table 1. It can be seen that TICAD proposed in this paper achieved the highest F1 score(0.974) and the highest recall(0.986). Although the precision is not the highest, it doesn't lag far behind other methods. In general, TICAD shows its availability and effectiveness in anomaly detection tasks at the level of trace.

Table 1. Trace-level Anomaly Detection Results

	Precision	Recall	F1 score
TICAD	0.961	**0.986**	**0.974**
TraceAnomaly	0.901	0.159	0.270
MEPFL-RF	**0.987**	0.953	0.970

5   Conclusion

In this paper, we propose an end-to-end trace anomaly detection method called TICAD. It has the ability to effectively learn the sequential patterns in the invocations and then infer the anomalies in the traces. TICAD will mine the inherent relationship between the current invocation and the previous ones through a neural network based on LSTM and self-attention.

References

1. Gan, Y., Zhang, Y., Hu, K., Cheng, D., Delimitrou, C.: Seer: leveraging big data to navigate the complexity of performance debugging in cloud microservices. In: the Twenty-Fourth International Conference (2019)
2. Li, Z., Chen, J., Jiao, R., Zhao, N., Wang, Z., Zhang, S., Pei, D.: Practical root cause localization for microservice systems via trace analysis. In: IWQOS (2021)
3. Liu, P., Xu, H., Ouyang, Q., Jiao, R., Pei, D.: Unsupervised detection of microservice trace anomalies through service-level deep Bayesian networks. In: 2020 IEEE 31st International Symposium on Software Reliability Engineering (ISSRE) (2020)
4. Nedelkoski, S., Cardoso, J., Kao, O.: Anomaly detection and classification using distributed tracing and deep learning. In: 2019 19th IEEE/ACM International Symposium on Cluster, Cloud and Grid Computing (CCGRID) (2019)
5. Nedelkoski, S., Cardoso, J., Kao, O.: Anomaly detection from system tracing data using multimodal deep learning. In: 2019 IEEE 12th International Conference on Cloud Computing (CLOUD) (2019)
6. Vaswani, A., et al.: Attention is all you need. arXiv (2017)
7. Zhou, X., et al.: Fault analysis and debugging of microservice systems: industrial survey, benchmark system, and empirical study. IEEE Trans. Softw. Eng. 1 (2018)
8. Zhou, X., et al.: Latent error prediction and fault localization for microservice applications by learning from system trace logs. In: 27th ACM Joint Meeting on European Software Engineering Conference (ESEC)/Symposium on the Foundations of Software Engineering (FSE) (2019)

Fusing Fine-Grained Information of Sequential News for Personalized News Recommendation

Jin-Cheng Zhang[1], Azlan Mohd Zain[1], Kai-Qing Zhou[2]([⊠]), Xi Chen[3], and Ren-Min Zhang[2]

[1] Faculty of Computing, Universiti Teknologi Malaysia, Skudai, Malaysia
zhangjincheng@graduate.utm.my, azlanmz@utm.my
[2] College of Information Science and Engineering, Jishou University, Jishou, China
{kqzhou,rzhang1981}@jsu.edu.cn
[3] Tencent Inc., Shenzhen, China
jasonxchen@tencent.com

Abstract. In this paper, we propose a novel method that fuses Fine-grained Information of Sequential News for personalized news recommendation (FISN). FISN comprises three primary modules: news encoder, clicked news optimizer and user encoder. The news encoder uses fine-grained information to learn accurate news representations. The clicked news optimizer introduces multi-headed self-attention and positional encoding techniques to optimize the clicked news representation. The user encoder uses news-level attention to learn user representations. Extensive experimental results demonstrate that FISN outperforms many baseline approaches in terms of metrics for real datasets.

Keywords: Fine-grained Information · News Recommendation · Personalized Attention

1 Introduction

With the exponential growth in the volume of news data, information overload has become a problem for users, and personalized news recommendation systems are employed to solve this problem. There are two central issues in this field, one is to fully utilize the rich information in content to accurately learn the news representation, and the other is to capture users' preferences through their behaviors to achieve a precise matching of users and candidate news. Traditional recommendation methods such as LibFM [5] and DeepFM [2] are classical recommendation models based on factorization machines. Recently, several studies have introduced deep learning techniques into news recommendations [9,11,14]. These existing methods have been remarkably successful, however, personalized news recommendation is a difficult task that still faces challenges. First, a news item usually contains much fine-grained information, including title, abstract, news category, and subcategory, which are very helpful to refine news representation. However, the majority of existing techniques fail to utilize fine-grained information [9,10]. Secondly, the sequence information of users' clicked history can also reflect users' preferences to some extent. Users' long-term preferences may change over time, and users' short-term interests may change suddenly due to recent hot events. Accurately obtaining this information from the sequence in which users have clicked on past news is also difficult.

Based on the above challenges in this field, This paper proposes a novel method that fuses Fine-grained Information of Sequential News for personalized news recommendation (FISN). This method exploits the fine-grained information of news and the sequential information of users' clicked history to provide personalized news recommendations. FISN contains three core components, of which the news encoder aims to extract fine-grained information such as news titles, abstracts, categories, and subcategories, and to fuse this information using a rational framework to obtain the final news representation. In addition, a novel clicked news optimization network for capturing the sequential information of clicked news is proposed. Finally, a user encoder is employed to learn user representations from the sequences of clicked news.

2   Related Work

Traditional news recommendation methods rely on manual feature engineering. Recently, numerous approaches based on deep learning have been presented [9–12,14]. For instance, Wang et al. [9] utilized knowledge graph and knowledge-aware CNN for news representation learning. Wu et al. [12] applied multi-headed self-attention to acquire news representations. However, most of them obtain news vectors based on a single piece of news information. Wu et al. [10] introduced a multi-view learning-based method that aggregates different types of information to enrich the news representation, but it does not capture the clicked news' sequential information and interaction between historical news in the user interest modeling stage.

3   Personalized News Recommendation

FISN is illustrated in Fig. 1. Our method has three core modules. i.e., the news encoder that learns news representations by incorporating fine-grained information, the clicked news optimizer that optimizes clicked news representations, and the user encoder that learns user representations by using news-level attention. We formulate the question in the following way. Denote the titles and abstracts of each news separately as $[w_1^t, w_2^t, \ldots, w_p^t]$ and $[w_1^a, w_2^a, \ldots, w_q^a]$, where p and q stand for the length of the title and abstract, respectively. Given the user's clicked news sequence $N_h = [N_h^1, N_h^2, \ldots, N_h^k]$ and the candidate news N_c, where k is the number of news articles that were clicked. We are attempting to calculate the probability of a user clicking on a piece of candidate news that this user has not browsed before.

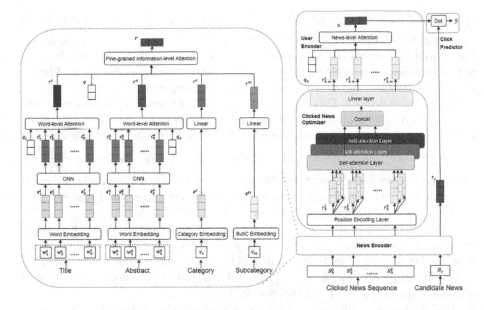

Fig. 1. The overall architecture of FISN.

3.1 News Encoder

This module is utilized to encode the fine-grained information of news to obtain the news representations. This fine-grained information is useful for generating news representations, as the title of the news often concisely conveys its content, and the abstract provides more detailed information. Furthermore, the category and subcategory are also instructive, and if a user clicks on numerous items in the same category and subcategory, the user's preferences can be inferred accordingly.

The initial stage is to obtain fine-grained information. The titles and abstracts can be represented as $[w_1^t, w_2^t, \ldots, w_p^t]$ and $[w_1^a, w_2^a, \ldots, w_q^a]$, respectively. Then using pre-trained Glove word embeddings [4], news titles and abstracts are converted to low-dimensional embedding vector sequences $[e_1^t, e_2^t, \ldots, e_p^t] \in \mathcal{R}^{p \times d_w}$ and $[e_1^a, e_2^a, \ldots, e_q^a] \in \mathcal{R}^{q \times d_w}$ where d_w denotes the dimension of word vector. Simultaneously, the ID of the category and subcategory can be converted into low-dimensional embeddings (convert to $e^c \in \mathcal{R}^{d_c}$ and $e^{sc} \in \mathcal{R}^{d_c}$, respectively) via category and subcategory embedding layer, where d_c indicates the dimension of the category embedding. Then a linear layer is employed to optimize these category embeddings into

$$r^c = \text{ReLu}\left(M_c \times e^c + b_c\right) \in \mathcal{R}^{d_w}$$
$$r^{sc} = \text{ReLu}\left(M_{sc} \times e^{sc} + b_{sc}\right) \in \mathcal{R}^{d_w}.$$

For news titles, the news encoder takes $[e_1^t, e_2^t, \ldots, e_p^t]$ as input, and obtains the contextual embedding c_i^t of each word by using a CNN layer to capture the local context, where i denotes the ith word in the title, which is computed as

$c_i^t = \text{ReLu}\left(H_t \times e_{(i-k,i+k)}^t + b_t\right) \in \mathcal{R}^{N_f}$, At this point, $\left[e_1^t, e_2^t, \ldots, e_p^t\right]$ is finally converted to $\left[c_1^t, c_2^t, \ldots, c_p^t\right] \in \mathcal{R}^{p \times N_f}$, where $H_t \in \mathcal{R}^{N_f \times (2K+1)d_w}$ and b_t denote the convolution kernel and bias parameters, respectively. N_f represents the number of convolution kernels equal to the dimension of word embedding. Then, a word-level attention network was applied to select important words in the title to model the news vectors. This is due to the fact that different words within a single news title have varying implications for modeling news representations. The word-level attention network can assign weights W_i^t to each word in news title based on the importance. The formula is $w_i^t = q_t^T \tanh\left(M_t \times c_i^t + b_t\right), W_i^t = \frac{\exp(w_i^t)}{\sum_{j=1}^{p} \exp(w_j^t)}$, in which q_t^T is the transpose of the attention query vector. A weighted total of the contextual embedding of each word in the title by attention weighting serves as the final representation of a news title $r^t = \sum_{j=1}^{p} W_j^t c_j^t \in \mathcal{R}^{N_f}$.

Similarly, news abstracts can be considered as longer titles with a larger number of words, hence the representation of news abstracts can be learned in a similar way to news title learning. The news encoder takes news abstract embedding sequence $\left[e_1^a, e_2^a, \ldots, e_q^a\right]$ as input, and obtains the final representation of a news abstract $r^a = \sum_{j=1}^{q} W_j^a c_j^a \in \mathcal{R}^{N_f}$.

Finally, because different kinds of fine-grained information are not equally important for modeling news representations. Hence, we take the final representation r^t of title, r^a of abstract, r^c of category, and r^{sc} of subcategory as inputs to the fine-grained information-level attention network, and assign attention weights W^t, W^a, W^c, and W^{sc}, respectively to each kind of information according to the importance. For example, the attention weight of the title W^t is calculated by $w^t = q^T \tanh\left(M^t \times r^t + b^t\right), W^t = \frac{\exp(w^t)}{\exp(w^t)+\exp(w^a)+\exp(w^c)+\exp(w^{sc})}$. The final news feature representation is a weighted sum of the fine-grained information by attention weights: $\boldsymbol{r} = \left(W^t r^t + W^a r^a + W^c r^c + W^{sc} r^{sc}\right)$.

3.2 Clicked News Optimizer and User Encoder

The clicked news optimizer fuses the position information of clicked news to optimize the feature representation of all the clicked news. Firstly, a learnable position encoding matrix $P \in \mathcal{R}^{N_f \times k}$ is randomly initialized by the location encoding layer to add position information for each news in the history sequence. This is used to assist the multi-headed self-attention layer [7] to learn the sequence information of clicked news. Where k represents the number of news in the user's history sequence. Then the representation of all clicked news is concatenated into $\boldsymbol{r_h} = \left[\boldsymbol{r_h^1}, \boldsymbol{r_h^2}, \ldots, \boldsymbol{r_h^k}\right] + P$ as the input to the multi-headed self-attention layer. Finally we get the output $\boldsymbol{r_{h\_m}} = \left[\boldsymbol{r_{h\_m}^1}, \boldsymbol{r_{h\_m}^2}, \ldots, \boldsymbol{r_{h\_m}^k}\right]^T \in \mathcal{R}^{k \times d_w}$.

The user encoder is designed to obtain the user's representation from the clicked news sequence. Different news in clicked news sequence are not equally important for modeling the user interest representation. Hence, a news-level attention network was applied to select important news to model the user representations. The news-level attention network can assign weights W_i^n to each news in clicked news sequence based on the importance. The formula is $w_i^n = q_n^T \tanh\left(M_n \times r_{h\_m}^i + b_n\right), W_i^n = \frac{\exp(w_i^n)}{\sum_{j=1}^k \exp(w_j^n)}$. The final representation of the user is a weighted sum of each clicked news in historical sequence by attention weighting $u = \sum_{j=1}^k W_j^n r_{h\_m}^j \in \mathcal{R}^{d_w}$.

3.3 Click Predictor and Model Training

The click predictor is applied to calculate the click probability. The candidate news N_c is denoted as r_c. The click probability $\hat{y}$ is calculated by the inner product of the user vector and the candidate news vector, i.e., $\hat{y} = u^T r_c$.

Our model was trained by using negative sampling techniques. For each clicked news(as a positive example), K news that not clicked are then randomly selected (as negative example). Then denote the click probabilities of positive example and K negative examples as $\hat{y}^+$ and $[\hat{y}_1^-, \hat{y}_2^-, \dots, \hat{y}_k^-]$, respectively. The posterior click probability of positive samples is determined by normalizing these click probability scores using softmax: $p_i = \frac{\exp(\hat{y}_i^+)}{\exp(\hat{y}_i^+) + \sum_{j=1}^K \exp(\hat{y}_{i,j}^-)}$. The loss function $\mathcal{L} = -\sum_{i \in \mathcal{S}} \log(p_i)$ in the model training is the negative log likelihood of all positive samples $\mathcal{S}$.

4 Experiments

Experiments are conducted on the real-world news recommendation dataset MIND-large and MIND-small. [13]. The hyperparameters are tuned on the validation set to optimize the performance of the FISN model. Word embedding is assigned a dimension of 300, and category and subcategory embeddings are assigned a dimension of 100. The attention queries have a dimension of 200, and the model utilizes 300 CNN filters with a window size of 3. We set the number of self-attention heads for multi-headed self-attention to 15. A dropout rate of 20% [6] is applied in the news encoder. We select a batch size of 128 and the Adam optimizer [3] with a learning rate of 0.0001.

By comparing with the following baseline approaches, we evaluate the performance of FISN. (1) LibFM [5], (2) DeepFM [2], (3) DKN [9], (4) NPA [11], (5) NAML [10], (6) LSTUR [1], (7) NRMS [12], (8) FIM [8]. The performance of these methods on two datasets is summarized in Table 1.

First, end-to-end learning techniques like DKN, NAML, LSTUR, NRMS, etc. outperform traditional methods that require human feature engineering (LibFM and DeepFM). The reason for this result may be those recommendation methods that rely on manual feature engineering have inconsistent training goals across modules, while the end-to-end approach avoids the inherent drawbacks of multiple modules. This finding supports the idea that neural networks are more effective at learning news and user representations from raw data than feature engineering. Second, in deep recommendation frameworks, using attention mechanisms(LSTUR, NRMS, NPA, and NAML) may assist recommendations to perform better to some degree, probably because attention mechanisms can model the relative importance of each word to learn news representations more accurately. In addition, attention mechanisms can also capture user preferences and thus model user representations more accurately. Third, FIM is superior to other methods because it can detect fine-grained matching signals for the interaction between each news pair at multiple levels of semantic granularity. Finally, across all metrics, our method FISN can consistently beat other baseline approaches. This result shows that our approach makes reasonable use of fine-grained information. Since the form of fine-grained information is usually different, a uniform approach to extracting features from this information is certainly not feasible. We use various components to extract different features, and finally, learn accurate representations for each piece of news through a fine-grained information-level attention network. In addition, our approach uses positional encoding techniques to capture sequential information of historical news and introduces multi-headed self-attention to learn interactions between clicked news. The user's representation is then learned through a news-level attention network, which enhances the user's representation learning. The results of the experiments validate the effectiveness of FISN.

Table 1. The overall performance of different methods on MIND

Method	MIND-small				MIND-large			
	AUC	MRR	nDCG@5	nDCG@10	AUC	MRR	nDCG@5	nDCG@10
LibFM	0.5974	0.2633	0.2795	0.3429	0.6185	0.2945	0.3145	0.3713
DeepFM	0.5989	0.2621	0.2774	0.3406	0.6187	0.2930	0.3135	0.3705
DKN	0.6175	0.2705	0.2890	0.3538	0.6407	0.3042	0.3292	0.3866
NPA	0.6321	0.2911	0.3170	0.3781	0.6592	0.3207	0.3472	0.4037
LSTUR	0.6438	0.2946	0.3189	0.3817	0.6708	03236	0.3515	0.4093
NRMS	0.6483	0.3001	0.3252	0.3892	0.6766	0.3325	0.3628	0.4198
NAML	0.6550	0.3039	0.3308	0.3931	0.6646	0.3275	0.3566	0.4140
FIM	0.6502	0.3026	0.3291	0.3910	0.6787	0.3346	0.3653	0.4221
FISN	**0.6658**	**0 .3211**	**0.3552**	**0.4161**	**0.6906**	**0.3379**	**0.3749**	**0.4369**
%Improv	1.08	1.72	2.24	2.30	1.19	0.33	0.96	1.48

5   Conclusion and Future Work

In this paper, we present the FISN for news recommendation. Our experimental results demonstrate that FISN outperforms several baseline approaches on real datasets based on various metrics. In future work, we aim to incorporate knowledge graphs as side information into news recommendations using knowledge graph embedding techniques.

Acknowledgment. This work was supported by the Fundamental Research Grant Scheme, Ministry of Higher Education Malaysia (R.J130000.7809.5F524) the National

Natural Science Foundation of China (62066016, 62061017), the Research Foundation of Education Bureau of Hunan Province, China (22B0549, 22C0282) the Xiangxi State Science Program Project-Collaborative control of UAV swarm system based on improved cuckoo algorithm([2022]5), the NSF of Hunan Province (No. 2021JJ30556, No. 2022JJ50310).

References

1. An, M., Wu, F., Wu, C., Zhang, K., Liu, Z., Xie, X.: Neural news recommendation with long-and short-term user representations. In: Proceedings of the 57th Annual Meeting of the Association for Computational Linguistics, pp. 336–345 (2019)
2. Guo, H., Tang, R., Ye, Y., Li, Z., He, X.: DeepFM: a factorization-machine based neural network for CTR prediction. arXiv preprint arXiv:1703.04247 (2017)
3. Kingma, D.P., Ba, J.: Adam: A method for stochastic optimization. arXiv preprint arXiv:1412.6980 (2014)
4. Pennington, J., Socher, R., Manning, C.D.: Glove: global vectors for word representation. In: Proceedings of the 2014 Conference on Empirical Methods in Natural Language Processing (EMNLP), pp. 1532–1543 (2014)
5. Rendle, S.: Factorization machines with libFM. ACM Trans. Intell. Syst. Technol. (TIST) **3**(3), 1–22 (2012)
6. Srivastava, N., Hinton, G., Krizhevsky, A., Sutskever, I., Salakhutdinov, R.: Dropout: a simple way to prevent neural networks from overfitting. J. Mach. Learn. Res. **15**(1), 1929–1958 (2014)
7. Vaswani, A., et al.: Attention is all you need. In: Advances in Neural Information Processing Systems, vol. 30 (2017)
8. Wang, H., Wu, F., Liu, Z., Xie, X.: Fine-grained interest matching for neural news recommendation. In: Proceedings of the 58th Annual Meeting of the Association for Computational Linguistics, pp. 836–845 (2020)
9. Wang, H., Zhang, F., Xie, X., Guo, M.: DKN: deep knowledge-aware network for news recommendation. In: Proceedings of the 2018 World Wide Web Conference, pp. 1835–1844 (2018)
10. Wu, C., Wu, F., An, M., Huang, J., Huang, Y., Xie, X.: Neural news recommendation with attentive multi-view learning. arXiv preprint arXiv:1907.05576 (2019)
11. Wu, C., Wu, F., An, M., Huang, J., Huang, Y., Xie, X.: NPA: neural news recommendation with personalized attention. In: Proceedings of the 25th ACM SIGKDD International Conference on Knowledge Discovery & Data Mining, pp. 2576–2584 (2019)
12. Wu, C., Wu, F., Ge, S., Qi, T., Huang, Y., Xie, X.: Neural news recommendation with multi-head self-attention. In: Proceedings of the 2019 Conference on Empirical Methods in Natural Language Processing and the 9th International Joint Conference on Natural Language Processing (EMNLP-IJCNLP), pp. 6389–6394 (2019)
13. Wu, F., et al.: Mind: a large-scale dataset for news recommendation. In: Proceedings of the 58th Annual Meeting of the Association for Computational Linguistics, pp. 3597–3606 (2020)
14. Zhang, Q., et al.: UNBERT: user-news matching BERT for news recommendation. In: IJCAI, pp. 3356–3362 (2021)

A Finite-Domain Constraint-Based Approach on the Stockyard Planning Problem

Sven Löffler, Ilja Becker[✉], and Petra Hofstedt[✉]

Brandenburg University of Technology Cottbus-Senftenberg,
Konrad-Wachsmann-Allee 5, Cottbus, Germany
{sven.loeffler,ilja.becker,hofstedt}@b-tu.de
https://www.b-tu.de/fg-programmiersprachen-compilerbau/

Abstract. Transport problems are a significant challenge for companies, i.e. due to concerns about climate change and the constant increase in raw material prices, such as for petrol. One issue in transporting bulk materials is the Stockyard Planning Problem (SPP), which plays a crucial role in mine production scheduling and bulk material transportation by using stockpiles to store and blend raw material. The SPP aims to: 1) blend superior and inferior components to achieve a desired quality, 2) transfer material from import (e.g. ocean-going ships) to interim storage (e.g. stockpiles) and then to export (e.g. docks at inland ports), and 3) find a time-/cost-optimized schedule for working steps of 1 and 2. This paper presents a novel constraint-based approach for solving the Stockyard Planning Problem (SPP) by utilizing new abstractions such as time, storage, and potential movements within a stockyard system.

Keywords: Planning and Scheduling · Decision Support Systems · Mining · Stockyard Planning Problem · SPP · Constraint Programming

1 Introduction

In a globalized world, raw materials are mined at one end of the world, refined and then distributed all over the world. In mining and transportation of bulk materials, there are certain transition points, often ports with stockpile systems, where much material of different qualities comes to a transition point where the materials need to be blended, stored, prepared and loaded for onward transport.

Each stockpile system can have unique and specific structures, components and constraints, like different import and export variants (e.g. ships or trains), different number of simultaneous imports and exports, materials, number of stockpiles, sizes of stockpiles, machines, conveyor belts and much more. This makes individual planning necessary. Mistakes in scheduling can lead to deliveries not being made (on time), agreed upon qualities not beeing achieved or ships docking longer than necessary, resulting in additional mooring fees.

Our industrial partner *ABB - Sales Minerals and Mining, Engineering Substation and Power Generation, Service Metals branch office Cottbus* implement

C. Strauss et al. (Eds.): DEXA 2023, LNCS 14147, pp. 126–133, 2023.
https://doi.org/10.1007/978-3-031-39821-6_10

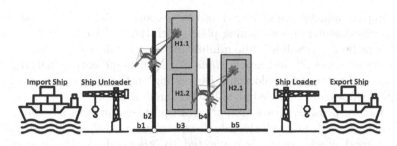

Fig. 1. An example stockyard system.

and operate stockpile monitoring and management systems all over the world. They also support their customer companies, e.g. ports, with the scheduling process. In doing so they recognized that a particularly high degree of optimization is possible here. Our goal is to create a constraint-based planning tool that takes into account all the requirements of the SPP and finds plans that minimize the idle time of import and export ships and relieve the human planner.

The rest of this paper is organized as follows. The second section introduces the stockyard planning problem (SPP) and its components and also gives an overview of previous research regarding SPP. Section 3 shows the basics of constraint programming. Section 4 describes the necessary abstractions to model the problem. The developed constraint problem for the presented SPP and possible improvements of it is described in Sect. 5. Finally, a conclusion is given and possible future improvements of the developed method are discussed.

2   The Stockyard Planning Problem

The Stockyard Planning Problem: The SPP comprises the unloading, transportation, blending and loading of bulk material at e.g. a port. In [6] a special SPP for the port of Newcastle is described. We created our own SPP problem, which better takes into account the limitations of the conveyor belt net.

Consider a conveyor belt net with import vehicles, storage places (stockpile areas), export vehicles and transportation vehicles. The goal is it to transport bulk materials from the import vehicles to the storage places (*import moves*), blend different qualities of the bulk materials to reach agreed qualities (*blending moves*), transport the bulk materials from a storage place to another storage place (*transport moves*) or transport the bulk materials from the storage places to the export vehicles (*export moves*), such that the import vehicles are all unloaded and the export vehicles are loaded correctly.

Figure 1 shows an example stockyard system. We see one import and one export ship, each with one ship unloader and one ship loader, respectively. There are five conveyor belts ($b1$ to $b5$), two stackers which are also reclaimers and two stockpiles with two ($H1.1$ and $H1.2$) respectively one stockpile areas ($H2.1$).

The import vehicles are unloaded and the export vehicles are loaded according to specific loading and unloading plans, with the goal to do the loading and unloading as fast as possible while minimizing vehicle effort (optional).

The *import plans* P^I and *export plans* P^E are lists indicating material, mass and order for loading and unloading. Depending on the system, several import and export ships can be loaded or unloaded simultaneously.

For the four possible moves (import, transport, export and blending) a system of *conveyor belts* and vehicles: *stackers, reclaimers, ship unloaders* and *ship loaders* is used. Each vehicle is connected to other vehicles via one or more conveyor belts. The conveyor belt net of the system decides which moves are possible separately and in parallel.

To store, refine or blend the imported materials separated areas on the stockpiles are used. Each stockpile can be accessed with a stacker or reclaimer. A stockyard plan contains information about the number of areas of each stockpile, the capacity of the stockpile areas, the stackers and reclaimers connectable to a certain stockpile and the type and quantity of material currently stored at each stockpile area.

To fulfill the export requirements, not only material transportation but also material refinement (or blending) is necessary. Each material category has a set of quality parameters. It is the scheduler's task to combine the materials provided from the import and stockpile areas in such a way that they meet the requirements of the export plan. To this end the blending ratio of the different materials must be determined and applied.

Related Aproaches for the Stockyard Planning Problem (SPP): Certain aspects of the SPP are well-studied, e.g., scheduling of trains and optimizing storage at seaport terminals. However, these papers usually only consider these aspects individually and not simultaneously in a single comprehensive problem setting. The problem described by Abdekhodaee et al. [3] considers a coal chain in Australia, but primarily focuses on train scheduling rather than stockyard management. They decompose the problem into several modules and solve each of the modules separately using a greedy heuristic.

The SPP also has elements of a 2D packing problem because a stockpile area takes up space and time in the stockyard, and no two stockpile areas can occupy the same space at the same time. Extensive work has been done on the 2D packing problem, see Lodi et al. [9] for a survey. However, the problem of locating stockpiles in the stockyard in the SPP is more complex than a traditional 2D packing problem, because the time dimension of a stockpile is not known in advance and depends on the material movements on the stockpiles and the import and export plans.

Wen et al. [13] focus mainly on the times and the places, when and where a stockpile area should be created and filled with material. For this they also use integer programming, in a similar way as we do for our SPP problem. Even if the description of the problem and the scenario presented in [6] looks very similar to ours at first, it is different in a few key points, such as: 1) in [6] it is possible to move materials from each input or reclaimer to each stacker or export, but this is not the case in our SPP; 2) in [6] the perfect stockpile areas are also planned (number, size, build and tear down time), whereas these are fixed in our SPP.

3   Basics of Constraint Programming

Constraint programming (CP) is a powerful way to declaratively model and solve NP problems. Typical research problems in CP are rostering, graph coloring, optimization, and satisfiability (SAT) problems [10]. In this section, we introduce basics of constraint programming needed for our approach.

The general procedure for constraint programming is divided into two parts: 1.: the declarative description/representation of a problem as a constraint model, 2.: the solving process of a constraint model using a constraint solver. The CP user is only responsible for modeling the application problem with constraints as well as setting up and starting the solver, which acts independently.

A *constraint satisfaction problem (CSP)* is defined as a 3-tuple $P = (X, D, C)$ where $X = \{x_1, x_2, \dots, x_n\}$ is a set of variables, $D = \{D_1, D_2, \dots, D_n\}$ is a set of finite-domains where D_i is the domain of x_i, and $C = \{c_1, c_2, \dots, c_m\}$ is a set of constraints each over between one and all variables of X [4].

A *constraint* is a tuple (X', R) of a relation R and an ordered set of variables $X' \subseteq X$ over which the relation R is defined [7]. A *solution* of a CSP is an instantiation of all variables x_i each with a value d_i of its corresponding domain D_i, such that all constraints are satisfied.

A *constraint optimization problem (COP)* is an extension of a CSP, where an optimization variable x_{opt} identified as such will be minimized or maximized.

To solve constraint problems (CSPs and COPs), so-called constraint solvers usually use backtracking search nested with propagation. Popular solvers are among others Google OR tools [1], Gecode [12], JaCoP [2], and the Choco-Solver [11], which is used here in the following. More information about solvers and how they work in detail can be found in [5,8].

4   Abstraction of the Problem

Abstraction of Time: In our modeling we want to optimize moves (import, transportation, blending, export) over the time, whereby multiple processing streams exist, such that moves typically happen in parallel. Each move runs in a time interval and can potentially start and end at any point of time.

One typical way to handle time is to consider time intervals of a fixed size. In our case this approach would mean to deal with a huge amount of relatively small time intervals because of the high flexibility of the start and end times of the moves. This inflates the problem size, and consequently increases the computation and optimization costs. Hence, we designed a more flexible interval partitioning.

We use $n \in \mathbb{N}$ time intervals $I = \{I_1, I_2, \dots, I_n\}$ which may be of different sizes. Each time interval $I_{i+1} = [b, c]$ starts when the previous time interval $I_i = [a, b]$ ends. Start and end times of the intervals I_i are start and end times

Table 1. The transition from one state to another by means of different moves.

	Mat_1^I	$Mass_1^I$	Mat_2^I	$Mass_2^I$	Mat_1^H	$Mass_1^H$		Mat_3^H	$Mass_3^H$	Mat_1^E	$Mass_1^E$	
State q_i	$Iron_{q1}$	15	$Iron_{q2}$	15	0	0	...	$Iron_{q3}$	15	0	0	...
Import	$Iron_{q1}$	-15	0	0	$Iron_{q1}$	+15	...	0	0	0	0	...
Transport	0	0	0	0	0	0	...	0	0	0	0	...
Export	0	0	0	0	0	0	...	$Iron_{q3}$	-10	$Iron_{q3}$	+10	...
State q_{i+1}	0	0	$Iron_{q2}$	15	$Iron_{q1}$	15		$Iron_{q3}$	5	$Iron_{q3}$	10	

of moves, such that each move can be represented by a sequence of successive intervals $[I_j, I_{j+1}, .., I_{j+k}]$. The start and the end time of each interval is handled as a constraint variable so that the interval sizes will separately be handeld as part of the solution process.

Abstraction of the Stockyard: We abstract the stockyard system by focusing on the main information about storage places which include the stockpiles, and import and export vehicles. Every storage place has a maximum capacity in tons it can store . We represent a storage place by a tuple $(Mat, Mass)$, with Mat being an integer assignment to the corresponding material or 0 if there is no material, and $Mass$ being the mass of the material in tons.

We abstract the state of a SPP by a three tuple (I, H, E) with import vehicles (resp. plans) $I = P^I = [(Mat_1^I, Mas_1^I), ..., (Mat_k^I, Mas_k^I)]$, storage places $H = [(Mat_1^H, Mas_1^H), ..., (Mat_m^H, Mas_m^H)]$, and output vehicles (or plans) $E = P^E = [(Mat_1^E, Mas_1^E), ..., (Mat_n^I, Mas_n^I)]$. For example, the stockard system of Fig. 1 is described accordingly, where $k = 2$ (2 hatches at the import ship), $m = 3$ (3 stockpile areas), and $n = 2$ (2 hatches at the export ship). The hatches must be unloaded resp. loaded in the given order. This means that the second hatch can not be unloaded or loaded until the first hatch is completely processed.

Abstraction of the Moves: Each move (import, transport, export) is represented by a vector $T = [(Mat_1^I, Mas_1^I), ..., (Mat_k^I, Mas_k^I), (Mat_1^H, Mas_1^H), ..., (Mat_m^H, Mas_m^H), (Mat_1^E, Mas_1^E), ..., (Mat_n^E, Mas_n^E)]$, where k is the number of hatches of the import ship, m is the number of storage places, and n is the number of hatches of the export ship. At this, exactly one mass $Mass_i$ has a value $a > 0, \in \mathbb{N}$, one mass $Mass_j$ has a value $-a$, and both corresponding materials Mat_i and Mat_j have the same value b, with all others material and mass values being 0. The negative value $-a$ for mass $Mass_j$ indicates that the corresponding storage place will be decreased by a thousand tons, while the corresponding storage place to $Mass_i$ will be increased by a thousand tons. The corresponding materials Mat_i and Mat_j show for the material b which is moved. For a move the mass values $Mass_i$ and $Mass_j$ correspond to storage places on stockyards, import or export ships. See Table 1 for an example of a parallel import and export move.

5 A Constraint Optimization Problem for the SPP

We use constraint programming to mathematically model and solve the SPP, in particular the choice and order of the moves (import, transport, blending, and export). We create a CSP that describes the conditions for the SPP, i.e. solutions of the CSP are all possible sequenzes of moves to unload the import ship, blend material and load the export ship in a way that the import $I = \{P_1^I, P_2^I, ...\}$ and export plans $E = \{P_1^E, P_2^E, ...\}$ (see Sect. 2) are satisfied. An optimization is then carried out by addition of an objective function $f = (minimize(x_n^{Time}))$. The aim of this COP is to achieve a minimum time duration for the planned moves to perform ship unloading, blending and loading.

Provided Starting Data: For the COP we assume a given fix number of time intervals n, a maximum time t_{max} how long the sequence to be planned may take, one import P^I, one export plan P^E, a stockyard system i.e. its structure, components, their connections, and dimensions. Furthermore, an initial state q_0 of the material types $Mat_0 = [mat_1, ..., mat_m]$ and masses $Mass_0 = [mass_1, ..., mass_m]$, where m is the number of storage places, of the whole system is given. Each state $q_i = [(mat_1, mass_1), (mat_2, mass_2), ..., (mat_m, mass_m)]$ is a tuple, where the first i^I values represent the hatches of the import ship, the next i^H values present the storage places of the stockyard and the last i^E values describe the hatches of the export ship (as schown e.g. in Table 1). The mixing ratios for blending two materials into another and the conveyor belt net are also provided.

Variables and Domains: The aim of the SPP is to model and optimally solve a sequence of moves to unload material from (an) import ship(s), transport and blend the material between stockpile areas, and finally load the material to (an) export ship(s).

Considering n time intervals means that the move sequence to be planned has $n + 1$ many states $q_0, ..., q_n$ with each m material variables $x_{i,j}^{Mat}$ and m mass variables $x_{i,j}^{Mass}$ with $i \in \{0, 1, ..., n\}$ and $j \in \{1, 2, ..., m\}$. The domains of the material variables $D_{i,j}^{Mat}$ contains all allowed materials (for example iron ore of different qualities), whereas the mass domains $D_{i,j}^{Mass}$ contain integer values between 0 and the capacity of the corresponding storage place in thousand tons.

To model state changes we use move vectors for the import, transportation, blending, and export moves (cf. Table 1). Each vector also consists of material variables x^{MatV} and mass variables x^{MassV}, which describe the value changes of a variable from one state to a subsequent state.

Finally, we also need variables X^{Time} to represent the time. Each state has a time variable x_i^{Time} (time stamp), $\forall i \in \{0, 1, ..., n\}$ with value between 0 and the maximum time $maxT$ available for the whole move sequence.

Constraints: We structure the constraints into basic constraints which realize basic rules of the stockyard system, all vector constraints which handle valid conditions for all move vectors, simple vector constraints (valid conditions for import, transportation and export move vectors), and blending vector constraints. Furthermore we consider vehicle constraints which realize vehicle dependencies like conveyor belt joints.

Basic Constraints: These realize basic conditions, e.g.: If there is no material at a storage places, then the corresponding material and the mass variable must be 0, otherwise both variables must be greater than 0 $((x_{i,j}^{Mat} = 0) \Leftrightarrow (x_{i,j}^{Mass} = 0))$. Other constraints realize that an import ship can only be unloaded, an export ship can only be loaded, you can not get different materials at the same storage place, you only can unload an import ship hatch if the previous hatch of the ship is empty, and you only can load an export ship hatch if the loading of the previous hatch is completed.

All Vector Constraints: If there is no material change for a storage place, then the material and mass variables in a vector must be assinged to 0. If material is moved, just as much material is removed from (one or two) storage places as is added to another storage place. The transported mass can not be higher than the flowrate of the conveyor belts and the duration of the move allow it.

Simple Vector Constraints: One simple move (import, transport, export) allows only the reduction of a mass variable with the simultaneous increase of another mass variable. So the mass values of an import, transportation or export move are either all 0 (no move happens) or exactly two of them are not 0. Furthermore, the two values different from 0 of a simple vector must be equal.

Blending Vector Constraints: In a blending process the mass of exactly two storage places is reduced while exactly one mass is increased. At the mass reducing storage places the material in the vector must be equal to the material at the storage in state before and in a state after or there is no material left afterwards. The same, but vice versa, applies to storage places, where the mass is increased when blending. At a blending move two materials are refined to one new material. For this the input and resulting materials as well as the blending ratio must be given and respected.

Vehicle Constraints: Each vehicle (stacker, reclaimer) can only reach stockpiles which are directly at the rail of the vehicle. We use table constraints with positive tuples to limit the pairing of vehicles and stockpiles in the conveyor belt network to specific combinations. We use table constraints with negative tuples to prevent vehicles from working simultaneously at neighboring stockpiles to avoid collisions.

Optimizations and Results: The solution speed of the presented COP can be improved by applying typical optimizations of constraint programming, such as parallelization, remodeling, (re)configuration of the solver settings, decomposition into small subproblems, or the use of implicit constraints. We are currently implementing some of these optimizations, and we have already achieved good results on smaller stockpile systems (e.g., 5 stockpiles, 20 stockpile areas, and 10 devices) for short periods of time (a few days). Our approach has been tested on both fictitious systems and the digital twin of a real system.

We expect to be able to further accelerate the solution process by implementing the optimizations mentioned above, which will enable us to plan ideal solutions for larger systems over longer periods of time. So far, we have been able to determine our results in a few minutes, which often represent optimal solutions for the mentioned problem size.

6   Conclusion and Future Work

We have presented and developed a new method for finding an optimal solution to the Stockyard Planning Problem (SPP) using finite-domain constraint programming. To achieve this, we abstracted various aspects of the stockyard planning process, such as time, storage, and possible moves in a stockyard system. We modeled and solved a representative Constraint Optimization Problem (COP) and explained methods to improve its solution process. Our approach was tested on simulated plants and the digital twin of a real plant operated by ABB.

Future work includes testing our method on more stockyard plants and integrating the optimization techniques presented in the COP to speed up the solution process and make it applicable for larger systems over longer periods. It would also be interesting to investigate whether the abstractions made in this paper can be used for other facilities, such as train-operated stockyard systems with different capacities, machines, and belt networks.

References

1. Google LLC, Google OR-Tools, 2019. http://developers.google.com/optimization/. Accessed 31 May 2023
2. Jacop solver 4.4. http://jacop.osolpro.com/. Accessed 24 Jan 2023
3. Abdekhodaee, A.H., Dunstall, S., Ernst, A.T., Lam, L.: Integration of stockyard and rail network: a scheduling case study. In: 5th Asia-Pacific Industrial Engineering and Management Systems Conference. Gold Coast, Australia, pp. 1–16 (2004)
4. Apt, K.: Constraint satisfaction problems: examples. In: [5] (2003), chapter 2
5. Apt, K.: Principles of Constraint Programming. Cambridge University Press, New York, NY, USA (2003)
6. Boland, N., Gulczynski, D., Savelsbergh, M.W.P.: A stockyard planning problem. EURO J. Transp. Logist. **1**(3) (2012). https://doi.org/10.1007/s13676-012-0011-z
7. Dechter, R.: Constraint networks. [8], chap. 2, pp. 25–49
8. Dechter, R.: Constraint Processing. Elsevier Morgan Kaufmann, Burlington (2003)
9. Lodi, A., Martello, S., Monaci, M.: Two-dimensional packing problems: a survey. Eur. J. Oper. Res. **141**(2) (2002). https://doi.org/10.1016/S0377-2217(02)00123-6
10. Marriott, K., Stuckey, P.J.: Programming with Constraints - An Introduction. MIT Press, Cambridge (1998)
11. Prud'homme, C., Fages, J.G., Lorca, X.: Choco documentation (2017). http://choco-solver.org/docs/considerations/. Accessed 31 May 2023
12. Schulte, C., Lagerkvist, M., Tack, G.: Gecode 6.2.0 (2019). www.gecode.org/. Accessed 31 May 2023
13. Wen, C., Zheng, L.: A constraint programming based method for stockyard management problem. In: Li, W., et al. (eds.) IDCS 2016. LNCS, vol. 9864, pp. 214–221. Springer, Cham (2016). https://doi.org/10.1007/978-3-319-45940-0_19

Data Analytics Framework for Smart Waste Management Optimisation: A Key to Sustainable Future for Councils and Communities

Sabbir Ahmed[1]($\boxtimes$) (iD), Sameera Mubarak[1], Santoso Wibowo[2], and Jia Tina Du[1]

[1] UniSA STEM, University of South Australia, Adelaide, Australia
`sabbir.ahmed@mymail.unisa.edu.au`, {`sameera.mubarak,tina.du`}`@unisa.edu.au`
[2] Central Queensland University, Melbourne, Australia
`s.wibowo1@cqu.edu.au`

Abstract. Smart waste management systems (SWMS), including various technologies, including routing, scheduling, infrastructure, and the Internet of Things (IoT), are used to enhance the efficiency and automation of waste management processes. The availability of big data generated by IoT sensors has the potential to significantly improve waste management systems by providing valuable insights and enabling automation. This study presents a data analytics framework that supports decision-makers in implementing, monitoring, and optimising SWMS. The framework utilises IoT sensor data and employs data analytic techniques to analyse and predict municipal bins' waste generation trends and patterns. Finally, the framework demonstrates the capability to forecast waste generation, leading to the development of a sustainable environment and efficient managerial administration in waste management.

Keywords: Smart waste management · Internet of Things (IoT) · councils · sustainability · data analytics

1 Introduction

Due to rapid urbanisation and population growth in urban areas, waste management has become one of the most challenging tasks for city councils [2,8]. Despite encompassing a wide range of technologies, waste management remains a significant global concern due to improper waste management [6,8]. Improper waste disposal systems, inadequate resources, and other factors have a negative impact on waste management systems. Improper waste disposal systems, inadequate resources, and other factors such as lack of policy-making knowledge and insufficient infrastructure have a detrimental effect on waste management systems.

To deal with waste management issues, several studies have been conducted on solid waste management, covering areas such as waste collection, classification, recycling, and resource allocation [1,3,10]. However, most of these studies

C. Strauss et al. (Eds.): DEXA 2023, LNCS 14117, pp. 134–139, 2023.
https://doi.org/10.1007/978-3-031-39821-6_11

have focused solely on the traditional system, rather than utilising the potential benefits of integrating computation, networking, and cyber-physical systems [1,12]. There is a limited study on the application of cyber-physical systems, which involve IoT, cloud computing, and data analytics techniques for waste management [6,13]. Furthermore, the complexity of data knowledge discovery, analysis, and computation can be effectively addressed by utilising machine-generated data from IoT devices, rather than relying solely on traditionally generated human data [8]. Therefore, to fully leverage the advantages of IoT and data analytics, it is essential to conduct an in-depth study on waste management for ensuring environmental awareness and public health through the adoption of more sustainable practices.

Currently, the smart waste management system (SWMS) incorporates IoT and data analytics tools [6,12]. The adoption of smart waste management solutions has been facilitated by enhanced machine communication, reduced costs of IoT sensors, and improved performance of data analytics tools. The implementation of IoT sensors in waste management enables effective solutions to challenges such as scheduling optimisation and trend analysis. By leveraging the integration of computation, networking, and physical systems enabled by IoT, city councils can achieve efficient and robust waste management [1]. However, the implementation of smart waste management systems faces challenges including limited knowledge among policymakers and a lack of standards and strategic guidelines [11]. Therefore, this study aims to address the main research question: *How can data analytics tools or algorithms utilise IoT sensor data to aid decision-makers in implementing, monitoring, and optimising SWMS?*

To address this research question, we propose a framework for a SWMS that leverages historical data from IoT sensors and employs data analytic techniques to achieve optimal waste management practices. This approach is significant as it aids decision-makers in implementing, monitoring, and optimising SWMS while ensuring efficient waste collection for public bins, including appropriate scheduling, bin deployment, and resource utilisation. Additionally, the SWMS utilises bin usage data and compartment filling levels for behavior analysis and forecasting. By applying data analytic techniques to analyse trends, patterns, frequency, and timing, a more optimised and sustainable waste management ecosystem can be achieved.

This paper is organised as follows: Sect. 2 introduces the proposed framework for SWMS. We then describe the implementation of the framework in Sect. 3. Finally, Sect. 4 concludes the study.

2   Proposed Data Analytics Framework

This section presents an overview of the proposed data analytics framework for the waste management system. Lin et al. [8] and Ihsanullah et al. [6] have recognised the need to adopt a data analytics framework for efficient waste administration. Therefore, our objective is to design an optimised system that utilises real-time IoT sensor data and data analytics to support a sustainable ecosystem.

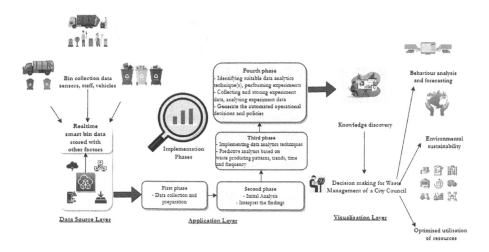

Fig. 1. Overview of the proposed data analytics framework for SWMS

Figure 1 presents the data analytics framework which provides an overview of the entire system for city council authorities to monitor and manage waste.

The waste data analytics framework comprises the data source layer (including physical infrastructure), the application layer (for data processing and analysis), and the visualisation layer (for knowledge discovery). By incorporating these layers, the framework enables efficient collection, analysis, and visualisation of waste data, leading to improved waste management strategies and decision-making. Moreover, the implementation of the framework can be divided into four phases: data collection and preparation from smart bins, initial dataset analysis, implementing of data analytics techniques for waste generation, and conducting experiments, collecting and analysing data to draw conclusions.

3   Implementation

3.1   Dataset

The smart bins in Wyndham city council, Melbourne, Australia have been generating data through IoT devices, including information on bin filling level, and geolocation. The structured data from public bins, collected since 2018 [4], was obtained from the data portal: https://data.gov.au. These bins are positioned in different public locations and the dataset spans for around four years. Table 1 shows the detailed description of each attribute.

The attributes are utilised to determine the appropriate bin for emptying based on the fill level detection. These attributes include the collection point, geographical position, status (empty/full), verbal status, serial number, description, position, and timestamp.

Table 1. Detail description of the collected data set

Coordinates0 Coordinates1	LatestFullness AgeThreshold	Reason	Serial Number	Description	Position	Fullness Threshold	Time stamp
Latitude Longitude	Numeric 0 to 10	Fullness, NoReady	Numeric value	Details for points	Where Locates	Numeric 6 or 8	Date

3.2 Waste Generation Forecasting

At this stage, data analytics techniques, specifically deep learning algorithms [7,8] have been applied using stochastic waste data. Convolution neural networks (CNN) and Recurrent neural network (RNN) have been applied to bigdata processing in different sectors in complex problems like waste management [9]. In this study, we have used 1D CNN, RNN and Long Short-Term Memory (LSTM) [9] to predict waste generation from smart bins.

Bin-level prediction plays a crucial role in waste management for municipalities. One approach to tackle this is by utilising CNN, RNN, and LSTM models, which are trained on historical data to forecast future waste generation trends. The dataset is split into training and testing sets, with 80% and 20% respectively, for the forecasting process. Python Deep Learning API Keras is used along with the Adam optimiser and 20 epochs for all models [5]. The prediction performance of the models is evaluated using three indices: Mean Absolute Error (MAE), Root Mean Square Error (RMSE), and Mean Absolute Percentage Error (MAPE) [6]. Lower values of MAE, RMSE, and MAPE indicate better prediction performance.

3.3 Results and Discussion

All experiment data is collected and stored during and after the experiments for further analysis. Table 2 shows the data using the MAE and RMSE performance indices to show the error values. The smallest value for MAE and RMSE is considered the best model. We found that LSTM is outperformed as it contains the lowest value.

Table 2. Error measures for deep learning models using performance indices

Model	MAE		RMSE		MAPE	
	Train	Test	Train	Test	Train	Test
1D CNN	0.651	0.779	1.758	2.026	1.054	1.277
RNN	0.818	0.951	1.863	2.124	1.317	1.536
LSTM	0.579	0.680	1.479	1.688	10.325	12.830

Figure 2(a) shows the train and test loss for the models. For 1D CNN, the trends of loss are getting higher after four epochs; however, it is close, starting from 2 to 4 epochs. A continuous gap of around 0.15 has been found, starting at 1.00 epoch to the end of the epochs for the RNN model. In terms of LSTM at the beginning, we found the loss containing greater value; however, after 2.75, the gap is too low. Hence, LSTM incorporates a better forecasting for waste generation procedures.

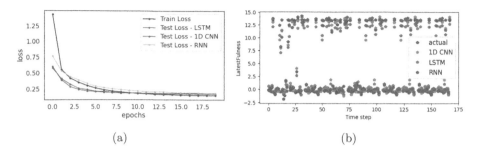

(a)   (b)

Fig. 2. Prediction results for models- a) epoch vs loss and b) actual vs prediction

The LSTM model outperforms CNN and RNN in forecasting waste generation, with higher accuracy and lower loss rates. LSTM demonstrates lower MAE and RMSE values for the training dataset, indicating better capture of trends and patterns. However, it has a higher MAPE value, suggesting less accuracy in predicting actual values. Overall, this study highlights the potential of LSTM deep learning models (Fig. 2(b)) for waste management authorities in accurately predicting waste generation patterns and optimising collection schedules.

The proposed techniques in this paper are applicable and reliable for waste forecasting. While previous studies have focused on different seasonal periods and total waste amounts, our study specifically focuses on daily waste prediction using real-time sensor data [6, 9, 10]. Lin et al. [9] achieved high correlation coefficients of 78, 86.6, and 90 using attention, one-dimensional CNN, and LSTM models for waste forecasting. Niu et al. [10] utilised two-year data and obtained an RMSE of 940 and a coefficient determination (R2) of 0.90 using the LSTM model. In our study, the LSTM model outperformed the CNN and RNN models, with the lowest MAE and RMSE values of 0.68 and 1.688, respectively, for both the training and test sets. The performance indicators confirm the accuracy and reliability of the proposed models in waste estimation, offering valuable insights for informed decision-making by waste management authorities.

4   Conclusion

This study developed a data analytics framework that supports decision-makers in implementing, monitoring, and optimising SWMS. The framework utilises IoT sensor data and employs data analytic techniques to analyse and predict municipal bins' waste generation trends and patterns. The system successfully imposed a platform incorporating the sensor data from the smart bins located in different public places. Finally, to implement the framework, we have devised the whole approach with four distinct phases, and the outcome has been measured through the evaluation of deep learning models to ensure a better predictive capability and is more accurate in forecasting waste generation patterns.

References

1. Anagnostopoulos, T., et al.: Challenges and opportunities of waste management in IoT-enabled smart cities: a survey. IEEE Trans. Sustain. Comput. **2**(3), 275–289 (2017). https://doi.org/10.1109/TSUSC.2017.2691049
2. Anagnostopoulos, T., et al.: A stochastic multi-agent system for internet of things-enabled waste management in smart cities. Waste Manage. Res. **36**(11), 1113–1121 (2018)
3. Bano, A., Ud Din, I., Al-Huqail, A.A.: AIoT-based smart bin for real-time monitoring and management of solid waste. Sci. Program. **2020**, 1–13 (2020)
4. Burton Watson, R., John Ryan, P.: Visualization and waste collection route heuristics of smart bins data using python big data analytics. In: 2021 The 4th International Conference on Software Engineering and Information Management, pp. 124–130 (2021)
5. Gulli, A., Kapoor, A., Pal, S.: Deep Learning with TensorFlow 2 and Keras: Regression, ConvNets, GANs, RNNs, NLP, and more with TensorFlow 2 and the Keras API, 2nd edn. Packt Publishing, Birmingham (2019)
6. Ihsanullah, I., Alam, G., Jamal, A., Shaik, F.: Recent advances in applications of artificial intelligence in solid waste management: a review. Chemosphere **309**, 136631 (2022). https://doi.org/10.1016/j.chemosphere.2022.136631
7. Jiang, P., Fan, Y.V., Zhou, J., Zheng, M., Liu, X., Klemeš, J.J.: Data-driven analytical framework for waste-dumping behaviour analysis to facilitate policy regulations. Waste Manage. **103**, 285–295 (2020)
8. Lin, K., et al.: Toward smarter management and recovery of municipal solid waste: a critical review on deep learning approaches. J. Clean. Prod. **346**, 130943 (2022)
9. Lin, K., Zhao, Y., Tian, L., Zhao, C., Zhang, M., Zhou, T.: Estimation of municipal solid waste amount based on one-dimension convolutional neural network and long short-term memory with attention mechanism model: A case study of Shanghai. Sci. Total Environ. **791**, 148088 (2021)
10. Niu, D., Wu, F., Dai, S., He, S., Wu, B.: Detection of long-term effect in forecasting municipal solid waste using a long short-term memory neural network. J. Clean. Prod. **290**, 125187 (2021)
11. Sharma, M., Joshi, S., Kannan, D., Govindan, K., Singh, R., Purohit, H.: Internet of things (IoT) adoption barriers of smart cities' waste management: an Indian context. J. Clean. Prod. **270**, 122047 (2020)
12. Wang, C., Qin, J., Qu, C., Ran, X., Liu, C., Chen, B.: A smart municipal waste management system based on deep-learning and internet of things. Waste Manage. **135**, 20–29 (2021)
13. Watson, R.B., Ryan, P.J.: Big data analytics in Australian local government. Smart Cities **3**(3), 657–675 (2020). https://doi.org/10.3390/smartcities3030034

Natural Language Processing

Hierarchy-Aware Bilateral-Branch Network for Imbalanced Hierarchical Text Classification

Jiangjiang Zhao[1,2], Jiyi Li[2(✉)], and Fumiyo Fukumoto[2]

[1] Hangzhou Dianzi University, Hangzhou, China
zhaojiangjiang@hdu.edu.com
[2] University of Yamanashi, Kofu, Japan
{jyli,fukumoto}@yamanashi.ac.jp

Abstract. Hierarchical text classification is an essential task in natural language processing. Existing studies focus only on label hierarchy structure, such as building classifiers for each level of labels or employing the label taxonomic hierarchy to improve the hierarchy classification performance. However, these methods ignore issues with imbalanced datasets, which present tremendous challenges to text classification performance, especially for the tail categories. To this end, we propose **Hierarchy-aware Bilateral-Branch Network (HiBBN)** to address this problem, where we introduce the bilateral-branch network and apply a hierarchy-aware encoder to model text representation with label dependencies. In addition, HiBBN has two network branches that cooperate with the uniform sampler and reversed sampler, which can deal with the data imbalance problem sufficiently. Therefore, our model handles both hierarchical structural information and modeling of tail data simultaneously, and extensive experiments on benchmark datasets indicate that our model achieves better performance, especially for fine-grained categories.

Keywords: Hierarchical Text Classification · Imbalanced Data

1 Introduction

As an important task in the natural language processing domain, multi-label text classification (MLC) is widely employed for various text classification procedures, such as news categorization, sentimental analysis and scientific paper classification. Hierarchical text classification (HTC) is an essential subtask of multi-label text classification, which introduces hierarchies to organize the category structure. As depicted in Fig. 1, the category hierarchical structure is commonly modeled as a tree or directed acyclic graph from coarse-grained categories (i.e., parent categories) to fine-grained categories (i.e., child categories), with the objective of predicting multiple labels in a given category hierarchy for a given text. In general, the fine-grained categories are the most appropriate labels to describe the input text, whereas the coarse-grained categories express more general concepts, represented as parent nodes of the fine-grained categories.

C. Strauss et al. (Eds.): DEXA 2023, LNCS 14147, pp. 143–157, 2023.
https://doi.org/10.1007/978-3-031-39821-6_12

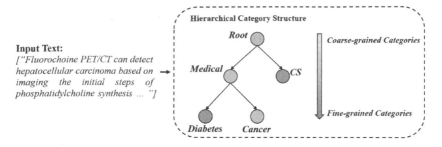

Fig. 1. A hierarchical text classification example from the WOS [12] dataset, tagged with category *Medical* and category *Cancer* from coarse-grained categories to fine-grained categories.

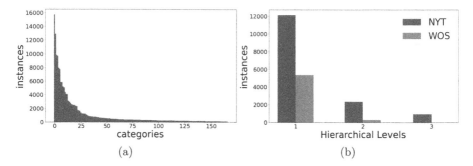

Fig. 2. The imbalance problem in the hierarchical text classification dataset. (a) In the distribution of instance numbers for different categories of NYT [5] training data, the number of instances is imbalanced, as many tail categories have only a small number of instances. (b) In the hierarchical text classification datasets, NYT [5] and WOS [12], as the hierarchy level deepens, the label frequency decreases, and the average number of coarse-grained categories significantly exceeds that of fine-grained categories.

In practice, however, in the real world, text classification datasets tend to exhibit imbalanced distributions, as depicted in Fig. 2(a), where the distribution of instances across the categories is biased or skewed. Specifically, a few categories dominate most of the data (i.e., head categories), while the majority are represented by few instances (i.e., tail categories), with the overall distribution varying between a slight and a severe imbalance. In a hierarchical text classification dataset, the number of coarse-grained categories may significantly exceed that of fine-grained categories, as depicted in Fig. 2(b), causing an imbalance between coarse-grained and find-grained categories. Imbalanced classification poses a tremendous challenge for predictive modeling, as most machine learning algorithms used for classification were designed around the assumption of an equal number of samples for each class. Consequently, models trained on these datasets exhibit relatively low representability on the learned deep features, resulting in an overall decrease in predictive performance, especially for the tail categories.

Existing HTC approaches can be categorized as local approaches and global approaches. Local approaches generally construct a classifier for each label or each level of the label hierarchy. Some local methods [18] have been proposed to overcome the data imbalance problem on child nodes by learning from the parent node. However, these methods require large quantities of parameters and lack holistic structural information, which can easily lead to exposure bias. Conversely, global approaches treat the HTC task as a flat multi-label classification (MLC) problem, and build a single classifier to simultaneously predict multiple labels for a given text. Early global methods [10] ignore the hierarchical structure of labels, instead assuming that no dependencies among labels exist. More recently, global approaches [15] have increasingly begun to employ various strategies to exploit information pertaining to the hierarchical structure of labels. However, these global methods ignore the data imbalance problem, which is crucial for accurate text classification. Therefore, these methods may be insufficient when modeling fine-grained category data and producing satisfactory classification performance.

To address the data imbalance problem in the hierarchical text classification task, we propose a hierarchy-aware bilateral-branch network (HiBBN) based on the conventional bilateral-branch network (BBN) [24], which was originally proposed to address the data imbalance problem in computer vision tasks. Specifically, HiBBN consists of two branches, a conventional learning branch, and a re-balancing branch, used for representation learning and modeling of tail data, respectively. Furthermore, we consider the uniform sampler and reversed sampler for each branch separately, wherein the uniform sampler retains the original data distribution and the reversed sampler is sampled by possibility. The conventional learning branch equipped with the uniform sampler is responsible for learning universal patterns, whereas the re-balancing branch coupled with the reversed sampler is designed to model the tail data. As an extension of a prior approach [25], we also introduce a hierarchy-aware encoder to model text features with hierarchical structure information. Experimental results confirm that HiBBN is more robust than previous models, and is able to alleviate the problem caused by imbalanced data. The contributions of this work can be summarized as follows:

- We address both the category hierarchical structure information and the data imbalance problem in HTC, and propose a novel end-to-end trainable model, called the hierarchy-aware bilateral-branch network (HiBBN), to address the data imbalance problem in the HTC task.
- We successfully adapted the bilateral-branch network proposed in the computer vision domain to the hierarchical text classification task. To our knowledge, this is the first research attempt to accomplish this.
- We evaluated our model on various hierarchical datasets with imbalanced data, with experimental results indicating that our model achieves better classification performance, especially for fine-grained categories.

2   Related Work

2.1   Hierarchical Text Classification

Hierarchical text classification is an important subtask of multi-label text classification, wherein classification results are assigned to one or more nodes within a taxonomic hierarchy. Although certain traditional methods exist for text classification [13,21], they do not consider the hierarchical structure information. Existing approaches for HTC can be categorized into two groups: local approaches and global approaches. Local approaches build a classifier for each label, or each level of the label hierarchy. For example, Shimura et al. [18] proposed a method to learn a local classifier per level that can effectively utilize data in the upper levels to contribute to categorization in the lower levels. Banerjee et al. [1] proposed a transfer learning strategy where classifiers at lower levels are initialized using parameters of the parent classifier and fine-tuned on the child category classification task. In contrast, global approaches treat the HTC task as a flat multi-label text classification problem. Early global methods effectively uses word order for text categorization, but ignore hierarchical label structures. Subsequently, many methods have been developed to extend flat multi-label classification models with hierarchical information. Peng et al. [16] proposed a GCN [22] model wherein the deep architecture is regularized with inter-label dependencies to improve large-scale HTC performance. Zhou et al. [25] formulated the hierarchy as a directed graph, and introduced a hierarchy-aware structure encoder to model label dependencies. Chen et al. [4] proposed HiMatch, which formulates HTC as a semantic matching problem with a hierarchy-aware text feature propagation module. However, the aforementioned methods ignore the data imbalance problem, which we believe must be addressed to ensure accurate classification performance.

2.2   Data Imbalance in Classification

The problem of imbalanced data among different categories is a common challenge in classification tasks. Various methods have been proposed to alleviate this problem. Some approaches [9,19] introduce re-sampling methods to over-sampling the tail categories or under-sampling the head categories. However, this incurs the problem of label co-occurrence in multi-label text classification, and simply increasing the quantity of tail data does not sufficiently address the data imbalance problem. Although some re-weight approaches [3,6] have been developed to assign a weight to each label, wherein labels with smaller sample sizes are assigned larger weights, these schemes are inconsistent with the principle of cross-entropy loss, and naturally result in conflict between learning generalizable representations and the facilitation of learning for tail categories. More recently, Yang et al. [20] proposed HSCNN with a multi-task architecture based on the Single and Siamese network [2] to improve performance on the tail or entire categories. Guo et al. [8] proposed an approach to train on both uniform and re-balanced samplings in a collaborative way, yielding performance improvement on both head and tail categories. Li et al. [14] proposed a novel balanced group softmax module to balance the classifiers, thereby ensuring that the head and tail categories are both sufficiently trained. However, these approaches ignore hierarchical label information, or are otherwise inapplicable to the HTC task. In contrast, we address the hierarchical structure information and data imbalance problem simultaneously.

3   Problem Definition

Hierarchical text classification (HTC) is an essential subtask of multi-label text classification (MLC). To an extent, the HTC problem is equivalent to the MLC problem, as both aim to predict multiple labels in a given text. The difference is that HTC organizes the label space with a predefined taxonomic hierarchy, which primarily contains a tree or a directed acyclic graph structure.

In general, we denote the dataset as $\mathcal{D}=(x,y)$, wherein $x = \{x_1, x_2, ..., x_n\}$ is the text content and n is the instance number of the dataset, $y_i \in \{0,1\}^{|\mathcal{C}|}$ is the corresponding label of instance x_i, and $|\mathcal{C}|$ is the size of the label set. Formulated, we write $y_i^j=1$ if label j belongs to instance x_i, else we set y_i^j to 0. Additionally, we formulate a hierarchical category structure as a directed graph $\mathcal{G} = (\mathcal{V}, \overrightarrow{E}, \overleftarrow{E})$, where $\mathcal{V} = (v_1, v_2, ..., v_{|\mathcal{C}|})$ indicates the set of hierarchical structure label nodes, $\overrightarrow{E}$ is the top-down hierarchy path, and $\overleftarrow{E}$ is built from the bottom-up hierarchy path to represent the connection relationship from child label nodes to parent label nodes.

4   Methodology

In this section, we introduce details pertaining to our **Hierarchy-aware Bilateral-Branch Network (HiBBN)** approach. As depicted in Fig. 3, our model is structured based on the BBN [24], with two branches for representation learning and modeling of tail data, equipped with the uniform sampler and reversed sampler respectively. Both branches use the same network structure and share all parameter weights except for the last classifier.

4.1   Data Sampler

The sampler is an important component of our method that controls a focus on the universal pattern or tail data in the learning process. For the bilateral branches, we considered the uniform sampler and reversed sampler separately. The uniform sampler focuses on universal pattern learning and retains the original data distribution so that each instance in the training dataset is sampled only once.

To allocate more attention to the tail data, we apply the reversed sampler, wherein the sampling possibility of each category is proportional to the reciprocal of its sample size. In formulation, we denote the number of samplers for label j

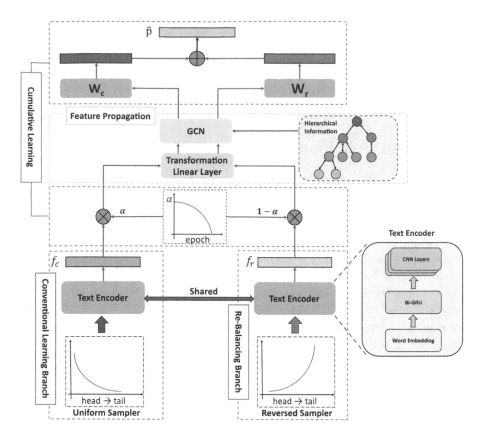

Fig. 3. Overall architecture of proposed model. The parameter weights of the *Text Encoder*, *Transformation Linear Layer*, and *Graph Convolution Network* are shared by two branches. (1) The conventional learning branch equipped with the uniform sampler is responsible for learning universal patterns; (2) The re-balancing branch coupled with the reversed sampler is designed for the tail data. (3) Feature propagation encodes the text features with the category hierarchy information. (4) Cumulative learning strategy shifts the learning focus between bilateral branches by controlling the feature weights generated by the two branches simultaneously with the classification loss.

as N_j, and the maximum number of all labels as N_{max}. The reversed sampling process is organized into the following steps:

- Step 1: calculate the ratio $w_j = \frac{N_{max}}{N_j}$;
- Step 2: calculate the sampling probability of label j based on w_j, the sampling probability $P_j = \frac{w_j}{\sum_{k=1}^{|\mathcal{C}|} w_k}$;
- Step 3: randomly select a label depending on the sampling probability;;
- Step 4: randomly select an instance that contains the selected label.

Iterations of this reversed sampling process produce training batch data of the reversed sampler is obtained.

4.2 Text Encoder

We use (x_c, y_c) and (x_r, y_r) to denote the input data sampled by the uniform sampler and reversed sampler respectively, where x_c and x_r are the text input sequence and y_c and y_r are the corresponding labels. Our proposed method can be applied to any existing text encoder. In this study, we implemented our model with two different encoders to verify our approach.

The first encoder is a traditional text encoder based on Bi-GRU and CNN, with word embeddings generated by Glove [17]. Concretely, the text input sequence is first fed into the embedding layer initialized by the Glove. Subsequently, a bidirectional GRU layer is employed to extract contextual features. Finally, CNN layers with top-k max-pooling generate key n-gram features $\mathbf{f_c} \in \mathbb{R}^{k \times d_{cnn}}$ and $\mathbf{f_r} \in \mathbb{R}^{k \times d_{cnn}}$, where d_{cnn} indicates the output dimension of the CNN layer.

The other encoder is based on BERT [7], a pre-trained language model. We employed the top-level representation h_{CLS} of BERT's special CLS token for text representation.

4.3 Cumulative Learning

The cumulative learning strategy was proposed in the BBN [24], which was designed to learn the universal patterns and then pay attention to the tail data gradually. During the training phase, there is an adaptive parameter α to shift the learning focus between the bilateral branches. The value of α is automatically generated according to the training epoch, and will gradually decrease with subsequent epochs.

$$\alpha = 1 - \left(\frac{T}{T_{max}}\right)^2 \tag{1}$$

where T is the current epoch and T_{max} is the total number of training epochs.

Concretely, the feature $\mathbf{f_c}$ of the conventional learning branch is multiplied by α, whereas the feature $\mathbf{f_r}$ of the re-balancing branch is multiplied by $1 - \alpha$. Then, the weighted features $\alpha \mathbf{f_c}$ and $(1 - \alpha)\mathbf{f_r}$ are sent into the feature propagation component to encode with the label hierarchy information.

In addition, the cumulative learning strategy is also applied to update the parameters and thereby control the weights for the classification losses of the two branches. This process is described in more detail in Subsect. 4.5.

4.4 Feature Propagation

Inspired by a previous approach [25], we introduce hierarchy-aware text feature propagation to model hierarchical category information. As illustrated in Fig. 4, the hierarchical category structure is a directed graph $\mathcal{G} = (\mathcal{V}, \overrightarrow{E}, \overleftarrow{E})$.

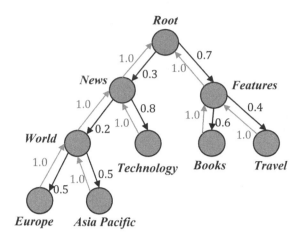

Fig. 4. Example of the hierarchical category structure. Each number indicates the prior statistical probability of label dependencies according to the training corpus.

Specifically, $\overrightarrow{E} \in \mathbb{R}^{|\mathcal{C}| \times |\mathcal{C}|}$ denotes the top-down hierarchy paths representing the prior statistical probability from parent nodes to child nodes, calculated as

$$P(U_b|U_a) = \frac{P(U_b \cap U_a)}{P(U_a)} = \frac{P(U_b)}{P(U_a)} = \frac{N_b}{N_a} \tag{2}$$

where U_a indicates the occurrence of parent label node v_a, U_b denotes that of child label node v_b, $P(U_b|U_a)$ represents the conditional probability of parent label node v_a given that child label node v_b occurs, $P(U_b \cap U_a)$ is the probability of $\{v_a, v_b\}$ occurring simultaneously, and N_a and N_b refer to the quantities of U_a and U_b in the training subset, respectively. Finally, we rescale and normalize the prior probability values of child nodes $v_{child(a)}$ to a sum total of 1.

Meanwhile, $\overleftarrow{E} \in \mathbb{R}^{|\mathcal{C}| \times |\mathcal{C}|}$ is constructed from the bottom-top hierarchy paths representing connectivity relationships from children nodes to parent nodes

$$P(U_a|U_b) = \frac{P(U_b \cap U_a)}{P(U_b)} = \frac{P(U_b)}{P(U_b)} = 1.0 \tag{3}$$

To obtain the aforementioned hierarchical category information, the text feature propagation component uses a transformation linear layer $\mathbf{W_p} \in \mathbb{R}^{|\mathcal{C}| \times d_{cnn} \times d_t}$ to project the text semantics features $\alpha \mathbf{f_c}$ and $(1-\alpha)\mathbf{f_r}$ into the node embedding space. Thus, we can obtain the text nodes $\mathbf{V_c}$ and $\mathbf{V_r}$ respectively

$$\mathbf{V_c} = \alpha \mathbf{W_p} \mathbf{f_c} \tag{4}$$

$$\mathbf{V_r} = (1 - \alpha)\mathbf{W_p}\mathbf{f_r} \tag{5}$$

where $\mathbf{V_c} \in \mathbb{R}^{|\mathcal{C}| \times d_t}$ and $\mathbf{V_r} \in \mathbb{R}^{|\mathcal{C}| \times d_t}$ are the text node features, and d_t represents the hierarchical structure node embedding dimension.

Finally, a graph convolutional network (GCN) is adopted to explicitly combine textual semantics with hierarchical structure information. Concretely, we take the text node feature and hierarchy path as input for the GCN. The hierarchy-aware text features $\mathbf{S_c} \in \mathbb{R}^{|\mathcal{C}| \times d_t}$ and $\mathbf{S_r} \in \mathbb{R}^{|\mathcal{C}| \times d_t}$ can be obtained by

$$\mathbf{S_c} = \sigma(\overrightarrow{E} \cdot \mathbf{V_c} \cdot \mathbf{W_{g1}} + \overleftarrow{E} \cdot \mathbf{V_c} \cdot \mathbf{W_{g2}}) \tag{6}$$

$$\mathbf{S_r} = \sigma(\overrightarrow{E} \cdot \mathbf{V_r} \cdot \mathbf{W_{g1}} + \overleftarrow{E} \cdot \mathbf{V_r} \cdot \mathbf{W_{g2}}) \tag{7}$$

where $\sigma(\cdot)$ is the ReLU activation function, and $W_{g1} \in \mathbb{R}^{d_t \times d_t}$ and $W_{g2} \in \mathbb{R}^{d_t \times d_t}$ are the trainable parameters of GCN.

4.5 Training and Inference

We flatten the hierarchy by regarding all nodes as leaf nodes for multi-label classification. Concretely, the hierarchy-aware text features $\mathbf{S_c}$ and $\mathbf{S_r}$ are sent into classifiers $\mathbf{W_c} \in \mathbb{R}^{|\mathcal{C}| \times d_t \times |\mathcal{C}|}$ and $\mathbf{W_r} \in \mathbb{R}^{|\mathcal{C}| \times d_t \times |\mathcal{C}|}$ respectively, and the outputs are integrated together by element-wise addition. The output logits are formulated as

$$\hat{y} = \mathbf{W_c S_c} + \mathbf{W_r S_r} \tag{8}$$

where $\hat{y} \in \mathbb{R}^{|\mathcal{C}|}$ denotes the prediction result.

During the training phase, the cumulative learning strategy is applied to control the weights for the classification losses of two branches, which shifts the learning focus between bilateral branches and ensures that both branches are constantly updated throughout the training process. Denoting $E(\cdot, \cdot)$ as the binary cross-entropy loss function, the weighted binary cross-entropy classification loss of our method can be written as

$$\mathcal{L} = \alpha E(\hat{y}, y_c) + (1 - \alpha) E(\hat{y}, y_r) \tag{9}$$

This loss function is minimized by gradient descent during training.

During the inference phase, we fix $\alpha = 0.5$ to predict the output logits $\hat{y}$, and apply the sigmoid activation function to calculate the probability of each label

$$p_j = \frac{1}{1 + e^{\hat{y}_j}} \tag{10}$$

If the value of p_j exceeds the classification threshold value λ, we append the label j to the instance.

5   Experiments

The following section presents the details of our experiments, including the dataset, implementation details, comparison, ablation study, and analysis of experimental results.

Table 1. Data Statics

Dataset	$|\mathcal{C}|$	Depth	Avg($|\mathcal{C}_i|$)	Avg(N_c)	N_c^{max}	N_c^{min}	Train	Valid	Test
NYT	166	8	7.6	1336	19610	121	23345	5834	7292
WOS	141	2	2.0	537	11783	34	30070	7518	9397

5.1 Datasets

To evaluate our model's effectiveness, we conducted experiments on two imbalanced datasets for hierarchical multi-label text classification. The statistical information of these datasets is listed in Table 1, where $|\mathcal{C}|$ is the size of label set, Avg($|\mathcal{C}_i|$) indicates the average number of labels of per sample, and *Depth* represents the maximum level of the hierarchy path. Furthermore, we define N_c as the number of labels c in the training data, Avg(N_c) as the mean of N_c and N_c^{max} and N_c^{min} as the maximum and minimum of N_c, respectively. We observe a significant difference between N_c^{max} and N_c^{min}, with the datasets exhibiting an imbalanced distribution.

NYT: The NYT [5] dataset is a news categorization corpus compiled and published by the New York Times between January 1, 1987 and June 19, 2007, with article metadata provided by the New York Times Newsroom. The dataset comprises 166 labels and 8 levels, with 4 labels in the first level.

WOS: The WOS [12] is the world's oldest, most widely used and authoritative database of research publications and citations, comprising 46,985 documents with 141 labels including 7 parent labels. This tree typical text classification datasets are all annotated with the ground truth of hierarchical taxonomic labels.

Note that the NYT dataset uses multi-path taxonomic labels, whereas the WOS dataset is employed for single-path hierarchical text classification.

5.2 Implementation Details

In our experimental setting, we initialized the embedding layer with 300 dimensional pre-trained word embeddings from Glove [17], and employed a one-layer bidirectional GRU with 100 hidden units. For the CNN layers, we applied 100 filters with a kernel size of [2–4], and set $k = 1$ for the top-k max-pooling. The dimension d_t of the text propagation feature and graph convolution weight matrix was 300, the classification threshold value $\lambda = 0.5$. All models were trained over 150 epochs with a batch size of 64. We used Adam [11] as our optimizer, and set the initial learning rate to 1e-4 with the warmup linear scheduler.

To comprehensively validate our model's performance, we further evaluated our method in a pre-trained language model scenario. Concretely, we applied BERT [7] as our encoder, and use the bert-base-cased version in all experiments. We also trained the model for 150 epochs with a fixed batch size of 16 and an initial learning rate of 2e–5.

We evaluated the experiment results by F1 score, which is a statistical measure that considers both the precision and recall of a classification model. The F1 score can further be divided into the micro-F1 and macro-F1 scores. The micro-F1 score accounts for the overall precision and recall of all instances, whereas the macro-F1 score represents the average F1 score of all categories. Thus, micro-F1 assigns higher weights to frequent categories, whereas macro-F1 weighs all categories equally. Because the tail categories are significantly underrepresented compared to the head categories, the micro average is somewhat biased. Accordingly, we employed the micro-F1 and macro-F1 scores simultaneously to ensure a more comprehensive analytical perspective.

5.3   Comparison Models

We compared our model with three types of existing works.

- Text classification baselines:
    - **TextCNN** [23] is the traditional deep learning method for text classification with a convolutional neural network (CNN).
    - **TextRCNN** [13] is based on TextCNN, and uses an extra recurrent structure to capture contextual information.
- Hierarchy-aware models:
    - **HiAGM** [25] is an end-to-end hierarchy-aware global method that extracts structural information to aggregate label-wise text features.
    - **HiMatch** [4] is a hierarchical text classification method that captures semantic relationships among texts and labels at different abstraction levels.
- Pre-trained language model:
    - **BERT** [7] is a powerful and broadly used pre-trained language model in the natural language processing domain, for the text classification task, which only uses text as input.

5.4   Results and Discussion

Table 2 reports the performance of our approach against existing methods on the NYT and WOS datasets. We observe that the traditional text classification methods TextCNN and TextRCNN did not perform as well as the hierarchy-aware models, as they fail to consider hierarchical structure information. Furthermore, our model consistently achieved the best performance, especially in terms of the macro-F1 score. Concretely, our model yielded an advantage of 0.68 on the micro-F1 score and 1.13 on the macro-F1 score, compared with HiMatch on the WOS dataset. Likewise, it exhibited an improvement of 1.84 in terms of macro-F1 score on the NYT dataset.

To further verify our model's performance, we evaluated it in a pre-trained language model scenario. BERT is an effective and powerful pre-trained language model for fine-tuning downstream tasks. However, the primary results indicate that BERT yields inferior performance compared to the other approaches, which

Table 2. The main experimental results.

Dataset	WOS		NYT	
Metric	Micro-F1	Macro-F1	Micro-F1	Macro-F1
Text classification baselines				
TextCNN [23]	83.32	76.41	70.25	54.27
TextRCNN [13]	83.02	76.65	70.57	54.68
Hierarchy-aware models				
HiAGM [25]	85.82	80.28	74.97	60.83
HiMatch [4]	86.00	80.11	75.00	59.42
HiBBN (Ours)	**86.68**	**81.24**	**75.39**	**61.26**
with Pre-trained language model				
BERT [7]	86.09	80.90	77.47	64.13
HiMatch+BERT [4]	86.70	81.06	78.39	66.00
HiBBN+BERT (Ours)	**86.73**	**81.28**	**78.53**	**66.57**

combine BERT with the hierarchy-aware method. In other words, our method brings improvements to BERT, thereby producing better performance compared to the other hierarchy-aware approach under the pre-trained language model scenario.

The experimental results indicate that our model successfully handles representation learning and modeling of tail data to achieve imbalanced HTC classification. Furthermore, the improvement in macro-F1 score demonstrates that our model significantly improves classification performance in the tail categories.

5.5   Label Granularity Performance Study

We analyzed the performance of different label granularities based on the label hierarchy levels. Figure 5 shows the level-based micro-F1 and macro-F1 scores obtained by TextCNN, TextRCNN, HiMatch, and our model. Because the second and third hierarchical levels contain some tail categories, their F1 scores are relatively low. From the figures, we observe that our method outperformed the other approaches on both datasets. Concretely, Fig. 5.(a) and Fig. 5.(c) show that our model achieves satisfactory performance scores in the coarse-grained categories, and outperforms in the fine-grained categories in terms of micro-F1 score. Figure 5.(b) and Fig. 5.(d) indicate that our model yields the best performance in terms of macro-F1 score, especially in the fine-grained categories, which demonstrates our model's ability to improve classification performance with imbalanced data.

5.6   Ablation Study

To further understand our proposed HiBBN model, we conducted an ablation study to evaluate the effectiveness of the bilateral-branch network component.

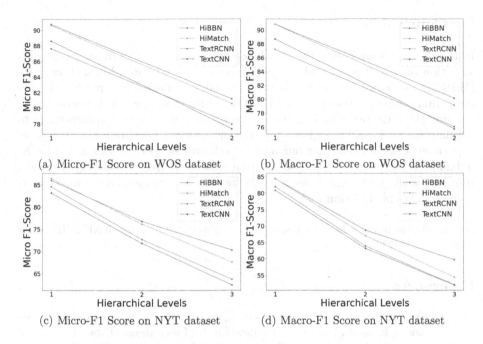

(a) Micro-F1 Score on WOS dataset   (b) Macro-F1 Score on WOS dataset

(c) Micro-F1 Score on NYT dataset   (d) Macro-F1 Score on NYT dataset

Fig. 5. Performance of different hierarchical levels

Table 3. Ablation study results.

Dataset	WOS		NYT	
Metric	Micro-F1	Macro-F1	Micro-F1	Macro-F1
HiBBN (Ours)	**86.68**	**81.24**	**75.39**	**61.26**
w/o BBN	86.24	80.15	75.02	58.74
with Pre-trained language model				
HiBBN+BERT (Ours)	**86.73**	**81.28**	**78.53**	**66.57**
w/o BBN	86.33	80.86	78.34	66.19

The ablation study results are reported in Table 3, showing that the micro-F1 score decreased by 0.44 and 0.37 when removing the bilateral-branch network on the WOS and NYT datasets respectively. These decreases are magnified to 1.09 and 2.52 on the macro-F1 score, as the bilateral-branch network exerts a significant influence on the tail categories. Under the pre-trained language scenario, the bilateral-branch network can still be effective in improving performance. These results also verify that our model can make better use of tail data to mitigate the imbalance problem in the hierarchical text classification task.

6 Conclusion

In this study, we addressed the imbalance problem in the hierarchical text classification task, and developed a novel hierarchy-aware bilateral-branch network (HiBBN) that considers hierarchical structural information and models the tail data simultaneously. We also considered the uniform sampler and reversed sampler for the bilateral branches. The network uses an adaptive parameter α to regulate model focus throughout the training process. By conducting extensive experiments, we verified that our method achieves superior performance in the imbalanced hierarchical text classification task compared to existing methods. In future work, we intend to research the zero-shot learning scenario in the hierarchical text classification task.

Acknowledgements. This work was partially supported by JSPS KAKENHI Grant Number 23H03402 and JKA.

References

1. Banerjee, S., Akkaya, C., Perez-Sorrosal, F., Tsioutsiouliklis, K.: Hierarchical transfer learning for multi-label text classification. In: Proceedings of the 57th Annual Meeting of the Association for Computational Linguistics, pp. 6295–6300 (2019)
2. Bromley, J., Guyon, I., LeCun, Y., Säckinger, E., Shah, R.: Signature verification using a "siamese" time delay neural network. Adv. Neural Inf. Process. Syst. **6**, 1–8 (1993)
3. Cao, K., Wei, C., Gaidon, A., Arechiga, N., Ma, T.: Learning imbalanced datasets with label-distribution-aware margin loss. Adv. Neural Inf. Process. Syst. **32**, 1–12 (2019)
4. Chen, H., Ma, Q., Lin, Z., Yan, J.: Hierarchy-aware label semantics matching network for hierarchical text classification. In: Proceedings of the 59th Annual Meeting of the Association for Computational Linguistics and the 11th International Joint Conference on Natural Language Processing, pp. 4370–4379 (2021)
5. Consortium, L.D., Company, N.Y.T.: The New York Times Annotated Corpus. Linguistic Data Consortium, LDC corpora (2008)
6. Cui, Y., Jia, M., Lin, T.Y., Song, Y., Belongie, S.: Class-balanced loss based on effective number of samples. In: Proceedings of the IEEE/CVF Conference on Computer Vision and Pattern Recognition, pp. 9268–9277 (2019)
7. Devlin, J., Chang, M.W., Lee, K., Toutanova, K.: Bert: pre-training of deep bidirectional transformers for language understanding (2018). arXiv preprint arXiv:1810.04805
8. Guo, H., Wang, S.: Long-tailed multi-label visual recognition by collaborative training on uniform and re-balanced samplings. In: Proceedings of the IEEE/CVF Conference on Computer Vision and Pattern Recognition, pp. 15089–15098 (2021)
9. He, H., Garcia, E.A.: Learning from imbalanced data. IEEE Trans. Knowl. Data Eng. **21**(9), 1263–1284 (2009)
10. Johnson, R., Zhang, T.: Effective use of word order for text categorization with convolutional neural networks (2014). arXiv preprint arXiv:1412.1058
11. Kingma, D.P., Ba, J.: Adam: a method for stochastic optimization (2014). arXiv preprint arXiv:1412.6980

12. Kowsari, K., Brown, D.E., Heidarysafa, M., Meimandi, K.J., Gerber, M.S., Barnes, L.E.: Hdltex: hierarchical deep learning for text classification. In: 2017 16th IEEE International Conference on Machine Learning and Applications, pp. 364–371 (2017)

13. Lai, S., Xu, L., Liu, K., Zhao, J.: Recurrent convolutional neural networks for text classification. In: Twenty-Ninth AAAI Conference on Artificial Intelligence (2015)

14. Li, Y., et al.: Overcoming classifier imbalance for long-tail object detection with balanced group softmax. In: Proceedings of the IEEE/CVF Conference on Computer Vision and Pattern Recognition, pp. 10991–11000 (2020)

15. Mao, Y., Tian, J., Han, J., Ren, X.: Hierarchical text classification with reinforced label assignment. In: Proceedings of the 2019 Conference on Empirical Methods in Natural Language Processing and the 9th International Joint Conference on Natural Language Processing, pp. 445–455 (2019)

16. Peng, H., et al.: Large-scale hierarchical text classification with recursively regularized deep graph-cnn. In: Proceedings of the 2018 World Wide Web Conference, pp. 1063–1072 (2018)

17. Pennington, J., Socher, R., Manning, C.D.: Glove: global vectors for word representation. In: Proceedings of the 2014 Conference on Empirical Methods in Natural Language Processing, pp. 1532–1543 (2014)

18. Shimura, K., Li, J., Fukumoto, F.: Hft-cnn: learning hierarchical category structure for multi-label short text categorization. In: Proceedings of the 2018 Conference on Empirical Methods in Natural Language Processing, pp. 811–816 (2018)

19. Wang, Y., Gan, W., Yang, J., Wu, W., Yan, J.: Dynamic curriculum learning for imbalanced data classification. In: Proceedings of the IEEE/CVF International Conference on Computer Vision, pp. 5017–5026 (2019)

20. Yang, W., Li, J., Fukumoto, F., Ye, Y.: Hscnn: a hybrid-siamese convolutional neural network for extremely imbalanced multi-label text classification. In: Proceedings of the 2020 Conference on Empirical Methods in Natural Language Processing, pp. 6716–6722 (2020)

21. Yang, Z., Yang, D., Dyer, C., He, X., Smola, A., Hovy, E.: Hierarchical attention networks for document classification. In: Proceedings of the 2016 Conference of the North American Chapter of the Association for Computational Linguistics: Human Language Technologies, pp. 1480–1489 (2016)

22. Yao, L., Mao, C., Luo, Y.: Graph convolutional networks for text classification. In: Proceedings of the AAAI Conference on Artificial Intelligence, pp. 7370–7377 (2019)

23. Zhang, Y., Wallace, B.: A sensitivity analysis of (and practitioners' guide to) convolutional neural networks for sentence classification (2015). arXiv preprint arXiv:1510.03820

24. Zhou, B., Cui, Q., Wei, X.S., Chen, Z.M.: Bbn: bilateral-branch network with cumulative learning for long-tailed visual recognition. In: Proceedings of the IEEE/CVF Conference on Computer Vision and Pattern Recognition, pp. 9719–9728 (2020)

25. Zhou, J., et al.: Hierarchy-aware global model for hierarchical text classification. In: Proceedings of the 58th Annual Meeting of the Association for Computational Linguistics, pp. 1106–1117 (2020)

Multi-Feature and Multi-Channel GCNs for Aspect Based Sentiment Analysis

Wenlong Xi[1] , Xiaoxi Huang[1] , Fumiyo Fukumoto[2] ,
and Yoshimi Suzuki[2(✉)]

[1] School of Computer Science and Technology, Hangzhou Dianzi University,
Hangzhou, China
{xwlcs,huangxx}@hdu.edu.cn
[2] Graduate Faculty of Interdisciplinary Research, University of Yamanashi,
Kofu, Japan
{fukumoto,ysuzuki}@yamanashi.ac.jp

Abstract. Aspect Based Sentiment Analysis (ABSA) is a challenging task in natural language processing that involves identifying the sentiment polarity of different aspects in a given text. While graph convolutional network (GCN) has been shown to achieve state-of-the-art results for ABSA, existing methods typically rely on investigating a single type of feature in the sentence to construct the graph representation, which may not capture enough relevant information. In this paper, we propose an approach that leverages multiple channels to extract multiple features, including aspect relations, word dependency relation types, and semantic information, to enhance the performance of the model. We evaluate our approach on four benchmark datasets and demonstrate its validity, achieving state-of-the-art results. We also conduct extensive ablation studies to analyze the contribution of different components of our model. Our findings suggest that the combination of multiple channels and multiple features GCN is crucial for achieving the best performance in ABSA tasks. In conclusion, our proposed approach provides a promising solution to the ABSA problem and contributes to advancing the field by highlighting the importance of considering multiple features and leveraging the power of multiple channels GCN to improve performance.

Keywords: Aspect Based Sentiment Analysis · GCN · Multiple features

1 Introduction

As the volume of online reviews and sentiment-expressing texts continues to increase, there is a growing need for algorithms that can automatically analyze sentiment. [18,31] Aspect-based sentiment analysis (ABSA) or Aspect-level sentiment analysis which is a fine-grained task that is more challenging to address. ABSA enables the extraction of detailed sentiment information from text documents or sentences. [2] Specifically, ABSA aims to infer the sentiment polarities

ⓒ The Author(s), under exclusive license to Springer Nature Switzerland AG 2023
C. Strauss et al. (Eds.): DEXA 2023, LNCS 14147, pp. 158–172, 2023.
https://doi.org/10.1007/978-3-031-39821-6_13

of the different aspects mentioned in a sentence. For example, the given sentence: *"The menu is limited but almost all of the dishes are excellent."* and the corresponding aspects of *"menu"* and *"dishes"*, the sentiment polarity of *"menu"* is negative, while the sentiment polarity of *"dishes"* is positive. Notably, these two aspects have opposite sentiment polarities, highlighting the need for aspect-level analysis, which can provide more nuanced insights than sentence-level analysis.

With the rapid development of deep learning, attention mechanisms have become popular in Aspect Based Sentiment Analysis (ABSA) research. Previous studies have applied various attention mechanisms to model the relationship between aspect words and opinion terms in the context [3,6,7,19,25,28]. However, simply applying the attention mechanism without thoroughly investigating the sentence information may lead to incorrect aspect-term associations and result in degrading the model's performance.

In recent years, Graph Neural Networks (GNNs) such as GCN [11] and GAT [27] have been introduced to the field of sentiment analysis. These models aim to better pair the aspect and corresponding opinion term, and have achieved state-of-the-art performance [1,13,14,16,21,23,30,32]. The overall procedure of this method can be summarized into several steps: *Graph construction*, which involves using a parser to analyze the text and construct a graph, and *Node representation*, which involves using an encoder to obtain hidden representations and then feeding them into a GNN-based model to extract more information and perform sentiment prediction. However, previous studies have only focused on one type of feature in the text, such as semantic information [16], aspect-based relation [14], or dependency relation type [1], without attempting to use multiple features simultaneously. This limitation may result in insufficient knowledge for the model to learn from.

Therefore, there is a need to explore and integrate multiple types of features in the GNN-based model to achieve more comprehensive and accurate sentiment prediction. In our approach, we propose to integrate multiple features into our model through a multiple channel network. Based on this, a Multi-Features and Multi-Channels Graph Convolution Networks (**MFMCGCN**) model is proposed to fit this need. By using this approach, we can successfully integrate different features of a sentence simultaneously for sentiment polarity prediction.

The main contributions of our work can be summarized as follows:

- We introduce a novel approach to sentiment analysis that integrates multiple features of a sentence simultaneously.
- Our approach uses a multiple channel network to effectively integrate different types of features in our model.
- Through extensive experiments on four benchmark datasets, we demonstrate that our proposed approach achieves the state-of-the-art result. This highlights the effectiveness of our approach in sentiment analysis tasks. The code and datasets involved in this paper are provided on Github.[1]

---

[1] https://github.com/xijuan-hdu/MFMCGCN.

2   Related Work

Early methods in ABSA relied on machine learning and used handcrafted features, which required significant human labor and failed to effectively model the correlation between aspects and their context [12, 24].

Recently, various attention mechanisms have been used in this field to automatically model the correlation between aspects and their context by assigning attention weights to relevant words in a sentence [3, 6, 7, 19, 22, 28]. For example, Wang [28] introduces an attention mechanism to dynamically weigh the importance of different parts of a sentence when predicting the sentiment of a particular aspect. Chen [3] uses a recurrent attention network to model interactions between different aspects and their associated sentiments. In addition, pre-trained models such as BERT [4] have been introduced into this field, involving the construction of an auxiliary sentence for each aspect to incorporate aspect information into the model [20].

More recently, researchers have focused on integrating GNN-based models into ABSA. Aspects can be represented as nodes in a graph, and the relationships between aspects can be represented as edges in the graph. For example, Sun [21] uses a convolution operation over a dependency tree to capture the relationships between different aspects and their associated sentiments. Liang [14] proposes a novel neural network architecture for ABSA that uses graph convolutional networks to learn aspect-focused and inter-aspect relations. Pang [16] uses a dynamic and multi-channel graph convolutional network to capture the relationships between aspects and their associated sentiments. Bai [1] uses typed syntactic dependencies and a graph attention neural network to capture relationships between aspects and their associated sentiments. Li [13] uses two parallel graph neural networks to capture both aspect-sentiment and aspect-term relations in a sentence. Zhang [32] combines both syntactic and semantic information to enhance a graph convolutional network for ABSA. However, none of these methods consider using multiple types of features, which could potentially enrich the information that the model can learn.

3   Proposed Approach

Figure 1 provides an overview of proposed model MFMCGCN. In this section, we will provide a detailed introduction to the components of MFMCGCN.

3.1   Embedding and Encoding Layer

Based on the previous studies [21], given an input sentence $s = \{w_1, w_2, \cdots, w_1^a, \cdots, w_m^a, \cdots, w_{n-1}, w_n\}$ with n words and a specific aspect $\{w_1^a, \cdots, w_m^a\}$ with m words. Each word in this sentence is fed into the embedding layer to obtain the corresponding word vector, which is retrieved from the embedding lookup table $E \in \mathbb{R}^{|V| \times d_e}$, where V is the vocabulary size and d_e is the dimension of the word embedding. After obtaining the word vectors, we get the corresponding

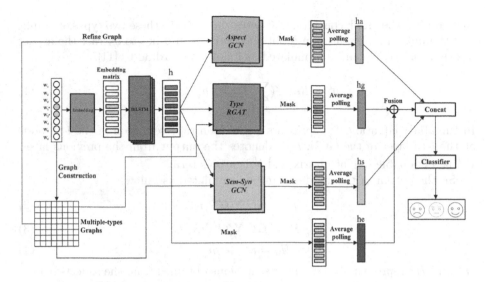

Fig. 1. Overview of proposed model MFMCGCN.

embedding matrix $v = [v_1, v_2, \cdots, v_n]$, where $v_i \in \mathbb{R}^{d_e}$ is the word embedding of w_i. Additionally, we concatenate POS-tag embeddings $t_i \in \mathbb{R}^{d_t}$ and position embeddings $p_i \in \mathbb{R}^{d_p}$ with the previous embedding matrix v. Here, d_t and d_p are the dimensions of the POS-tag and position embeddings, respectively. The final representation of the sentence s is $x = [x_1, x_2, \cdots, x_1^a, \cdots, x_m^a, \cdots, x_{n-1}, x_n]$, where $x_i = [v_i; t_i; p_i]$ represents the concatenation of the three embeddings.

We then input x to a bidirectional LSTM to learn the hidden representation of the given sentence. Specifically, a forward $\overrightarrow{LSTM}$ generates the forward hidden representation $\overrightarrow{h} = \{\overrightarrow{h}_1, \overrightarrow{h}_2, \cdots, \overrightarrow{h}_1^a, \cdots, \overrightarrow{h}_m^a, \cdots, \overrightarrow{h}_{n-1}, \overrightarrow{h}_n\}$, while a backward $\overleftarrow{LSTM}$ generates the backward hidden representation $\overleftarrow{h} = \{\overleftarrow{h}_1, \overleftarrow{h}_2, \cdots, \overleftarrow{h}_1^a, \cdots, \overleftarrow{h}_m^a, \cdots, \overleftarrow{h}_{n-1}, \overleftarrow{h}_n\}$. Finally, we concatenate the $\overrightarrow{h}$ and $\overleftarrow{h}$ to obtain the hidden representation $h = [\overrightarrow{h}; \overleftarrow{h}]$.

3.2 Aspect GCN Module

Follows the previous study [21,30], We can represent the dependency tree G of an sentence as an adjacency matrix D of size $n \times n$. Each entry D_{ij} in the matrix indicates whether node i is connected to node j by a single dependency path in G. More specifically, $D_{ij} = 1$ if node i is connected to node j, and $D_{ij} = 0$ if there is no connection between them.

Our Aspect GCN module mainly focuses on the aspect relation of a sentence, which requires two types of graphs as inputs: an aspect-focused graph and an inter-aspect graph. The module aims to enhance the aspect-related representation of the sentence. Following the approach in the previous work [14], we constructed the aspect-focused graph G^F and the inter-aspect graph G^I. We then applied the graph convolutional network(GCN) to these two types of graphs and the hidden representation h obtained earlier, in order to extract the aspect information. GCN can be formulated as follows, according to [11]:

$$h_i^l = ReLU(\sum_{j=1}^{n} D_{ij}W^l h_j^{l-1} + b^l) \tag{1}$$

In the above equation, h_i^l represents the hidden representation of the i-th node at the l-th layer of the GCN. h_j^{l-1} denotes the output from the previous layer. W^l is a trainable weight matrix and b^l is a bias term.

So the output of this module can be formulated as follows:

$$H^f = GCN(G^F, h) \tag{2}$$

$$H^i = GCN(G^I, h) \tag{3}$$

$$H_a = h^F + \gamma h^I \tag{4}$$

H^f and H^a represent the final representations obtained from the aspect-focused and inter-aspect GCN layers, respectively. The variable γ is a tunable parameter, as used in previous work [14]. H_a is the final output of this module.

3.3   Type RGAT Module

Our Type RGAT module focuses mainly on the dependency relations between words in a sentence, following previous study [1]. This module requires a label matrix R_{ij}, which records the corresponding label of D_{ij} if $D_{ij} = 1$, otherwise $R_{ij} = 0$. R_{ij} is then transformed into a vector $r_{ij} \in \mathbb{R}^{d_r}$, where d_r is the dimension of the relation matrix. The procedure to obtain the hidden representation of this module can be summarized as the following equation:

$$h_i^l = \|_{z=1}^{Z}\sigma(\sum_{j \in N(i)} \hat{\alpha}_{ij}^{lz}(W_V^{lz} h_j^{l-1} + W_{Vr}^l r_{ij})) \tag{5}$$

In the equation, Z represents the number of attention heads, $\|$ denotes vector concatenation, and σ is the sigmoid function. $W_V^{lz} \in \mathbb{R}^{\frac{d}{Z} \times d}$ represents the parameter matrices of the z-th head at layer l, while $W_{Vr}^l \in \mathbb{R}^{\frac{d}{Z} \times d_r}$ represents a parameter matrix for the relation matrix. In addition, $j \in N(i)$ denotes the neighbor word index, where $N(i)$ is induced from the adjacent matrix D. $\hat{\alpha}_{ij}^{lz}$ represents a weight that models the extent to which h_i^l depends on h_j^{l-1}. The weight is determined by the following equation:

$$\hat{\alpha}_{ij} = \frac{exp(e_{ij}^N + e_{ij}^R)}{\sum_{j \in N(i)} exp(e_{ij}^N + e_{ij}^R)} \tag{6}$$

e_{ij}^N is the node-aware attention, and e_{ij}^R is the relation-aware attention, as given by the following equations:

$$e_{ij}^N = \begin{cases} f(h_i^{(l-1)}, h_j^{(l-1)}) & \text{if } j \in N(i) \\ -inf & \text{otherwise} \end{cases} \tag{7}$$

$$e_{ij}^R = \begin{cases} f(h_i^{(l-1)}, r_{ij}) & \text{if } j \in N(i) \\ -inf & \text{otherwise} \end{cases} \tag{8}$$

f is a scaled dot-product attention function [26], given by the following equation:

$$f(h_i^{l-1}, h_j^{l-1}) = \frac{(W_Q^{lz} h_i^{l-1})^T (W_K^{lz} h_J^{l-1})}{\sqrt{d/Z}} \tag{9}$$

For simplicity, we represent the graph representation H_g as:

$$H_g = RGAT(D, h, r_{ij}) \tag{10}$$

which is the output of this module.

3.4 Sem-Syn GCN Module

The Sem-Syn module addresses the problem of sentences lacking remarkable syntactic structure, as described in the previous research [16]. To enhance semantic information, we generate a training matrix D_{se} following the previous work's approach. The Sem-Syn module consists of three parts: Syntactic GCN, Semantic GCN, and Common-GCN, which we have simplified based on the previous study [16].

$$H_{sy}^l = SYGCN(D, h, W_{sy}) \tag{11}$$

$$H_{se}^l = SEGCN(D_{se}, h, W_{se}) \tag{12}$$

and the common representation H_c^l:

$$H_c^l = \frac{\mathcal{A}H_{c-sy}^l + \mathcal{B}H_{c-se}^l}{2} \tag{13}$$

where $\mathcal{A}$ and $\mathcal{B}$ are trainable parameters, H_{c-sy}^l and H_{c-se}^l is given by the following equations:

$$H_{c-sy}^l = SYGCN(D, h, W_c) \tag{14}$$

$$H_{c-se}^l = SEGCN(D_{se}, h, W_c) \tag{15}$$

where $W_c \in \mathbb{R}^{d_{lstm}+l \times d_{gcn}}$ is the l-th layer learnable weight matrix.

The final output of this module is denoted as H_s, which is obtained by concatenating H_{sy}, H_{se} and H_c:

$$H_s = [H_{sy}; H_c; H_{se}] \tag{16}$$

3.5 Feature Fusion

Before fusing the representations extracted by our three modules, we employ a mask mechanism and average pooling to capture the aspect representation, following the previous studies [1, 14, 16, 21, 30].

$$h_a = AveragePooling(mask(H_a)) \tag{17}$$

$$h_g = AveragePooling(mask(H_g)) \qquad (18)$$

$$h_s = AveragePooling(mask(H_s)) \qquad (19)$$

$$h_e = AveragePooling(mask(h)) \qquad (20)$$

h_e will be used to fuse with the h_g as the previous work [1]:

$$h_t = g \times h_g + (1 - g) \times h_e \qquad (21)$$

where g is a gate computed by:

$$g = \sigma(W_g[h_g; h_e] + b_g) \qquad (22)$$

$[;]$ is concatenation operation, W_g and b_g are model learnable parameters. σ is the sigmoid function. The final aspect representation is obtained by concatenating the output vectors h_a, h_t, and h_s. The formulas used are as follows:

$$h_f = [\alpha * h_a, \beta * h_t, \gamma * h_s] \qquad (23)$$

where α, β, γ are tunable parameters.

$$mask = \begin{cases} 1 & \text{if word index} \in aspect\ words\ indexes \\ 0 & \text{otherwise} \end{cases} \qquad (24)$$

3.6   Classifier

The classifier in our model is implemented as a fully connected neural network. It takes the fused final representation h_F as input and computes the probability of each sentiment class c using the softmax function. Specifically, the probability $P(y = c)$ is calculated as:

$$P(y = c) = \frac{exp(Wh_f + b)_c}{\sum_{c' \in C} exp(Wh_f + b)_{c'}} \qquad (25)$$

where W and b are trainable parameters and C is the set of sentiment classes.

3.7   Training

To train the classifier, we define the objective as minimizing the cross-entropy loss between the predicted probabilities and the ground-truth distribution. Specifically, let y_i be the ground-truth label for the i-th example and $p_i(c)$ be the predicted probability for class c. The cross-entropy loss $\mathcal{L}$ is computed as:

$$\mathcal{L} = -\frac{1}{N} \sum_{i=1}^{N} \sum_{c \in C} y_i(c) \log p_i(c) + \lambda \|\Theta\|^2 \qquad (26)$$

where N is the total number of examples and $y_i(c)$ is an indicator function that equals 1 if the label of example i is c, and 0 otherwise. λ is the weight of L_2 regularization and Θ is the trainable parameters in the model.

4   Experiments

4.1   Dataset

Table 1. Statistics of experimental datasets.

Dataset	Positive		Neutral		Negative	
	Train	Test	Train	Test	Train	Test
MAMS	3380	400	5042	607	2764	329
Twitter	1507	172	3016	336	1528	169
Restaurant	2164	727	637	196	807	196
Laptop	976	337	455	167	807	196

We used four datasets to evaluate our model. These four datasets include the MAMS dataset [9], the ACL14 Twitter dataset [5], the Restaurant dataset and Laptop dataset from SemEval 2014 [17]. It is worth noting that each data may have multiple aspects. For each aspect, there is a corresponding sentimental polarity including positive, neutral, and negative. The statistics of each dataset is given in Table 1.

4.2   Experimental Setting

Our experimental setup is similar to that of previous work, such as RGAT [1] and Inter-GCN [14]. We investigated two types of models: one based on GloVe and the other based on BERT. For the GloVe-based model, we utilized 840 billion tokens with 300-dimensional GloVe vectors[2] for word embedding. Additionally, we set the position, POS, and dependency embeddings to 30. To train the models, we used the AdaMax [10] optimizer with a learning rate of 10^{-2}. The train epoch is set to 100. For the BERT-based model, we employ the bert-base-uncased pre-trained model[3]. We fine-tuned a pre-trained BERT and used the Adam [10] optimizer for training. The train epochs is set to 15. We evaluated the models using two metrics: accuracy and Macro-Averaged F1 score. And we obtained the α, β, and γ parameters by trying different combinations. The final parameter settings for each dataset can be found in our code. To ensure a fair comparison between models, we selected the best results achieved by each model for comparison.

4.3   Baseline Methods

We compared our proposed method with the following methods.

---

[2] https://github.com/stanfordnlp/GloVe.
[3] https://github.com/huggingface/transformers.

GloVe-Based Models: SVM [12], ATAE-LSTM [28], IAN [15], AOA [8], AEN [19], ASGCN [30], CDT [21], InterGCN [14], SD-GCN-G [33], DGEDT [23], RGAT [1], DM-GCN [16],

BERT-Based Models: RGAT+BERT [1], BERT4GCN [29], DualGCN+BERT [13], SSEGCN+BERT [32].

Table 2. Main experimental results. Acc means accuracy rate. F1 means Macro-F1 score. Best results are in bold. Results underlined indicate that it is the second best result. The symbol '-' indicates this result is not available.

Model	MAMS		Twitter		Restaurant		Laptop	
	Acc	F1	Acc	F1	Acc	F1	Acc	F1
SVM	-	-	63.30	64.40	80.16	-	70.49	-
ATAE-LSTM	-	-	-	-	78.60	67.02	68.88	63.93
IAN	76.60	-	-	-	78.60	-	72.10	
AOA	77.26	-	-	-	79.97	70.42	72.62	67.52
AEN	66.72	-	72.83	69.81	80.98	72.14	73.51	69.04
ASGCN	-	-	72.15	70.40	80.77	72.02	75.55	71.05
CDT	80.70	79.79	74.66	73.66	82.30	74.02	77.19	72.99
InterGCN	-	-	-	-	82.23	74.01	77.86	74.32
SD-GCN-G	-	-	-	-	82.95	75.79	75.55	71.35
DGEDT	-	-	74.8	73.4	83.9	75.1	76.8	72.3
RGAT	<u>81.75</u>	<u>80.87</u>	75.36	74.15	83.55	<u>75.99</u>	78.02	74.00
DM-GCN	-	-	**76.93**	**75.9**	<u>83.98</u>	75.59	<u>78.48</u>	<u>74.9</u>
MFMCGCN	**82.76**	**81.91**	<u>76.51</u>	<u>75.31</u>	**83.99**	**77.07**	**78.65**	**75.19**
RGAT+BERT	84.52	83.74	76.28	75.25	86.68	80.92	80.94	78.20
BERT4GCN	-	-	74.43	73.76	84.75	77.11	77.49	73.01
DualGCN+BERT	-	-	<u>77.40</u>	<u>76.02</u>	<u>87.13</u>	<u>81.16</u>	<u>81.80</u>	<u>78.10</u>
SSEGCN+BERT	-	-	<u>77.40</u>	<u>76.02</u>	**87.31**	81.09	81.01	77.96
MFMCGCN+BERT	**84.97**	**84.43**	**77.44**	**76.53**	<u>87.13</u>	**81.76**	**82.03**	**78.86**

4.4   Main Results

Our main experimental results of all the models are shown in Table 2. We compare the GloVe-based models and BERT-based models respectively.

We observed that the MFMCGCN model showed accuracy improvement on the MAMS dataset, achieving a 1.01 percent gain compared to the performance of RGAT [1]. For the Restaurant dataset, while the accuracy gain was relatively modest, our model achieved a noteworthy improvement of 1.41 percent in F1 score compared to DM-GCN [16]. This is particularly meaningful since the

Table 3. Experimental results of the ablation study. The w/o means remove one module from MFMCGCN.

Model	MAMS		Twitter		Restaurant		Laptop	
	Acc	F1	Acc	F1	Acc	F1	Acc	F1
MFMCGCN w/o h_A	80.13	79.42	75.00	72.90	83.36	76.00	76.15	72.17
MFMCGCN w/o h_T	81.36	80.80	74.52	72.95	83.00	75.57	75.89	72.03
MFMCGCN w/o h_S	81.43	80.62	72.53	70.96	82.57	74.29	77.34	73.05
MFMCGCN w/o h_A and h_T	80.23	79.21	74.01	71.79	82.38	74.00	75.94	72.10
MFMCGCN w/o h_A and h_S	79.95	79.32	73.10	71.86	81.30	72.97	75.52	71.21
MFMCGCN w/o h_T and h_S	79.85	78.83	73.51	72.24	78.26	73.09	75.55	71.29
MFMCGCN	**82.76**	**81.91**	**76.51**	**75.31**	**83.99**	**77.07**	**78.65**	**75.19**

Restaurant dataset is unbalanced, making F1 score a more relevant and desirable metric for evaluation. We suspect that the reason why our method did not outperform DM-GCN on the Twitter dataset is because the Twitter data is less structured and more informal. Since the data of this dataset is collected from the Internet, it may contain a higher degree of noise, making it more challenging for our proposed model to handle. Specifically, our model may have struggled to effectively process the informal language used in the Twitter data, which DM-GCN may have been better equipped to handle.

Our MFMCGCN+BERT model has achieved state-of-the-art results. Compared to current state-of-the-art results, the MFMCGCN+BERT model's accuracy improvement is relatively small, with the biggest gain being 0.45 percent on the MAMS dataset, as compared to RGAT+BERT [1]. However, there is a more significant improvement in F1 score, with a 0.76 percent increase on the Laptop dataset, as compared to DualGCN+BERT [13].

4.5   Ablation Study

To further investigate the role of modules in the MFMCGCN model, we conduct extensive ablation study. The results are shown in Table 3. It is obvious that MFMCGCN obtains the best result on all dataset, which means that all three modules contribute to the performance of our proposed model.

In the first three rows of our experimental results, we evaluated the performance of our MFMCGCN model by removing one module at a time. The results showed that the Aspect GCN module (h_A) played a crucial role in achieving the best performance on the MAMS dataset. Similarly, the Sem-Syn GCN module (h_S) was the most important part of our model for the Twitter and Restaurant datasets. In particular, the MFMCGCN model without h_S achieved the worst result on the Twitter and Restaurant datasets. For the Laptop dataset, we found that the Type-GCN module (h_T) was also a critical component of our model, as the MFMCGCN model without h_T had the worst performance. These observations suggest that certain datasets may have more pronounced or hidden features, highlighting the importance of extracting and utilizing multiple features of data simultaneously.

We have conducted experiments where we removed two modules at a time, and the results are shown in the fourth to sixth row of the experimental results table. We observed that MFMCGCN without both h_T and h_S performed poorly on the MAMS and Laptop datasets, indicating that this combination is more critical for these datasets. On the other hand, for the Twitter and Laptop datasets, MFMCGCN without both h_A and h_S yielded worse results, suggesting that this combination is more important for these two datasets. We have thus identified the relative importance of different module combinations for each dataset.

It is noteworthy that in some cases, removing two modules at once can result in better performance than removing just one module. For instance, we observed that the performance of MFMCGCN w/o h_S was worse than that of MFMCGCN w/o h_A and h_S, as well as w/0 h_T and h_S. This finding suggests that if the combination of different modules is not well-matched, it may have a negative impact on the overall performance of the model. The specific reasons behind this phenomenon will require further investigation in future work. Despite this, we can draw preliminary conclusions. It is important to take into account the interactions between different modules and their compatibility during the design and optimization of GCN-based models.

4.6   Case Study

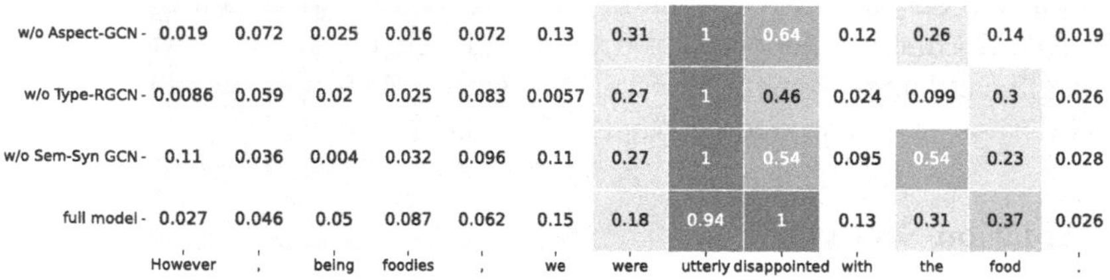

Fig. 2. Visualization of attention weights assigned to each word by our model, as well as the ablation models.

To gain a better understanding of the different modules in our model, we will use a sentence from the Restaurant dataset as a case study. Using the methodology employed in CDT [21], we created heatmaps that display the weight assigned by our model and ablation models to each word in the given sentence, as shown in Fig. 2.

Comparing the first row result obtained by the w/o Aspect-GCN model with the fourth row result obtained by our full model, it is evident that our model pays more attention to the aspect term *"food"*. This indicates that the Aspect-GCN module has a positive effect on helping our model learn the knowledge related to aspects. Moreover, it can be observed that the remaining two ablation models pay more attention to *"food"* than the first one.

In the second row, comparing the result obtained by the w/o Type-RGCN model with that of the full model, we notice that the w/o Type-RGCN model almost does not pay any attention to the word *"the"* which is related to *"food"* and it pays less attention to the opinion term *"disappointed"*.

Finally, comparing the results of the last two rows, we can conclude that the w/o Sem-Syn GCN model cannot pay more attention to the opinion term and wrongly pays more attention to *"the"* than *"food"*.

From the findings above, it is evident that each module in our model plays an important role.

5   Conclusion

In this paper, we proposed a multiple features and multiple channels GCN model for Aspect Based Sentiment Analysis. Our experiments on four benchmark datasets showed that introducing multiple features through a multiple channels model is useful for ABSA task. Moreover, the ablation study justified the influence of each module, and the case study visualized the attention weight, indicating that our model can pay more attention to the corresponding word based on the condition. In future investigations, we will explore the factors contributing to the incompatibility between different modules in specific scenarios.

Acknowledgements. This work is supported by the Kajima Foundation's Support Program for International Joint Research Activities and JSPS KAKENHI (No.22K12146, 23H03402).

References

1. Bai, X., Liu, P., Zhang, Y.: Investigating typed syntactic dependencies for targeted sentiment classification using graph attention neural network. IEEE/ACM Trans. Audio, Speech, Lang. Proc. **29**, 503–514 (2021). https://doi.org/10.1109/TASLP. 2020.3042009
2. Brauwers, G., Frasincar, F.: A survey on aspect-based sentiment classification. ACM Comput. Surv. **55**(4), 1–37 (2022)
3. Chen, P., Sun, Z., Bing, L., Yang, W.: Recurrent attention network on memory for aspect sentiment analysis. In: Proceedings of the 2017 Conference on Empirical Methods in Natural Language Processing, pp. 452–461. Association for Computational Linguistics, Copenhagen, Denmark (2017). https://doi.org/10.18653/v1/ D17-1047, http://aclanthology.org/D17-1047
4. Devlin, J., Chang, M.W., Lee, K., Toutanova, K.: BERT: pre-training of deep bidirectional transformers for language understanding. In: Proceedings of the 2019 Conference of the North American Chapter of the Association for Computational Linguistics: Human Language Technologies, Volume 1 (Long and Short Papers), pp. 4171–4186. Association for Computational Linguistics, Minneapolis, Minnesota (2019). https://doi.org/10.18653/v1/N19-1423, http://aclanthology.org/N19-1423

5. Dong, L., Wei, F., Tan, C., Tang, D., Zhou, M., Xu, K.: Adaptive recursive neural network for target-dependent Twitter sentiment classification. In: Proceedings of the 52nd Annual Meeting of the Association for Computational Linguistics (Volume 2: Short Papers), pp. 49–54. Association for Computational Linguistics, Baltimore, Maryland (2014). https://doi.org/10.3115/v1/P14-2009, http://aclanthology.org/P14-2009

6. Fan, F., Feng, Y., Zhao, D.: Multi-grained attention network for aspect-level sentiment classification. In: Proceedings of the 2018 Conference on Empirical Methods in Natural Language Processing, pp. 3433–3442. Association for Computational Linguistics, Brussels, Belgium (2018). https://doi.org/10.18653/v1/D18-1380, http://aclanthology.org/D18-1380

7. Hu, M., et al.: CAN: constrained attention networks for multi-aspect sentiment analysis. In: Proceedings of the 2019 Conference on Empirical Methods in Natural Language Processing and the 9th International Joint Conference on Natural Language Processing (EMNLP-IJCNLP), pp. 4601–4610. Association for Computational Linguistics, Hong Kong, China (2019). https://doi.org/10.18653/v1/D19-1467, http://aclanthology.org/D19-1467

8. Huang, B., Ou, Y., Carley, K.M.: Aspect level sentiment classification with attention-over-attention neural networks (2018). https://doi.org/10.48550/ARXIV.1804.06536, http://arxiv.org/abs/1804.06536

9. Jiang, Q., Chen, L., Xu, R., Ao, X., Yang, M.: A challenge dataset and effective models for aspect-based sentiment analysis. In: Proceedings of the 2019 Conference on Empirical Methods in Natural Language Processing and the 9th International Joint Conference on Natural Language Processing (EMNLP-IJCNLP), pp. 6280–6285. Association for Computational Linguistics, Hong Kong, China (2019). https://doi.org/10.18653/v1/D19-1654, http://aclanthology.org/D19-1654

10. Kingma, D.P., Ba, J.: Adam: a method for stochastic optimization (2014). https://doi.org/10.48550/ARXIV.1412.6980, http://arxiv.org/abs/1412.6980

11. Kipf, T.N., Welling, M.: Semi-supervised classification with graph convolutional networks. In: International Conference on Learning Representations (2017). http://openreview.net/forum?id=SJU4ayYgl

12. Kiritchenko, S., Zhu, X., Cherry, C., Mohammad, S.: NRC-Canada-2014: detecting aspects and sentiment in customer reviews. In: Proceedings of the 8th International Workshop on Semantic Evaluation (SemEval 2014), pp. 437–442. Association for Computational Linguistics, Dublin, Ireland (2014). https://doi.org/10.3115/v1/S14-2076, http://aclanthology.org/S14-2076

13. Li, R., Chen, H., Feng, F., Ma, Z., Wang, X., Hovy, E.: Dual graph convolutional networks for aspect-based sentiment analysis. In: Proceedings of the 59th Annual Meeting of the Association for Computational Linguistics and the 11th International Joint Conference on Natural Language Processing (Volume 1: Long Papers), pp. 6319–6329. Association for Computational Linguistics, Online (2021). https://doi.org/10.18653/v1/2021.acl-long.494, http://aclanthology.org/2021.acl-long.494

14. Liang, B., Yin, R., Gui, L., Du, J., Xu, R.: Jointly learning aspect-focused and inter-aspect relations with graph convolutional networks for aspect sentiment analysis. In: Proceedings of the 28th International Conference on Computational Linguistics, pp. 150–161. International Committee on Computational Linguistics, Barcelona, Spain (Online) (2020). https://doi.org/10.18653/v1/2020.coling-main.13, http://aclanthology.org/2020.coling-main.13

15. Ma, D., Li, S., Zhang, X., Wang, H.: Interactive attention networks for aspect-level sentiment classification (2017). https://doi.org/10.48550/ARXIV.1709.00893, http://arxiv.org/abs/1709.00893
16. Pang, S., Xue, Y., Yan, Z., Huang, W., Feng, J.: Dynamic and multi-channel graph convolutional networks for aspect-based sentiment analysis. In: Findings of the Association for Computational Linguistics: ACL-IJCNLP 2021, pp. 2627–2636 (2021)
17. Pontiki, M., Galanis, D., Pavlopoulos, J., Papageorgiou, H., Androutsopoulos, I., Manandhar, S.: SemEval-2014 task 4: aspect based sentiment analysis. In: Proceedings of the 8th International Workshop on Semantic Evaluation (SemEval 2014), pp. 27–35. Association for Computational Linguistics, Dublin, Ireland (2014). https://doi.org/10.3115/v1/S14-2004, http://aclanthology.org/S14-2004
18. Schouten, K., Frasincar, F.: Survey on aspect-level sentiment analysis. IEEE Trans. Knowl. Data Eng. **28**(3), 813–830 (2016). https://doi.org/10.1109/TKDE.2015.2485209
19. Song, Y., Wang, J., Jiang, T., Liu, Z., Rao, Y.: Attentional encoder network for targeted sentiment classification. arXiv:1902.09314 (2019)
20. Sun, C., Huang, L., Qiu, X.: Utilizing BERT for aspect-based sentiment analysis via constructing auxiliary sentence. In: Proceedings of the 2019 Conference of the North American Chapter of the Association for Computational Linguistics: Human Language Technologies, Volume 1 (Long and Short Papers), pp. 380–385. Association for Computational Linguistics, Minneapolis, Minnesota (2019). https://doi.org/10.18653/v1/N19-1035, http://aclanthology.org/N19-1035
21. Sun, K., Zhang, R., Mensah, S., Mao, Y., Liu, X.: Aspect-level sentiment analysis via convolution over dependency tree. In: Proceedings of the 2019 Conference on Empirical Methods in Natural Language Processing and the 9th International Joint Conference on Natural Language Processing (EMNLP-IJCNLP), pp. 5679–5688. Association for Computational Linguistics, Hong Kong, China (2019). https://doi.org/10.18653/v1/D19-1569, http://aclanthology.org/D19-1569
22. Tan, X., Cai, Y., Zhu, C.: Recognizing conflict opinions in aspect-level sentiment classification with dual attention networks. In: Proceedings of the 2019 Conference on Empirical Methods in Natural Language Processing and the 9th International Joint Conference on Natural Language Processing (EMNLP-IJCNLP), pp. 3426–3431. Association for Computational Linguistics, Hong Kong, China (2019). https://doi.org/10.18653/v1/D19-1342, http://aclanthology.org/D19-1342
23. Tang, H., Ji, D., Li, C., Zhou, Q.: Dependency graph enhanced dual-transformer structure for aspect-based sentiment classification. In: Proceedings of the 58th Annual Meeting of the Association for Computational Linguistics, pp. 6578–6588. Association for Computational Linguistics, Online (2020). https://doi.org/10.18653/v1/2020.acl-main.588, http://aclanthology.org/2020.acl-main.588
24. Titov, I., McDonald, R.: Modeling online reviews with multi-grain topic models. In: Proceedings of the 17th International Conference on World Wide Web, pp. 111–120. WWW 2008, Association for Computing Machinery, New York, NY, USA (2008). https://doi.org/10.1145/1367497.1367513
25. Truşcă, M.M., Wassenberg, D., Frasincar, F., Dekker, R.: A hybrid approach for aspect-based sentiment analysis using deep contextual word embeddings and hierarchical attention. In: Bielikova, M., Mikkonen, T., Pautasso, C. (eds.) ICWE 2020. LNCS, vol. 12128, pp. 365–380. Springer, Cham (2020). https://doi.org/10.1007/978-3-030-50578-3_25
26. Vaswani, A., et al.: Attention is all you need (2017). http://arxiv.org/pdf/1706.03762.pdf

27. Velickovic, P., Cucurull, G., Casanova, A., Romero, A., Liò, P., Bengio, Y.: Graph attention networks. In: International Conference on Learning Representations (2018). http://openreview.net/forum?id=rJXMpikCZ

28. Wang, Y., Huang, M., Zhu, X., Zhao, L.: Attention-based LSTM for aspect-level sentiment classification. In: Proceedings of the 2016 Conference on Empirical Methods in Natural Language Processing, pp. 606–615. Association for Computational Linguistics, Austin, Texas (2016). https://doi.org/10.18653/v1/D16-1058, http://aclanthology.org/D16-1058

29. Xiao, Z., Wu, J., Chen, Q., Deng, C.: BERT4GCN: using BERT intermediate layers to augment GCN for aspect-based sentiment classification. In: Proceedings of the 2021 Conference on Empirical Methods in Natural Language Processing, pp. 9193–9200. Association for Computational Linguistics, Online and Punta Cana, Dominican Republic (2021). https://doi.org/10.18653/v1/2021.emnlp-main.724, http://aclanthology.org/2021.emnlp-main.724

30. Zhang, C., Li, Q., Song, D.: Aspect-based sentiment classification with aspect-specific graph convolutional networks (2019). https://doi.org/10.48550/ARXIV.1909.03477, http://arxiv.org/abs/1909.03477

31. Zhang, W., Li, X., Deng, Y., Bing, L., Lam, W.: A survey on aspect-based sentiment analysis: tasks, methods, and challenges. IEEE Trans. Knowl. Data Eng., 1–20 (2022). https://doi.org/10.1109/TKDE.2022.3230975

32. Zhang, Z., Zhou, Z., Wang, Y.: SSEGCN: syntactic and semantic enhanced graph convolutional network for aspect-based sentiment analysis. In: Proceedings of the 2022 Conference of the North American Chapter of the Association for Computational Linguistics: Human Language Technologies, pp. 4916–4925. Association for Computational Linguistics, Seattle, United States (2022). https://doi.org/10.18653/v1/2022.naacl-main.362, http://aclanthology.org/2022.naacl-main.362

33. Zhaoa, P., Houb, L., Wua, O.: Modeling sentiment dependencies with graph convolutional networks for aspect-level sentiment classification (2019). https://doi.org/10.48550/ARXIV.1906.04501, http://arxiv.org/abs/1906.04501

Knowledge Injection for Aspect-Based Sentiment Classification

Romany Dekker, Danae Gielisse, Chaya Jaggan, Sander Meijers,
and Flavius Frasincar[✉][iD]

Erasmus University Rotterdam, Burgemeester Oudlaan 50,
3062 PA Rotterdam, The Netherlands
{529424rd,539908dg,538604aj,530192sm}@student.eur.nl,
frasincar@ese.eur.nl

Abstract. Since the increase of Web reviews of products and services,
Aspect-Based Sentiment Classification (ABSC) has become more important to determine the sentiment of online opinions. Useful information
extracted from these reviews can then be used by companies themselves,
but can also be applicable by consumers. In the recent literature on
ABSC, hybrid methods, which combine knowledge-based and machine
learning approaches, are becoming more popular as well. However, in this
work, instead of following a two-step procedure, we attempt to improve
the model accuracy by proposing to directly inject the information from
a domain ontology in a state-of-the-art neural network model, more precisely LCR-Rot-hop++. Furthermore, by using soft-positioning and visible matrices we aim to prevent that the injected knowledge hinders
the semantics of the original sentences. To evaluate the accuracy of our
model, LCR-Rot-hop-ont++, we use the standard SemEval 2015 and
SemEval 2016 datasets for ABSC. We conclude that knowledge injection
in the neural network is effective for sentiment classification, especially
if the amount of labeled data is limited.

Keywords: LCR-Rot-hop-ont++ · ABSC · Knowledge Injection

1 Introduction

In an era where life seems to be moving more and more towards the online
dimension, information has as well moved to the online domain. Taking form in
Web reviews of products and services, enormous amounts of useful information
are up for grabs. When the size of this information gets enormously large, the
sheer size of data requires automatic approaches in order to retrieve people's
opinion from their reviews. One proposal for an automatic approach for gathering
these opinions is called sentiment analysis [6], which determines the sentiment
expressed in a text.

A discipline within sentiment analysis is Aspect-Based Sentiment Analysis
(ABSA) [15]. Here, the overall sentiment of an entity is determined by identifying the sentiment of the aspects of that entity. ABSA usually consists of two

C. Strauss et al. (Eds.): DEXA 2023, LNCS 14147, pp. 173–187, 2023.
https://doi.org/10.1007/978-3-031-39821-6_14

steps: finding the aspects present in the sentences and determining the sentiments associated to these aspects. The sentiments are often defined as positive, neutral, or negative. For example, in the sentence "I was very disappointed with this restaurant." the aspect is "restaurant" and the sentiment is classified as negative. In this research, we assume the aspect to be given and therefore focus only on the sentiment classification part of the ABSA, also known as Aspect-Based Sentiment Classification (ABSC) [3]. To efficiently and accurately predict the sentiment of the given aspects, knowledge-based methods and machine learning approaches are generally used. Because a combination of both techniques outperforms the models that only use one technique [16], a hybrid approach is considered [2].

One hybrid approach is HAABSA++ [17] which increased the accuracy of sentiment classification w.r.t. its predecessor the Hybrid Approach for Aspect-Based Sentiment Analysis (HAABSA) [19] by introducing two extensions. First, in HAABSA++ the non-contextual GloVe [11] word embeddings are replaced by deep contextual word embeddings like BERT [4]. Furthermore, a hierarchical attention is implemented ensuring that the high-level input sentence representations are tuned to each other. HAABSA++ is a two-step model which first uses a domain ontology to estimate the sentiment. When this turns out to be inconclusive, a back-up algorithm is used taking form as the state-of-the-art neural network LCR-Rot-hop++ which has a rotatory attention mechanism based on LCR-Rot [23] and a hopping (repetition) mechanism defined by its successor LCR-Rot-hop [19].

Different than in [17] where the ontology is used before the back-up neural network, in this work we plan to inject the knowledge of the domain ontology in the neural network itself. We call the new model LCR-Rot-hop-ont++, referring to the ontology now being incorporated in the neural network LCR-Rot-hop++. We know that such an approach has been used for contextual word embeddings, i.e., K-BERT [8], for language representation, but very little is known about applying this method to the neural models (in our case LCR-Rot-hop++) in the context of ABSC. We approach our knowledge injection inspired by K-BERT, where the target text is first augmented with domain knowledge before being passed on to the neural network, providing for a richer context. Furthermore, our research contributes to the previous work in the following way. We make use of a domain ontology instead of a knowledge graph to inject knowledge. Furthermore, we follow a multi-hop approach where different concepts, which are semantically close to the original words, from the domain ontology are injected into the sentences. The Python source code of our model has been made publicly available at https://github.com/DanaeGielisse/LCR-Rot-hop-ont-plus-plus.

The rest of the paper is organized as follows. In Sect. 2 relevant related works regarding sentiment analysis and knowledge injection are discussed. Next, Sect. 3 presents the used data for experiments and Sect. 4 provides a detailed description of the proposed method. Furthermore, in Sect. 5 the results are presented and discussed. Last, Sect. 6 gives the conclusions of our research and provides suggestions for future work.

2   Related Work

In this section we discuss the research that is relevant to our work. First, Sect. 2.1 presents hybrid methods for ABSC. Second, in Sect. 2.2 the literature regarding knowledge injection in neural models is outlined.

2.1   Hybrid Models for ABSC

Hybrid models [2] combine the symbolic with sub-symbolic approaches in classifying sentiment. The former makes use of ontologies and lexicons to determine if words are positive, neutral, or negative, and the latter includes machine learning techniques that perform sentiment classification [5]. Several researches concluded that the use of hybrid models for ABSC leads to an increase in the performance of sentiment prediction [9,10,16,19].

ALDONA [9] and its extension ALDONAr [10] are among those researches which use a hybrid approach of ABSC on a sentence-level. Both models determine the polarity of the aspect by using a lexicalized domain ontology and the statistical relations captured by a regularized neural attention model. ALDONAr improves the results of ALDONA by using BERT word embeddings among other things. Furthermore, these models also outperform state-of-the-art models such as CABASC [7].

[17] introduced the HAABSA++ model which is a two-step hybrid model for ABSC. In the first step, a domain sentiment ontology is employed in order to predict the target polarities. If this ontology leads to inconclusive results (e.g., due to conflicting sentiments), the neural network LCR-Rot-hop++ is used. LCR-Rot-hop++ extends the neural network LCR-Rot-hop of HAABSA [19] in two ways. First, deep contextual word embeddings are taken into account, namely ELMo [12] and BERT [4] (better results are obtained with BERT), in order to better deal with word semantics in text. Second, hierarchical attention is used by adding an extra attention layer to the HAABSA high-level representations. Therefore, the method is more flexible to model the input data. It was proved that the fourth method, out of three other methods, gives the highest accuracy, where for each iteration the rotatory attention weighting was applied separately on the intermediate context and target vectors pairs.

In this work, we directly include the information from the domain ontology in the neural network, which results in the model LCR-Rot-hop-ont++. Hence, the ontology is now part of the neural network and thus our model does not follow a two-step approach. More precisely, we account for our whole domain knowledge by replacing the BERT word embeddings with the word embeddings from K-BERT [8]. We also devise a novel multi-hop approach for embedding domain knowledge using K-BERT into the neural network. Last, we compare the performance of our model with HAABSA++ and LCR-Rot-hop++.

2.2   Knowledge Injection in Neural Networks

One of the first works to inject knowledge in a deep neural network is the knowledge-enhanced language representation model K-BERT [8]. This is an extension of the BERT model which is a language representation model that takes contextual information into account [4]. Both the architectures of BERT and K-BERT are based on the Transformer [18]. However, K-BERT outperforms BERT on domain-specific tasks by injecting knowledge obtained from a Knowledge Graph (KG), which could be described as a graph structure modeling a domain, into sentences. Although, this has to be done with care as too much knowledge could be injected which could change the meaning of the original input sentence. This undesired information is called Knowledge Noise (KN). K-BERT could be made more robust to KN in two ways. First, soft-positioning retains the sequential structure of the original sentence. Second, visible matrices control the visible area of each word in the sentence by blocking the words from the injected knowledge except from the corresponding subject.

[22] proposed to use a knowledge-enhanced BERT model for ABSC. The authors utilize a Sentiment Knowledge Graph (SKG) as an external source to increase the accuracy of sentiment detection by injecting domain knowledge into BERT. Like our research, [22] employs the BERT component and injects the information from the SKG into the input sentences and then uses these knowledge enhanced embeddings for ABSC. However, in [22] the knowledge is injected in the same way as in the K-BERT model, namely, in the form of candidate triples which are pairs of concepts and their associated relation. Differently, in our approach we inject the lexical representations obtained from a domain ontology. After comparing the model BERT+SKG to another external knowledge-enhanced model and several self-attention based models, the authors concluded that the BERT+SKG is effective for the ABSC task.

[21] also makes use of a KG to enhance the language representation model. The authors use the model SAKG-BERT, which utilizes the information from Sentiment Analysis Knowledge Graph (SAKG) and the language model BERT. Contrary to our research, the authors focus on the extraction of knowledge triples from the SAKG. Then, the triples are injected as domain knowledge into the input sentences. Also different to our research, the authors have constructed their own KG that is related to online reviews. It was found that the use of a KG shows promising results for sentiment analysis tasks and sentence completion.

In our research, we use an approach based on K-BERT to inject knowledge into a neural network for ABSC instead of a general language model. We, however, inject the information from a domain ontology in the form of lexical representations (words) associated to ontology concepts. Furthermore, compared to previous work, we follow a multi-hop approach for knowledge injection. We only opt to perform knowledge injection at test time. This is because employing such an approach at training time might result in semantic loss, i.e., the model not capturing the meaning conveyed in the original sentences.

3   Data

To evaluate the performance of our model, we have used the SemEval 2015 Task 12 [14] and SemEval 2016 Task 5 [13] datasets. These datasets are often used in research relating to ABSC which makes them convenient for comparing the performance of our model to that of other models. The data contains customers reviews about restaurants which can consist of multiple sentences. Furthermore, the sentences express one or more opinions about the provided aspects. Each aspect is assigned to a predefined category that is labeled as positive, neutral, or negative. In Fig. 1 an example of a customer review from the SemEval 2016 dataset is presented. It is shown that a sentence can have multiple targets, here "food", "drinks", and "atmosphere". Furthermore, the polarities and categories for each target are listed.

```
<sentence id="1609375:1">
    <text>We love the food, drinks, and atmosphere!</text>
    <Opinions>
        <Opinion from="12" to="16" polarity="positive"
            category="FOOD#QUALITY" target="food"/>
        <Opinion from="18" to="24" polarity="positive"
            category="DRINKS#QUALITY" target="drinks"/>
        <Opinion from="30" to="40" polarity="positive"
            category="AMBIENCE#GENERAL" target="atmosphere"/>
    </Opinions>
</sentence>
```

Fig. 1. Example of a sentence from the SemEval 2016 dataset.

For training we have used 1278 and 1880 reviews and for testing 597 and 650 reviews of the SemEval 2015 and SemEval 2016 dataset, respectively. Table 1 shows the frequencies of the positive, neutral, and negative targets of the SemEval 2015 Task 12 and SemEval 2016 Task 5 datasets. For both datasets it holds that the targets are valued the most as positive, followed by negative, and neutral.

Table 1. Polarity frequencies of the training and test data of the SemEval 2015 and SemEval 2016 datasets for ABSC.

	Training Data			Test Data		
	Positive	Neutral	Negative	Positive	Neutral	Negative
SemEval 2015	75.3%	2.8%	21.9%	59.1%	6.1%	34.8%
SemEval 2016	70.2%	3.8%	26.0%	74.3%	4.9%	20.8%

To inject knowledge into our input sentences we have used the information obtained from a restaurant domain ontology [16]. An ontology describes a set of concepts and the relations between these concepts. This ontology is divided in multiple classes and subclasses regarding the sentiments and targets. The subclasses of the targets are, for example, service, restaurant, and ambience. The sentiment class only consists of the subclasses positive, neutral, and negative. We have especially focused on the synonyms or different spellings of a certain concept (called concept lexical representations) that the ontology provides. For instance, for the concept "Outside" the ontology gives the lexical representations "outdoor", "outside", and "outdoors". However, related concepts could also be found in the ontology, such as "chef" being a direct subclass of "staff".

Furthermore, in our research we make use of the pre-trained BERT word embeddings [4]. In particular, we use the uncased BERT base model which has 12 transformer layers, 768 as hidden size, and 12 self-attention heads and was pre-trained on BookCorpus and English Wikipedia.

4   Methodology

In this section we explain our model. As an extension of the LCR-Rot-hop++ model we propose to use the information from a domain ontology to enrich the word embeddings. We call this model LCR-Rot-hop-ont++. First, in Sect. 4.1 we describe how the LCR-Rot-hop++ model works. Furthermore, in Sect. 4.2 we explain how we inject the knowledge from the domain ontology in the test sentences using K-BERT. The methodology was implemented in Python (version 3.7) with the HuggingFace's Transformer [20] and Google's TensorFlow [1] packages.

4.1   LCR-Rot-hop++

LCR-Rot-hop++ builds further on the model proposed in [17] which uses contextual word embeddings as input for three bidirectional Long Short-Term Memory (Bi-LSTM) models. Bi-LSTMs are a particular Recurrent Neural Network (RNN) that are able to incorporate long-term dependencies in their learning. In Fig. 2 a visualization of the LCR-Rot-hop-ont++ model is presented. This figure is split into two parts. The lower part of this figure is inspired by K-BERT. On the left side the training process is displayed and on the right side the testing. For training we just use the input sentence. However, for testing we also use the input sentence and the information from the domain ontology to create a sentence tree and contextualized word embeddings. We describe this step more extensively in Sect. 4.2.

The procedure described in the upper part of Fig. 2 goes as follows. First the sentence $s = [w_1, w_2, ..., w_N]$ is split into three parts: the left context $s^l = [w_1^l, w_2^l, ..., w_L^l]$, the target phrase $s^t = [w_1^t, w_2^t, ..., w_T^t]$, and the right context $s^r = [w_1^r, w_2^r, ..., w_R^r]$. Hereby, N is the number of words in sentence s, and L, T, and R are the number of tokens in each part of the sentence, respectively. These three parts are then embedded together using BERT for training data and K-BERT for test data.

The embeddings are then used as input for three separate Bi-LSTMs and give three hidden states h for each part, namely: $[h_1^l, h_2^l, ..., h_L^l]$ for the left context,

$[h_1^t, h_2^t, ..., h_T^t]$ for the target phrase, and $[h_1^r, h_2^r, ..., h_R^r]$ for the right context. Next, the three Bi-LSTMs hidden states are used in a two-step rotatory mechanism which creates in the first step the new context representations using target information. In the first iteration, we use an average pooling mechanism to obtain the target representation $r^{t_p} = \text{pooling}[h_1^t, h_2^t, ..., h_T^t]$.

Furthermore, we calculate the attention scores for each word using a softmax function in order to compute the new context and target representations. We compute the representation of the left context for $i = 1, ..., L$ as follows:

$$f(h_i^l, r^{t_p}) = \tanh(h_i^{l'} \times W_c^l \times r^{t_p} + b_c^l), \tag{1}$$

where W_c^l is a weight matrix and b_c^l is the bias for the left context l.

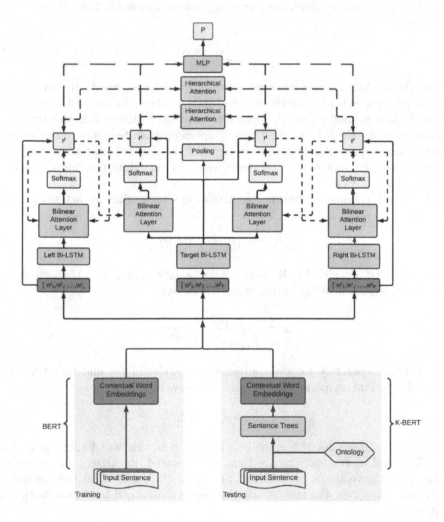

Fig. 2. Visualization of LCR-Rot-hop-ont++.

Next, we normalize the attention scores by using a softmax function, where $i = 1, ..., L$:

$$\alpha_i^l = \frac{\exp(f(h_i^l, r^{t_p}))}{\sum_{j=1}^{L} \exp(f(h_j^l, r^{t_p}))}. \tag{2}$$

Furthermore, we calculate the representation of the left context as:

$$r^l = \sum_{i=1}^{L} \alpha_i^l \times h_i^l. \tag{3}$$

In a similar way, we obtain the representation of the right context. In the second step of this rotary mechanism, we determine the target representations w.r.t the left and right context. Here, the target representation w.r.t. the left context is given by:

$$r^{t_l} = \sum_{i=1}^{T} \alpha_i^{t_l} \times h_i^t, \tag{4}$$

where the attention scores are computed between target words and previously computed representations of the left and right context. In the first iteration of the multi-hop method we use the pooled target vector. However, in the iterations thereafter we use the left and right target representations to compute the left and right context representations, respectively. We then use a hierarchical attention model to further improve the representations by including the information on sentence-level as well.

First, we calculate the normalized attention scores using the attention function in (1) for $i = 1, ..., 4$:

$$f(v^i) = \tanh(v^{i'} \times W + b), \tag{5}$$

where $v^i \in \{r^l, r^{t_l}, r^{t_r}, r^r\}$, W is a weight matrix, and b is a bias. Next, we compute the normalized attention scores as follows:

$$\alpha_i = \frac{\exp(f(v^i))}{\sum_{j=1}^{2} \exp(f(v^j))}. \tag{6}$$

This is done separately for the context vectors (r^l and r^r) and target vectors (r^{t_l} and r^{r_l}). Furthermore, we determine the new representations:

$$v_{new}^i = v_{old}^i \times \alpha^i. \tag{7}$$

We use $n = 3$ hops for the multi-hop approach as this was found optimal in [17]. The final representations are then concatenated, linearized, and squashed to compute the vector of probabilities of length $|C|$, where $|C|$ is the amount of sentiment categories. For this we use the vector v obtained by concatenation of four vectors:

$$v = [r^l; r^{t_l}; r^{t_r}; r^r], \tag{8}$$

Then, after applying a linear layer this is sent into the softmax function in order to predict the sentiment probability:

$$p = \text{softmax}(W_c \times v + b_c), \tag{9}$$

where p is the conditional probability distribution, W_c is a weight matrix, and b_c is the bias for class c.

To obtain the model parameters that best capture the semantics in the training data, we minimize the cross-entropy loss function which measures the difference in the estimated probability and the true value of the sentiment classification.

In addition, we penalize the complexity of the system by using L_2-regularization. The loss formula is:

$$Loss = -\sum_j y_j \cdot \log(\hat{p}_j) + \lambda ||\theta||_2^2, \tag{10}$$

where y_j is a $|C| \times 1$ vector of the true sentiment classification values, $\hat{p}_j$ is the $|C| \times 1$ vector of sentiment classification predictions, λ is the L_2-regularization term that penalizes the squared values of the model parameters, and θ is the vector containing all the model parameters, these are the LSTM parameters and $\{W_c^l, b_c^l, W_c^r, b_c^r, W_t^l, b_t^l, W_t^r, b_t^r, W_c, b_c, W, b\}$. The squaring of the model parameters ensures that we penalize the large model parameters more than the small model parameters.

4.2   Knowledge Injection

We inject knowledge from a domain ontology which represents concepts and relations between these concepts. Furthermore, a concept has one or more lexical representations attached which denote synonyms. We define a k-hop as navigating k classes from the current class in the ontology in order to extract lexical representations, for $k = 0, ..., N - 1$ with N being the total number of classes. In this case, performing a 0-hop points to synonyms which are then injected into the test input sentence, because we do not have to navigate through the ontology.

We inject the information from the ontology according to the following procedure. First in the K-Query step, we determine for each k-hop all the lexical representations. Furthermore, $w_{i,k}$ denotes the word i in the sentence for which the lexical representations of the subclasses were obtained by navigating k-hops from the current class in ontology O (domain-specific relations from the ontology are ignored in this study due to the possibility of divergent semantics w.r.t. the current concept). Second, the words corresponding to the concepts are added to sentence s. For sentence s, the selection of these words could then be written as:

$$E = \text{K-Query}(s, O, k). \tag{11}$$

The collection of a k-hop is then:

$$E = \cup_{w_i \in s}[(w_i \cup \{w_{i,0}\} \cup ...\{w_{i,k}\})], \tag{12}$$

whereby $\{w_{i,n}\}$ represents the set of all words obtained from word i using n hops in the ontology.

The second step is called K-Inject, where we create a sentence tree t by injecting the queried concepts of the k-hops to the corresponding positions. Next, we enter the embedding layer where we input the new sentence tree and get as output an embedding representation. Since the sentence tree is incompatible with the embedding layer of BERT, we first have to create a new sentence tree. We inject the words, obtained through the concepts by navigating k-hops, directly after the corresponding words in the sentence. Note that if a word from the original sentence is also listed in the ontology, we do not inject this word again in the sentence. We consider injecting synonyms by adding "soft branches" to our sentence tree. We define "hard branches" for the words obtained through the concepts of the other k-hops. Hereby, we only add branches of length one. In Fig. 3 an example of a sentence tree is shown. The original word in the sentence "staff" has a synonym, namely "crew". Furthermore, the words attached to the "hard branches" which represents the 1-hops, are "Chef", "Waiter", and "Host". Also, the words obtained by performing a 1-hop could have synonyms themselves given by corresponding "soft branches".

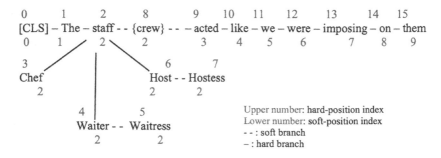

Fig. 3. Example of a sentence tree.

The embedding representation is a sum of three parts: the token embedding, soft-position embedding, and the segment embedding. The token and segment embeddings are the same as in BERT. However, the main difference is in the soft-position embedding, because we cannot input the new sentence tree into the embedding layer. As mentioned, we have to place the words obtained through the concepts of performing k-hops in the original sentence at specific places which ruins the structure of the original sentence. This problem could be solved by using soft-positioning in which the lexical representations are given the same soft-position as the original word in the sentence, displayed by the red numbers (below text numbers for black and white printing) in Fig. 3. This way the structural information of the sentence is retained, however, this might also lead to false semantic information.

To solve the previously identified problem, we use the visible matrix M. This makes sure that the k-hop of the original word does not affect the hidden state value of this word that is not in the same branch of the sentence tree. We create a visible matrix by using the hard-position indexes, displayed by the gray numbers (above text numbers for black and white printing) in Fig. 3. We define the visible matrix as:

$$M_{ij} = \begin{cases} 0, & \text{if } w_i \oplus w_j \\ -\infty, & \text{if } w_i \otimes w_j \end{cases}, \tag{13}$$

where $\oplus$ means that the words are in the same branch and thus are visible to each other and $\otimes$ means the words are not visible to each other. Furthermore, i and j are the hard-position indexes. Thus, it holds that all words in the original sentence are visible to each other. Furthermore, the synonyms are only visible to the words they belong to and the words obtained from performing the k-hops are not able to see each other unless they are in the same branch. In Fig. 4 an example of a visible matrix is shown which has as input the sentence tree of Fig. 3. The numbers on the sides are the hard-position indexes, the red dots (gray dots for black and white printing) indicate that the words are visible to each other and the white dots indicate that they are invisible. For example, the word "staff" of the original sentence with hard-position 2, is visible to the words obtained by performing k-hops and the other words of the original sentence. It is, however, unable to see the synonyms "Waitress" and "Hostess" with hard-positions 5 and 7, respectively.

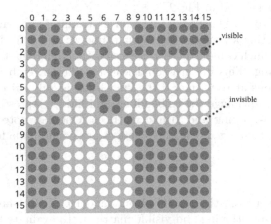

Fig. 4. Example of a visible matrix.

Next, we apply the BERT-encoder where we change the multi-head attention block by adding the visible matrix M in (15). The input of the first multi-head attention block of the first layer is $h^0 = \text{sum}(E_t, E_s, E_{seg})$, where E are the embeddings of the tokens, soft-positions, and segments, respectively. The input plus output of this block is then the input for the first normalization layer in the BERT-encoder. Furthermore, the output of this first normalization layer is then the input of the feed-forward layer. Then, the input plus output of the feed-forward layer is used as the input of the second normalization layer. The output of the second normalization layer is then used as the input for the next encoder layer. This process is repeated until we arrive at the last encoder layer and set that output as hidden states for all tokens. These steps are repeated for all twelve encoder layers. In the next three equations the multi-head attention block is given where we add the visible matrix M (13).

$$Q^{i+1}, K^{i+1}, V^{i+1} = h^i W_q, h^i W_k, h^i W_v, \tag{14}$$

$$S^{i+1} = \text{softmax}\left(\frac{Q^{i+1}K^{i+1^\top} + M}{\sqrt{d_k}}\right), \tag{15}$$

$$h^{i+1} = S^{i+1}V^{i+1}, \tag{16}$$

where Q, K, and V are matrices of queries, keys, and values, respectively. Furthermore, h^i and h^{i+1} denote the input and output of the multi-head attention block, respectively. In addition, W_q, W_k, and W_v are trainable model parameters and d_k is a scaling factor. Now, we are able to identify that if the words k and j are not in the same branch, S_{kj}^{i+1} is set to zero. This means that word k does not make a contribution to the hidden state of word j. This process is repeated for the whole sentence. Next, the sum of the outputs of the last four encoder layers are split into the three parts that correspond to the left context, target, and right context of the sentence. These are then each used as input for the Bi-LSTMs as was showed in Fig. 2.

It is possible to do knowledge injection at training time and at test time. However, by injecting knowledge at training time we add lexical representations to the sentences which could lead to sentences with a different structure and thus a different meaning. This in turn could lead to different embeddings and thus different predictions at test time. If the knowledge is only injected at test time then we just have to change the embeddings and hidden states in K-BERT with the use of soft-positioning and visible matrices. One can note that the model training remains unchanged, the only changes are being made for testing.

5 Results

In the section we report the results of our model LCR-Rot-hop-ont++ which makes use of soft-positioning and visible matrices. To evaluate the performance of our model we have used the SemEval 2015 Task 12 and SemEval 2016 Task 5 datasets. Furthermore, we compare the accuracy of our model with the accuracy of two benchmarks. These are HAABSA++ and LCR-Rot-hop++. In this paper we adopt a conservative approach and focus on 0-hops as these affect the least the semantics of the original sentences. By performing 0-hops, synonyms are injected into the input sentences. In Table 2 the test accuracies of the LCR-Rot-hop-ont++ model and its benchmarks are presented for both the SemEval 2015 and SemEval 2016 datasets. We do not replicate the HAABSA++ and LCR-Rot-hop++ work and thus give the results as reported in the previous work [17].

In Table 2 is given that for the SemEval 2015 dataset LCR-Rot-hop-ont++ outperforms HAABSA++ and LCR-Rot-hop++ in terms of test accuracy by 0.2% and 0.8%, respectively. In addition, for the SemEval 2016 dataset LCR-Rot-hop-ont++ outperforms LCR-Rot-hop++ by 0.2% points, but LCR-Rot-hop-ont++ has a lower performance than HAABSA++ (0.1% points). As SemEval 2015 is a smaller dataset than SemEval 2016 we conclude that our proposed approach helps boost the performance of a neural model where for smaller datasets knowledge injection is preferred over the hybrid approach.

Table 2. Comparison of LCR-Rot-hop-ont++ to its benchmarks using accuracy (the bold font indicates the best results).

	SemEval 2015	SemEval 2016
LCR-Rot-hop-ont++	**81.9%**	86.9%
HAABSA++	81.7%	**87.0%**
LCR-Rot-hop++	81.1%	86.7%

6   Conclusion

In this research, we propose the LCR-Rot-hop-ont++ model for performing Aspect-Based Sentiment Classification (ABSC) on a sentence-level of restaurant reviews. The model is formed by injecting the information from a domain ontology into the state-of-the-art neural network model LCR-Rot-hop++ in our test instances. This approach of knowledge injection is based on the language representation model K-BERT which also makes use of soft-positioning and visible matrices. These are used to control the scope of the injected knowledge and thus help to retain the meaning of the original sentences. Furthermore, the performance of the LCR-Rot-hop-ont++ model was evaluated by using the standard SemEval 2015 and SemEval 2016 datasets for ABSC. We have compared our model to its benchmarks, namely HAABSA++ (ont+LCR-Rot-hop++) and LCR-Rot-hop++.

Hereby, we conclude that LCR-Rot-hop-ont++ boosts the testing accuracy for the SemEval 2015 dataset, namely from 81.7% for HAABSA++ and 81.1% for LCR-Rot-hop++ to 81.9%. Furthermore, for the SemEval 2016 dataset we conclude that LCR-Rot-hop-ont++ outperforms LCR-Rot-hop++, but does not outperform HAABSA++. Therefore, we conclude that knowledge injection is effective for the neural network to perform sentiment classification, especially for smaller datasets.

For future work, we propose to further explore the amount of information that can be injected from the ontology into the neural network. We have only researched knowledge injection for a 0-hop, but perhaps injecting more information could boost the performance of the neural network more for sentiment classification. In addition, we could try different percentages of knowledge injection, e.g., 20%, 40%, 60%, 80%, and 100%. Furthermore, we could analyze the characteristics of sentences that can benefit the most from knowledge injection. For example, it is expected that shorter sentences could benefit the most from knowledge injection for building additional context information.

References

1. Abadi, M., et al.: TensorFlow: large-scale machine learning on heterogeneous systems. In: 12th USENIX Symposium on Operating Systems Design and Implementation (OSDI 2016), pp. 265–283. USENIX Association (2016)
2. van Bekkum, M., de Boer, M., van Harmelen, F., Meyer-Vitali, A., ten Teije, A.: Modular design patterns for hybrid learning and reasoning systems. Appl. Intell. **51**(9), 6528–6546 (2021)
3. Brauwers, G., Frasincar, F.: A survey on aspect-based sentiment classification. ACM Comput. Surv. **55**(4), 65:1–65:37 (2023)
4. Devlin, J., Chang, M.W., Lee, K., Toutanova, K.: BERT: pre-training of deep bidirectional transformers for language understanding. In: 17th Conference of the North American Chapter of the Association of Computational Linguistics: Human Language Technologies (NAACL-HLT 2019), pp. 4171–4186. ACL (2019)
5. Dragoni, M., Poria, S., Cambria, E.: OntoSenticNet: a commonsense ontology for sentiment analysis. IEEE Intell. Syst. **33**(3), 77–85 (2018)
6. Liu, B.: Sentiment Analysis: Mining Opinions, Sentiments, and Emotions. Cambridge University, 2nd edn. (2020)
7. Liu, Q., Zhang, H., Zeng, Y., Huang, Z., Wu, Z.: Content attention model for aspect based sentiment analysis. In: 27th International World Wide Web Conference (WWW 2018), pp. 1023–1032. ACM (2018)
8. Liu, W., et al.: K-BERT: enabling language representation with knowledge graph. In: 34th AAAI Conference on Artificial Intelligence (AAAI 2020), vol. 34, pp. 2901–2908. AAAI (2020)
9. Meškelė, D., Frasincar, F.: ALDONA: a hybrid solution for sentence-level aspect-based sentiment analysis using a lexicalised domain ontology and a neural attention model. In: 34th ACM Symposium on Applied Computing (SAC 2019), pp. 2489–2496. ACM (2019)
10. Meškelė, D., Frasincar, F.: ALDONAr: a hybrid solution for sentence-level aspect-based sentiment analysis using a lexicalized domain ontology and a regularized neural attention model. Inf. Process. Manage. **57**(3), 102211 (2020)
11. Pennington, J., Socher, R., Manning, C.D.: GloVe: global vectors for word representation. In: 2014 Conference on Empirical Methods in Natural Language Processing (EMNLP 2014), pp. 1532–1543. ACL (2014)
12. Peters, M.E., et al.: Deep contextualized word representations. In: 2018 Conference of the North American Chapter of the Association for Computational Linguistics-Human Language Technologies (NAACL-HLT 2018), pp. 2227–2237. ACL (2018)
13. Pontiki, M., et al.: SemEval-2016 task 5: aspect-based sentiment analysis. In: 10th International Workshop on Semantic Evaluation (SemEval 2016), pp. 19–30. ACL (2016)
14. Pontiki, M., Galanis, D., Papageorgiou, H., Manandhar, S., Androutsopoulos, I.: SemEval-2015 task 12: aspect-based sentiment analysis. In: 9th International Workshop on Semantic Evaluation (SemEval 2015), pp. 486–495. ACL (2015)
15. Schouten, K., Frasincar, F.: Survey on aspect-level sentiment analysis. IEEE Trans. Knowl. Data Eng. **28**(3), 813–830 (2015)

16. Schouten, K., Frasincar, F., de Jong, F.: Ontology-enhanced aspect-based sentiment analysis. In: Cabot, J., De Virgilio, R., Torlone, R. (eds.) ICWE 2017. LNCS, vol. 10360, pp. 302–320. Springer, Cham (2017). https://doi.org/10.1007/978-3-319-60131-1_17

17. Truşcă, M.M., Wassenberg, D., Frasincar, F., Dekker, R.: A hybrid approach for aspect-based sentiment analysis using deep contextual word embeddings and hierarchical attention. In: Bielikova, M., Mikkonen, T., Pautasso, C. (eds.) ICWE 2020. LNCS, vol. 12128, pp. 365–380. Springer, Cham (2020). https://doi.org/10.1007/978-3-030-50578-3_25

18. Vaswani, A., et al.: Attention is all you need. In: 31st Annual Conference on Neural Information Processing Systems (NIPS 2017), vol. 30, pp. 5998–6008. Curran Associates (2017)

19. Wallaart, O., Frasincar, F.: A hybrid approach for aspect-based sentiment analysis using a lexicalized domain ontology and attentional neural models. In: Hitzler, P., et al. (eds.) ESWC 2019. LNCS, vol. 11503, pp. 363–378. Springer, Cham (2019). https://doi.org/10.1007/978-3-030-21348-0_24

20. Wolf, T., et al.: Transformers: state-of-the-art natural language processing. In: 2020 Conference on Empirical Methods in Natural Language Processing: System Demonstrations (EMNLP 2020), pp. 38–45. ACL (2020)

21. Yan, X., Jian, F., Sun, B.: SAKG-BERT: enabling language representation with knowledge graphs for Chinese sentiment analysis. IEEE Access 9, 101695–101701 (2021)

22. Zhao, A., Yu, Y.: Knowledge-enabled BERT for aspect-based sentiment analysis. Knowl.-Based Syst. 227, 107220 (2021)

23. Zheng, S., Xia, R.: Left-center-right separated neural network for aspect-based sentiment analysis with rotatory attention. arXiv preprint arXiv:1802.00892 (2018)

Towards Ensemble-Based Imbalanced Text Classification Using Metric Learning

Takahiro Komamizu[✉][ID]

Nagoya University, Nagoya, Japan
taka-coma@acm.org

Abstract. This paper reports s series of ensemble approaches for imbalance text classification. All the approaches utilize a metric learning technique for obtaining better representations of texts to train weak classifiers. Each approach deals with the class imbalance problem with an undersampling-based ensemble approach, because metric learning techniques also suffer from this problem. In this paper, four ensemble approaches (namely, MLBagging, MLBoosting, MLStacking, and MLBoostacking) are proposed, three of which are corresponding to ensemble frameworks (namely, bagging, boosting, and stacking), and the other is a combination of boosting and stacking. MLBagging, MLBoosting, and MLStacking train metric learners on the individual undersampled dataset and combine them, while MLBoostacking trains metric learners in a step-by-step manner; that is, a metric learner learns a feature transformation so that failed-to-classify samples in the previous step should be correctly classified. The experimental evaluation on three imbalanced text classification datasets (namely, unfair statement classification in terms of service, hate speech detection in a forum, and hate speech tweet detection) shows that the proposed approaches lift classification performance from BERT-based approaches, by improving the representations of texts through metric learning.

Keywords: Imbalanced Classification · Metric Learning · Ensemble

1 Introduction

Text classification [16,22] is a fundamental problem in a wide range of research areas. This problem is to assign classes defined in advance for given texts. Text classification has been studied for a while, and its major issue is representations of texts. A breakthrough of the representations of texts is the appearance of contextualized embeddings of texts, especially, Transformer [30] has a great impact and is used for a wide variety of approaches such as BERT (Bidirectional Encoder Representations from Transformers) [4], RoBERTa (Robustly Optimized BERT Pretraining Approach) [19], and DeBERTa (Decoding-enhanced BERT with Disentangled Attention) [10]. These approaches realize to represent semantic relationships among words on an embedding space.

C. Strauss et al. (Eds.): DEXA 2023, LNCS 14147, pp. 188–202, 2023.
https://doi.org/10.1007/978-3-031-39821-6_15

From the viewpoint of classification, the class imbalance is a crucial problem in real-world applications, which degrades classification performance, especially on minority classes. The class imbalance refers to a situation with datasets in which the number of instances in a class is much larger than that in other classes. This large difference of the numbers of instances causes classifiers to be biased toward the majority class. Text classification tasks also suffer from the class imbalance problem, such as the unfair statement prediction in terms of service [17], the hate speech detections [8,32].

In this paper, learning representations of texts is a focused issue in the imbalanced text classification. One straightforward text classification method is to fine-tune the aforementioned Transformer-based models using class-imbalanced datasets. However, it does not show good results, that will be shown in Sect. 4. Another approach is to train classifiers modeled for the class imbalance (e.g., EasyEnsemble [18]) based on the representations obtained from these models. Section 4 shows that this approach does not show good results either. These two results indicate that classification methodology is not the main issue for the imbalanced text classification problem. Instead, representations obtained from pre-trained models can be a major issue for the imbalanced text classification.

In this paper, a series of ensemble approaches (called **MLEnsemble**) is proposed to tackle imbalanced text classification for incorporating metric learning to learn appropriate representations of texts. The basic idea of MLEnsemble inherits that of existing approaches [13,33], that is an undersampling-based ensemble with metric learning. MLEnsemble includes four approaches, namely, **MLBagging**, **MLBoosting**, **MLStacking**, and **MLBoostacking**, each of which corresponds with the ensemble frameworks; MLBagging is based on a bagging ensemble framework; MLBoosting is based on a boosting ensemble framework; MLStacking is based on a stacking ensemble framework; and MLBoostacking is a combination of boosting and stacking frameworks.

In an experimental evaluation, MLEnsemble is examined on three imbalanced text classification datasets (namely, unfair statement classification in terms of service, hate speech detection in a forum, and hate speech tweet detection). It is compared with the Transformer-based approaches including BERT [4], advanced models like DeBERTa [10] and DistilRoBERTa (distilled version of RoBERTa) [19,26], and pretrained models for dedicated domains such as Legal-BERT [3]. The experiment shows that MLEnsemble outperforms the Transformer-based approaches in terms of precision, recall, f-measure, and geometric mean (*gmean* for short) of true positive rate and true negative rate. This superiority of MLEnsemble to the existing approaches means that it lifts up classification performance from Transformer-based approaches, by improving the representations of texts through metric learning.

In summary, this paper includes the following contributions.

- **Feasibility of Ensemble Frameworks**: The proposed method, MLEnsemble, is a series of approaches each of which takes advantage of an undersampling-based ensemble and metric learning. In contrast to the existing approaches, MLEnsemble covers wide variety of ensemble frameworks, namely, bagging, boosting, and stacking.

- **Lifting-up Classification Performance of Imbalanced Text Classification**: Transformer-based models (e.g., BERT) have been succeeded in many natural language processing tasks, however, their classification performance is limited in imbalanced text classification. To cope with this difficulty, MLEnsemble transforms representations of texts obtained from these models by using the metric learning technique. In an experiment in this paper, MLEnsemble shows better performances than Transformer-based models. This fact indicates that MLEnsemble successfully lifts the classification performance via transforming representations of texts so that texts in majority classes and those in minority classes are separable.

The rest of this paper is organized as follows. Section 2 gives a brief review of related work to this paper. Section 3 introduces the proposed method, named MLEnsemble, and individual approaches (namely, MLBagging, MLBoosting, MLStacking, and MLBoostacking) in detail. Section 4 shows the experimental results and findings. Finally, Sect. 5 concludes this paper.

2   Related Work

A key to text classification is to obtain the best representations of texts, and this is the same for imbalanced text classification. Therefore, this section first introduces the current state of representations of texts, and the following section introduces related work of imbalanced classification.

2.1   Text Representations

To obtain representations of texts, there have been a large variety of approaches, and they can be roughly grouped into three categories; bag-of-words (BoW), word representation learning, and contextualized embedding. The BoW is a primitive representation of a text that is a count vector of words in a sentence, and TFIDF (term frequency-inverse document frequency) is a popular extended approach of BoW. Word representation learning (a.k.a. Word2Vec) is a data-oriented approach to obtaining representations of a word by using the surrounding words as its context. There are many variations of Word2Vec, such as SkipGram [21] and FastText [12]. Based on these word representations, representations of texts can be obtained in various ways such as using recurrent neural network, long short-term memory, and smooth inverse frequency [1].

BERT [4] is a neural language model that captures semantic relationships among text tokens (i.e., word or subword) by learning from vast amounts of text data. In addition to the success of BERT in various natural language processing tasks, Transformer-based approaches have been proposed such as RoBERTa [19], DeBERTa [10], and ALBERT [15]. An important feature of these Transformer-based models is that they can be pretrained on a vast amount of text data and applications can use the pretrained models for their tasks by fine-tuning the models in addition to training parameters of additional layers.

2.2   Imbalanced Classification

To deal with class imbalance, resampling is widely accepted preprocessing. It samples datasets so that the number of instances in different classes are balanced. Resampling can be divided into two major approaches; one is (synthetic) oversampling (such as SMOTE [5] and PolynomFitSMOTE [7]), and the other is undersampling (such as EasyEnsemble [18] and MUEnsemble [14]). The former generates synthetic instances of minority classes, and the latter samples instances of majority classes.

For imbalanced text classification, there are few attempts, and their main focus is oversampling [9,29]. [9] explores possibilities of SMOTE and its variants for text classification. MISO (Mutual Information constrained Semantically Oversampling framework) [29] is an oversampling-based approach for imbalanced text classification. It re-embeds difficult-to-classify texts by generating anchor instances. Though data augmentation methods by using text style transfer [6] and text generation [24] can be used as oversampling, their performances on imbalanced text classification are not revealed. [2] explores classification performances when using random under- and over-sampling for fake review detection, which is an imbalanced text classification task. However, it only applies simple random undersampling which discards a large portion of texts of the majority classes. In consequence, imbalanced text classification has been scarcely studied.

3   MLEnsemble: Metric Learning-Based Ensemble for Imbalanced Text Classification

MLEnsemble is a series of ensemble approaches that integrate metric learning into undersampling-based ensemble frameworks. In this paper, it consists of four approaches: **MLBagging**, **MLBoosting**, **MLStacking**, and **MLBoostacking**. MLBagging is based on bagging framework, MLBoosting is based on boosting framework, MLStacking is based on stacking framework, and MLBoostacking is based both on boosting and stacking frameworks

3.1   MLBagging

An overview of MLBagging is shown in Fig. 1, where it integrates metric learning into a *balanced bagging ensemble*. The basic idea of MLBagging is as follows. To deal with the class imbalance problem in metric learning, MLBagging utilizes undersampling to balance instances fed to metric learning models. A simple RUS throws many instances of majority classes away; therefore, MLBagging borrows the idea of EasyEnsemble. EasyEnsemble is based on a technique called balanced bagging. Balanced bagging is a bagging-based ensemble technique by which each batch of samples is balanced by undersampling. The light grey box in Fig. 1 is the balanced bagging part of MLBagging. It illustrates that given input data is copied several times, and each piece of copied data is randomly undersampled, and these individual pieces of undersampled data are used for learning different

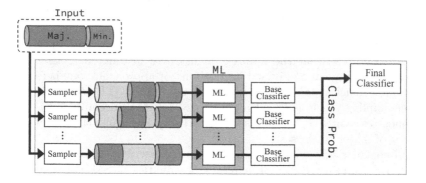

Fig. 1. Overview of MLBagging

spaces in metric learning and base classifiers on the basis of data transformed by metric learning. Each base classifier predicts a label with a probability (or confidence score) for a text and pushes that label to a merger of the ensemble. The merger collects the predictions from the base classifiers, and finalizes the prediction for the label with the highest probability.

3.2 MLBoosting

An overview of MLBoosting is shown in Fig. 2, where it integrates metric learning into a boosting framework. MLBoosting is based on boosting framework with undersampling (i.e., RUSBoost [27]); that is, for each iteration of learning a base classifier including a metric learning module, difficult-to-classify texts are determined, and MLBoosting gives higher weights to these texts for a sampling of the next iteration. The light grey box in Fig. 2 is the boosting part of MLBoosting. It illustrates that given input data is sampled for each iteration. Weights for sampling to sample difficult-to-classify texts are calculated by a weak classifier in the preceding iteration. In addition to the weighted sampling, undersampling is also applied to balance the data to be fed to metric learning and base classifier in that iteration.

3.3 MLStacking

An overview of MLStacking is shown in Fig. 3, where it integrates metric learning into a stacking framework. MLStacking applies undersampling to deal with the class imbalance problem in metric learning in a similar way to balanced bagging, and, it combines representations of input texts with prediction probabilities from individual base classifiers, instead of passing prediction results of base classifiers to the merger of the ensemble. A basic idea of the stacking framework is to integrate different characteristics of base classifiers by training a final classifier based on predictions of these base classifiers. Therefore, the stacking framework expects that base classifiers are in a wide variety, and MLStacking expects these

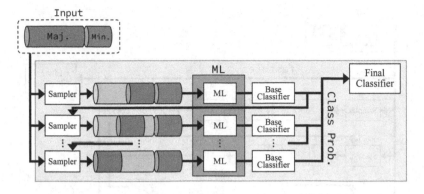

Fig. 2. Overview of MLBoosting

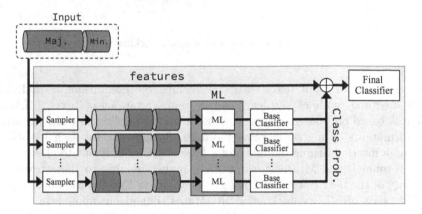

Fig. 3. Overview of MLStacking

base classifiers trained on different pieces of data with different transformations by metric learning are of a wide variety. The light grey box in Fig. 3 is the stacking part of MLStacking. It illustrates that given input data is copied several times, and each piece of copied data is randomly undersampled, and these individual pieces of undersampled data are used for learning different spaces in metric learning and base classifiers on the basis of transformed data by metric learning. The original representations of texts and predictions from base classifiers are concatenated and are fed into the final classifier.

3.4 MLBoostacking

An overview of MLBoostacking is shown in Fig. 4, where it integrates metric learning into a combination of boosting and stacking frameworks. The basic idea of MLBoostacking is close to MLBoosting except for learning the final classifier. MLBoostacking applies undersampling to deal with the class imbalance problem in metric learning in a similar way to MLBoosting, and, it combines

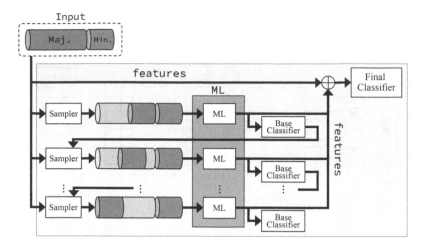

Fig. 4. Overview of MLBoostacking

representations of input texts with transformed representations by individual metric learning modules. To learn better representations of texts, MLBoostacking is based on boosting framework with undersampling to train variational transformations by metric learning modules. Specifically, for each iteration of learning a metric learning module and a base classifier, difficult-to-classify texts are determined, and MLBoostacking gives higher weights to these texts for a sampling of the next iteration. The light grey box in Fig. 4 is the boosting-and-stacking part of MLBoostacking. It illustrates that given input data is copied several times, and each piece of copied data is undersampled with weights from the previous iteration, and these individual pieces of undersampled data are used for learning different spaces in metric learning and base classifiers based on data transformed by metric learning. The original representations of texts and transformed representations from metric learning modules are concatenated and are fed into the final classifier.

4 Experimental Evaluation

In this experiment, MLEnsemble approaches were compared with Transformer-based classifiers and undersampling-based ensemble classifiers on three imbalanced text classification tasks. In the recent literature on text classification, Transformer-based approaches have achieved better performances than traditional approaches. A major contribution to this achievement is a learning procedure of representation of text. However, the learning procedure requires a vast amount of text, therefore, the classification performance of Transformer-based approaches is limited for domains that have a potentially small amount of texts, especially, for those whose amounts of texts in a domain are highly imbalanced. MLEnsemble, proposed in this paper, deals with this problem by using metric learning. To show the capability of MLEnsemble, in this experiment, representations obtained by using Transformer-based pretrained models were used for MLEnsemble.

4.1 Settings

Evaluation Metrics: The evaluation metrics were *precision, recall, F_1,* and *Gmean.* Let *TP, FN, TN,* and *FP* be true positives, false negatives, true negatives, and false positives. *Precision* $= TP/(TP + FP)$ measures how many positive instances are correctly among instances predicted as positive. *Recall* $= TP/(TP + FN)$ measures how many positive instances are correctly classified among all true positive instances. $F_\beta = ((1 + \beta^2)Recall \cdot Precision)/(Recall + \beta^2 Precision)$ is the harmonic mean of recall and precision, where β determines the weight on recall. In this experiment, β was set to 1, where $\beta = 1$ means that precision and recall were considered equally. *Gmean* $= \sqrt{Recall \cdot TNR}$ is the geometric mean of the recalls of both classes, where $TNR = TN/(TN + FP)$.

Table 1. Statistics of Datasets: the numbers of positive and negative sentences (#positives, #negatives) and imbalance ratio which is #negatives/#positives

Dataset	# positives	# negatives	Imbalance Ratio
claudette	1,032	8,382	8.12
hate-speech18	1,914	15,210	7.95
tweets-hate-speech-detection	2,242	29,720	13.26

Datasets and Tasks: In this experiment, three datasets were used: claudette dataset[1], hate-speech18 dataset[2], and tweets-hate-speech-detection dataset[3]. The claudette dataset is a dataset for detecting unfair sentences in a terms of service. The hate-speech18 dataset is for detecting hate speech in a Stormfront, a white supremacist forum, and the tweets-hate-speech-detection dataset is for detecting hate speech on Twitter. These three datasets are potentially imbalanced; that is, they are not controlled (or sampled) to make imbalanced datasets. Statistics of these datasets are summarized in Table 1.

The claudette dataset [17] consists of 50 documents (i.e., terms of service) and 9,414 sentences. The classification task in this experiment was a binary classification setting, where an unfair label was considered to belong to the positive class and the others the negative class. In the dataset, 1,032 sentences belonged to the positive class, and 8,382 sentences belonged to the negative class. The

---

[1] http://claudette.eui.eu/ToS.zip.
[2] https://huggingface.co/datasets/hate_speech18.
[3] https://huggingface.co/datasets/tweets_hate_speech_detection.

imbalance ratio (which is a ratio of negative instances to positive instances) of this dataset was 8.12; in other words, the number of negative texts was 8.12 times larger than that of positive texts.

In contrast, the hate speech detection tasks (hate-speech18 and tweets-hate-speech-detection) were sentence-level prediction. In these datasets, sentences of hate speech were considered to belong to the positive class and the others to the negative class. hate-speech18 dataset [8] is a set of sentences from forum posts in the Stormfront forum, and those have been manually labeled as containing hate speech or not. This dataset contains 17,124 sentences, and 1,914 sentences out of them belong to the positive class, therefore, the imbalance ratio of this dataset was 7.95. tweets-hate-speech-detection [28] dataset also consists of sentences (i.e., tweets), and "hate speech" in this dataset means that a tweet has a racist or sexist sentiment associated with it. This dataset contains 31,962 sentences, and 2,242 sentences out of them belong to the positive class, therefore, the imbalance ratio of this dataset was 13.26.

In this experiment, for claudette dataset, the experimental task followed that in [17]; that is, leave-one-document-out cross validation was applied, meaning that 1 out of 50 documents was used for testing and the rest for training. This setting follows a real situation; that is, given a document (a terms of service), a classifier should find unfair sentences in the document. Therefore, the validation process was repeated for each document being tested, and scores of evaluation metrics were summarized as macro averages over the 50 documents. In contrast, for hate speech detection tasks (hate-speech18 and tweets-hate-speech-detection), a sentence-level cross validation was applied. To prepare datasets for trials, a set of sentences was randomly split into 8:2 for training and testing, and this random split was repeated 10 times to generate 10 trials for each task. The scores of evaluation metrics were summarized as macro averages over the 10 trials.

Baseline Methods: The baseline methods in this experiment were selected from undersampling-based ensemble classification method, and neural language model-based representations of text. Approaches on the former category were bagging-based and boosting-based ensemble methods; namely, EasyEnsemble [18] and RUSBoost [27]. Those on the latter category were Transformer-based models, BERT [4] and its extensions for dedicated tasks.

For the input to the former category approaches, representations of texts were generated by using sentence-transformers library [25]. In addition, to compare MLEnsemble with deep metric learning approaches, representations of texts were re-trained by using training data with triplet loss [31].

For the latter category, the following approaches were used in this experiment.

- BERT[4] for all tasks, which is pretrained by using the BooksCorpus [34] and English Wikipedia.
- LegalBERT[5] for the claudette task, which is a pretrained BERT by using English legal texts from several fields (e.g., legislation and court cases).

---

[4] https://huggingface.co/bert-large-uncased.

[5] https://huggingface.co/nlpaueb/legal-bert-base-uncased.

- DeBERTa[6] for the hate-speech18 task, which is a fine-tuned model of DeBERTa V3 [11] on the dataset of this task.
- DiRoBERTa[7] for the tweets-hate-speech-detection task, which is a fine-tuned model of a distilled version of RoBERTa by using the dataset of this task.

The classifiers based on them were implemented by adding a linear layer on top of the pooled output of each model. Their optimizers were the Adam weighted decay (AdamW) optimizer [20]. In addition, to deal with imbalanced text classification, the balanced sampling (BS) and a weighted loss function (WCE) were also used.

- **BS**: The sampling weights for instances of positive and negative classes were set to $\left|D^{(+)}\right|^{-1}$ and $\left|D^{(-)}\right|^{-1}$, respectively, where $D^{(+)}$ and $D^{(-)}$ denote sets of positive and negative pieces of training data, respectively.
- **WCE**: In this experiment, weighted cross entropy loss was used. The weight of weighted cross entropy loss was calculated as follows. $w^{(+)}$ and $w^{(-)}$ denote weights for positive and negative classes, respectively, and they are computed as $w^{(+)} = \log\left(|D| / \left|D^{(+)}\right|\right)$ and $w^{(-)} = \log\left(|D| / \left|D^{(-)}\right|\right)$, where $D = D^{(+)} \cup D^{(-)}$ is a total set of instances.

4.2   Results

Table 2, 3, and 4 showcase the comparison results. The tables are vertically separated into three groups: (a) baseline methods including BERT-based classifier, RUSBoost, and EasyEnsemble, (b) extended BERT model-based classifier with balanced sampling and weighted loss function, and (c) the proposed method, MLEnsemble, is based on representations from extended BERT models. In each table, the best scores for each metric (i.e., column) in the table are boldfaced, and the best score of each metric in each group is underlined.

Results on claudette Task: In Table 2, MLBoostacking, one of the framework in MLEnsemble, performed the best in terms of precision, Gmean and F_1 metrics. It outperformed the best Transformer-based model (i.e., LegalBERT+BS) in all the metrics with large gaps, especially the precision score of MLBoostacking was .289 higher than that of LegalBERT+BS, while, in contrast, the gap of recall scores between them was not large (i.e., .029). MLBoostacking also showed superior performance to the best imbalanced classification methods (i.e., EasyEnsemble) in all the metrics.

In terms of Transformer-based classifiers, pretrained model trained by using dedicated corpus (i.e., LegalBERT) performed better than that by using general-purpose corpus (i.e., BERT), and the improvement by using balanced sampling (BS) and weighted loss function (WCE) was limited. BERT classifier achieved the highest recall score, however, its precision score was low, and LegalBERT classifiers achieved higher precision scores and their recall scores were still high. Legal-BERT with balanced sampling performed (LegalBERT+BS) the best among

---

[6] https://huggingface.co/Narrativaai/deberta-v3-small-finetuned-hate_speech18.
[7] https://huggingface.co/mrm8488/distilroberta-finetuned-tweets-hate-speech.

Table 2. Comparison for `claudette` dataset.

Model	Feature	Precision	Recall	Gmean	F_1
BERT		.244	**.944**	.754	.382
LegalBERT		<u>.361</u>	.899	.844	.508
LegalBERT+BS		.356	.910	<u>.848</u>	<u>.509</u>
LegalBERT+WCE		.327	.907	.824	.474
LegalBERT+BS+WCE		.338	.931	.842	.492
RUSBoost	LegalBERT	.388	.752	.788	.503
EasyEnsemble	LegalBERT	<u>.451</u>	.840	<u>.844</u>	<u>.579</u>
EasyEnsemble	LegalBERT+Triplet	.432	<u>.854</u>	<u>.844</u>	.565
MLBagging	LegalBERT	**.636**	.894	.910	**.736**
MLBoosting	LegalBERT	.554	.883	.890	.672
MLStacking	LegalBERT	.582	.902	.905	.702
MLBoostacking	LegalBERT	.629	<u>.939</u>	**.919**	**.736**

LegalBERT-based approaches, however, gaps between LegalBERT+BS and others were not significant.

Comparison between MLEnsemble with imbalanced classification methods shows the effectiveness of involvement of metric learning into undersampling-based ensemble classification frameworks. In particular, in the bagging framework, MLBagging was superior to EasyEnsemble in all metrics, and, in the boosting, MLBoosting also totally outperformed RUSBoost. In addition, EasyEnsemble trained on re-trained representations of text by re-training text encoder of LegalBERT by using triplet loss did not show clear superiority to that trained on the original representations by using the LegalBERT model. This observation indicates that deep metric learning did not always help learn suitable representations for classification as complained in [23]. In contrast, a fact that approaches in MLEnsemble showed their superiority over EasyEnsemble indicates that the non-deep metric learning (i.e., LMNN) performed better.

Results on `hate-speech18` Task: In Table 3, MLBoostacking performed the best in terms of recall and Gmean metrics. It outperformed Transformer-based models in terms of recall and Gmean, while these models achieved higher precision and F_1. MLBoostacking outperformed the best imbalanced classifier (i.e., EasyEnsemble on DeBERTa+Triplet encoding). In terms of Transformer-based classifiers, fine-tuned models (i.e., DeBERTa) outperformed the general model (i.e., BERT), and balanced sampling (BS) and weighted loss function (WCE) improve the base model in terms of recall. In terms of imbalanced classifiers, re-trained encoder by using triplet loss for representation of text contributed to improving classification performance.

Similar to the result of `claudette` task, a comparison between MLEnsemble with imbalanced classification methods shows the effectiveness of involvement of metric learning in undersampling-based ensemble classification frameworks. A

Table 3. Comparison for `hate-speech18` dataset.

Model	Feature	Precision	Recall	Gmean	F_1
BERT		.856	.727	.845	.784
DeBERTa		**.898**	.825	.902	.857
DeBERTa+BS		.890	.885	.934	**.886**
DeBERTa+WCE		.841	.876	.926	.857
DeBERTa+BS+WCE		.791	.916	.942	.847
RUSBoost	DeBERTa	.595	.822	.872	.688
EasyEnsemble	DeBERTa	.670	.921	.932	.775
EasyEnsemble	DeBERTa+Triplet	.683	.937	.941	.790
MLBagging	DeBERTa	.713	.947	.949	.813
MLBoosting	DeBERTa	.724	.957	.956	.824
MLStacking	DeBERTa	.733	.967	.961	.834
MLBoostacking	DeBERTa	.745	**.969**	**.963**	.841

fact that all the approaches in MLEnsemble outperformed the imbalanced classifiers (RUSBoost and EasyEnsemble) indicates that involvement of metric learning contributes to improve precision. Another fact that approaches in MLEnsemble showed their superiority to EasyEnsemble based on DeBERTa+Triplet representations indicates that the non-deep metric learning (i.e., LMNN) performed better.

Results on `tweets-hate-speech-detection` Task: In Table 4, MLBoostacking performed the best in terms of Gmean and F_1 metrics. It outperformed Transformer-based approaches in all the metrics except precision, and it achieved superior performance to the best imbalanced classifier (i.e., EasyEnsemble+Triplet) except recall. For the DistilRoBERTa approach, balanced sampling and weighted loss function assisted to improve recall scores, however, its precision scores dropped. In terms of imbalanced classifiers, deep metric learning by triplet loss slightly improved classification performance.

Comparison among MLBagging, EasyEnsemble, and EasyEnsemble+Triplet shows that deep/non-deep metric learning improves classification performance. A fact that MLBoostacking performed superior to them indicates that stacking transformed representations of texts can construct more suitable representations for imbalanced text classification. Another fact that MLBoostacking outperformed MLBoosting also supports this observation, that is, stacking transformed representations of texts is a better strategy for imbalanced text classification than stacking classification probabilities of weak classifiers.

4.3   Lessons Learned

MLEnsemble Confirms Effectiveness of Metric Learning for Undersampling-Based Ensemble Classifiers. This experiment showed that the ensemble

Table 4. Comparison for `tweets-hate-speech-detection` dataset.

Model	Feature	Precision	Recall	Gmean	F_1
BERT		.780	.730	.847	.752
DiRoBERTa		**.847**	.547	.733	.655
DiRoBERTa+BS		.718	.704	.827	.700
DiRoBERTa+WCE		.712	.595	.757	.634
DiRoBERTa+BS+WCE		.485	.840	.882	.607
RUSBoost	DiRoBERTa	.547	.840	.889	.660
EasyEnsemble	DiRoBERTa	.647	.959	.961	.776
EasyEnsemble	DiRoBERTa+Triplet	.658	.964	.963	.782
MLBagging	DiRoBERTa	.704	.958	.964	.812
MLBoosting	DiRoBERTa	.625	.864	.910	.723
MLStacking	DiRoBERTa	.673	**.967**	.966	.794
MLBoostacking	DiRoBERTa	.722	.964	**.967**	**.823**

approaches in MLEnsemble outperformed the imbalanced classifiers (namely, EasyEnsemble and RUSBoost), and a deep metric learning approach using triplet loss also improves classification performance. It is noteworthy that MLEnsemble outperformed EasyEnsemble with representations by using a deep metric learning. This indicates that the quality of batch-level metric learning depends on the batch size. Since GPU memory was limited, the batch size in the experiment was 16. In contrast, due to larger RAM for CPU, MLEnsemble can utilize a larger subset of training data for each metric learning model. This fact indicates that larger GPU memory or memory-effective deep metric learning method can improve deep metric learning. To the best of our knowledge, such a deep metric learning approach does not exist.

Stacking Transformed Representations is the Best Approach Among the Ensemble Frameworks in MLEnsemble. A major stream of undersampling-based ensemble approaches is based on bagging framework [13,14,18,33]. This paper shows another direction of ensemble for imbalanced text classification, the combination of boosting and stacking framework. MLBagging, MLBoosting and MLStacking showed their superiority to the undersampling-based imbalanced classifiers (i.e., RUSBoost and EasyEnsemble), besides that, MLBoostacking outperformed them. Stacking transformed representations can be regarded as a feature expansion, thus MLBoostacking realizes it based on classification difficulties on weak classifiers.

5   Conclusion

This paper reports series of ensemble approaches, MLEnsemble, for imbalance text classification. Each approach deals with the class imbalance problem with an undersampling-based ensemble approach, because metric learning techniques also suffer from class imbalance problems. In this paper, four ensemble approaches (namely, MLBagging, MLBoosting, MLStacking, and MLBoostacking) are proposed, three of which are corresponding to ensemble frameworks (namely, bagging, boosting, and stacking), and the other is a combination of boosting and stacking. The experimental evaluation on three imbalanced text classification datasets (namely, unfair statement classification in terms of service, hate speech detection in a forum, and hate speech tweet detection) shows that the proposed approaches lift classification performance from BERT-based approaches, by improving the representations of texts through metric learning. MLBoostacking achieves the best among the approaches in MLEnsemble, and it indicates that stacking transformed representations of texts is an effective approach for imbalanced text classification.

Acknowledgement. This work was partly supported by JSPS KAKENHI JP21H0355.

References

1. Arora, S., Liang, Y., Ma, T.: A simple but tough-to-beat baseline for sentence embeddings. In: ICLR 2017 (2017)
2. Budhi, G.S., Chiong, R., Wang, Z.: Resampling imbalanced data to detect fake reviews using machine learning classifiers and textual-based features. Multimed. Tools Appl. **80**(9), 13079–13097 (2021)
3. Chalkidis, I., Fergadiotis, M., Malakasiotis, P., Aletras, N., Androutsopoulos, I.: LEGAL-BERT: the Muppets straight out of law school. In: EMNLP 2020, pp. 2898–2904 (2020)
4. Devlin, J., Chang, M., Lee, K., Toutanova, K.: BERT: pre-training of deep bidirectional transformers for language understanding. In: NAACL-HLT 2019, pp. 4171–4186 (2019)
5. Fernández, A., García, S., Herrera, F., Chawla, N.V.: SMOTE for learning from imbalanced data: progress and challenges, marking the 15-year anniversary. J. Artif. Intell. Res. **61**, 863–905 (2018)
6. Fu, Z., Tan, X., Peng, N., Zhao, D., Yan, R.: Style transfer in text: exploration and evaluation. In: McIlraith, S.A., Weinberger, K.Q. (eds.) AAAI 2018, pp. 663–670 (2018)
7. Gazzah, S., Amara, N.E.B.: New oversampling approaches based on polynomial fitting for imbalanced data sets. In: DAS 2008, pp. 677–684 (2008)
8. de Gibert, O., Perez, N., García-Pablos, A., Cuadros, M.: Hate speech dataset from a white supremacy forum. In: ALW2@EMNLP 2018, pp. 11–20 (2018)
9. Glazkova, A.: A comparison of synthetic oversampling methods for multi-class text classification. CoRR abs/2008.04636 (2020)
10. He, P., Liu, X., Gao, J., Chen, W.: DeBERTa: decoding-enhanced BERT with disentangled attention. CoRR abs/2006.03654 (2020)
11. He, P., Liu, X., Gao, J., Chen, W.: DeBERTa: decoding-enhanced BERT with disentangled attention. In: International Conference on Learning Representations (2021). http://openreview.net/forum?id=XPZIaotutsD

12. Joulin, A., Grave, E., Bojanowski, P., Douze, M., Jégou, H., Mikolov, T.: Fast-Text.zip: compressing text classification models. CoRR abs/1612.03651 (2016)
13. Komamizu, T.: Combining multi-ratio undersampling and metric learning for imbalanced classification. J. Data Intell. **2**(4), 462–474 (2021)
14. Komamizu, T., Ogawa, Y., Toyama, K.: An ensemble framework of multi-ratio undersampling-based imbalanced classification. JDI **2**(1), 30–46 (2021)
15. Lan, Z., Chen, M., Goodman, S., Gimpel, K., Sharma, P., Soricut, R.: ALBERT: a lite BERT for self-supervised learning of language representations. In: ICLR 2020 (2020)
16. Li, Q., et al.: A survey on text classification: from traditional to deep learning. ACM Trans. Intell. Syst. Technol. **13**(2), 31:1–31:41 (2022)
17. Lippi, M., et al.: CLAUDETTE: an automated detector of potentially unfair clauses in online terms of service. Artif. Intell. Law **27**(2), 117–139 (2019)
18. Liu, X., Wu, J., Zhou, Z.: Exploratory undersampling for class-imbalance learning. IEEE Trans. Syst. Man Cybern. Part B **39**(2), 539–550 (2009)
19. Liu, Y., et al.: RoBERTa: a robustly optimized BERT pretraining approach. CoRR abs/1907.11692 (2019)
20. Loshchilov, I., Hutter, F.: Decoupled weight decay regularization. In: ICLR 2019 (2019)
21. Mikolov, T., Chen, K., Corrado, G., Dean, J.: Efficient estimation of word representations in vector space. In: ICLR 2013 (2013)
22. Minaee, S., Kalchbrenner, N., Cambria, E., Nikzad, N., Chenaghlu, M., Gao, J.: Deep learning-based text classification: a comprehensive review. ACM Comput. Surv. **54**(3), 62:1–62:40 (2022)
23. Musgrave, K., Belongie, S., Lim, S.-N.: A metric learning reality check. In: Vedaldi, A., Bischof, H., Brox, T., Frahm, J.-M. (eds.) ECCV 2020. LNCS, vol. 12370, pp. 681–699. Springer, Cham (2020). https://doi.org/10.1007/978-3-030-58595-2_41
24. Nie, W., Narodytska, N., Patel, A.: RELGAN: relational generative adversarial networks for text generation. In: ICLR 2019. OpenReview.net (2019)
25. Reimers, N., Gurevych, I.: Sentence-BERT: sentence embeddings using Siamese BERT-networks. In: EMNLP-IJCNLP 2019, pp. 3980–3990 (2019)
26. Sanh, V., Debut, L., Chaumond, J., Wolf, T.: DistilBERT, a distilled version of BERT: smaller, faster, cheaper and lighter. CoRR abs/1910.01108 (2019)
27. Seiffert, C., Khoshgoftaar, T.M., Hulse, J.V., Napolitano, A.: RUSBoost: a hybrid approach to alleviating class imbalance. IEEE Trans. Syst. Man Cybern. Part A **40**(1), 185–197 (2010)
28. Sharma, R.: Twitter-Sentiment-Analysis (2019). http://github.com/sharmaroshan/Twitter-Sentiment-Analysis
29. Tian, J., et al.: Re-embedding difficult samples via mutual information constrained semantically oversampling for imbalanced text classification. In: EMNLP 2021, pp. 3148–3161 (2021)
30. Vaswani, A., et al.: Attention is all you need. In: NIPS 2017, pp. 5998–6008 (2017)
31. Wang, J., et al.: Learning fine-grained image similarity with deep ranking. In: CVPR 2014, pp. 1386–1393 (2014)
32. Waseem, Z., Hovy, D.: Hateful symbols or hateful people? Predictive features for hate speech detection on Twitter. In: SRW@NAACL 2016, San Diego, California, pp. 88–93 (2016)
33. Yin, J., Gan, C., Zhao, K., Lin, X., Quan, Z., Wang, Z.: A novel model for imbalanced data classification. In: AAAI 2020, pp. 6680–6687 (2020)
34. Zhu, Y., et al.: Aligning books and movies: towards story-like visual explanations by watching movies and reading books. In: ICCV 2015, pp. 19–27 (2015)

Target and Precursor Named Entities Recognition from Scientific Texts of High-Temperature Steel Using Deep Neural Network

M. Saef Ullah Miah[1,2] ⓘ, Junaida Sulaiman[2] ⓘ, Talha Bin Sarwar[1,2(✉)] ⓘ,
Imam Ul Ferdous[2] ⓘ, Saima Sharleen Islam[3] ⓘ, and Md. Samiul Haque[4] ⓘ

[1] American International University-Bangladesh, Dhaka, Bangladesh
talhasarwar40@gmail.com
[2] Universiti Malaysia Pahang, Pekan, Malaysia
[3] University of Trento, Trento, Italy
[4] University of Dhaka, Dhaka, Bangladesh

Abstract. Named Entity Recognition (NER) is an essential task in natural language processing, especially in the domain of scientific texts. This paper presents a study of NER for scientific texts in high-temperature steel, a type of alloy used in various applications where high temperatures prevail. We propose a NER system using Bi-LSTM with a domain-specific embedding approach and evaluate its performance on a test dataset. The study results show that the proposed NER system achieves an F1 score of 0.99, indicating that it can accurately identify and classify named entities in scientific texts about high-temperature steel with high precision and recall. The proposed approach was more effective than the classical machine learning-based approach. Our results suggest that the domain-specific embedded Bi-LSTM technique can be an effective approach for NER in scientific texts, especially in specialized domains such as high-temperature steel.

Keywords: Material Named Entity Recognition · High-temperature steel · Bi-LSTM · Deep learning in industry · Material Natural Language Processing

1 Introduction

Named entity recognition (NER) is a critical task in natural language processing, with applications ranging from information extraction and text summarization to machine translation and question answering. In the scientific text, NER is particularly important for extracting and organizing information about specific concepts, techniques, and tools, often mentioned using technical terms and abbreviations.

High-temperature steel [9] is a type of alloy steel used in various applications where high temperatures are encountered, such as power generation, aerospace, and automotive engineering. Accurate identification and classification of named entities in the scientific text on high-temperature steel are important for tasks such as information extraction and text summarization and for enabling the automated analysis of scientific literature in this domain.

However, NER in scientific texts can be challenging due to the named entities' complexity and the lack of annotated data for training and evaluating NER systems [14]. In this paper, we present a study of NER for scientific texts in the domain of high-temperature steel, focusing on the key challenges and approaches to this task, as well as the recent advances and areas for future research. We also propose a NER system that utilizes Bi-LSTM [4] with a domain-specific embedding approach and evaluates its performance on a domain expert-curated test dataset [10].

To conduct the study, we have reviewed the existing literature on NER for scientific texts in the domain of high-temperature steel, using a variety of databases such as the ACM Digital Library, IEEE Xplore, and PubMed. We have analyzed the methods used for identifying and classifying named entities, the performance of different NER systems, and the challenges and limitations of these systems. Our study has identified several key challenges in NER for scientific texts in high-temperature steel. These include the complexity of the named entities, the lack of annotated data, the ability to handle synonyms and variations of named entities, and named entities mentioned in different ways in the text. To address these challenges, we propose the NER system, which utilizes a domain-specific embedding with the Bi-LSTM technique.

We evaluate the performance of the proposed NER system on a test dataset and compare it to several classic machine learning-only approaches [7]. Pre-trained models fine-tuned with the test dataset, namely BERT [3] and sciB-ERT [1].

2   Related Works

Kim et al. investigate the use of neural networks in predicting synthesis conditions for inorganic materials [8]. The Conditional Variational Autoencoder (CVAE) model identifies influential parameters and suggests changes for optimal results. It learns from literature data and generalizes to new materials, enabling diverse applications. The model achieves accuracy of 59%, 67%, and 91% on training datasets with 100, 1000, and 10,000 precursors, respectively. Increasing latent dimensions to 64 improves accuracy to 98%, but affects conditioning similarly to the synthesis action CVAE.

Cruse et al. present a dataset extracted from the scientific literature on gold nanoparticle synthesis [2]. The text-mining algorithms accurately capture information on synthesis methodologies, nanoparticle morphologies, and sizes. The dataset enables the study of synthesis-process associations and facilitates the design of new materials.

Wang et al. propose an automated NLP pipeline for superalloy data, extracting chemical composition and property data from 14425 journal articles [16].

Their synthesis and characterization of three alloys confirm the pipeline's accuracy. The recovered database through text mining provides a valuable superalloy production and global analysis resource.

Gupta et al. compares MatSciBERT to SciBERT, showing MatSciBERT's superior performance in recognizing named entities, relation classification, and abstract classification [5]. MatSciBERT achieves an 82% Macro-F1 score on the SOFC-Slot dataset using five cross-validation splits.

3 Methodology

Several steps are performed to create the proposed deep neural network model. First, pertinent scientific publications are collected in PDF format and converted to plain text with appropriate sentence boundaries [11]. The dataset is then annotated and validated by domain experts. The dataset is produced in IOB format. After preparing the dataset, the model TP-NER is created, trained, tested, and fine-tuned to ensure it reliably recognizes named entities in unseen text. Finally, the proposed approach extracts named entities from scientific articles.

3.1 Data Collection and Processing

Through the electronic resource of Universiti Malaysia Pahang's library, scientific papers are downloaded [15]. Using the search term "high temperature steel AND P91 P92", the top 500 most-cited scholarly articles are collected and analyzed for use in this investigation.

3.2 Dataset Preparation for TP-NER Model Training

Experts in the domain have used 50 relevant papers to compile the dataset. 1551 sentences from these articles are annotated with the labels B-target, I-target, B-precursor, I-precursor, and O. The output of the annotation process is saved in IOB format. The annotated dataset contains 426 tokens with the target class, which combines B-target and I-target labels, and 486 tokens with the precursor class, which combines B-precursor and I-precursor labels. The inter-annotator agreement (IAA) score for this dataset is 94% of Cohen's Kappa, indicating a high level of agreement between the annotators as they assigned labels to the texts. The dataset can be accessed via Mendeley data [10].

3.3 TP-NER: Deep Neural Network Model Development

There are 5 layers in the proposed model, with 3 of them being concealed. The first layer is an input layer, the second is a domain-specific integer-coded word embedding layer, the third is a dropout layer, the fourth is a Bi-LSTM layer with TanH activation and 200 neurons, and the fifth and final layer is a time-distributed dense layer with softmax activation. Algorithm 1 illustrates the architecture and layer-by-layer setup.

Algorithm 1 Architecture of the Proposed TP-NER DNN model

1: Input: Sequence with token and labels: SL
2: num_words = unique words in dataset, num_tags = unique labels in dataset
3: set $maxLen$ = 60
4: For $\{S$ in $sent\_list\}\{$ X = pad sequences words $\}$
5: For $\{S$ in $sent\_list\}\{$ y = pad sequences labels, y = hot encode y $\}$
6: Split data into training and testing set
7: $input\_word = (maxLen)$
8: Embedding($input\_dim$ = num_words, $output\_dim$ = $maxLen$, $input\_length$ = $maxLen$)
9: SpatialDropout1D(0.2)
10: Bidirectional(LSTM($units$ = 200,$return\_sequences$ = True, $recurrent\_dropout$ = 0.2))
11: out=TimeDistributed(Dense($activation$ = 'softmax'))
12: $matrec$ = Model($input\_word,out$)
13: $matrec$.compile($optimizer$ = 'adam', $loss$ = 'categorical_crossentropy', $metrics$ = [accuracy, precision_mt, recall_mt, f1_mt])
14: $matrec$.fit(train-test data, $batch\_size$ = 16, $verbose$ = 1, $epochs$ = 50,
15: $validation\_split$ = 0.2, $callbacks$=[tensorboard_cbk, es])
16: $model$.save('tpner.h5')
17: Return $medner.h5$

3.4   Evaluation Metrics

The proposed TP-NER model is evaluated utilizing Precision(Pr), Recall(Rcl), and $F1$ metrics. If each token is identified correctly during entity prediction, the entity is marked as true. True Positives (TPs) are calculated when the entities are accurately predicted. However, False Positives (FPs) are indicated when the predicted beginning token does not match the marked token of the entity. False Negatives (FNg) are recorded whenever the system wrongly forecasts the initial token of expected entities. The equations in Eq. (1) calculate the evaluation metrics.

$$Pr = \frac{TPs}{TPs + FPs}, \quad Rcl = \frac{TPs}{TPs + FNg}, \quad F1 = \frac{2 * Pr * Rcl}{Pr + Rcl} \qquad (1)$$

4   Result and Discussion

This study examines the efficacy of machine learning methods for recognising target and precursor-named entities for high-temperature steel from scientific publications. On a dataset of published scientific publications, the study compares the performance of numerous algorithms, including BERT [3], sci-BERT [1], Decision Tree [13], Random Forest [6], and Gaussian Naive Bayes [12]. Table 1 presents the outcomes of the experiments.

The proposed TP-NER model based on Bi-LSTM achieves the highest F1 score of 0.99, followed by Decision Tree with an F1 score of 0.92 and Random Forest with an F1 score of 0.91. In addition, the performance of the

Table 1. Precision, recall and F1 scores for different models

Model Name	Precision	Recall	F1
Proposed TP-NER	0.9972	0.9964	0.9968
BERT	0.81	0.865	0.8365
sci-BERT	0.88	0.81	0.85
Decision Tree	.89	.95	.923
Random Forest	.91	0.92	.91
Gaussian Naive Bayes	.85	.88	.86

approaches improves when the training dataset is domain-specific. Still, larger BERT and sci-BERT models perform less adequately despite being fine-tuned using the same dataset. This observation presents an opportunity to explore larger language models with a growing volume of domain-specific data to determine whether their performance may be enhanced. In contrast, the experimental results demonstrate that a domain-specific embedded neural network model without pre-trained embedding and language models outperforms finely tuned large pre-trained models for target and precursor-named entity recognition in high-temperature steel-related scientific papers.

The experimental results show that the proposed model for target and precursor named entity recognition from scientific text is the most successful algorithm among those compared. However, an additional study is required to evaluate the effectiveness of the suggested model on various forms of texts, such as patent records and technical reports.

5 Conclusion and Future Work

This study demonstrates the effectiveness of the proposed deep learning-based method for recognising target and precursor named entities. The experimental results show that the proposed method outperforms conventional machine learning techniques and pre-trained models and achieves high precision, recall, and F1 score of 0.9972, 0.9964, and 0.9968, respectively, in recognizing different types of named entities related to high-temperature steel, such as targets and precursors.

To further validate the proposed deep learning model, it is planned to extend the study to other material domains and languages in the future. In addition, the proposed model can combine the named entity identification system with different natural language processing approaches to improve the system's overall performance.

Acknowledgment. This research was supported by the Post Graduate Research Scheme of Universiti Malaysia Pahang entitled: An automated knowledge-based framework for EDLC using deep learning approach, PGRS220337.

References

1. Beltagy, I., Lo, K., Cohan, A.: SciBERT: a pretrained language model for scientific text. arXiv preprint arXiv:1903.10676 (2019)
2. Cruse, K., et al.: Text-mined dataset of gold nanoparticle synthesis procedures, morphologies, and size entities. Sci. Data **9**(1), 234 (2022)
3. Devlin, J., Chang, M.W., Lee, K., Toutanova, K.: BERT: pre-training of deep bidirectional transformers for language understanding. In: Proceedings of the 2019 Conference of the North American Chapter of the Association for Computational Linguistics: Human Language Technologies, vol. 1 (Long and Short Papers), pp. 4171–4186. Association for Computational Linguistics, Minneapolis, Minnesota (2019). https://doi.org/10.18653/v1/N19-1423. https://aclanthology.org/N19-1423
4. Graves, A., Schmidhuber, J.: Framewise phoneme classification with bidirectional LSTM and other neural network architectures. Neural Netw. **18**(5–6), 602–610 (2005)
5. Gupta, T., Zaki, M., Krishnan, N.A.: MatSciBERT: a materials domain language model for text mining and information extraction. NPJ Comput. Mater. **8**(1), 102 (2022)
6. Ho, T.K.: Random decision forests. In: Proceedings of 3rd International Conference on Document Analysis and Recognition. vol. 1, pp. 278–282 (1995)
7. Islam, S.S., Haque, M.S., Miah, M.S.U., Sarwar, T.B., Nugraha, R.: Application of machine learning algorithms to predict the thyroid disease risk: an experimental comparative study. PeerJ Comput. Sci. **8**, e898 (2022)
8. Kim, E., et al.: Inorganic materials synthesis planning with literature-trained neural networks. J. Chem. Inf. Model. **60**(3), 1194–1201 (2020)
9. Meetham, G.W.: High-temperature materials - a general review. J. Mater. Sci. **26**(4), 853–860 (1991). https://doi.org/10.1007/BF00576759
10. Miah, M.S.U., Sulaiman, J., Ferdous, I.U., Sarwar, T.B.: TP-NER: a named entity recognition dataset of target and precursor named entities for high-temperature steel (2022). https://doi.org/10.17632/5ZNG6KHY9H.1. https://data.mendeley.com/datasets/5zng6khy9h/1
11. Miah, M.S.U., et al.: Sentence boundary extraction from scientific literature of electric double layer capacitor domain: tools and techniques. Appl. Sci. **12**(3), 1352 (2022)
12. Rish, I., et al.: An empirical study of the Naive Bayes classifier. In: IJCAI 2001 Workshop on Empirical Methods in Artificial Intelligence, vol. 3, pp. 41–46 (2001)
13. Rokach, L.: Decision forest: twenty years of research. Inf. Fusion **27**, 111–125 (2016)
14. Sharma, A., Amrita, Chakraborty, S., Kumar, S.: Named entity recognition in natural language processing: a systematic review. In: Gupta, D., Khanna, A., Kansal, V., Fortino, G., Hassanien, A.E. (eds.) Proceedings of Second Doctoral Symposium on Computational Intelligence. Advances in Intelligent Systems and Computing, vol. 1374, pp. 817–828. Springer, Singapore (2022). https://doi.org/10.1007/978-981-16-3346-1_66
15. UMP Library: e-Resource UMP Lib (2022). https://login.ezproxy.ump.edu.my/login
16. Wang, W., et al.: Automated pipeline for superalloy data by text mining. NPJ Comput. Mater. **8**(1), 9 (2022)

Enabling PII Discovery in Textual Data via Outlier Detection

Md. Rakibul Islam[1]([✉]), Anne V. D. M. Kayem[2], and Christoph Meinel[2]

[1] Department of Computational Science, University of Potsdam, Potsdam, Germany
md.rakibul.islam@uni-potsdam.de
[2] Hasso-Plattner-Institute for Digital Engineering, University of Potsdam, Potsdam, Germany
Anne.Kayem@hpi.de, Christoph.Meinel@hpi.de

Abstract. Discovering Personal Identifying Information (PII) in textual data is an important pre-processing step to enabling privacy preserving data analytics. One approach to PII discovery in textual data is to characterise the PII as abnormal or unusual observations that can potentially result in privacy violations. However, discovering PII in textual data is challenging because the data is unstructured, and comprises sparse representations of similar text elements. This limits the availability of labeled data for training and the accuracy of PII discovery. In this paper, we present an approach to discovering PII in textual data by characterising the PII as outliers. The PII discovery is done without labelled data, and the PII are identified using named entities. Based on the recognised named entities, we then employ five (5) unsupervised outlier detection models (LOF, DBSCAN, iForest, OCSVM, and SUOD). Our performance comparison results indicate that iForest offers the best prediction accuracy with an ROC AUC value of 0.89. We employ a masking mechanism, to replace discovered PII with semantically similar values. Our results indicate a median semantic similarity score of 0.461 between original and transformed texts which results in low information loss.

Keywords: Outlier Detection · Named Entity Recognition · Data Masking · Personal Identifying Information (PII)

1 Introduction

Growing instances of information and data sharing abound on the Internet, with an increasing representation in the form of free text on social media, forums, blogs, and wikis. According to Gandomi and Haider [9], textual data makes up to 95% of all unstructured data online. Sharing textual data can inadvertently lead to sensitive information disclosure, without either the subjects concerned or the data owners being aware of it.

Discovering personal identifying information (PII) in unstructured textual data is challenging because the data does not lend itself well to labelling. This is

C. Strauss et al. (Eds.): DEXA 2023, LNCS 14147, pp. 209–216, 2023.
https://doi.org/10.1007/978-3-031-39821-6_17

mainly because unstructured textual data is comprised of sparse representations of similar text elements, that do not necessarily obey grammatical structures. This as such limits the availability of labeled data for training and the accuracy of PII discovery. PII discovery is also typically followed by masking and/or deletion which results in high information loss.

In this paper, we present an approach to discovering and masking PII in textual data by characterising PII as outliers. Our results show that iForest predicts outliers with ROC AUC value of 0.89, confirming that iForest performs well for large datasets. Detected outliers are masked to anonymise but preserve semantic similarity. Our similarity scores comparing the original and anonymised text show a median score of 0.461.

The rest of the paper is structured as follows, Sect. 2 presents related work and Sect. 3 presents our outlier detection and masking approach. Section 4 presents our results and Sect. 5, concludes the paper.

2   Related Work

Outlier detection has been researched primarily with respect to structured data [1,10,12]. Recent work also shows that approaches such as deep feature extraction using neural networks [5] and generative neural networks [17] can also be used to predict outliers. However, the correctness of labeled data impacts significantly on the performance and accuracy of these models. Unsupervised approaches such as proximity-based, density, and cluster-based methods [4,10,11] handle low dimensional numerical data well but are prone to overfitting on textual data due to assumptions about data format and distance differences [15,20]. Angle-based vector similarity is useful in estimating divergence in textual documents that are represented as feature vectors based on word occurrences, and vector cosine similarity but is not scalable to large datasets [21]. While cluster-based approaches handle large datasets well by emphasising cluster tightness but are dependent on threshold values and so are not suited to textual data [7]. Furthermore, identifying outliers in textual data using distance and density-based approaches are processing intensive in terms of similarity calculations [1]. Dimension reduction can address this problem, but incurs high information loss when applied to identifying sensitive data [3]. Alternatively, outlier identification approaches based on subspaces can address this issue by integrating pattern analysis of local data with analysis of subspaces [2], but are processing intensive [1,13]. Other work on PII discovery, focuses either on structured data [6,18] or semi-structured data [18] but assumes the availability of labelled data to support training PII discovery models, which is impractical for unstructured textual data.

We present an approach to solving the problem of PII discovery in unstructured textual data in the next section.

3 PII Discovery and Masking

Our PII discovery and masking mechanism operates in three (3) steps, namely: (1.) Named entity recognition to support feature generation, (2.) Using the named entities to support PII discovery and (3.) Replacing the identified PII with semantically similar but different values.

Table 1. Named entity categories based on spaCy NER system

Feature	Description
PERSON	People, including fictional
EMAIL	Any valid email
PHONE	Any valid phone number
NORP	Nationalities or religious or political groups
FAC	Buildings, airports, highways, bridges, etc
ORG	Companies, agencies, institutions, etc
GPE	Countries, cities, states
LOC	Non-GPE locations, mountain ranges, bodies of water
PRODUCT	Objects, vehicles, foods, etc
EVENT	Named hurricanes, battles, wars, sports events, etc
WORK_OF_ART	Titles of books, songs, etc
LAW	Named documents made into laws
LANGUAGE	Any named language
DATE	Absolute or relative dates or periods
TIME	Times smaller than a day
PERCENT	Percentage, including "%"
MONEY	Monetary values, including unit
QUANTITY	Measurements, as of weight or distance
ORDINAL	"first", "second", etc
CARDINAL	Numerals that do not fall under another type

We define an outlier as the occurrence of PII in a text. To detect outliers (PII), We are only interested in phrases that contain PII such as *name, date of birth, address, etc.*. Typically, these sensitive phrases form named entities, thus requiring the use of Named Entity Recognition (NER) [19]. Most NER systems are largely dependant on plain features and domain-specific information to learn reliably from already available supervised training corpora. We address this issue by identifying named entities (NE) using a pre-trained transition-based parser model [14]. The model constructs portions of the input sequentially using a stack data structure. To generate representations of the stack required for prediction, our NER model employs the Stack-LSTM, which augments the LSTM model with a stack pointer [8]. NER is done by detecting a single word or a collection of words that comprise an entity and classifying them into different categories. So, given a collection of comments $C_1, C_2, ..., C_n$, we want to locate all named entities and calculate their frequency count by category. Each of the named entity (NE) categories is considered a feature for detecting outliers. We selected 20 categories that can represent most of the known named entities. Table 1 describes the NE categories that we used as features to represent documents. Among them, 18 of these categories were selected based on the NER implementation of spaCy library. The remaining two (i.e., EMAIL and PHONE) were manually annotated. Thus the feature extraction process of finding PII in unstructured data reduces to locating named entities for each document and creating a feature matrix with the frequency count by each named entity category. This process gives us a concise representation of a textual document compared to the traditional bag of words model, which requires a large representational space.

Five unsupervised outlier detection models (LOF, DBSCAN, iForest, OCSVM, SUOD) were then employed for outlier detection. The PIIs (outliers) were then transformed by substituting named entities with pseudo-values. Pseudo-values are created as comparable replacement types for the named entities based on the types of named entities in the text. For instance, when an `EMAIL, PHONE,` or `DATE` is discovered as a named entity, the masking algorithm generates entities of a similar kind. We maintain a hash-table lookup approach to produce consistent masking values that translate to the same masking value each time a particular type of named item is discovered. In terms of content replacement, we used pre-defined pseudo-values to replace PII realistically without mapping to a real person. Semantic similarity, based on comparing word embeddings, is used to evaluate the distance between the original and anonymised textual data elements rather than their lexicographical similarity [16]. We trained our Word2Vec embeddings on the Common Bag of Words (CBOW) pre-trained model for performance efficiency and accuracy for representations of more frequently occurring words. The resulting word embeddings are used to calculate document similarity by measuring the cosine angle.

4 Experimental Evaluation and Results

Code for our implementation can be found at[1]. We used AirBnB review data for Berlin, Germany, compiled on 17 December 2021 containing $410, 291$ reviews including spam[2]. We considered comments written in English only, for a total of $253, 908$ reviews.

Using the named entities in Table 1, we pre-trained an NER system to identify named entities and calculated their frequency count by category, giving a $253, 908 \times 20$ initial feature matrix. We applied dimension reduction using Principal Component Analysis (PCA) and Singular Value Decomposition (SVD)to reduce the sparsity of the feature vectors.

---

[1] Github Code.

[2] AirBnB Dataset.

Since we do not have any ground truth about what an outlier (PII) looks like in our context, we used domain knowledge and data analysis, to make two assumptions for labeling comments as an outlier (PII). (1) If EMAIL or PHONE is present as a named entity, we consider the comment to be an outlier (PII), and (2) likewise, for the presence of PERSON or ORG with other named entities.

$$v_i > 0; \ i \in \{EMAIL, PHONE\}$$

$$v_i > 0 \ and \ v_j > 0; \ i \in \{PERSON, ORG\}; \ j \notin \{PERSON, ORG\}$$

Here v_i is an element of the feature vector and i represents a category of named entities. We take the same sample of 50,782 reviews that are used in the model implementation part. After labeling, we get 42% outlier reviews and 58% not outlier reviews.

Table 2. Execution time comparison for base models

Model	n_jobs = 1	n_jobs = −1
LOF	**3.51 s ± 274** ms	**2.32 s ± 92.9** ms
DBSCAN	8.54 s ± 799 ms	3.74 s ± 346 ms
iForest	5.23 s ± 485 ms	4.2 s ± 109 ms
OCSVM	7 min 23 s ± 8.99 s	7 min 2 s ± 24.8 s
SUOD	5 min 32 s ± 1 min 19 s	4 min 46 s ± 16.6 s

Table 2 shows the execution time of the five models. As the density calculation depends on the dimension of the dataset, dimension reduction helps run LOF and DBSCAN faster, but both do not scale well for PII discovery in unstructured textual data. iForest is slower, but scales well with growing data sizes and, due to the isolation property, is faster than density-based approaches. Also, iForest has linear time complexity and requires low memory, while SVM is based on a nonlinear kernel function which can have a complexity of up to $O(n_{features} \times n_{samples}^3)$. Table 3 illustrates the outlier score threshold, precision, recall, F1-score, ROC AUC, and PR AUC score for five models. For model evaluation, recall is the most important metric as we interested in reducing the false negative value. The table shows that based on the recall and F1-score value, SUOD, iForest, and OCSVM perform well with recall values of 0.70, 0.69, and 0.68, respectively. On the other hand, LOF and DBSCAN perform worst, with 0.33 and 0.18 recall values, respectively. Figure 1 shows the TPR (True Positive Rate)/recall versus the FPR(False Positive Rate) at various outlier score thresholds. In this case, iForest performs best with a ROC AUC of 0.86, followed by SUOD and OCSVM. LOF and DBScan performed worst, which is aligned with our previous result based on recall and F1-score (Table 3). Figure 2 and Table 3 show results of the PR curve, indicating that iForest performs best with a PR AUC of 0.78, followed by SUOD, OCSVM, LOF, and DBSCAN.

Table 3. Evaluation matrices for the models

Model	Threshold	Precision	Recall	F1-score	ROC AUC	PR AUC
LOF	1.00	0.84	0.33	0.47	0.61	0.55
DBSCAN	–	0.90	0.18	0.30	0.58	0.51
iForest	0.00	0.73	0.69	0.71	**0.86**	**0.78**
OCSVM	500.77	0.74	0.68	0.71	0.79	0.73
SUOD	−0.16	0.75	**0.70**	**0.72**	0.84	0.77

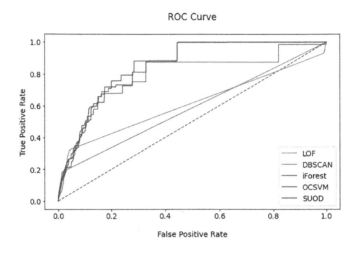

Fig. 1. ROC curve

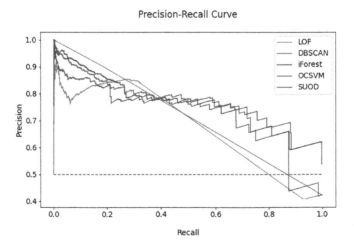

Fig. 2. Precision-Recall curve

During the data masking step, we only substitute the named entities from the outlier comments and use these named entities to generate document embeddings. This avoids the remaining terms from affecting the embeddings that are unchanged in both original and transformed comments. The results show that 50% of the anonymised comments have a similarity score between 0.357 to 0.554 with a median score of 0.461 while only 7% of the transformed comments have a similarity score less than or equal to 0. As the majority of the similarity score has a value greater than 0, we can conclude that our proposed data masking approach preserves most of the semantic properties of the original comments. LOF performs best after tuning with a recall value of 0.74. The Receiver Operating Characteristic (ROC) curves of the tuned iforest model performs best with a ROC AUC of 0.89, followed by SUOD, LOF, and OCSVM. Furthermore, iforest performs best with Precision-Recall (PR) AUC of 0.81, followed by SUOD, OCSVM, LOF, and DBSCAN.

5   Conclusion

We presented an approach to discovering personal identifying information (PII) in unstructured textual data, by characterising PIIs as outliers. We show that by using named entities it is possible to detect outliers (PIIs) using traditional unsupervised outlier detection models. Our experiments show that iForest predicts outliers with a ROC AUC score of 0.86 and a recall value of 0.69. Detected outliers are masked to anonymise but preserve semantic similarity. Our similarity scores comparing the original and anonymised text show a median score of 0.461.

References

1. Aggarwal, C.C.: An introduction to outlier analysis. In: Aggarwal, C.C. (ed.) Outlier Analysis, pp. 1–34. Springer, Cham (2017). https://doi.org/10.1007/978-3-319-47578-3_1
2. Aggarwal, C.C., Sathe, S.: Theoretical foundations and algorithms for outlier ensembles. ACM SIGKDD Explor. Newsl. **17**(1), 24–47 (2015)
3. Blouvshtein, L., Cohen-Or, D.: Outlier detection for robust multi-dimensional scaling. IEEE Trans. Pattern Anal. Mach. Intell. **41**(9), 2273–2279 (2019)
4. Breunig, M.M., et al.: LOF: identifying density-based local outliers. In: Proceedings of the 2000 ACM SIGMOD International Conference on Management of Data, SIGMOD 2000, pp. 93–104. Association for Computing Machinery (2000)
5. Chakraborty, D., Narayanan, V., Ghosh, A.: Integration of deep feature extraction and ensemble learning for outlier detection. Pattern Recogn. **89**, 161–171 (2019)
6. Domingo-Ferrer, J., Sánchez, D., Soria-Comas, J.: Database Anonymization: Privacy Models, Data Utility, and Microaggregation-Based Inter-model Connections, vol. 8. Morgan & Claypool Publishers (2016)
7. Duan, L., et al.: Cluster-based outlier detection. Ann. Oper. Res. **168**(1), 151–168 (2009)
8. Dyer, C., et al.: Transition-based dependency parsing with stack long short-term memory. arXiv preprint arXiv:1505.08075 (2015)

9. Gandomi, A., Haider, M.: Beyond the hype: big data concepts, methods, and analytics. Int. J. Inf. Manag. **35**(2), 137–144 (2015)
10. Hautamaki, V., Karkkainen, I., Franti, P.: Outlier detection using k-nearest neighbour graph. In: Proceedings of the 17th International Conference on Pattern Recognition (ICPR 2004), vol. 3, pp. 430–433 (2004)
11. Knorr, E.M., Ng, R.T.: Algorithms for mining distance-based outliers in large datasets. In: Proceedings of the 24rd International Conference on Very Large Data Bases, VLDB 1998, pp. 392–403 (1998)
12. Kokkula, S., Musti, N.M.: Classification and outlier detection based on topic based pattern synthesis. In: Perner, P. (ed.) MLDM 2013. LNCS (LNAI), vol. 7988, pp. 99–114. Springer, Heidelberg (2013). https://doi.org/10.1007/978-3-642-39712-7_8
13. Kriegel, H.-P., Kröger, P., Schubert, E., Zimek, A.: Outlier detection in axis-parallel subspaces of high dimensional data. In: Theeramunkong, T., Kijsirikul, B., Cercone, N., Ho, T.-B. (eds.) PAKDD 2009. LNCS (LNAI), vol. 5476, pp. 831–838. Springer, Heidelberg (2009). https://doi.org/10.1007/978-3-642-01307-2_86
14. Lample, G., et al.: Neural architectures for named entity recognition. In: Proceedings of the 2016 Conference of the North American Chapter of the Association for Computational Linguistics: Human Language Technologies, pp. 260–270. Association for Computational Linguistics (2016)
15. Liu, H., et al.: Efficient outlier detection for high-dimensional data. IEEE Trans. Syst. Man Cyberne.: Syst. **48**(12), 2451–2461 (2018)
16. Liu, Y., et al.: Computing semantic text similarity using rich features. In: Proceedings of the 29th Pacific Asia Conference on Language, Information and Computation, pp. 44–52 (2015)
17. Liu, Y., et al.: Generative adversarial active learning for unsupervised outlier detection. IEEE Trans. Knowl. Data Eng. **32**(8), 1517–1528 (2020)
18. Mrabet, A., Bentounsi, M., Darmon, P.: SecP2I a secure multi-party discovery of personally identifiable information (PII) in structured and semi-structured datasets. In: 2019 IEEE International Conference on Big Data (Big Data), pp. 5028–5033 (2019). https://doi.org/10.1109/BigData47090.2019.9006605
19. Nadeau, D., Sekine, S.: A survey of named entity recognition and classification. Lingvisticae Investigationes **30**(1), 3–26 (2007)
20. Sutanto, T., Nayak, R.: Semi-supervised document clustering via Loci. In: Wang, J., et al. (eds.) WISE 2015. LNCS, vol. 9419, pp. 208–215. Springer, Cham (2015). https://doi.org/10.1007/978-3-319-26187-4_16
21. Wong, S.K.M., Ziarko, W., Wong, P.C.N.: Generalized vector spaces model in information retrieval. In: Proceedings of the 8th Annual International ACM SIGIR Conference on Research and Development in Information Retrieval, SIGIR 1985, pp. 18–25. Association for Computing Machinery (1985)

Deep Learning

An Efficient Embedding Framework for Uncertain Attribute Graph

Ting Jiang[1] , Ting Yu[1] , Xueting Qiao[1] , and Ji Zhang[2](✉)

[1] Zhejiang Lab, Hangzhou, China
{jiangting,yuting}@zhejianglab.com, xueting.qiao@connect.polyu.hk
[2] University of Southern Queensland, Toowoomba, Australia
zhangji77@gmail.com

Abstract. Graph data with uncertain connections between entities is commonly represented using uncertain graphs. This paper tackles the challenge of graph embedding within such uncertain attribute graphs. Current graph embedding techniques are typically oriented towards deterministic graphs, or uncertain graphs that lack attribute data. Furthermore, the majority of studies on uncertain graph learning simply adapt conventional algorithms for deterministic graphs to handle uncertainty, leading to compromised computational efficiency. To address these issues, we introduce an optimized embedding framework $UAGE$ for uncertain attribute graphs. In $UAGE$, nodes are represented within a Gaussian distribution space to learn node attributes. We also propose a Probability Similarity Value (PSV) to manage relationship uncertainty and ensure that nodes with higher-order similar structures are located more closely in the latent space. Real-world dataset experiments confirm that $UAGE$ surpasses contemporary methods in performance for downstream tasks.

Keywords: Uncertain graph · Gaussian embedding · encoder-decoder

1 Introduction

Graphs have been widely used to model various real-world networks, such as social networks [15] and traffic networks [21]. However, uncertainties caused by data collection defects, fuzzy data sources, and inaccurate data [18], can pose challenges in effective network analysis. Moreover, the links in real-world networks may be derived from probabilistic techniques, resulting in uncertain graphs.

An uncertain graph [1] is a type of graph where nodes are connected with probability values. The interpretation of these probabilities may vary from the context. In social networks, the relationships between users, such as influence, are not directly visible but instead inferred from user activities, such as paper citations. In PPI networks, an edge between two proteins corresponds to an

C. Strauss et al. (Eds.): DEXA 2023, LNCS 14147, pp. 219–229, 2023.
https://doi.org/10.1007/978-3-031-39821-6_18

interaction that may be observed in a noisy experiment with uncertainty. Thus, the probability value of an edge represents the probability of such an interaction existing. Graph embedding [30] is a technique used to learn effective representations of nodes in a lower-dimensional space while preserving structural and semantic information. It plays a vital role in network analysis [29] and enables many downstream machine-learning tasks.

However, most existing graph embedding methods, such as DeepWalk [22] and SDNE [28], don't consider uncertainty during the modeling process and are highly susceptible to noise from node attributes and graph structure [8]. Recently, some studies [2,32] have leveraged the Gaussian embedding technique to obtain better representations with uncertainty quantification, but they have not considered the attribute noise of the graph and are designed for deterministic graphs and are unable to support uncertain graphs. For studies on uncertain graph learning [1], most of them simply modify existing algorithms for deterministic graphs to handle the uncertain information of graph edges, and there are few studies on embedding uncertain graphs, such as URGE [11] and DGCU [5]. However, both of them only learn from the topological structure of the graph and are designed for uncertain graphs without attributes. Many real-world networks have both uncertainties on relationships and attributes on objects. Learning node attribute information at the same time can lead to more effective embedding [25].

In this paper, we propose an efficient embedding framework *UAGE* for uncertain attribute graphs. To the best of our knowledge, *UAGE* is the first method that uses deep learning to embed uncertain attribute graphs. Specifically, 1) PSV is proposed and utilized in our model to handle the relationship uncertainty and preserve the higher-order structural information in the embedding space; 2) the Gaussian embedding method is adopted to capture the structural uncertainty and node attributes, and an encoder-decoder structure is designed to reconstruct node attributes; 3) our model is evaluated on several real-world networks and experiment results show that *UAGE* outperforms state-of-the-art methods.

2   Related Works

In this section, we provide a brief overview of recent works on graph embedding learning and uncertain graph learning.

Graph Embedding Learning. Graph embedding learning aims to project nodes into a lower-dimensional latent space while preserving the original graph structure properties. There is some deterministic vector-based graph embedding methods. Several popular methods leverage random walks, such as DeepWalk [22] and node2vec [6], to learn embeddings. Other approaches, such as LINE [26] and SDNE [28], learn from first/second-order proximity, respectively.

Inspired by word2Gauss [27], which embeds words into an infinite-dimensional Gaussian distributional functional space, several studies have proposed the learning of the uncertainty of embeddings using Gaussian distribution-based graph embedding methods. VGAE [13] embeds nodes into a standard normal distribution space and learns embeddings by reconstructing the graph structure. G2G [2] attempts to preserve the graph structural proximity by controlling the distance between probability distribution embeddings. DVNE [32] learns Gaussian embeddings using the deep variational mode. The Gaussian embedding technique enables better node representations with quantification of uncertainty. However, current approaches can only capture the structural noise without considering the attributes noise on nodes. For a more comprehensive introduction to graph embedding learning, we refer readers to [30].

Uncertain Graph Learning. Several tasks, such as clustering [12], classification [4], and kNN [23], have been studied for uncertain graphs. However, most of these studies just modify deterministic graph models to handle the uncertainty of graph edges, leading to computational inefficiencies. Taking clustering, for example, most methods are based on the possible worlds model. For a graph with m edges, the number of possible worlds is 2^m and grows exponentially. Consequently, many approximate algorithms have been developed to efficiently cluster uncertain graphs. pKwikCluster [14] is a random 5-approximation algorithm, which extends the edit-distance-based definition of graph clustering to uncertain graphs. EA-CPG [7] uses a multi-population evolutionary algorithm (EA) for clustering, guided by pKwikCluster. USCAN [24] adopts a method based on structural clustering, establishing a structural clustering model based on reliable structural similarity obtained from the possible world of the uncertain graph. However, the computational efficiency is low, and ProbSCAN [17] attempts to improve the time complexity of this part. URGE [11] is the first to adopt the embedding method for uncertain graphs, using the matrix factorization method to calculate the embedding vector and preserve both structural and uncertainty information of the relationship in the embedding space. DGCU [5] is the first to use the deep learning method to learn uncertain graphs for node clustering tasks. However, both methods only learn the structural information of graphs and ignore the attribute information of nodes. For a more comprehensive introduction to uncertain graphs learning, we refer readers to [1].

3   The Proposed Methods

In this section, we formally present the problem definition and introduce the framework of our proposed model.

3.1   Problem Definition

Given a uncertain attribute graph $\mathcal{G} = \{V, E, X, P\}$, where $V = \{v_i | i = 1, ..., N\}$, $E = \{(v_i, v_j) | v_i, v_j \in V\}$ are the node and edge set of graph $\mathcal{G}$. $X = [x_i] \in \mathbb{R}^{N \times F}$ is a attribute matrix and $x_i \in \mathbb{R}^F$ is a F-dimension attribute vector of node v_i. The transition probability matrix $P = [p_{ij}] \in [0, 1]^{N \times N}$, represents the connect probability between N nodes, where $p_{ij} > 0$ if and only if there is an edges $(v_i, v_j) \in E$. In this paper, we propose an embedding framework that can handle uncertain graphs and preserve both node attributes and graph structure.

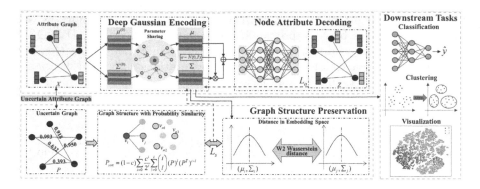

Fig. 1. The framework of *UAGE*.

3.2 Embedding Framework for Uncertain Attribute Graph

As Fig. 1 shown, our embedding framework *UAGE* consists of three main components: deep Gaussian encoding, graph structure preservation, and node attribute decoding. We describe each of these components in detail below.

Deep Gaussian Encoding. We use a deep feed-forward network to learn the attributes and uncertainty of nodes, which are used as inputs to two encoders: f_μ and f_σ, which aggregate and transform information while propagating the uncertainty in node attributes. They can be implemented using either a multi-layer Graph Neural Network or an MLP. In this paper, We utilize two L-layer MLPs to encode the attributes and uncertainty of nodes and learn the vector values of nodes in the Gaussian distribution space, where $\mu \in \mathbb{R}^{N \times emb}$ and $\Sigma \in \mathbb{R}^{N \times emb}$ are the embedded mean and covariance vectors calculated by:

$$\mu = f_\mu \left(ReLU(XW_\mu^{(0)} + b_\mu^{(0)})|P \right), \Sigma = f_\sigma \left(ELU(XW_\sigma^{(0)} + b_\sigma^{(0)}) + 1|P \right) \quad (1)$$

where X are attribute vectors, $W_\mu^{(0)}, W_\sigma^{(0)} \in \mathbb{R}^{F \times F^{(0)}}, b_\mu^{(0)}, b_\sigma^{(0)} \in \mathbb{R}^{F^{(0)}}$ are parameters shared across instances. Functions $ReLU$ and ELU are defined as:

$$ReLU(x) = max(0,x), ELU(x) = \begin{cases} e^x - 1, & x < 0 \\ x\ \ , & x \geq 0 \end{cases} \quad (2)$$

Graph Structure Preservation. Inspired by *SimRank**** [31], we propose the Probability Similarity Value (PSV) to estimate the probability of similarity between node pairs, maintain the structural information of the graph, and ensure that nodes with similar structures are mapped closer together in the latent space:

$$P_{sim} = (1 - c) \sum_{i=0}^{t} \frac{c^i}{2^i} \sum_{l=0}^{i} \binom{i}{l} (P)^l (P^T)^{i-l} \quad (3)$$

where P is the transition probability matrix, t is the number of iterations, representing the highest order of neighbors, and $c \in (0,1)$ is the damping factor.

To ensure that nodes with a higher probability of similarity to node v_i are mapped closer to v_i in the embedding space, we try to satisfy the following set of constraints, where the symmetric 2nd Wasserstein distance is used:

$$\forall v_i, v_j, v_k \in V : \quad P_{sim}[i,j] > P_{sim}[i,k] \Leftrightarrow dist(v_i, v_j) > dist(v_i, v_k) \quad (4)$$

To preserve high-order similarity and satisfy the constraints of Eq. 4, We treat the probability similarity between nodes as a distribution and minimize the distance between the empirical and conditional probability distribution Eq. 5:

$$\hat{p}(i,j) = \frac{P_{sim}[i,j]}{\sum_{k \in V} P_{sim}[i,k]}, p(i,j) = \frac{exp(-dist(v_i, v_j))}{\sum_{k \in V} exp(-dist(v_i, v_k))} \quad (5)$$

$$L_s = - \sum_{(v_i, v_j) \in E} P_{sim}[i,j] \, log p(i,j) \quad (6)$$

Node Attribute Decoding. The values of node vectors z can be sampled from the Gaussian distribution $N(\mu, \Sigma)$, which are then used as inputs to an MLP decoder for reconstructing the node attributes Z:

$$Z = MLP(z) \in \mathbb{R}^{N \times F}, z \sim N(\mu, \Sigma) \quad (7)$$

The entire model cannot perform Back Propagation (BP) due to the non-differentiable sampling operation described in Eq. 7. To overcome this limitation, we utilize the reparametrization technique. Specifically, we first sample ϵ from a standard Gaussian distribution $N(0, I)$, and then transform it through scaling and shifting, which makes BP possible.

To mitigate the impact of attribute uncertainty on the model's error, nodes with greater uncertainty are given lower reconstruction error weights:

$$L_a = \sum_{i=0}^{N-1} e^{-\lambda \sigma_{mean}[i]} \|X_i - Z_i\|_F^2 \quad (8)$$

where $\lambda > 0$ is a hyper-parameter and $\sigma_{mean}[i] = Mean(\Sigma_i) \in \mathbb{R}$ and $e^{-\lambda \sigma_{mean}[i]}$ is the reconstruction error weight of the ith node.

Model Optimization. To preserve node attributes and graph structure, we create a loss function that combines Eq. 6 and Eq. 8, which is represented as:

$$L = L_s + \alpha L_a \quad (9)$$

where $\alpha > 0$ is a hyper-parameter that balances structure and attribute losses. Optimizing L_s can be computationally expensive, as evaluating the entire node set is required to calculate the conditional probability in Eq. 5. To address this issue, we adopt the negative sampling method:

$$L_s = \sum_{P_{sim}[i,j]>0} \left(log\sigma(-dist(v_i, v_j)) + \sum_{n=1}^{K} \mathbb{E}_{v_n \sim P_n(v)} log\sigma(dist(v_i, v_n)) \right) \quad (10)$$

Table 1. The statistics of social networks and PPI networks.

Datasets	#Nodes	#Edges	#Dimension	#Labels
Cora_ml	2,995	8,416	2,879	7
Citeseer	4,230	5,358	602	6
PUBMED	19,717	88,648	500	3
–	#Nodes	#Edges	Min. Pro	Max. Pro
Collins	1,622	9,074	0.48	0.99
Gavin	1,855	7,669	0.24	1
Krogan-core	2,708	7,123	0.27	0.99
Krogan-extended	3,672	14,317	0.1	0.99

where $\sigma(x)$ is the *Sigmoid* function, (v_i, v_j) represents the observed edge, and (v_i, v_n) denotes the negative edge sampled from the noise distribution $P_n(v)$ [19]. K is the number of negative edges for node v_i. Similarly, we sample a batch of nodes to compute the attribute reconstruction loss in each batch to optimize L_a.

4   Experiments

In this section, we conduct experiments on several real-world graphs for node classification, clustering, and visualization tasks.

4.1   Datasets and Experiment Setup

Datasets and Baselines. We evaluated our model for node classification, clustering, and visualization tasks on 3 widely-used social networks, including *Cora_ml*, *Citeseer* and *PUBMED*. An obfuscating algorithm [3] is used to obtain uncertain graphs. Additionally, we use 4 classical PPI networks collated by Nepusz et. al [20] for node clustering tasks, including *Collins*, *Gavin*, *Krogan-core* and *Krogan-extended*. Table 1 provides detailed statistical information on all datasets. We compare the performance of our model *UAGE* with state-of-the-art methods, including: *Attributes* directly uses original node attributes as node features; *LINE*[1] respectively preserves the first and second-order proximity between nodes, and directly concatenates the representations; *VGAE*[2] embeds node into a standard normal distribution space by an encoder and reconstructs the graph structure by a decoder; *uBayes+*[3] is a Bayes-based classification algorithm customized for uncertain graphs; *DGCU* is a Gaussian embedding-based clustering method customized for uncertain graphs; *G2G*[4] and *GLACE* [9][5] learn the lower-dimensional Gaussian distribution embedding of graphs.

---

[1] https://github.com/tangjianpku/LINE.
[2] https://github.com/tkipf/gae.
[3] https://helios2.mi.parisdescartes.fr/~themisp/collectiveclassification/.
[4] https://www.cs.cit.tum.de/daml/g2g/.
[5] https://github.com/bhagya-hettige/GLACE.

Table 2. The results of classification with 10% of labeled nodes.

Datasets	Metric	LINE	Attributes	VGAE	uBayes+	GLACE	G2G	**OURS**
Cora_ml	F1	0.6125	0.7222	0.6583	0.7024	0.7494	0.8041	**0.8175**
	ACC	0.6391	0.7296	0.6684	0.7149	0.7685	0.8149	**0.8315**
	Prec	0.6584	0.7852	0.7107	0.7075	0.7522	0.8021	**0.8202**
	Recall	0.5932	0.6855	0.6318	0.6989	0.7549	0.8080	**0.8176**
	AUPRC	0.6839	0.8024	0.7230	0.7518	0.8036	0.8690	**0.8910**
	AUROC	0.8867	0.9412	0.9111	0.9304	0.9454	0.9662	**0.9737**
Citeseer	F1	0.4492	0.7397	0.5832	0.6821	0.7739	0.7911	**0.8278**
	ACC	0.4610	0.7389	0.5868	0.6833	0.7780	0.7943	**0.8278**
	Prec	0.4487	0.7473	0.5842	0.6841	0.7744	0.7943	**0.8276**
	Recall	0.4516	0.7377	0.5870	0.6829	0.7750	0.7948	**0.8281**
	AUPRC	0.4853	0.5895	0.6490	0.7425	0.8432	0.8707	**0.8957**
	AUROC	0.759	0.8426	0.8590	0.8462	0.9372	0.9559	**0.9655**
PUBMED	F1	0.5282	0.8416	0.8204	0.6451	0.8358	0.8065	**0.8454**
	ACC	0.5910	0.8414	0.8293	0.6488	0.8416	0.8171	**0.8470**
	Prec	0.5615	0.8428	0.8236	0.646	0.8367	0.8107	**0.8448**
	Recall	0.5318	0.8405	0.8176	0.6447	0.8353	0.8033	**0.8465**
	AUPRC	0.6006	0.9119	0.8852	0.7187	0.9061	0.8791	**0.9132**
	AUROC	0.7537	0.9520	0.9413	0.8551	0.9503	0.9354	**0.9537**

Experimental Setup. In all models, the dimension of the embedding vector *emb* is set to 128 by default. For *UAGE*, we set the initially learned intermediate vector dimension $F^{(0)}$ to 215. We set the iteration number t and the sampling factor c to 2 and 0.6, respectively. The reconstruction error parameter λ and loss weight α are set to 0.05 and 0.1. The number of negative edges for each node K is set to 10, and all model parameters are optimized using the Adam optimizer with a fixed learning rate of 0.001.

4.2 Performance Comparisons

Node Classification. In this task, we use the learned embeddings (mean vectors μ) as input to an *MLP* to classify nodes. Table 2 summarizes the average classification results of ten randomly sampled training runs on 3 datasets with 10% of labeled nodes as the training set. The results indicate that *UAGE* outperforms the baseline methods in all metrics. We also evaluate the performance of our model on datasets with more labeled nodes. Figure 2 shows the performance curves of F1 and AUPRC scores for each algorithm as the proportion of labeled nodes in all datasets increased from 10% to 50%. The results show that *UAGE* is more effective than other methods, which is particularly evident for the Citeseer and PUBMED datasets as the proportion of labeled nodes increases.

Node Clustering. In this task, we use the learned embeddings (mean vectors μ) for clustering using the k-means algorithm, where the cluster number is set

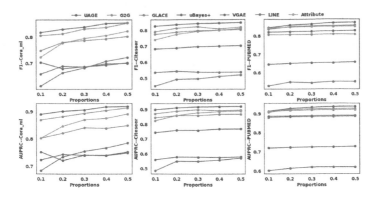

Fig. 2. Classification results with different proportions of labeled nodes.

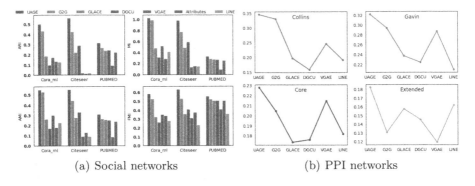

(a) Social networks   (b) PPI networks

Fig. 3. Clustering results on real-world networks.

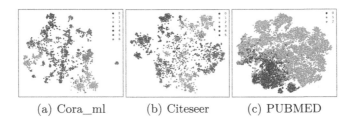

(a) Cora_ml   (b) Citeseer   (c) PUBMED

Fig. 4. The 2-dimensional visualizations of the latent representations.

to the number of categories in each dataset. As shown in Fig. 3(a), our method *UAGE* achieves the highest scores in all metrics and shows the best clustering performance, which is particularly evident in Citeseer. For PPI networks, we use a benchmark dataset of 323 authentic complexes from SGD [10] for evaluation and compare the F1 scores of our model with other algorithms. The results show that our model outperforms other algorithms, as shown in Fig. 3(b).

Visualization. In this task, we visualize the learned latent representation of the Cora_ml, Citeseer, and PUBMED datasets by mapping embeddings into 2-dimensional space using the *t-SNE* algorithm [16]. Figure 4 shows the *t-SNE* visualization results of our *UAGE* model, where the same color indicates the same class. The visualization results are visually appealing, with points of the same color forming segmented clusters and the boundaries of each class being clearer. This is particularly evident in dataset Citeseer. The results demonstrate that our model successfully captures the underlying structure of the graph and preserves the node attributes in the lower-dimensional space.

5   Conclusion

In this study, we proposed an efficient embedding framework *UAGE* for uncertain attribute graphs, which embeds nodes into Gaussian distribution space to account for the uncertainty present in the graph, preserving the graph's structural and attribute information. We evaluated *UAGE* on several real-world networks, and the results demonstrate the effectiveness of *UAGE* for downstream tasks such as node classification, clustering, and visualization. In conclusion, *UAGE* provides a promising direction for uncertain attribute graph embedding, improving downstream task performance by better handling uncertainty.

Acknowledgement. The authors would like to thank the support from the National Natural Science Foundation of China under Grant (No. 62172372), the Natural Science Foundation of Zhejiang Province, China (No. LZ21F030001, No. LQ22F020033), and the Exploratory Research project of Zhejiang Lab (No. 2022KG0AN01).

References

1. Banerjee, S.: A survey on mining and analysis of uncertain graphs. Knowl. Inf. Syst. **64**, 1–37 (2022)
2. Bojchevski, A., Günnemann, S.: Deep gaussian embedding of graphs: unsupervised inductive learning via ranking. In: ICLR (2018)
3. Boldi, P., Bonchi, F., Gionis, A., Tassa, T.: Injecting uncertainty in graphs for identity obfuscation. arXiv preprint arXiv:1208.4145 (2012)
4. Dallachiesa, M., Aggarwal, C., Palpanas, T.: Node classification in uncertain graphs. In: SSDBM (2014)
5. Danesh, M., Dorrigiv, M., Yaghmaee, F.: DGCU: a new deep directed method based on gaussian embedding for clustering uncertain graphs. Comput. Electr. Eng. **101**, 108066 (2022)
6. Grover, A., Leskovec, J.: node2vec: Scalable feature learning for networks. In: SIGKDD, pp. 855–864 (2016)
7. Halim, Z., Waqas, M., Hussain, S.F.: Clustering large probabilistic graphs using multi-population evolutionary algorithm. Inf. Sci. **317**, 78–95 (2015)
8. Hamilton, W., Ying, Z., Leskovec, J.: Inductive representation learning on large graphs. In: Advances in Neural Information Processing Systems, vol. 30 (2017)

9. Hettige, B., Li, Y.-F., Wang, W., Buntine, W.: Gaussian embedding of large-scale attributed graphs. In: Borovica-Gajic, R., Qi, J., Wang, W. (eds.) ADC 2020. LNCS, vol. 12008, pp. 134–146. Springer, Cham (2020). https://doi.org/10.1007/978-3-030-39469-1_11

10. Hong, E.L., et al.: Gene ontology annotations at SGD: new data sources and annotation methods. Nucleic Acids Res. 36(suppl_1), D577–D581 (2007)

11. Hu, J., Cheng, R., Huang, Z., Fang, Y., Luo, S.: On embedding uncertain graphs. In: CIKM, pp. 157–166 (2017)

12. Hussain, S.F., Maab, I.: Clustering probabilistic graphs using neighbourhood paths. Inf. Sci. 568, 216–238 (2021)

13. Kipf, T.N., Welling, M.: Variational graph auto-encoders. arXiv preprint arXiv:1611.07308 (2016)

14. Kollios, G., Potamias, M., Terzi, E.: Clustering large probabilistic graphs. TKDE 25(2), 325–336 (2011)

15. Kumar, S., Mallik, A., Khetarpal, A., Panda, B.: Influence maximization in social networks using graph embedding and graph neural network. Inf. Sci. 607, 1617–1636 (2022)

16. Laurens, V.D.M., Hinton, G.: Visualizing data using t-SNE. J. Mach. Learn. Res. 9(2605), 2579–2605 (2008)

17. Liang, Y., Hu, T., Zhao, P.: Efficient structural clustering in large uncertain graphs. In: ICDE, pp. 1966–1969 (2020)

18. Liu, H., Zhang, X., Zhang, X., Li, Q., Wu, X.M.: RPC: representative possible world based consistent clustering algorithm for uncertain data. Comput. Commun. 176, 128–137 (2021)

19. Mikolov, T., Sutskever, I., Chen, K., Corrado, G., Dean, J.: Distributed representations of words and phrases and their compositionality. In: NIPS, pp. 3111–3119 (2013)

20. Nepusz, T., Yu, H., Paccanaro, A.: Detecting overlapping protein complexes in protein-protein interaction networks. Nat. Methods 9(5), 471–472 (2012)

21. Peng, H., et al.: Dynamic graph convolutional network for long-term traffic flow prediction with reinforcement learning. Inf. Sci. 578, 401–416 (2021)

22. Perozzi, B., Al-Rfou, R., Skiena, S.: Deepwalk: online learning of social representations. In: SIGKDD, pp. 701–710 (2014)

23. Potamias, M., Bonchi, F., Gionis, A., Kollios, G.: K-nearest neighbors in uncertain graphs. Proc. VLDB Endowment 3(1–2), 997–1008 (2010)

24. Qiu, Y.X., et al.: Efficient structural clustering on probabilistic graphs. TKDE 31(10), 1954–1968 (2019)

25. Sun, G., Zhang, X.: A novel framework for node/edge attributed graph embedding. In: Advances in Knowledge Discovery and Data Mining, pp. 169–182 (2019)

26. Tang, J., Qu, M., Wang, M., Zhang, M., Yan, J., Mei, Q.: Line: large-scale information network embedding. In: WWW (2015)

27. Vilnis, L., McCallum, A.: Word representations via gaussian embedding. arXiv preprint arXiv:1412.6623 (2014)

28. Wang, D., Cui, P., Zhu, W.: Structural deep network embedding. In: SIGKDD, pp. 1225–1234 (2016)

29. Wang, X., Bo, D., Shi, C., Fan, S., Ye, Y., Yu, P.S.: A survey on heterogeneous graph embedding: methods, techniques, applications and sources. IEEE Trans. Big Data 1-1 (2022)

30. Xu, M.: Understanding graph embedding methods and their applications. SIAM Rev. **63**(4), 825–853 (2021)
31. Yu, W., Lin, X., Zhang, W., Pei, J., McCann, J.: Simrank*: effective and scalable pairwise similarity search based on graph topology. VLDB J. **28**, 401–426 (2019)
32. Zhu, D., Cui, P., Wang, D., Zhu, W.: Deep variational network embedding in Wasserstein space. In: SIGKDD, pp. 2827–2836 (2018)

Double-Layer Attention for Long Sequence Time-Series Forecasting

Jiasheng Ma, Xiaoye Wang$^{(\boxtimes)}$, and Yingyuan Xiao

School of Computer Science and Engineering, Tianjin University of Technology,
Tianjin 300384, China
jiaoliu456@163.com, yyxiao@tjut.edu.cn

Abstract. Time series forecasting (TSF) is crucial in many real-world applications. This paper studies the long-term forecasting problem of time series. Recent research has demonstrated that Transformer-based forecasting models can enhance forecasting accuracy, but their computational demands present a significant challenge for Long Sequence Time-series Forecasting (LSTF). To mitigate this, some researchers propose using a sparse attention network to reduce computational costs, but this approach can result in low information utilization and hinder long-term forecasting performance. This limitation impacts the overall effectiveness of the forecasting model. To address this issue, a new approach called Double-layer Efficient ProbSparse self-attention (DEPformer) is proposed in this paper for Long Sequence Time-series Forecasting. It combines a sparse attention network with an attention network that extracts global context vectors. This approach improves upon the low information utilization of sparse attention alone and enhances long-term forecasting performance. Experiments using standard and real datasets show that DEPformer outperforms the previous mainstream models.

Keywords: Long sequence time-series forecasting · ProbSparse self-attention · Efficient self-attention

1 Introduction

TSF has become increasingly prevalent in diverse fields including energy consumption, voltage sensor network detection, and weather prediction. In such practical scenarios, the ability to accurately forecast long-term future trends (LSTF) is crucial for enabling effective strategic planning and providing advanced warning. Traditional TSF methods, including the Autoregressive Integrated Moving Average (ARIMA) model [1], Support Vector Machine (SVM) [2] and the Holt-Winters seasonal method [3], offer theoretical guarantees. However, their application is primarily limited to univariate forecasting problems, which restricts their practicality when dealing with complex, real-world time series data. While conventional approaches have centered on parametric models grounded in field-specific knowledge, such as autoregressive (AR) [4] and exponential smoothing [5,6], contemporary machine learning techniques offer a means of acquiring knowledge of

C. Strauss et al. (Eds.): DEXA 2023, LNCS 14147, pp. 230–244, 2023.
https://doi.org/10.1007/978-3-031-39821-6_19

temporal dynamics through purely data-driven means [7]. As data access and computing capacity have increased in recent times, machine learning has emerged as an essential component of the upcoming cohort of time series forecasting models. However, these methods exhibit limitations in effectively handling forecasting problems with long time series. In recent years, the field of TSF has seen a significant advancement through the development of deep learning techniques. These methods have demonstrated superior forecasting accuracy when compared to traditional forecasting technologies. Among the various deep learning models, the Transformer-based model has emerged as the most prominent. Its effectiveness can be attributed to the incorporation of a self-attention mechanism, which allows the model to capture remote dependencies more effectively than previous approaches.

However, the Transformer's complexity grows exponentially with the length of the time series, rendering direct modeling of long time series infeasible. To address this challenge, Zhou and Peng et al. proposed a novel model called Informer [8], which is based on the Transformer architecture and designed to solve LSTF problem. However, due to the sparse point-by-point connection mode, using the sparse attention network alone may lead to the problem of low information utilization, which leads to the bottleneck of long-term prediction of time series and further affects the prediction effect. To solve this problem, this paper proposes a new model that combines sparse attention network with an efficient attention network that computes attention by extracting global context vectors. For simplicity, name this model DEPformer. The main contributions of this work are as follows:

- The global context vector extraction is added to the ProbSparse attention network, to calculate attention through the extracted global context vector to further improve the forecasting accuracy.
- We propose a new model, named DEPformer, which combines the ProbSparse attention network with an efficient attention network that computes attention by extracting global context vectors. DEPformer model makes up for the defect of using single sparse attention network through the combination of dual attention networks.
- Our proposed model has been extensively experiments on the public standard data set and the real data set, and the experimental results show that our model is superior to the previous mainstream models, in which the Double-layer Efficient ProbSparse self-attention has played an important role. The experimental results confirm the superiority of our proposed model.

2 Related Work

In this section, a review is presented on conventional TSF techniques as well as Transformer-based methods for time series analysis.

2.1   Traditional Time Series Forecasting Models

Because of the great importance of TSF, people have proposed various methods to solve this problem. Classical time-series models are a reliable and indispensable tool in the field of time-series forecasting, offering attractive features such as interpretability and theoretical guarantees [9,10]. ARIMA [1] solves the forecasting problem by transforming the non-stationary process into a stationary process by difference. In addition, the recurrent neural network (RNN) model is used to model the time correlation of time series [11–14]. DeepAR [15] models the probability distribution of future series by combining autoregressive method and RNN. LSTNet [16] captures the short-term and long-term time models by introducing the convolutional neural network (CNN) with recurrent-skip connection. The attention-based RNN [17–19] explores the long-term dependence of forecasting by introducing temporal attention. In addition, there are many studies based on temporal convolution network (TCN) [20–22] trying to model temporal causality by using causal convolution.

2.2   Transformer-Based Time Series Models

Inspired by the success of Transformers in CV and NLP, the TSF model based on Transformer has been actively studied recently. LogTrans [23] introduces the local context into the Transformer model by querying the causal convolution in the key projection layer, and proposes the LogSparse focus to reduce the complexity to $O(LlogL)$. Informer [8] extends Transformer by proposing ProbSparse attention and distillation operations to achieve $O(LlogL)$ complexity, and alleviates the problem that the complexity of traditional attention in Transformer increases twice with the length of the sequence. However, using a sparsely processed attention network alone may result in low information utilization, which in turn can become a bottleneck for long-term time series forecasting and ultimately impact the accuracy of the forecast. In order to solve this problem, this paper not only integrates a special attention with a global mechanism into sparse attention, but also integrates a layer of this attention network with a global mechanism on the updated sparse attention to alleviate the problem of low information utilization.

3   The Proposed Model: DEPformer

3.1   Problem Definition

Firstly, we introduce the problem of time series prediction, which is defined as follows: in a sliding window prediction scenario, at time t, the input is represented by $X^t = \{x_1^t, .., x_{L_x}^t | x_i^t \in R^{d_x}\}$ and the output is a predicted sequence $Y^t = \{y_1^t, .., y_{L_y}^t | y_i^t \in R^{d_y}\}$. Long-term time series forecasting aims to predict a longer horizon, that is,the longer L_y, and encourage the prediction variable to not limit the situation of single variable ($d_y \geq 1$).

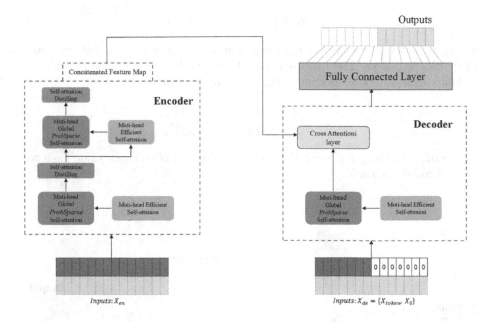

Fig. 1. Schematic diagram of DEPformer model

3.2   DEPformer Model

The DEPformer model is shown in Fig. 1. The DEPformer model consists of two key components, encoder and decoder. The encoder converts the input information into a dense vector with fixed dimensions, and extracts features from elements to generate feature maps. On the contrary, the decoder combines information and feature mapping to jointly predict the output. The DEPformer model proposed in this paper is an encoder-decoder architecture for LSTF problems, and has good information utilization and predictive performance.

Efficient Self-attention. Efficient self-attention is shown in Fig. 2. Efficient Attention [24] is a mechanism different from the traditional attention network. This mechanism is equivalent to the click attention mechanism, but it is faster and more efficient. It is still to map the input into three eigenvectors through three linear layers, namely $Q \in R^{L \times d_q}$, $K \in R^{L \times d_k}$, $V \in R^{L \times d_v}$. What is different from the traditional attention mechanism is that Efficient attention no longer takes the dot product of Q and $K^{\top}$ calculated by the normalization function as the weight of all positions, but uses each feature in these feature maps as the weight of all positions, and all eigenvalues are aggregated to form a global context vector by weighted summation. This also reflects a phenomenon that the vector formed by normalization does not correspond to a specific position, but represents the global description of input features. The effective attention formula is as follows:

$$E(Q, K, V) = \rho_q(Q)(\rho_k(K)^\top V) \tag{1}$$

where ρ_q and ρ_k represent the normalization functions of the two feature vectors of queries and keys. The main normalization function used in this paper is the softmax function. The specific functions are as follows:

$$Softmax : \rho_q(Y) = \sigma_{row}(Y) \tag{2}$$
$$\rho_k(Y) = \sigma_{col}(Y)$$

where σ_{row} and σ_{col} represent the function mapping of softmax along each row or column of matrix Y.

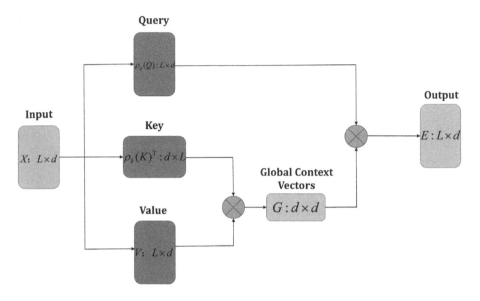

Fig. 2. Efficient Self-attention

Global ProbSparse Self-attention. After analyzing the Efficient attention, we propose Global ProbSparse self-attention, which is an attention mechanism that integrates the global mechanism of Efficient attention into sparse attention. This novel sparse attention network not only has a sparse mechanism, but instead of calculating the weights of all positions according to the normalization of traditional attention, it uses each feature in the feature map as a weight for all positions, and aggregates all features into a sparse global context vector by weighted summation. Global ProbSparse attention is shown in Fig. 3. The Global ProbSparse self-attention formula is as follows:

$$A(Q, K, V) = \frac{Softmax(\bar{Q}) \times (Softmax(K^\top)V)}{\sqrt{d}} \tag{3}$$

where $\bar{Q}$ is the filtered sparse matrix, which is the top u Qs with a higher score calculated by the sparsity measurement formula, and $Q \in R^{u \times d}$, $K \in R^{L \times d}$, $V \in R^{L \times d}$. The mapping matrix of global attention score is $d \times d$. Its meaning is no longer the similarity matrix between input sequences, but the global description of input characteristics. Finally, the global attention map and the filtered Q are weighted and summed to obtain the final output of Global ProbSparse attention. The activity of input elements needs to be calculated before calculating the attention of Global ProbSparse self-attention. The activity of elements is calculated through sparsity measurement. The sparsity measurement method of Global ProbSparse self-attention is as follows:

$$\bar{M}(q_i, K) = \max_j \left\{ \frac{q_i k_j^\top}{\sqrt{d}} \right\} - \frac{1}{L_K} \sum_{j=1}^{L_K} \frac{q_i k_j^\top}{\sqrt{d}} \qquad (4)$$

where L_K is $1/q(k_j|q_i)$; q_i is the i times query measured by sparsity, and q_i and k_j are from Q and K, respectively. The Global ProbSparse self-attention selects the Q with higher score in the sparsity measurement, and uses these screened Q to calculate the corresponding attention.

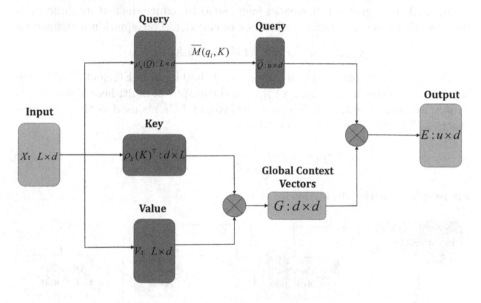

Fig. 3. Global ProbSparse Self-attention

Encoder. The experiments in this paper show that Global ProbSparse Self-attention has a better prediction effect than the original sparse attention network, but it also has the problem of low information utilization. Therefore, in this paper, a layer of Efficient attention is integrated in all places where Global ProbSparse self-attention is used in the encoder and decoder to alleviate this situation. First, the encoder receives the encoded input matrix, and then calculates through the double-layer attention network. The double-layer attention network D is defined as follows:

$$D(Q, K, V) = A(Q, K, V) \otimes Softmax(E(Q, K, V)) \qquad (5)$$

where A is formula (3) and E is formula (1). From the formula, we can see that the output of the double-layer attention network is to calculate two kinds of attention respectively, and then calculate the output of Efficient Self-attention through $softmax$, and then multiply it with the output of the element at the corresponding position of Global ProbSparse self-attention to obtain the final Output. The purpose of this method is to make the information representing the global feature description extracted from Efficient Self-Attention can be integrated into Global ProbSparse Self-Attention, so as to make up for the problem of low information utilization caused by only using sparse attention, so as to improve the prediction effect.

Self-attention Distilling. Self-attention distilling is shown in Fig. 4. After passing through the double-layer attention network, it will go through the self-attention distilling layer, which adds convolution, activation and maximum pooling operations between each encoder layer, so as to reduce the feature dimension, thereby improving the stacking efficiency of encoder. The equation is defined as:

$$X_{j+1}^t = MaxPool(ELU(Conv1d([X_j^t]_{AB}))) \qquad (6)$$

where X_{j+1}^t is the output of the multi-head double-layer Efficient ProbSparse self-attention layer in this layer; $[X_j^t]_{AB}$ is the output of multi-head double-layer ProbSparse self-attention in the previous layer; $ELU(.)$ is used as the activation function. The ELU function is as follows:

$$ELU(x) = \begin{cases} x, & x > 0 \\ a(e^x - 1), & x \leq 0 \end{cases} \qquad (7)$$

where a is a positive decimal close to 0.

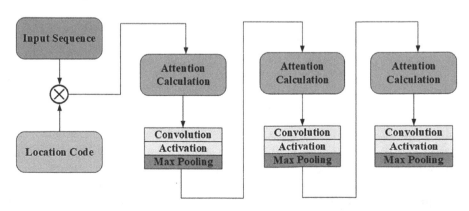

Fig. 4. Self-attention distilling

Decoder. The decoder part uses the method of batch generation to directly output the multi-step prediction results at one time, thus improving the efficiency of long series prediction. The equation is as follows:

$$X_{de}^t = Concat(X_{token}^t, X_0^t) \in R^{(L_{token}+L_y) \times d_{model}}$$ (8)

where X_{de}^t is the predicted output of the decoder, $X_{token}^t \in R^{L_{token} \times d_{model}}$ representations a known sequence prior to the predicted time, X_0^t indicates the timestamp of prediction results, L_{token} is the length of the sequence of start tokens, L_y is the length of the sequence to be predicted, and d_{model} is the dimension of the model. Unlike the traditional coder, it needs to input a period of time series X_{token}^t in advance, and X_0^t uses 0 to fill in and mask, so as to predict this part of the hidden content. For example, if we want to predict the temperature data of the next 7 days, we will use the known data of the past 7 days as "start token", set it as X_{7d}, and enter it as $X_{de} = X_{7d}, X_0$. It can be understood that the input part contains a part of known information, which is used as prompt information to jointly assist the prediction. Finally, the prediction results are obtained through a full connection layer.

4 Experiments

4.1 Datasets

In our experiment, we used ETT(Electric Transformer Temperature) dataset and Weather dataset.

ETT (Electricity Transformer Temperature) [8]: ETT is important indicators for long-term deployment of electricity. It collects two years of data from two different regions in one province of China. In order to explore the granularity of the LSTF problem, this article selected datasets ETTm_1 for 15-minute-level and {ETTh_1, ETTh_2} for 1-hour-level. Each sample has seven features, including oil temperature and six different types of external power load features.

Weather: This is a dataset containing the local climate data of nearly 1600 regions in the United States from 2010 to 2013. The data is recorded on an hourly basis, with each data point consisting of the target variable "wet bulb" and 11 other climate parameters.

4.2 Baselines

This paper has selected 7 baseline methods, including Informer [8], Reformer [25], LogTrans [23], LSTnet [16], LSTMa [26], Prophet [27], DeepAR [15] and ARIMA [28].

4.3 Hyper-parameter Tuning

In the experiments in this paper, hyperparameters are optimized by performing a grid search on the dataset. DEPformer contains a 3−layer stack and a 1−layer stack (1/4 input) in the encoder, and a 2−layer decoder. The optimizer is Adam and learning rate starts from le^{-4}, decaying twice every epoch. For the double-layers Efficient ProbSparse self-attention mechanism, this paper sets the same settings for Global ProbSparse self-attention and Efficient self-attention, both set to $d = 32$, $n = 16$. In addition, a position-wise feed-forward network layer (inner-layer dimension is 2048) and a dropout layer ($p = 0.1$) likewise. All experiments are conducted at 5 random train/val shifting selection along time, and the results are averaged over five runs. The total number of epochs is 6 with proper early stopping. This paper sets up the comparison methods as suggested, with a batch size is 32.

Setup: The input of each dataset is zero-mean normalized. Under the LSTF settings, this paper gradually extends the prediction windows size L_y progressively, i.e., {6 h, 12 h, 24 h, 72 h, 168 h} in ETTm, {1d, 2d, 7d, 14d, 30d, 40d} in {ETTh, Weather}. In addition, this paper uses the following two evaluation metrics:

$$MSE = \frac{1}{n}\sum_{i=1}^{n}(y - \hat{y})^2 \tag{9}$$

$$MAE = \frac{1}{n}\sum_{i=1}^{n}|y - \hat{y}| \tag{10}$$

Use these two indicators on each prediction window (averaging for multivariate prediction), and roll the whole set with $stride = 1$. And for all methods, the input length of recurrent component is chosen from $\{24, 48, 96, 168, 336, 720\}$ for the $ETTh_1$, $ETTh_2$ and Weather dataset, and chosen from $\{24, 48, 96, 192, 288, 672\}$ for the $ETTm_1$ dataset.

4.4   Results and Analysis

Table 1 and Table 2 summarize the univariate/multivariate evaluation results of all the methods on 4 datasets. Best results are highlighted in bold.

Univariate Time-Series Forecasting. Table 1 shows the comparison between the experimental results of the seven baseline models of univariate prediction and DEPformer model on four datasets. From the table, it can be found that DEPformer achieves the best performance on all datasets, which also shows the advantages of the DEPformer model. In the experiment, the dynamic decoding maintained by the Reformer does not perform well in LSTF, while the effect of other methods is better than that of it, which may be the reason why the generative style decoder as non-autoregressive predictors. Informer model is obviously superior to recurrent neural network LSTMa, which shows that the shorter network path in the self-attention mechanism can obtain better prediction effect than the model based on RNN. And from the experimental results, it can be seen that as the prediction length increases, the prediction error of our model increases steadily and slowly, which shows that DEPformer is successful in the prediction ability of LSTF.

Table 1. Univariate long sequence time-series forecasting results on four datasets.

Methods		DEPformer	Informer	LogTrans	Reformer	LSTMa	DeepAR	ARIMA	Prophet
Metric		MSE MAE	MSE MAE	MSE MAE	MSE MAE	MSE MAE	MSE MAE	MSE MAE	MSE MAE
$ETTh_1$	24	**0.091 0.233**	0.098 0.247	0.103 0.259	0.222 0.389	0.114 0.272	0.107 0.280	0.108 0.284	0.115 0.275
	48	**0.155 0.314**	0.158 0.319	0.167 0.328	0.284 0.445	0.193 0.358	0.162 0.327	0.175 0.424	0.168 0.330
	168	**0.162 0.322**	0.183 0.346	0.207 0.375	1.522 1.191	0.236 0.392	0.239 0.422	0.396 0.504	1.224 0.763
	336	**0.207 0.361**	0.222 0.387	0.230 0.398	1.860 1.124	0.590 0.698	0.445 0.552	0.468 0.593	1.549 1.820
	720	**0.248 0.408**	0.269 0.435	0.273 0.463	2.112 1.436	0.683 0.768	0.658 0.707	0.659 0.766	2.735 3.253
$ETTh_2$	24	**0.087 0.219**	0.093 0.240	0.102 0.255	0.263 0.437	0.155 0.307	0.098 0.263	3.554 0.445	0.199 0.381
	48	**0.139 0.297**	0.155 0.314	0.169 0.348	0.458 0.545	0.190 0.348	0.163 0.341	3.190 0.474	0.304 0.462
	168	**0.221 0.382**	0.232 0.389	0.246 0.422	1.029 0.879	0.385 0.514	0.255 0.414	2.800 0.595	2.145 1.068
	336	**0.261 0.412**	0.263 0.417	0.267 0.437	1.668 1.228	0.558 0.606	0.604 0.607	2.753 0.738	2.096 2.543
	720	**0.265 0.423**	0.277 0.431	0.303 0.493	2.030 1.721	0.640 0.681	0.429 0.580	2.878 1.044	3.355 4.664
$ETTm_1$	24	**0.027 0.132**	0.030 0.137	0.065 0.202	0.095 0.228	0.121 0.233	0.091 0.243	0.090 0.206	0.120 0.290
	48	**0.061 0.184**	0.069 0.203	0.078 0.220	0.249 0.390	0.305 0.411	0.219 0.362	0.179 0.306	0.133 0.305
	96	**0.186 0.367**	0.194 0.372	0.199 0.386	0.920 0.767	0.287 0.420	0.364 0.496	0.272 0.399	0.194 0.396
	288	**0.398 0.547**	0.401 0.554	0.411 0.572	1.108 1.245	0.524 0.584	0.948 0.795	0.462 0.558	0.452 0.574
	672	**0.496 0.629**	0.512 0.644	0.598 0.702	1.793 1.528	1.064 0.873	2.437 1.352	0.639 0.697	2.747 1.174
$Weather$	24	**0.104 0.234**	0.117 0.251	0.136 0.279	0.231 0.401	0.131 0.254	0.128 0.274	0.219 0.355	0.302 0.433
	48	**0.171 0.316**	0.178 0.318	0.206 0.356	0.328 0.423	0.190 0.334	0.203 0.353	0.273 0.409	0.445 0.536
	168	**0.263 0.381**	0.266 0.398	0.309 0.439	0.654 0.634	0.341 0.448	0.293 0.451	0.503 0.599	2.441 1.142
	336	**0.291 0.409**	0.297 0.416	0.359 0.484	1.792 1.093	0.456 0.554	0.585 0.644	0.728 0.730	1.987 2.468
	720	**0.324 0.445**	0.359 0.466	0.388 0.499	2.087 1.534	0.866 0.809	0.499 0.596	1.062 0.943	3.859 1.144

Table 2. Multivariate long sequence time-series forecasting results on four datasets.

Methods		DEPformer	Informer	LogTrans	Reformer	LSTMa	LSTnet
Metric		MSE MAE	MSE MAE	MSE MAE	MSE MAE	MSE MAE	MSE MAE
$ETTh_1$	24	**0.510 0.529**	0.577 0.549	0.686 0.604	0.991 0.754	0.650 0.624	1.293 0.901
	48	**0.653 0.607**	0.685 0.625	0.766 0.757	1.313 0.906	0.702 0.675	1.456 0.960
	168	**0.929 0.746**	0.931 0.752	1.002 0.846	1.824 1.138	1.212 0.867	1.997 1.214
	336	**1.022 0.863**	1.128 0.873	1.362 0.952	2.117 1.280	1.424 0.994	2.655 1.369
	720	**1.197 0.891**	1.215 0.896	1.397 1.291	2.415 1.520	1.960 1.322	2.143 1.380
$ETTh_2$	24	**0.706 0.659**	0.720 0.665	0.828 0.750	1.531 1.613	1.143 0.813	2.742 1.457
	48	**1.433 0.954**	1.457 1.001	1.806 1.034	1.871 1.735	1.671 1.221	3.567 1.687
	168	3.466 **1.508**	3.489 1.515	4.070 1.681	4.660 1.846	4.117 1.674	**3.242** 2.513
	336	2.711 **1.324**	2.723 1.340	3.875 1.763	4.028 1.688	3.434 1.549	**2.544** 2.591
	720	**3.443 1.468**	3.467 1.473	3.913 1.552	5.381 2.015	3.963 1.788	4.625 3.709
$ETTm_1$	24	**0.315 0.368**	0.323 0.369	0.419 0.412	0.724 0.607	0.621 0.629	1.968 1.170
	48	**0.472 0.483**	0.494 0.503	0.507 0.583	1.098 0.777	1.392 0.939	1.999 1.215
	96	**0.657 0.605**	0.678 0.614	0.768 0.792	1.433 0.945	1.339 0.913	2.762 1.542
	288	**0.991 0.764**	1.056 0.786	1.462 1.320	1.820 1.094	1.740 1.124	1.257 2.076
	672	**1.095 0.859**	1.192 0.926	1.669 1.461	2.187 1.232	2.736 1.555	1.917 2.941
$Weather$	24	**0.316 0.372**	0.335 0.381	0.435 0.477	0.655 0.583	0.546 0.570	0.615 0.545
	48	**0.376 0.417**	0.395 0.459	0.426 0.495	0.729 0.666	0.829 0.677	0.660 0.589
	168	**0.587 0.542**	0.608 0.567	0.727 0.671	1.318 0.855	1.038 0.835	0.748 0.647
	336	**0.644 0.597**	0.702 0.620	0.754 0.670	1.930 1.167	1.657 1.059	0.782 0.683
	720	**0.807 0.718**	0.831 0.731	0.885 0.773	2.726 1.575	1.536 1.109	0.851 0.757

Multivariate Time-Series Forecasting. Table 2 shows the comparison between the experimental results of multivariate prediction in five baseline models and our model in four data sets. Compared with RNN-based LSTMa and CNN-based LSTnet, transformer-based methods (such as LogTrans, Reformer and Informer) have better prediction effect, and our method is better than the above transformer-based methods, which also shows that our method has better ability to capture long-term potential laws in the entire historical data, thus reducing the prediction error in the future.

4.5   Model Analysis and Discussion

In this section, we will discuss the proposed model in depth, and the experiment is only for multivariate time-series forecasting. The purpose is to better understand our model behavior and compare the effects of different hyperparameters on the model.

Table 3. Influence of different attention modules on the effect of the model.

Methods		DEPformer	Informer	DEPformer-P
Metric		MSE MAE	MSE MAE	MSE MAE
Weather	24	**0.316 0.372**	0.335 0.381	0.326 0.379
	48	**0.376 0.417**	0.395 0.459	0.389 0.431
	168	**0.587 0.542**	0.608 0.567	0.619 0.582
	336	**0.644 0.597**	0.702 0.620	0.681 0.610
	720	**0.807 0.718**	0.831 0.731	0.819 0.722

The Effect of the Attention Module. The model proposed in this paper contains two parts: encoder and decoder, and each part has two attention mechanisms. In this experiment, we not only want to analyze the effectiveness of the Global ProbSparse multi-headed attention network with a global mechanism added to sparse attention compared to ordinary sparse attention networks, but also want to prove whether adding efficient multi-headed attention to the Global ProbeSparse multi-headed attention network can actually improve the prediction effect. This paper divides it into three models: DEPformer-P is a model that uses the Global ProbSparse multi-headed attention network only, Informer model, and DEPformer is a double-layer network model that uses two kinds of attention networks at the same time.

Table 3 shows the performance of the three models on the Weather dataset. It can be seen from the table that when the model uses a separate Global ProbSparse multi-head attention network, it is better than the traditional Informer model that uses sparse attention. This demonstrates the effectiveness of extracting global feature descriptions in sparse attention networks. However, the effect of DEPformer-P is lower than that of DEPformer. This may be due to the sparse effect of DEPformer-P, which results in some valuable data that may exist not being utilized, making its effect not as good as that of DEPformer using double-layer attention network. Compared with the other two models, the DEPformer model using a double-layer attention network has improved, which also shows the effectiveness of the double-layer attention network model proposed in this paper.

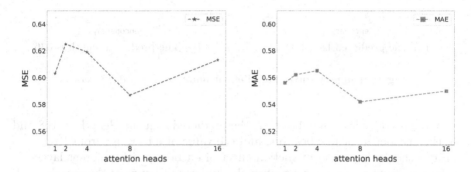

Fig. 5. The performance of different number of attention heads

The Effect of the Number of Attention Heads. The multi-head attention mechanism used in the model can expand the model to focus on different positions, allowing the model with this mechanism to better extract potential rules from longer sequences. However, if there are too many attention heads, the dimension of a single attention may become too small, which can limit the attention performance. Therefore, this paper conducted an experiment to investigate whether the number of heads in the double-layer attention network of the model has an impact on the model's prediction performance. In the experiment, we found that the number of heads in Global ProbSparse attention and Efficient attention had the same effect on the experimental results. Therefore, this paper will not analyze these two types of attention networks separately in this part of the experiment.

Figure 5 shows the effect of different number of attention heads on the effect of the model. This paper conducted experiments on the Weather dataset. The predicted sequence length was set 168. This paper compares the number of attention heads for five settings of 1, 2, 4, 8 and 16, respectively. From the results in the table, we can find that when the number of attention heads is 8, the model works best on the Weather dataset.

The Effect of the Number of Encoder Layers. The Encoder extracts long-range correlations to improve prediction accuracy in the decoder, so the number of encoder layers affects prediction performance significantly. To investigate the influence of different encoder layer numbers on the model's prediction effect, we conducted a comparative experiment on the Weather dataset. In this experiment,

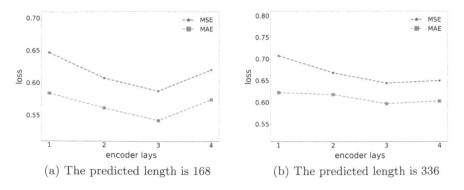

(a) The predicted length is 168   (b) The predicted length is 336

Fig. 6. The performance of different number of encoder layers

we set the decoder layer number to 2, the predicted sequence length to 168 and 336, the encoder layer number to e, and the value of e to range from 1 to 4.

Fig. 6 shows the model prediction effect of encoders with different layers. It can be seen from the figure that when the number of layers of the encoder is 3, the prediction effect of the model on the Weather dataset is the best.

5 Conclusion

In the context of long sequence time series forecasting (LSTF), using a traditional Transformer model may result in significant computational cost issues. On the other hand, the Transformer model using sparse attention network can greatly improve the computational efficiency and prediction accuracy. However, due to the use of sparse attention mechanisms, such models may have the problem of low information utilization. To alleviate this problem, this study proposes a double-layer attention mechanism. Experiments were conducted on public and real datasets, and good predictive performance was obtained.

Acknowledgments. This work is supported by "Tianjin Project + Team" Key Training Project under Grant No. XC202022

References

1. Box, G.E., Jenkins, G.M.: Some recent advances in forecasting and control. J. Roy. Stat. Soc. Ser. C (Appl. Stat.) **17**(2), 91–109 (1968)
2. Smola, A.J., Schölkopf, B.: A tutorial on support vector regression. Stat. Comput. **14**, 199–222 (2004)
3. Holt, C.C.: Forecasting seasonals and trends by exponentially weighted moving averages. Int. J. Forecast. **20**(1), 5–10 (2004)
4. Box, G.E., Jenkins, G.M.: Time Series Analysis: Forecasting and Control. Holden-Day, San Francisco (1976)

5. Gardner, E.S., Jr.: Exponential smoothing: the state of the art. J. Forecast. **4**(1), 1–28 (1985)
6. Winters, P.R.: Forecasting sales by exponentially weighted moving averages. Manag. Sci. **6**(3), 324–342 (1960)
7. Ahmed, N.K., Atiya, A.F., Gayar, N.E., El-Shishiny, H.: An empirical comparison of machine learning models for time series forecasting. Econ. Rev. **29**(5–6), 594–621 (2010)
8. Zhou, H., Zhang, S., Peng, J., Zhang, S., Li, J., Xiong, H., Zhang, W.: Informer: beyond efficient transformer for long sequence time-series forecasting. In: Thirty-Fifth AAAI Conference on Artificial Intelligence, AAAI 2021, Thirty-Third Conference on Innovative Applications of Artificial Intelligence, IAAI 2021, The Eleventh Symposium on Educational Advances in Artificial Intelligence, EAAI 2021, Virtual Event, 2–9 February 2021, pp. 11106–11115. AAAI Press (2021). https://ojs.aaai.org/index.php/AAAI/article/view/17325
9. Box, G.E., Jenkins, G.M., Reinsel, G.C., Ljung, G.M.: Time Series Analysis: Forecasting and Control. John Wiley & Sons, Hoboken (2015)
10. Ray, W.: Time Series: Theory and Methods (1990)
11. Wen, R., Torkkola, K., Narayanaswamy, B., Madeka, D.: A multi-horizon quantile recurrent forecaster (2017). arXiv preprint arXiv:1711.11053
12. Rangapuram, S.S., Seeger, M.W., Gasthaus, J., Stella, L., Wang, Y., Januschowski, T.: Deep state space models for time series forecasting. Adv. Neural Inf. Process. Syst. **31**, 1–10 (2018)
13. Yu, R., Zheng, S., Anandkumar, A., Yue, Y.: Long-term forecasting using higher order tensor rnns (2017). arXiv preprint arXiv:1711.00073
14. Maddix, D.C., Wang, Y., Smola, A.: Deep factors with gaussian processes for forecasting (2018). arXiv preprint arXiv:1812.00098
15. Salinas, D., Flunkert, V., Gasthaus, J., Januschowski, T.: Deepar: probabilistic forecasting with autoregressive recurrent networks. Int. J. Forecast. **36**(3), 1181–1191 (2020)
16. Lai, G., Chang, W.C., Yang, Y., Liu, H.: Modeling long-and short-term temporal patterns with deep neural networks. In: The 41st International ACM SIGIR Conference on Research & Development in Information Retrieval, pp. 95–104 (2018)
17. Qin, Y., Song, D., Chen, H., Cheng, W., Jiang, G., Cottrell, G.: A dual-stage attention-based recurrent neural network for time series prediction (2017). arXiv preprint arXiv:1704.02971
18. Shih, S.Y., Sun, F.K., Lee, H.Y.: Temporal pattern attention for multivariate time series forecasting. Mach. Learn. **108**, 1421–1441 (2019)
19. Song, H., Rajan, D., Thiagarajan, J., Spanias, A.: Attend and diagnose: clinical time series analysis using attention models. In: Proceedings of the AAAI Conference on Artificial Intelligence (2018)
20. Borovykh, A., Bohte, S., Oosterlee, C.W.: Conditional time series forecasting with convolutional neural networks (2017). arXiv preprint arXiv:1703.04691
21. Bai, S., Kolter, J.Z., Koltun, V.: An empirical evaluation of generic convolutional and recurrent networks for sequence modeling (2018). arXiv preprint arXiv:1803.01271
22. Sen, R., Yu, H.F., Dhillon, I.S.: Think globally, act locally: a deep neural network approach to high-dimensional time series forecasting. Adv. Neural Inf. Process. Syst. **32**, 1–10 (2019)
23. Li, S., et al.: Enhancing the locality and breaking the memory bottleneck of transformer on time series forecasting. Adv. Neural Inf. Process. Syst. **32**, 1–11 (2019)

24. Shen, Z., Zhang, M., Zhao, H., Yi, S., Li, H.: Efficient attention: attention with linear complexities. In: Proceedings of the IEEE/CVF Winter Conference on Applications of Computer Vision, pp. 3531–3539 (2021)
25. Kitaev, N., Kaiser, Ł., Levskaya, A.: Reformer: the efficient transformer (2020). arXiv preprint arXiv:2001.04451
26. Bahdanau, D., Cho, K., Bengio, Y.: Neural machine translation by jointly learning to align and translate (2014). arXiv preprint arXiv:1409.0473
27. Taylor, S.J., Letham, B.: Forecasting at scale. Am. Stat. **72**(1), 37–45 (2018)
28. Ariyo, A.A., Adewumi, A.O., Ayo, C.K.: Stock price prediction using the arima model. In: 2014 UKSim-AMSS 16th International Conference on Computer Modelling and Simulation, pp. 106–112. IEEE (2014)

Multi-core Adaptive Merging of the Secondary Index for LSM-Based Stores

Wojciech Macyna[✉], Michal Kukowski, and Michal Zwarzko

Department of Computer Science, Faculty of Information and Communication Technology,
Wrocław University of Science and Technology, Wrocław, Poland
{wojciech.macyna,michal.kukowski}@pwr.edu.pl,
261753@student.pwr.edu.pl

Abstract. NoSQL databases have gained great popularity recently. Most of them use the Log Structured Merge (LSM) tree which provides fast write throughput and fast lookup of primary keys. Nevertheless, searching by non-key attributes is very slow because the entire LSM-tree must be scanned. To overcome this problem, the secondary index can be used. Typically, all items in the database are equally covered by the secondary index. However, this is not effective in big data stores where some items are queried very often and some never. To solve this problem, adaptive merging has been introduced. The key idea is to create a secondary index adaptively as a side-product of query processing. Consequently, the database is indexed partially depending on the query workload.

The paper considers the adaptive merging of the secondary index in LSM-based stores. In this approach, the secondary index can be initiated at an arbitrary moment. Thereafter, only the requested data are inserted into the secondary index. They are retrieved from the independent immutable files created during the index initialization in a parallel way. The method can work in the dynamic database environment where database modifications interleave with user queries. The experiments show that the proposed approach outperforms traditional methods by about 30%.

Keywords: LSM-tree · adaptive indexing · NoSQL database

1 Introduction and Motivation

NoSQL databases have reached great popularity due to their scalability, simplicity, and flexibility. Nowadays, they are used in various areas such as social networks, sensor networks, and other big data applications. The most popular NoSQL database systems are HBase [1], Cassandra [2], AsterixDB [3], MongoDB [4], BigTable [5], LevelDB [6], and RocksDB [7]. Typically, NoSQL database systems store the data in Log Structured Merge (LSM) tree [8]. The LSM-tree consists of a few levels. Each level holds many key-value entries that are sorted within the level. A new entry can be inserted only into the top level. When the number of entries in the level exceeds the predefined limit, this level is merged with the level below.

In general, LSM-based stores support fast write throughput and fast lookups on primary keys. However, this is not enough to support effective query processing on non-key attributes. Let us consider a shopping database containing sale transactions defined

© The Author(s), under exclusive license to Springer Nature Switzerland AG 2023
C. Strauss et al. (Eds.): DEXA 2023, LNCS 14147, pp. 245–257, 2023.
https://doi.org/10.1007/978-3-031-39821-6_20

as $e = <id, \{user, product, date, quantity, price\}>$. It means that the *user* bought a *product* with a specific *quantity* and *price* on a particular *date*. In this case, a sale transaction identifier (id) is the key of entry e. Unfortunately, to find all transactions for the particular *product*, the whole LSM-tree must be scanned. Clearly, the secondary index on the *product* would improve the query execution significantly. However, secondary indexes cannot be used without any limits. When the database size grows the maintaining of the secondary index on the entire dataset may spoil system performance. Let us assume that the secondary index is imposed on the attribute *date*. If only records from the last year are queried, it is not necessary to index the whole table by *date*. In this case, it would be sufficient to index only the entries of the last year and leave the older ones not indexed. To cope with this problem, two main approaches of partial indexing were introduced: database cracking and adaptive merging.

This paper faces a problem of adaptive merging for secondary indexing in the LSM-tree storage. The secondary index is created adaptively as a consequence of range query processing. In the initial phase, the files (SSTables) of the main database are copied to the Adaptive Log. Then, when a range query arrives, the files of the Adaptive Log are scanned separately using different threads. To skip the irrelevant files, each file in the Adaptive Log is equipped with a bitmap and a range filter. The query results are inserted into the partial secondary index. In summary, the contribution of the paper is as follows:

- We propose the multi-core adaptive merging of the secondary index for LSM-tree. Although we assume that the secondary index is based on the LSM-tree, the approach is flexible enough to utilize the other index structures (for example B+ tree).
- Unlike most of the approaches, our framework is very efficient in the dynamic database environment where updates interleave with range queries.
- We did a real implementation of the framework and integrate it into LevelDB key-value storage [6] in a non-intrusive way.

The paper is organized as follows. Section 2 contains the related work. In Sect. 3, we first describe the adaptive merging of the LSM tree without dataset modification. Then, we extend the method so that it could consider modified data. In Sect. 4, we present several experiments that confirm the efficiency of the proposed method.

2   Related Work

NoSQL databases adopt two different strategies for secondary indexing: stand-alone index table and embedded secondary indexing (see [9]). In the first strategy, the secondary index is separated from the data. It uses B+-tree (MongoDB [4]) or LSM index table (BigTable [5], Cassandra [2], AsterixDB [3], Spanner [10]). The second strategy relies on storing the attribute information inside the original data blocks. In [11], the authors analyze the utilization of auxiliary structures like secondary indexes, range filters, primary key indexes, and bitmaps in the LSM-stores. They propose some validation and query processing strategies that use these structures. In [12], the authors explore the challenges associated with indexing modern distributed data stores and propose two different approaches for secondary indexing on HBase [1].

The first approach to partial indexing is described in [13]. The authors present a cracking architecture in the context of a full-fledged relational system. This concept is extended in [14]. The paper proposes novel algorithms for database cracking in a dynamic environment where updates interleave with queries. In [15], adaptive merging is introduced. In that approach, the data are first split into n sorted partitions. Then, a range query fetches the data from these partitions and inserts them into the secondary index. As a consequence, the process considers only such data that are relevant to actual range queries, leaving all other data in their initial places. Both approaches have some drawbacks. Database cracking has a low initialization cost but converges relatively slowly. On the other hand, adaptive merging enjoys a high initialization cost but converges rapidly. To overcome such limitations, a hybrid approach is proposed (see [16, 17]). The paper [18] concentrates on tuning the merge policy, the buffer size, and the Bloom filters across the LSM-tree's different levels. The authors in [19] analyze the index cracking algorithms in a parallel environment. They mainly check how the increasing number of threads impacts the efficiency of index cracking. The paper [20] concentrates on the adaptive merging for phase change memory (PCM). The authors apply several techniques to optimize this process taking into account PCM limitations.

3 Adaptive Indexing

This section presents the adaptive merging for LSM-based stores. We assume that the main database holds $<key, value>$ entries in the LSM-tree. When a range query on $value$ is executed, the main database is scanned and the required entries are fetched. Then, these entries are inserted into the secondary index. So, the secondary index is updated as a response to user queries. The architecture of our system consists of the following components:

- *Main LSM database.* The main data storage. The data are held in the LSM-tree which consists of several independent levels. Each level has many files called SSTables. Each file contains $<key, value>$ entries.
- *Partial Secondary Index.* The structure for storing the secondary index entries. An entry is of the form $<value, key>$. Please note that $value$ can be duplicated while $<value, key>$ cannot be.
- *Adaptive Log.* A set of files copied from the Main LSM database. Generally speaking, the files hold such entries $<value, key>$ that have not yet been added to the partial secondary index. Each file f is equipped with a range filter and a bitmap. The range filter: $<f.min, f.max>$ holds the minimum and maximum values in f. Each entry of the file f has its bit in the bitmap. It shows whether the entry is already in the partial secondary index or not. The range filter and the bitmap for f can be stored in RAM.
- *Index Range.* The tree-like structure stored in RAM that contains already indexed query ranges.

3.1   Adaptive Merging Without Database Modification

This subsection presents an intuitive description and basic algorithms of the approach. First, we assume that the database is not modified during the adaptive merging process (see Fig. 1).

Initially (step 1), the Main LSM database contains three levels and each level has a fixed number of SSTables. In the example, the SSTable can hold two key-value entries.

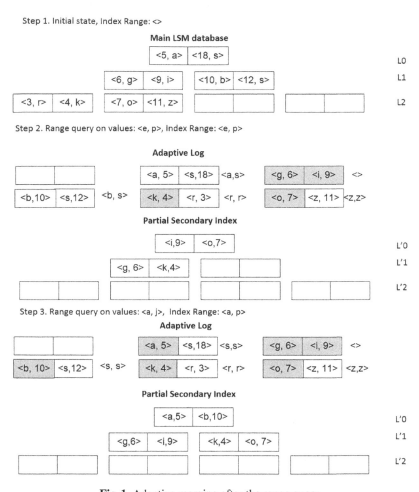

Fig. 1. Adaptive merging after the range query

The adaptive index is initiated when the range query arrives (step 2). In this case, the Adaptive Log is created. In the beginning, the Adaptive Log is a simple copy of all SSTables ordered by time. It means that the SSTables from the top level of the Main LSM database are at the front of the Adaptive Log and the SSTables from the oldest level are placed at the end of the Adaptive Log. The entries inside the file are sorted by *value* and each file holds a filter range. To answer the range query $q = <e, p>$, the system must iterate over all files in the Adaptive Log and fetch the entries that satisfy the query condition. It is easy to see that query q retrieves the following entries: $<g, 6>$, $<i, 9>$, $<k, 4>$ and $<o, 7>$. Then, these entries are inserted into the partial secondary index. In this example, the secondary index is based on LSM-tree, but this assumption is not obligatory. After that, the Adaptive Log is updated by setting the bitmap values to 0 for the already indexed entries (they are marked with the shadowed rectangle) and by changing the range filter of each file. Please note that the range filter only considers the entries with a bitmap value set to 1. So, the range filter of the third file is empty. At last, the query range $<e, p>$ is added to the Index Range.

In step 3, a range query $q_1 = <a, j>$ arrives. Obviously, the following entries fulfill the search condition: $<g, 6>$, $<i, 9>$, $<a, 5>$ and $<b, 10>$. As q_1 partially overlaps the actual Index Range ($<e, p>$), all entries in $<e, j>$ can be retrieved from the partial secondary index. To fetch entries of the range $<a, e>$, the Adaptive Log must be traversed. Fortunately, each file in the Adaptive Log is equipped with the bitmap and the range filter. So, the iteration process can skip the irrelevant files and entries. As a result, only two files in the Adaptive Log must be accessed. The entries fetched from the Adaptive Log are inserted into the partial secondary index. It is carried out in the same way as in the case of the "normal" LSM-tree. So, the entries $<a, 5>$ and $<b, 10>$ are added to $L'0$. If $L'0$ is full, the LSM-tree merge process must be invoked. At the end of the method, the range filters of all files in the Adaptive Log and the Index Range must be updated.

Code Description. When the Adaptive Log (L) is initiated, Algorithm 1.1 traverses all the levels of the Main LSM database and copies its files into Adaptive Log L. For each file f in L, the bitmap is created and the range filter $<f.min, f.max>$ is estimated. It denotes the minimum and maximum values in f. When a range query $r = <v_1, v_2>$ arrives, Algorithm 1.2 is invoked. The entry e with value v that fulfills the condition r can be found either in the partial secondary index or in the Adaptive Log. If v is already in the Index Range, it means that the value has been queried before. In this case, the entries can be directly fetched from the partial secondary index (line 6). Otherwise, Algorithm 1.3 is invoked. It handles the files of the Adaptive Log in parallel. When the condition in line 6 is fulfilled, the iterator seeks the entries for which the value is equal to v and its bit in the bitmap are set to 1 (line 11). If the entry is found, it is added to the result set, and its bit is set to 0 (lines 12 to 13). After that, the $f.min$ and $f.max$ values of file f are updated (lines 15 and 16). The entries returned by Algorithm 1.3 are then inserted into the partial secondary index and the $IndexRange$ is updated (see lines 10 and 12 of Algorithm 1.2).

3.2 Database Modification

In the previous subsection, we consider adaptive merging without database modification. Obviously, the database can be changed after adaptive merging initiation.

Algorithm 1.1: initAdaptiveLog(input: Main LSM database T)

---

1 Let *height* be the height of T
2 Let $SSTables[\text{k}]$ denote the SSTable set in level k of T
3 Let L denote the Adaptive Log
4 Let f denote a file in the Adaptive Log
5 Let $f.min$ and $f.max$ denote the minimum and maximum value of file f in L
6 $lvl := 1$
7 **while** $lvl <= height$ **do**
8   **foreach** *SSTable* $s \in SSTables[lvl]$ **do**
9       $f := \text{createFile}(s)$
10       $f.min := \text{getMinValue}(f)$
11       $f.max := \text{getMaxValue}(f)$
       // setting all bitmap values to 1
12       $\text{createBitmap}(f)$
13       $\text{addFileToAdaptiveLog}(L,f)$
14    $lvl := lvl + 1$
15 $IndexRange := \emptyset$

Algorithm 1.2: userQuery(input: QueryRange q, Main LSM database T, Partial Secondary Index S, output: Entry $resultSet[\ ]$)

---

1 Entry $resultSet[\ ] := \emptyset$
2 **if** S *is Empty* **then**
3   initAdaptiveLog(T)
4 **foreach** *Value* $v \in q$ **do**
5   **if** $v \in IndexRange$ **then**
      // fetching directly from the partial secondary index
6   Entry[] $e := \text{findInPartialSecondaryIndex}(v, S)$
7   **else**
      // fetching from the Adaptive Log
8   Entry[] $e := \text{findInAdaptiveLog}(v, L)$
9   **if** e *is not null* **then**
         // inserting into the partial secondary index
10          $\text{insertIntoPartialSecondaryIndex}(S, e)$
11    $resultSet := resultSet \cup e$
12 $IndexRange := IndexRange \cup q$
13 **return** $resultSet$

Typically, each modification of the LSM-tree is performed as an insertion. The new entry is simply inserted into the MemTable. When an entry $e = <key, val>$ must be deleted, a new deleted entry $e_d = <key, del>$ with the same key is inserted into

---

Algorithm 1.3: findInAdaptiveLog(input: Value v, AdaptiveLog L, output: Entry $resultSet[\,]$)

---
1 Entry $resultSet[\,] := \emptyset$
2 Let $f.min$ and $f.max$ denote the minimum and maximum value of file f in L
3 Let $files$ be an array of files in Adaptive Log L
4 Let $entries$ denote an array of entries in file f
5 **foreach** File $f \in files$ **do**
6   **if** $v \geq f.min$ and $v \leq f.max$ **then**
       // taking all entries in f
7         $entries[\,]:=$getEntries(f)
8   Int posEntry:=0
9   **while** $posEntry \leq entries.size$ **do**
10   Entry e := entries[posEntry]
           // when the entry is not in the partial secondary
             index (bitmap=1)
11   **if** $bitmap[posEntry]=1$ and $value(e) = v$ **then**
12              $resultSet:= resultSet \cup e$
13   bitmap[posEntry]:=0
14   posEntry: = posEntry +1
15        $f.min:=$getMinValue(f)
16        $f.max:=$getMaxValue(f)

17 return $resultSet$

---

the MemTable. Then, when the query requires entry e, it comes across e_d at first. In this case, entry e would not be considered in the query result, because it is marked as deleted. A similar situation happens when the entry $e = <key, val>$ must be updated. In that case, the old version of the entry is deleted (by insertion of e_d) and a new version $e_n = <key, val_n>$ is inserted into the MemTable.

In our approach, the database modifications are treated a similar. The insertion into the Main LSM database must be reflected in the adaptive merging. For that, we consider the following cases.

- *New entry.* When a new entry $e = <key, val>$ is inserted into the Main LSM database, a corresponding secondary index entry $e' = <val, key>$ is created. If val is inside the Index Range, entry e' is inserted directly into the partial secondary index. Otherwise, it is inserted at the beginning of the Adaptive Log.
- *Deleted entry.* When an entry $e = <key, val>$ is deleted from the Main LSM database, a new deleted entry $e'_d = <val, key, del>$ is created. If val is inside the Index Range, entry e'_d is inserted directly into the partial secondary index. Otherwise, it is inserted at the beginning of the Adaptive Log. Then, even when val fulfills a query condition, the corresponding entry would not be fetched by this query.
- *Updated entry.* As we mentioned before, when an entry $e = <key, val>$ is updated in the Main LSM database, the old version of the entry is deleted (by inserting of e_d) and a new version $e_n = <key, val_n>$ is inserted into the Main LSM database. Consequently, a new secondary entry $e'_n = <val_n, key>$ and a new deleted entry

$e'_d = <val, key, del>$ are created. The entries e'_n and e'_d are inserted into the partial secondary index or the Adaptive Log. It depends on whether the values val and val_n are inside the Index Range or not.

In the example on Fig. 2, we insert two new entries $<15, m>$ and $<19, x>$ into level L_0 of the Main LSM database. As a consequence, two new secondary index entries are created: $e_1 = <m, 15>$ and $e_2 = <x, 19>$ Please note that the Index Range is $<a, p>$. So, entry e_1 is added to L'_0 in the partial secondary index and entry e_2 to the Adaptive Log.

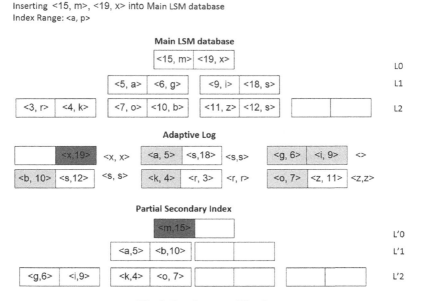

Fig. 2. Database modification

4 Experiments

In this section, we present some experiments that confirm the effectiveness of the method. We implement the adaptive merging framework as an extension to LevelDB, which is a very popular key-value storage used as a persistent level in many commercial database systems. The experiments are conducted on Dell PowerEdge R910 equipped with 4 CPU and 256 GB RAM. Each CPU has 20 cores and the cache lines: L1, L2, and L3 of 640 KB, 2560 KB, and 24 MB, respectively.

In each experiment, we create the Main LSM database with 10 million entries of the form $<key, value>$. Then, we generate a set of range queries. To do this, we fetch the minimum (v_{min}) and maximum (v_{max}) values from all entries. Then, we randomly choose v_1 with a uniform distribution from the range $<v_{min}, v_{max}>$. After that, we calculate v_2 so that the fixed query selectivity is obtained. By selectivity we mean which fraction of the total database entries must be queried. A range query fetches all entries for which *value* is between v_1 and v_2. We repeat this process until the whole set of range queries is generated. In the experiments, we execute the same set of range queries for the following methods.

1. *Full scan* - only the Main LSM database is used. The data are stored as $<key, value>$. To answer the range query with a predicate on *value*, the whole Main LSM database must be traversed.
2. *Secondary scan* - a secondary index is created for all entries of the Main LSM database before each experiment. The entries in this index are of the form: $<value, key>$ In this paper, the secondary index relies on the LSM-tree, so the entries are sorted by *value* within each level. A range query on *value* seeks the appropriate *key* in the secondary index and then, using this *key*, fetches the corresponding entry in the Main LSM database.
3. *Adaptive merging* - the approach described in this paper. When a secondary index is initiated, the SSTables of the Main LSM database are copied to the Adaptive Log. When a range query arrives, the Adaptive Log is scanned in parallel and the data fetched by the query are added to the partial secondary index. Then, the next queries can benefit from this index if their range overlaps with the range of the already stored data.

In the first experiment, we consider the creation time of the secondary index for the whole Main LSM database and compare it with the creation time of the Adaptive Log in the adaptive merging. Figure 3 presents elapsed time depending on the thread number. As the Adaptive Log is a set of independent files, it may be created separately by several threads. On the contrary, the secondary index used in the secondary scan must be created sequentially since the sorting order of the keys has to be retained. It explains the big overhead difference between both methods. Figure 4 shows index creation time depending on the data size. Clearly, when the data size grows, creation time gets larger as well. All the subsequent experiments include the construction time in the runtime.

In the next experiment, we measure the time elapsed after 100 range queries with different selectivity (see Fig. 5). We can observe that the adaptive merging shows better performance than the secondary scan. When the selectivity increases, elapsed time gets higher. We can also see that elapsed time of the full scan is independent of the selectivity. It comes from the fact that the whole Main LSM database must be traversed for each query. In Fig. 6, we consider a different data size. Clearly, elapsed time grows proportionally with the data size.

In Fig. 7 we measure time for different query numbers. We clearly show that the time increases with the increasing number of queries. The adaptive merging outperforms the secondary scan by about 30%. We can also see that the adaptive merging performance gets closer to the performance of the secondary scan with the increasing number of queries. Clearly, both methods are drastically better than the full scan. Figure 8 compares a different dataset range used in the queries. It means that the query set fetches the data not from the whole database but only from a fixed fraction of the dataset. When

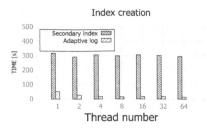

Fig. 3. Different thread number

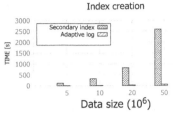

Fig. 4. Different data size

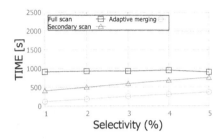

Fig. 5. 100 queries, 10 million entries, different selectivity

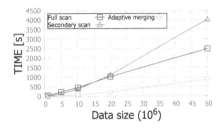

Fig. 6. 50 queries, 1% selectivity, different data size (10^6)

the system uses a small dataset range (for example 5%), the performance of the adaptive merging is better since more data is indexed in the partial secondary index and fewer data must be fetched from the Adaptive Log.

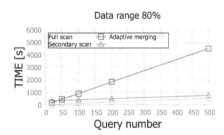

Fig. 7. 10 million entries, data range: 80%, selectivity: 1%, different query number

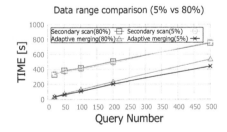

Fig. 8. 10 million entries, selectivity: 1%, different data range

In the next experiment (see Fig. 9), we execute the same query 100 times and compare it with the execution of different queries 100 times. We see that when the data retrieved by the query are already in the partial secondary index, the system does not have to traverse through the Adaptive Log and better performance is obtained.

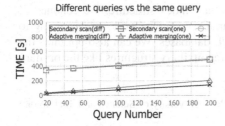

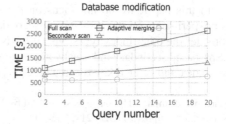

Fig. 9. 10 million entries, selectivity: 1%, different queries vs the same query

Fig. 10. 10 batches, different range query number in each batch

The last experiment assumes that the database is modified after the partial secondary index initialization. In this example, we execute 10 batches. Each batch consists of a fixed number of range queries, insert and delete operations. Figure 10 shows an example where the number of range queries in each batch is different (1, 5, 10, 20) and the number of inserts and deletes is equal to 10 and 5, respectively. As in the previous experiments, we clearly show that the adaptive merging outperforms the other methods.

From all the experiments, we can draw the following conclusions. The full scan is very effective for queries on *key* but not effective for queries on *value*. The most time-consuming part of the secondary scan is its initiation phase, i.e. a secondary index creation for the entire dataset. Clearly, when the secondary index is already created, its effectiveness is very high. On the other hand, the adaptive merging creates the secondary index partially. It can use threads both when creating the Adaptive Log and when executing the query. Thus, although the query execution is slower on the adaptive merging, the total performance of this approach is better since it avoids the creation of the secondary index for the whole database at the beginning.

5 Conclusion

The paper faces adaptive merging of the secondary index in LSM-based stores. Traditionally, the secondary index covers all records equally. The key idea of this paper is to insert into the partial secondary index only the data required by the user queries. In this way, we avoid indexing unnecessary and unused records. Such an approach drastically decreases the creation and maintenance cost of the secondary index. When we initialize a new secondary index, we copy the SSTable files and fetch the data requested by the user from them independently, in a parallel way. Worth mentioning is the fact that the Main LSM database can be modified during the adaptive merging. The experiments clearly show that the method is very useful in the situation when the limited subset of the values is frequently requested by the user. Moreover, modern SSD disks exploit multi-threading and parallel access very well. Thus, the multi-core approach will gain even more advantages over the single-thread secondary index creation in the future.

In this paper, we do not focus on the secondary index structure. We conduct all the experiments on the LSM-based secondary index. In future work, we are going to consider the other index types typical for NoSQL databases. We can also optimize the LSM-based secondary indexes by applying several auxiliary structures like the primary key index and deleted key B+ tree (see [11]).

Acknowledgment. The paper is supported by Wroclaw University of Science and Technology (subvention number: IDUB/8211204601).

References

1. George, L.: HBase: The Definitive Guide, 1st edn. O'Reilly Media, Sebastopol (2011)
2. Lakshman, A., Malik, P.: Cassandra: a decentralized structured storage system. ACM SIGOPS Oper. Syst. Rev. **44**(2), 35–40 (2010)
3. Alsubaiee, S., et al.: Storage management in AsterixDB. Proc. VLDB Endow. **7**(10), 841–852 (2014)
4. Chodorow, K., Dirolf, M.: MongoDB - The Definitive Guide: Powerful and Scalable Data Storage. O'Reilly, Sebastopol (2010)
5. Chang, F.W., et al.: Bigtable: a distributed storage system for structured data. ACM Trans. Comput. Syst. **26**, 4:1–4:26 (2008)
6. Google: LevelDB. https://github.com/google/leveldb
7. Cao, Z., Dong, S., Vemuri, S., Du, D.H.C.: Characterizing, modeling, and benchmarking RocksDB key-value workloads at Facebook. In: 18th USENIX Conference on File and Storage Technologies, FAST 2020, Santa Clara, CA, USA, 24–27 February 2020, pp. 209–223. USENIX Association (2020)
8. O'Neil, P.E., Cheng, E., Gawlick, D., O'Neil, E.J.: The log-structured merge-tree (LSM-tree). Acta Informatica **33**(4), 351–385 (1996)
9. Qader, M.A., Cheng, S., Hristidis, V.: A comparative study of secondary indexing techniques in LSM-based NoSQL databases. In: Proceedings of the 2018 International Conference on Management of Data, SIGMOD Conference 2018, Houston, TX, USA, 10–15 June 2018, pp. 551–566. ACM (2018)
10. Corbett, J.C., et al.: Spanner: Google's globally distributed database. ACM Trans. Comput. Syst. **31**(3), 8:1–8:22 (2013)
11. Luo, C., Carey, M.J.: Efficient data ingestion and query processing for LSM-based storage systems. Proc. VLDB Endow. **12**(5), 531–543 (2019)
12. D'silva, J.V., Ruiz-Carrillo, R., Yu, C., Ahmad, M.Y., Kemme, B.: Secondary indexing techniques for key-value stores: two rings to rule them all. In: Proceedings of the Workshops of the EDBT/ICDT 2017 Joint Conference (EDBT/ICDT 2017), Venice, Italy, 21–24 March 2017. CEUR Workshop Proceedings, vol. 1810. CEUR-WS.org (2017)
13. Idreos, S., Kersten, M.L., Manegold, S.: Database cracking. In: CIDR (2007)
14. Idreos, S., Kersten, M.L., Manegold, S.: Updating a cracked database. In: Proceedings of the 2007 ACM SIGMOD International Conference on Management of Data, SIGMOD 2007, pp. 413–424. ACM, New York (2007)
15. Graefe, G., Kuno, H.: Self-selecting, self-tuning, incrementally optimized indexes. In: Proceedings of the 13th International Conference on Extending Database Technology, EDBT 2010, pp. 371–381. ACM, New York (2010)
16. Idreos, S., Manegold, S., Kuno, H.A., Graefe, G.: Merging what's cracked, cracking what's merged: adaptive indexing in main-memory column-stores. PVLDB **4**(9), 585–597 (2011)

17. Xue, Z., Qin, X., Zhou, X., Wang, S., Yu, A.: Optimized adaptive hybrid indexing for in-memory column stores. In: Hong, B., Meng, X., Chen, L., Winiwarter, W., Song, W. (eds.) DASFAA 2013. LNCS, vol. 7827, pp. 101–111. Springer, Heidelberg (2013). https://doi.org/10.1007/978-3-642-40270-8_9

18. Dayan, N., Athanassoulis, M., Idreos, S.: Optimal bloom filters and adaptive merging for LSM-trees. ACM Trans. Database Syst. **43**(4), 16:1–16:48 (2018)

19. Alvarez, V., Schuhknecht, F.M., Dittrich, J., Richter, S.: Main memory adaptive indexing for multi-core systems. In: Tenth International Workshop on Data Management on New Hardware, DaMoN 2014, Snowbird, UT, USA, 23 June 2014, pp. 3:1–3:10. ACM (2014)

20. Macyna, W., Kukowski, M.: Adaptive merging on phase change memory. Fundamenta Informaticae **188**(2) (2023)

CAGAIN: Column Attention Generative Adversarial Imputation Networks

Jun Kawagoshi[1], Yuyang Dong[2]([✉]), Takuma Nozawa[2], and Chuan Xiao[1,3]

[1] Osaka University, Suita, Japan
{kawagoshi.jun,chuanx}@ist.osaka-u.ac.jp
[2] NEC Corporation, Tokyo, Japan
{dongyuyang,nozawa-takuma}@nec.com
[3] Nagoya University, Nagoya, Japan

Abstract. Imputation for missing values is a key operation in building data analysis models. In this paper, we target numerical and categorical values in tabular data. While previous studies have demonstrated the effectiveness of state-of-the-art methods, a major limitation is that these methods lack robustness and their performance significantly varies across datasets and the missing rate of values, hence posing considerable overhead of selecting and tuning models in a real-world scenario. To tackle this problem, we propose a Column Attention Generative Adversarial Imputation Network (CAGAIN), an imputation model which employs a generative adversarial network (GAN) and the attention mechanism. The generator of CAGAIN mimics the distribution of original data and generates imputed samples similar to real ones. The discriminator of CAGAIN distinguishes real and generated samples, so as to improve the quality of the imputed data. At the same time, the attention mechanism captures the correlation between attributes and focuses on the most significant attributes that determine the values of the missing positions. By inheriting the advantages of GAN and the attention mechanism, our model is endowed with robustness to shifting datasets and missing rates, which is demonstrated by experiments using 9 real datasets.

Keywords: data imputation · GAN · attention mechanism

1 Introduction

With the explosion of data, one of the major issues is dealing with missing values in a dataset. This problem can be caused by a variety of factors, including human error during data manipulation, machine error due to mechanical breakdowns, and discrepancies in merging datasets. Data imputation is the process of filling missing data with substituted values. It is a key operation in data analytics because missing data can cause severe problems for data analysis models, such as reduction in accuracy and non-convergence. In data management, data imputation has also been extensively studied as a data cleaning problem.

In this paper, we study the problem of data imputation for tabular data, and focus on numerical values and categorical values, which are among the most

C. Strauss et al. (Eds.): DEXA 2023, LNCS 14147, pp. 258–273, 2023.
https://doi.org/10.1007/978-3-031-39821-6_21

common data types in tabular data. While early attempts at this problem are based on simple statistical techniques [10] such as filling by mean or mode, more advanced approaches resort to machine learning to impute missing values by creating either a discriminative (e.g., decision trees [5,23]) or a generative (e.g., autoencoders [7,14]) model from the dataset. Whereas state-of-the-art approaches have been shown to deliver high-quality imputation results by previous studies, a main drawback is that their performance lacks robustness and significantly varies when applied to different datasets or missing rates of values. For this reason, data analysts tend to spend considerable effort and computing resources in selecting and tuning data imputation models. For example, even if high accuracy has been reported by previous studies, we still need to try many hyperparameter combinations to reproduce the results on the same datasets, and sometimes domain knowledge is required for feature selection.

To address the above limitation, we propose a Column Attention Generative Adversarial Imputation Network (CAGAIN) for imputing numerical and categorical values in tabular data. CAGAIN features a **generative adversarial network (GAN)** [8] and the **attention mechanism** [26]. CAGAIN is composed of a generator network and a discriminator network trained in an adversarial manner. The generator imputes missing values by generating samples that are close to real complete values. The discriminator distinguishes generated values from real ones, thereby improving the quality of imputed values from the generator. At the same time, the attention mechanism, which devotes more focus to a subset of attributes (i.e., columns), is used to discover the correlation between the missing value and those of other attributes.

To use the attention mechanism together with the GAN, we develop a network architecture consisting of an embedding layer, an attention layer, and a prediction layer. Since GANs are often unstable in training [12] due to gradient vanishing, the overhead of model tuning tends to be increased, and thus the usability in real-world imptation tasks is compromised. This occurs when the discriminator mistakes a false sample for a real sample, and the problem is also observed when one simply integrates the attention mechanism into a GAN. To address this issue, based on our network architecture, we propose a loss function for the generator, which balances the similarities of generated data to original input and complete data, and a least-squared loss for the discriminator, which aims to prevent gradient vanishing.

CAGAIN inherits the advantages of both GAN and the attention mechanism. By adversarial training, the imputed values keep approaching the distribution in the original dataset, and the attention mechanism helps infer the correct values of the missing positions by focusing on the most determinant subset of attributes. Moreover, since CAGAIN's embedding layer transforms each input value to a vector, it is enabled to model complex correlations such that the determinant attributes are conditional on the observed values (e.g., risk is highly correlated to age if COVID-19 is "positive"). As such, CAGAIN is endowed with the ability to deliver *robust imputation performance* to shifting datasets and missing rates. Despite the existence of approaches to tabular data imputation using either

GANs [19,28] or the attention mechanism [24], or both for other data types [18], to the best of our knowledge, we are the first of combining them to effectively cope with the imputation for numerical and categorical values in tabular data.

We perform experiments on 9 real datasets with missing rates from 20% to 80%. The results showcase the superiority of CAGAIN in a diversified real-world scenario. The accuracy of CAGAIN, without feature selection and laborious tuning (we tune one hyperparameter of 3 options and report the best results, while fixing other hyperparameters), always ranks in the top-3 among 9 competitors and outperforms the runner-up baselines by up to 15% RMSE for numerical values and 3% accuracy for categorical values, while the performance of existing approaches drastically varies across these datasets and missing rate settings. Moreover, even trained without the discriminator, CAGAIN is still robust and outperforms some state-of-the-art approaches. A post-imputation prediction experiment demonstrates that the imputation by CAGAIN results in higher accuracy in classification tasks. We would also like to mention that the performance of CAGAIN can be further improved by more effort of hyperparameter tuning, yet this is not our goal in this paper.

Our contributions are summarized as follows. (1) We propose a new solution to data imputation for numerical and categorical values in tabular data. Our solution, which integrates a GAN and the attention mechanism, is robust against shifting datasets and missing rates. (2) We develop a network architecture which consists of an embedding layer, an attention layer, and a prediction layer. We design a series of loss functions for model training, which balances the similarities of generated data to original input and complete data. (3) We conduct experiments on 9 real datasets with missing rates varying from 20% to 80%. The experiments demonstrate that with simple hyperparameter tuning, our solution reports high accuracy of data imputation across these datasets and missing rate settings.

2   Related Work

We focus on tabular data and learning-based methods, the state-of-the-art methods for data imputation. The methods targeting different data types (e.g., time series [18]) and those based on different methodologies (e.g., constraints [3,20,22]) are omitted due to the page limitation.

Discriminative methods formulate data imputation as a classification or a regression problem and train a learning model to solve it. A widely used method is KNN [9], which assigns the missing value to the most common class or the average value of the sample's k nearest neighbors. Another series of popular solutions is tree-based methods (MissForest, XGBoost, etc.) [4,5,23]. While tree-based models have demonstrated outstanding performance for classification and regression problems, they suffer from poor generalizability of imputation when missing values occur systematically or explanatory variables for prediction are misaligned between observed and missing values [2]. Multivariate imputation by chained equations (MICE) [25] is a popular tool for data imputation. It is valid only under the assumption that missing values are dependent on other values, and does not converge if the assumption does not hold. AimNet [27] is a state-of-the-art method based on the attention mechanism, with attention weights indicating the importance of the features used for imputation. Despite superior performance on some datasets, it is less effective for the case of complex data distributions or limited training data.

Generative methods model the joint probability distribution of observed values and missing values. Most methods in this category employ autoencoders or GANs. Denoising autoencoder (DAE) [7,14,21,24] is a representative method in this category. By adding a noise process to the standard autoencoder, a DAE takes a noisy input and attempts to recover the original data without noise. Among various DAEs, DAEMA [24] features a DAE with the attention mechanism and trains the model to focus on observed values rather than on missing values. It demonstrates competitive performance on numerical values, whereas it often overfits on categorical values and reports inferior accuracy. Another notable autoencoder-based method is HI-VAE [16]. It uses reconstruction error for observed values and takes as input missing values filled with zeros. While HI-VAE is able to correct data with a mixture of numerical and categorical values, it is difficult to train, and the predicted values may significantly diverge in some cases. For GANs, a representative method is GAIN [28]. Its generator takes as input the cell values of a record, along with a mask matrix to indicate whether a cell is missing or observed, and outputs the predicted values. Its discriminator aims to distinguish true values from the generated values. By training them in an adversarial way, GAIN is enabled to generate values following a conditional distribution of the observed data. Yet, mode collapse, a common problem for GANs, is sometimes observed in GAIN, and it is consistently outperformed by AimNet [27]. Other notable GAN-based methods include IFGAN [19], which proposes the notion of feature-specific GANs to preserve inter-feature correlations, as well as WGAIN [6] and WSGAIN [17], which employ the Wasserstein distance in the loss function to stabilize GAN training and tackle mode collapse.

3   Problem Statement

Given a data space $\mathcal{X} = \mathcal{X}_1 \times \cdots \times \mathcal{X}_d$, where each $\mathcal{X}_j, j \in [1, d]$ refers to either a numerical or a categorical column, we consider n i.i.d. random variables $\mathbf{X} = \{\mathbf{X}_1, \ldots, \mathbf{X}_n\}$. Each variable $\mathbf{X}_i$ is a vector $\{X_{i,1}, \ldots, X_{i,d}\}$, where each $X_{i,j}$ is either an observed value in $\mathcal{X}_j$, or a missing value, denoted by $*$. A matrix $\mathbf{M} \in \{0, 1\}^{n \times d}$ indicates whether a value is observed or missing; i.e., $M_{i,j} = 1$, if $X_{i,j}$ is observed, or 0, if $X_{i,j}$ is missing. Each sample $\mathbf{x}$ in $\mathcal{X}$ is called a data vector, and each $\mathbf{m} = \{0, 1\}^d$ is called a mask vector.

For data imputation, our task is to impute the missing values in $\mathbf{X}$. Specifically, for each (i, j) such that $M_{i,j} = 0$, we predict the value of $X_{i,j}$ upon observing $\mathbf{X}_{i,-j}$ and $\mathbf{M}_{i,-j}$, where $-j$ denotes all the columns except j.

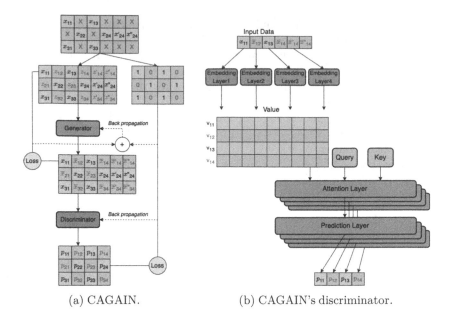

(a) CAGAIN.   (b) CAGAIN's discriminator.

Fig. 1. Model architecture.

4   Method

4.1   The CAGAIN Model

The architecture of CAGAIN, our imputation model, is shown in Fig. 1a (notations $'$ and $''$ denote the bits in the one-hot encoding of categorical variables). CAGAIN employs a GAN, which consists of a generator G and a discriminator D. The input of G is a vector that preserves the observed values of a data vector $\mathbf{x}$ and fills each missing value of $\mathbf{x}$ with a random noise z taken from a Gaussian distribution where $\mu = 0$ and $\sigma = 0.01$. G outputs $\bar{\mathbf{x}}$, a predicted variable for imputation. Note that each column has an output value from G, even if it is observed. The input of D is a completed vector $\hat{\mathbf{x}}$, which completes a data vector $\mathbf{x}$ by replacing missing values $*$ with the corresponding values in $\bar{\mathbf{x}}$. D tries to distinguish between generated values and observed (i.e., real) values by outputting a d-dimensional vector that indicates the probability of each position being observed. Note that the input of D is different from the traditional setting of a GAN, which takes real and generated data as input separately. Inspired by AimNet [27], we design both G and D as a network consisting of an embedding layer, an attention layer, and a prediction layer.

Embedding Layer. The embedding layer transforms each input feature x_j to a k-dimensional embedding vector $\mathbf{V}_j \in \mathbb{R}^k$, implemented as a two-layer neural network with ReLU activation function. Before the input, observed numerical values are standardized by mean and variance, and observed categorical values are transformed into vectors by one-hot encoding.

Attention Layer. To effectively capture the correlation between the missing value and the observed values in other columns, we employ the attention mechanism [26], which enhances some observed values while diminishing others. First, we are given the output of the embedding layer, i.e., the embedding vectors $\mathbf{V} = [\mathbf{V}_1, \ldots, \mathbf{V}_d]$ of the input features. To apply the attention mechanism, we have a set of query vectors $\mathbf{Q} \in \mathbb{R}^{d \times d}$ and key vectors $\mathbf{K} \in \mathbb{R}^{d \times d}$. $\mathbf{Q}$ indicate which column we shall focus on. $\mathbf{K}$ are unit vectors and they are used to select from $\mathbf{Q}$ to produce a set of weights to represent the correlation between columns. The softmax function normalizes the weights, so they sum up to 1. The normalized weights are then multiplied to $\mathbf{V}$ to output context vectors $\mathbf{S}$. Formally, for column j, $\mathbf{S}_j = \mathrm{softmax}(\mathbf{K}_j^{\mathrm{T}} \mathbf{Q}) \mathbf{V}$. Moreover, the weight of a column to itself is masked to be 0 for G.

Prediction Layer. The prediction layer transforms the context vectors $\mathbf{S}$ to the output of either G or D. It is a two-layer neural network for G and a three-layer neural network for D, with ReLU activation function. For categorical values, G also learns target embedding vectors. It is unique for target domain values in a categorical column. G computes the inner product between the output of a two-layer neural network and the categorical target embedding vectors, and then, the softmax function is applied to generate prediction probabilities for each domain value.

Let $\bar{\mathbf{x}}$ be the output of G, i.e., the data vector $\mathbf{x}$ with imputed values. Since each column has an output value from G even if it is observed, we use the following equation to obtain the completed vector, which serves as the result of data imputation: $\hat{\mathbf{x}} = \mathbf{m} \odot \mathbf{x} + (\mathbf{1} - \mathbf{m}) \odot \bar{\mathbf{x}}$, where $\mathbf{m}$ is the mask vector, $\mathbf{1}$ is a vector of 1s, and $\odot$ denotes element-wise multiplication.

D takes $\hat{\mathbf{x}}$ as input and outputs the probability of each value being observed. Figure 1b shows its architecture. Since G is very similar to this (only differ in the attention and prediction layers, as mentioned above), we do not repeatedly illustrate the architecture of G.

4.2 Loss Functions

We design loss functions to measure the degree of imputation fitness. For G, the loss function consists of two parts: observation loss L_G^{Obs} and adversarial loss L_G^{Adv}. Minimizing the observation loss has the effect of making generated samples similar to incomplete input samples, while minimizing the adversarial loss has the effect of making generated samples similar to real complete samples. In essence, the objective of G tries to strike a balance between the original input and the complete data. The two loss functions are defined in the following equations. Specifically, for categorical data, $\| \cdot \|$ denotes 1-norm, $\cdot$ denotes the inner product, and log returns the logarithmic value for each element of a one-hot encoding vector.

$$L_G^{Obs}(G) = \begin{cases} \mathbb{E}[\mathbf{m}^{\mathrm{T}}(\mathbf{x} - \bar{\mathbf{x}})^2], & \text{if numerical} \\ \mathbb{E}[\mathbf{m}^{\mathrm{T}} \| \mathbf{x} \cdot \log \bar{\mathbf{x}} \|_1], & \text{if categorical.} \end{cases}$$

Table 1. Dataset statistics.

Dataset	# samples	# columns	# numerical	# categorical
Spam	4601	58	57	1
Letter	20000	17	16	1
Balance	625	5	4	1
Breast	569	31	30	1
Eye EEG	14980	15	14	1
Credit	30000	24	14	10
Solar Flare	1066	13	3	10
Contraceptive	1473	10	2	8
Thoracic	470	17	3	14

$$L_G^{Adv}(D) = \frac{1}{2}\mathbb{E}[(\mathbf{1} - \mathbf{m})^{\mathrm{T}}(D(\hat{\mathbf{x}}))^2],$$

where $D(\cdot)^2$ returns the square value of each element in D's output. Then, we have the loss function for G:

$$L_G(D) = \lambda L_G^{Obs}(G) + L_G^{Adv}(D). \tag{1}$$

λ is a tunable hyperparameter to balance the impact of the observation loss and the adversarial loss. When λ is large, G is trained to minimize the error to the original input. When λ is small, G is trained to output imputed values similar to the distribution of real data.

For D, to stable the GAN training so as to achieve more robustness, we use a least-squared loss as the adversarial loss, in line with LSGAN [13]. The loss function for D is given as follows.

$$L_D(G) = \frac{1}{2}\mathbb{E}[\mathbf{m}^{\mathrm{T}}(D(\hat{\mathbf{x}}) - \mathbf{1})^2] + \frac{1}{2}\mathbb{E}[(\mathbf{1} - \mathbf{m})^{\mathrm{T}}(D(\hat{\mathbf{x}}) + \mathbf{1})^2]. \tag{2}$$

With the above loss functions, G and D are trained in an alternate manner, by minimizing Eq. 1 for G and minimizing Eq. 2 for D.

5   Experiments

5.1   Experimental Setup

Datasets. We choose a variety of 9 real-world datasets from the UCI Machine Learning Repository [1]. Dataset statistics is given in Table 1. Raw data are used, except Eye EEG, for which we remove the outliers beyond three standard deviations from the mean. We randomly split each dataset into training, validation, and testing datasets with a 80-10-10 ratio. To generate missing values, we use a missing completely at random (MCAR) strategy by randomly removing 20% cell values from each dataset, unless otherwise noted.

Baselines. We compare the following methods, including 7 existing ones as baselines. We also show the hyperparameter settings, of which the best is used for reporting the experimental results: (1) Mean/mode imputation, which fills with the mean value for numerical data and the mode for categorical data; (2) KNN [9], k = 3, 5, or 7; (3) MICE [25], number of iteration = 1, 5, or 10; (4) MissForest [23], number of trees = 50, 100, or 300; (5) GAIN [28], α = 0.1, 1, or 10; (6) AimNet [27], embedding size = 16, 64, or 128; (7) DAEMA [24], dropout rate = 0.2, 0.3, or 0.4; (8) CAGAIN (our model), λ = 0.1, 1, or 10; (9) CAGAIN w/o D, CAGAIN evaluated without training its discriminator, in order to show the effectiveness of the balanced design for the generator's loss function. Its hyperparameter setting is the same as CAGAIN.

Other hyperparameters of CAGAIN are as follows. Embedding size k is 64 for both D and G. We train CAGAIN by Adam [11], with $\beta_1 = 0$ and $\beta_2 = 0.9$. We train D for two epochs and G for one epoch in each iteration. The learning rate is 0.004 for D and 0.001 for G. We use spectral normalization [15] for D except its output layer. We set batch size to 128 for the four large datasets (Spam, Letter, Eye EEG, and Credit), and 64 for the five small ones.

Due to the versioning of TensorFlow and other issues (e.g., unable to retrieve the original hyperparameter settings), the observed results in our experiments may differ from what have been reported in the original papers of the baselines. Nonetheless, many of our observed results are even better than in the original papers. For example, with the same missing rate for categorical data, out of the six datasets which were also used in the AimNet [27] paper, here we observe higher accuracy for AimNet on three of them and equal accuracy on two of them.

Metrics. We measure root mean square error (RMSE) for numerical data and accuracy for categorical data: $RMSE = \sqrt{\frac{\sum_1^N (\hat{X}-X)^2}{N}}$, where $\hat{X}$ is the imputed value and X is the true value. Accuracy (ACC) measures the percentage of correctly imputed values: $ACC = \frac{\sum_1^N 1(\hat{X}=X)}{N}$, where $1(\cdot)$ is the indicator function.

Environments. Experiments are run on a server with a 2.20 GHz Intel Xeon Gold CPU and 630 GB RAM. Models are trained and run on a NVidia Tesla V100 Tensor Core. All the competitors are implemented in Python 3.8.

5.2 Experimental Results

Varying Datasets. We first evaluate the performance for numerical data. Table 2 shows the results of the competitors on the 9 datasets. For each competitor, we report the average RMSE, standard deviation, and the rank among all the competitors, with the best one highlighted in boldface. Mean imputation has an overall worst rank as it always fills with the average value. For existing learning methods, the performance significantly varies across datasets. Many of them report high ranking for a few datasets, but the performance is mediocre on others. For example, KNN is the best on Letter and Eye EEG, but ranks from 5th to 9th for the other datasets. The most robust existing method is AimNet,

Table 2. Varying datasets. Numerical, missing rate = 20%. RMSE ± std (rank), lower is better.

	Spam	Letter	Balance	Breast	Eye EEG
Mean Imputation	0.99±0.04 (9)	1.01±0.01 (9)	1.01±0.06 (6)	1.06±0.02 (9)	1.00±0.01 (9)
KNN	0.95±0.04 (7)	**0.41±0.01** (1)	1.10±0.10 (9)	0.68±0.04 (7)	**0.40±0.01** (1)
MICE	0.93±0.03 (6)	0.65±0.01 (7)	0.93±0.05 (2)	0.65±0.04 (5)	0.49±0.01 (7)
MISSFOREST	0.95±0.04 (7)	0.51±0.01 (5)	1.08±0.03 (7)	0.67±0.06 (6)	0.46±0.01 (6)
GAIN	**0.85±0.01** (1)	0.80±0.01 (8)	1.08±0.06 (7)	0.73±0.03 (8)	0.61±0.02 (8)
AimNet	0.87±0.05 (2)	0.43±0.01 (2)	0.97±0.04 (4)	0.62±0.06 (2)	0.41±0.01 (2)
DAEMA	0.91±0.01 (5)	0.58±0.01 (6)	0.94±0.02 (3)	**0.58±0.04** (1)	0.45±0.01 (5)
CAGAIN w/o D	0.90±0.04 (4)	0.44±0.01 (3)	0.99±0.03 (5)	0.64±0.06 (4)	0.43±0.01 (4)
CAGAIN	0.88±0.03 (3)	0.44±0.01 (3)	**0.90±0.05** (1)	0.62±0.04 (2)	0.42±0.01 (3)

	Credit	Solar Flare	Contraceptive	Thoracic
Mean Imputation	1.00±0.04 (9)	0.87±0.41 (4)	0.99±0.04 (7)	0.83±0.14 (3)
KNN	0.76±0.04 (6)	0.95±0.29 (7)	1.09±0.11 (9)	0.88±0.17 (5)
MICE	0.76±0.04 (6)	1.00±0.36 (8)	0.93±0.04 (5)	1.02±0.12 (8)
MISSFOREST	0.65±0.03 (4)	1.27±0.14 (9)	0.99±0.09 (7)	**0.76±0.17** (1)
GAIN	0.80±0.05 (8)	0.86±0.36 (3)	0.93±0.04 (5)	2.68±0.89 (9)
AimNet	0.61±0.03 (3)	0.92±0.35 (6)	0.91±0.04 (3)	0.91±0.19 (7)
DAEMA	0.68±0.04 (5)	0.88±0.33 (5)	0.87±0.05 (2)	0.83±0.17 (3)
CAGAIN w/o D	**0.59±0.04** (1)	0.76±0.31 (2)	0.92±0.04 (4)	0.88±0.14 (5)
CAGAIN	**0.59±0.03** (1)	**0.71±0.32** (1)	**0.83±0.04** (1)	0.77±0.18 (2)

which ranks from 2th to 4th on most datasets, yet ends up with 6th on Solar Flare. In contrast, CAGAIN is always in the top-3 for all the datasets, showcasing its robustness to various datasets. In particular, CAGAIN performs the best on Balance, Credit, Solar Flare, and Contraceptive, and the gap to the runner-up baseline is up to 15%, observed on Solar Flare. We also observe that even without the discriminator, the performance of CAGAIN is still robust, ranking in the 4th to 5th for 6 out of 9 datasets and top-3 for the others. This result demonstrates the robustness of our model design in generating imputed values.

We then evaluate the competitors on categorical data. Table 3 shows their average accuracies, standard deviations, and ranks. Mode imputation is overall the worst. Similar to what we have witnessed in numerical data, the performance of existing methods exhibits a considerable variation across datasets, though AimNet is still competitive on some datasets. In contrast, CAGAIN still delivers a rank in top-3 across all the datasets, and is the winner on Balance, Breast, Eye EEG, Credit, and Solar Flare. The gap to the runner-up baseline is up to 3%, observed on Balance. Its robustness is also demonstrated when we remove its discriminator, with top-3 ranking for all the datasets except Breast and Thoracic. We also observe that GAIN does not perform well, ranking at 4th on Breast and 7th to 9th on the other datasets. In addition, along with the results on numerical data, the results show that the methods based on the attention mechanism (AimNet, DAEMA, and CAGAIN) generally perform better than the other competitors. We think the reason is that the attention mechanism correctly captures

Table 3. Varying datasets. Categorical, missing rate = 20%. ACC ± std (rank), higher is better.

	Spam	Letter	Balance	Breast	Eye EEG
Mode Imputation	0.59±0.02 (9)	0.04±0.01 (8)	0.38±0.09 (9)	0.69±0.15 (9)	0.54±0.02 (9)
KNN	0.84±0.02 (8)	**0.87±0.01** (1)	0.63±0.04 (6)	**1.00±0.00** (1)	**0.85±0.02** (1)
MICE	0.89±0.03 (6)	0.27±0.01 (6)	0.83±0.06 (3)	0.98±0.03 (4)	0.61±0.02 (7)
MISSFOREST	**0.92±0.02** (1)	0.76±0.01 (4)	0.67±0.12 (5)	0.94±0.09 (7)	0.78±0.02 (4)
GAIN	0.86±0.06 (7)	0.03±0.01 (9)	0.52±0.16 (8)	0.98±0.04 (4)	0.56±0.01 (8)
AimNet	0.90±0.03 (5)	0.64±0.02 (5)	0.81±0.05 (4)	0.98±0.04 (4)	0.71±0.01 (6)
DAEMA	0.91±0.02 (2)	0.12±0.01 (7)	0.57±0.15 (7)	**1.00±0.00** (1)	0.72±0.02 (5)
CAGAIN w/o D	0.91±0.02 (2)	0.77±0.01 (2)	0.85±0.04 (2)	0.98±0.04 (4)	0.80±0.01 (3)
CAGAIN	0.91±0.02 (2)	0.77±0.01 (2)	**0.86±0.05** (1)	**1.00±0.00** (1)	**0.85±0.02** (1)

	Credit	Solar Flare	Contraceptive	Thoracic
Mode Imputation	0.56±0.01 (9)	0.67±0.03 (9)	0.61±0.02 (5)	0.83±0.05 (6)
KNN	0.73±0.01 (6)	0.75±0.02 (6)	0.60±0.03 (7)	0.83±0.05 (6)
MICE	0.74±0.01 (5)	0.76±0.02 (4)	0.61±0.03 (5)	**0.86±0.03** (1)
MISSFOREST	0.79±0.01 (3)	0.76±0.03 (4)	0.62±0.02 (4)	0.84±0.05 (5)
GAIN	0.65±0.01 (8)	0.68±0.04 (8)	0.58±0.04 (9)	0.80±0.04 (9)
AimNet	0.78±0.01 (4)	0.78±0.02 (2)	**0.66±0.01** (1)	0.85±0.05 (3)
DAEMA	0.71±0.01 (7)	0.75±0.03 (6)	0.59±0.01 (8)	**0.86±0.05** (1)
CAGAIN w/o D	**0.80±0.01** (1)	0.78±0.02 (2)	0.64±0.02 (2)	0.82±0.02 (8)
CAGAIN	**0.80±0.01** (1)	**0.79±0.01** (1)	0.64±0.01 (2)	0.85±0.01 (3)

the correlation between attributes and focuses on the more correlated attributes to produce a more accurate imputed value.

Varying Missing Rates. We investigate how missing rate affects the performance of data imputation, and report the results in Tables 4, 5, 6 and 7, by varying missing rates from 20% to 80% on Credit and Solar Flare. Other datasets, on which similar results are observed, are omitted here due to the page limitation. When the missing rate increases, almost all the competitors have their RMSE in a rising trend and accuracy in a falling trend. Some are also very sensitive and even cannot converge (e.g., MICE). This is expected, because a higher missing rate yields a dirtier and thus harder dataset. Again, we observe that CAGAIN is in top-3 across all the settings, showcasing its robustness to missing rates.

Performance Analysis. To explain why CAGAIN performs better than existing GAN-based methods and attention-based methods, we show in Fig. 2 a toy example of 50% missing values on a moon dataset generated using scikit-learn. CAGAIN is compared with GAIN and AIMNET. GAIN uses an autoencoder as its generator, which reduces the dimensions and recovers them. However, dimension reduction may lose some necessary information, and thus result in inaccurate imputation results for many values, as shown in the figure. Moreover, the training of GAIN is unstable and requires considerable effort in tuning the model. For AIMNET, which uses a least-squares loss function, often imputes with a point in the middle which is far from the original data. In contrast, CAGAIN imputes the missing value with a latent variable, and the use of an adversarial loss

Table 4. Varying missing rates (MR). Numerical, Credit. RMSE ± std (rank), lower is better, N/C for non-convergence.

	MR = 20%	MR = 40%	MR = 60%	MR = 80%
Mean Imputation	1.00 ± 0.04 (9)	1.01 ± 0.02 (8)	0.99 ± 0.01 (7)	1.00 ± 0.01 (6)
KNN	0.76 ± 0.04 (6)	0.91 ± 0.02 (7)	1.06 ± 0.04 (8)	1.06 ± 0.02 (7)
MICE	0.76 ± 0.04 (6)	N/C (9)	N/C (9)	N/C (9)
MISSFOREST	0.65 ± 0.03 (4)	0.77 ± 0.03 (5)	0.92 ± 0.03 (6)	1.06 ± 0.06 (7)
GAIN	0.80 ± 0.05 (8)	0.83 ± 0.03 (6)	0.83 ± 0.02 (5)	0.92 ± 0.01 (5)
AimNet	0.61 ± 0.03 (3)	0.69 ± 0.02 (2)	**0.72 ± 0.02 (1)**	**0.83 ± 0.01 (1)**
DAEMA	0.68 ± 0.04 (5)	0.72 ± 0.03 (4)	0.74 ± 0.02 (2)	0.84 ± 0.01 (3)
CAGAIN w/o D	**0.59 ± 0.04** (1)	0.69 ± 0.02 (2)	0.75 ± 0.02 (4)	0.84 ± 0.01 (3)
CAGAIN	**0.59 ± 0.03** (1)	**0.67 ± 0.02** (1)	0.73 ± 0.01 (2)	**0.83 ± 0.01** (1)

Table 5. Varying missing rates (MR). Categorical, Credit. ACC ± std (rank), higher is better.

	MR = 20%	MR = 40%	MR = 60%	MR = 80%
Mode Imputation	0.56 ± 0.01 (9)	0.56 ± 0.01 (9)	0.56 ± 0.01 (8)	0.56 ± 0.01 (8)
KNN	0.73 ± 0.01 (6)	0.69 ± 0.01 (7)	0.62 ± 0.01 (7)	0.59 ± 0.01 (6)
MICE	0.74 ± 0.01 (5)	0.71 ± 0.01 (5)	0.53 ± 0.01 (9)	0.40 ± 0.15 (9)
MISSFOREST	0.79 ± 0.01 (3)	0.76 ± 0.01 (3)	0.71 ± 0.01 (6)	0.64 ± 0.01 (4)
GAIN	0.65 ± 0.01 (8)	0.65 ± 0.01 (8)	0.63 ± 0.01 (9)	0.59 ± 0.01 (6)
AimNet	0.78 ± 0.01 (4)	0.75 ± 0.01 (4)	0.72 ± 0.01 (3)	0.66 ± 0.01 (3)
DAEMA	0.71 ± 0.01 (7)	0.69 ± 0.01 (6)	0.66 ± 0.01 (5)	0.62 ± 0.01 (5)
CAGAIN w/o D	**0.80 ± 0.01** (1)	0.77 ± 0.01 (2)	**0.73 ± 0.01** (1)	0.67 ± 0.01 (2)
CAGAIN	**0.80 ± 0.01** (1)	**0.78 ± 0.01** (1)	**0.73 ± 0.01** (1)	**0.68 ± 0.01** (1)

Table 6. Varying missing rates (MR). Numerical, Solar Flare. RMSE ± std (rank), lower is better, N/C for non-convergence.

	MR = 20%	MR = 40%	MR = 60%	MR = 80%
Mean Imputation	0.87 ± 0.41 (4)	0.72 ± 0.18 (2)	0.87 ± 0.08 (1)	0.84 ± 0.08 (2)
KNN	0.95 ± 0.29 (7)	0.75 ± 0.18 (3)	0.94 ± 0.11 (7)	0.93 ± 0.08 (7)
MICE	1.00 ± 0.36 (8)	0.87 ± 0.24 (8)	1.63 ± 0.38 (9)	N/C (9)
MISSFOREST	1.27 ± 0.14 (9)	1.12 ± 0.20 (9)	1.42 ± 0.43 (8)	1.55 ± 0.71 (8)
GAIN	0.86 ± 0.36 (3)	0.79 ± 0.17 (5)	0.88 ± 0.06 (2)	0.89 ± 0.06 (6)
AimNet	0.92 ± 0.35 (6)	0.85 ± 0.25 (6)	0.92 ± 0.07 (5)	0.87 ± 0.10 (5)
DAEMA	0.88 ± 0.33 (5)	0.77 ± 0.17 (4)	0.89 ± 0.06 (4)	0.85 ± 0.07 (3)
CAGAIN w/o D	0.76 ± 0.31 (2)	0.86 ± 0.22 (7)	0.92 ± 0.07 (5)	0.85 ± 0.08 (3)
CAGAIN	**0.71 ± 0.32** (1)	**0.71 ± 0.15** (1)	0.88 ± 0.10 (2)	**0.83 ± 0.08** (1)

function makes it possible to generate values that follow the same distribution as the original data. When the adversarial loss function is not used, CAGAIN behaves similarly to AIMNET.

Model Training. The training time of CAGAIN varies from 3 min to 97 min across the 9 datasets. It is fast on Balance and Contraceptive, but slow on Spam

Table 7. Varying missing rates (MR). Categorical, Solar Flare. ACC ± std (rank), higher is better.

	MR = 20%	MR = 40%	MR = 60%	MR = 80%
Mode Imputation	0.67 ± 0.03 (9)	0.67 ± 0.01 (8)	0.67 ± 0.02 (6)	0.68 ± 0.01 (1)
KNN	0.75 ± 0.02 (6)	0.70 ± 0.02 (7)	0.66 ± 0.01 (8)	0.67 ± 0.01 (5)
MICE	0.76 ± 0.02 (4)	0.71 ± 0.01 (5)	0.69 ± 0.02 (4)	0.62 ± 0.06 (9)
MISSFOREST	0.76 ± 0.03 (4)	0.71 ± 0.02 (5)	0.67 ± 0.02 (6)	0.65 ± 0.01 (7)
GAIN	0.68 ± 0.04 (8)	0.65 ± 0.01 (9)	0.64 ± 0.02 (9)	0.64 ± 0.01 (8)
AimNet	0.78 ± 0.03 (2)	**0.75 ± 0.01** (1)	**0.72 ± 0.01** (1)	**0.68 ± 0.01** (1)
DAEMA	0.75 ± 0.03 (6)	0.73 ± 0.01 (4)	0.69 ± 0.01 (4)	0.67 ± 0.01 (5)
CAGAIN w/o D	0.78 ± 0.02 (2)	0.74 ± 0.02 (2)	0.70 ± 0.01 (3)	**0.68 ± 0.01** (1)
CAGAIN	**0.79 ± 0.01** (1)	0.74 ± 0.01 (2)	0.71 ± 0.01 (2)	**0.68 ± 0.01** (1)

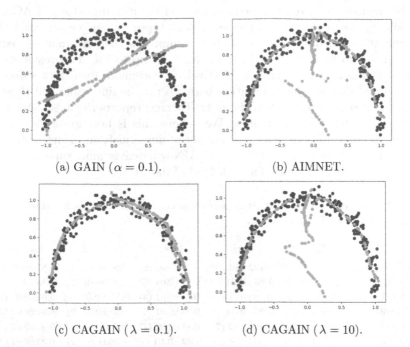

(a) GAIN ($\alpha = 0.1$).   (b) AIMNET.

(c) CAGAIN ($\lambda = 0.1$).   (d) CAGAIN ($\lambda = 10$).

Fig. 2. Toy example: 50% missing values on a scikit-learn moon dataset. Blue points have no missing values. Orange points have one missing value on either coordinate. (Color figure online)

Table 8. Effect of attention. Numerical, missing rate = 20%, RMSE, lower is better.

	Credit	Solar Flare	Contraceptive	Thoracic
CAGAIN	0.59 ± 0.03	0.71 ± 0.32	0.83 ± 0.04	0.77 ± 0.18
CAGAIN w/o Attention in D	0.59 ± 0.05	0.76 ± 0.41	0.94 ± 0.05	0.99 ± 0.27

and Letter, approximately proportional to the dataset size reported in Table 1. Compared to AimNet, another attention-based method, CAGAIN spends about twice training time. Nonetheless, considering the advantage of CAGAIN in

Table 9. Effect of attention. Categorical, missing rate = 20%, ACC, higher is better.

	Credit	Solar Flare	Contraceptive	Thoracic
CAGAIN	0.80 ± 0.01	0.79 ± 0.01	0.64 ± 0.01	0.85 ± 0.01
CAGAIN w/o Attention in D	0.79 ± 0.01	0.77 ± 0.01	0.61 ± 0.01	0.79 ± 0.03

robustness, CAGAIN saves the effort of users in selecting and tuning models for data imputation.

Ablation Study. To evaluate the impact of using the attention mechanism, we conduct an ablation study and report the results in Table 8 for numerical data and Table 9 for categorical data on Credit, Solar Flare, Contraceptive, and Thoracic. We remove the attention mechanism from the discriminator of CAGAIN, and comapre the performance. For both data types, it can be observed that using the attention mechanism consistently improves the performance, reducing RMSE by up to 25% and raising accuracy by up to 8%. This result shows that the attention mechanism can effectively devote more focus to a subset of attributes and discover the correlation between the missing value and those of other attributes. Among the four datasets, Thoracic reports the most significant impact of the attention mechanism. We believe this is because of its dataset statistics and distributions, e.g., less number of numericals and more number of categoricals. As such, we expect that CAGAIN will deliver more superior results when imputing missing values for such kind of datasets.

Table 10. Post-imputation prediction performance. Numerical, missing rate = 20%. ACC $\pm$ std (rank), higher is better.

	Spam	Letter	Balance	Breast	Eye EEG
Mean Imputation	0.92±0.01 (8)	0.78±0.01 (8)	0.79±0.01 (8)	0.92±0.01 (8)	0.87±0.01 (8)
KNN	0.96±0.01 (6)	**0.95±0.01** (1)	0.87±0.01 (6)	0.98±0.01 (2)	**0.96±0.01** (1)
MICE	0.98±0.01 (2)	0.85±0.01 (6)	0.91±0.01 (3)	**0.99±0.01** (1)	0.90±0.01 (6)
MISSFOREST	0.97±0.01 (2)	**0.95±0.01** (1)	0.90±0.01 (5)	0.98±0.01 (2)	0.94±0.02 (2)
GAIN	0.96±0.01 (6)	0.80±0.02 (7)	0.81±0.02 (7)	0.97±0.02 (7)	0.89±0.02 (6)
AimNet	0.97±0.01 (2)	0.93±0.01 (4)	**0.92±0.01** (1)	0.98±0.01 (2)	0.93±0.01 (4)
DAEMA	0.97±0.01 (2)	0.91±0.01 (5)	0.90±0.01 (3)	0.98±0.02 (2)	0.93±0.01 (4)
CAGAIN	**0.98±0.01** (1)	0.94±0.01 (3)	**0.92±0.01** (1)	0.98±0.01 (2)	0.94±0.01 (2)

	Credit	Solar Flare	Contraceptive	Thoracic
Mean Imputation	0.85±0.01 (8)	0.83±0.01 (7)	0.76±0.01 (6)	**0.93±0.01** (1)
KNN	**0.86±0.01** (1)	**0.86±0.01** (1)	0.78±0.01 (4)	**0.93±0.01** (1)
MICE	**0.86±0.01** (1)	0.85±0.01 (3)	**0.79±0.01** (1)	**0.93±0.01** (1)
MISSFOREST	**0.86±0.01** (1)	0.83±0.01 (7)	0.74±0.01 (8)	0.92±0.01 (8)
GAIN	**0.86±0.01** (1)	**0.86±0.01** (1)	0.77±0.02 (5)	**0.93±0.01** (1)
AimNet	**0.86±0.01** (1)	0.84±0.01 (5)	**0.79±0.01** (1)	**0.93±0.01** (1)
DAEMA	**0.86±0.01** (1)	0.84±0.01 (5)	0.76±0.01 (6)	**0.93±0.01** (1)
CAGAIN	**0.86±0.01** (1)	0.85±0.01 (3)	**0.79±0.01** (1)	**0.93±0.01** (1)

Table 11. Post-imputation prediction performance. Numerical, missing rate = 80%. ACC ± std (rank), higher is better.

	Spam	Letter	Balance	Breast	Eye EEG
Mean Imputation	0.67±0.01 (7)	0.05±0.01 (8)	0.47±0.01 (6)	0.68±0.01 (8)	0.59±0.01 (8)
KNN	0.67±0.01 (7)	0.18±0.01 (5)	0.61±0.02 (3)	0.82±0.03 (6)	0.63±0.01 (4)
MICE	0.74±0.03 (5)	0.08±0.01 (6)	**0.63±0.02** (1)	0.80±0.02 (7)	0.60±0.01 (7)
MISSFOREST	0.81±0.01 (3)	0.23±0.02 (3)	0.53±0.03 (5)	0.89±0.01 (2)	0.62±0.02 (5)
GAIN	0.73±0.01 (6)	0.06±0.01 (7)	0.34±0.09 (8)	0.84±0.02 (4)	0.62±0.02 (5)
AimNet	0.80±0.02 (4)	0.27±0.01 (2)	0.62±0.02 (2)	0.82±0.07 (5)	0.66±0.01 (2)
DAEMA	0.82±0.01 (2)	0.20±0.01 (4)	0.34±0.05 (7)	0.89±0.02 (2)	0.65±0.01 (3)
CAGAIN	**0.83±0.01** (1)	**0.30±0.01** (1)	0.57±0.02 (4)	**0.91±0.01** (1)	**0.67±0.01** (1)

	Credit	Solar Flare	Contraceptive	Thoracic
Mean Imputation	0.79±0.01 (7)	0.80±0.01 (3)	0.42±0.03 (5)	0.86±0.01 (4)
KNN	0.80±0.01 (4)	0.77±0.02 (5)	**0.50±0.01** (1)	**0.87±0.01** (1)
MICE	0.52±0.07 (8)	0.56±0.23 (8)	0.48±0.05 (2)	0.86±0.01 (4)
MISSFOREST	0.80±0.01 (4)	0.62±0.11 (7)	0.41±0.05 (6)	0.85±0.01 (7)
GAIN	0.80±0.01 (4)	0.81±0.03 (2)	0.39±0.01 (7)	0.85±0.02 (7)
AimNet	**0.82±0.01** (1)	0.78±0.02 (4)	0.47±0.03 (3)	**0.87±0.01** (1)
DAEMA	**0.82±0.01** (1)	0.77±0.02 (5)	0.37±0.02 (8)	0.86±0.01 (4)
CAGAIN	0.81±0.01 (3)	**0.82±0.02** (1)	0.47±0.03 (3)	**0.87±0.01** (1)

Post-imputation Performance To show how data imputation affects downstream tasks, we perform a post-imputation prediction with a random forest to predict the class labels in the datasets. Tables 10 and 11 report the classification accuracy with pre-imputation missing rate of 20% and 80%, respectively. When missing rate is 20%, CAGAIN is always in the top-3 on all the datasets, and is the best on 5 of them. When the missing rate is 80%, CAGAIN is the winner of 6 datasets and in the top-3 on 8 datasets, showcasing its robustness and advantage in a high missing rate scenario. In contrast, the performance of other competitors drastically varies across the datasets, and many recently proposed methods are even worse than KNN. This experiment demonstrates the superiority of CAGAIN in imputation quality.

6 Conclusion and Future Work

Seeing the weakness of existing approaches in producing robust results for dataset or missing rate shift, we proposed CAGAIN, a new model for tabular data imputation. CAGAIN employs a GAN along with the attention mechanism, with former mimicking real complete data, and the latter capturing the correlation between columns by focusing on the most important attributes that determine the true values of the missing positions. The experiments showed that while the performance of existing approaches significantly varies across datasets and missing rates, our model is always in the top-3, thereby demonstrating its robustness. A post-imputation prediction test reported higher and more robust accuracy using CAGAIN. Future work includes further improving the imputation accuracy and training time.

Acknowledgements. This work is mainly supported by NEC Corporation, and partially supported by JSPS Kakenhi 22H03903 and CREST JPMJCR22M2.

References

1. Asuncion, A., Newman, D.: UCI machine learning repository (2007). http://archive.ics.uci.edu/ml
2. Belkin, M., Hsu, D.J., Mitra, P.: Overfitting or perfect fitting? risk bounds for classification and regression rules that interpolate. In: NeurIPS, pp. 2306–2317 (2018)
3. Breve, B., Caruccio, L., Deufemia, V., Polese, G.: RENUVER: a missing value imputation algorithm based on relaxed functional dependencies. In: EDBT, pp. 1:52–1:64. OpenProceedings.org (2022)
4. Chen, T., Guestrin, C.: XGBoost: a scalable tree boosting system. In: KDD, pp. 785–794 (2016)
5. D'Ambrosio, A., Aria, M., Siciliano, R.: Accurate tree-based missing data imputation and data fusion within the statistical learning paradigm. J. Classif. **29**(2), 227–258 (2012)
6. Friedjungová, M., Vašata, D., Balatsko, M., Jiřina, M.: Missing features reconstruction using a wasserstein generative adversarial imputation network. In: ICCS, pp. 225–239 (2020)
7. Gondara, L., Wang, K.: MIDA: multiple imputation using denoising autoencoders. In: PAKDD, pp. 260–272 (2018)
8. Goodfellow, I.J., et al.: Generative adversarial nets. In: NIPS, pp. 2672–2680 (2014)
9. Jonsson, P., Wohlin, C.: An evaluation of k-nearest neighbour imputation using likert data. In: METRICS, pp. 108–118 (2004)
10. Kalton, G., Kasprzyk, D.: Imputing for missing survey responses. In: ASA-SRMS, vol. 22, p. 31 (1982)
11. Kingma, D.P., Ba, J.: Adam: a method for stochastic optimization. In: ICLR (2015)
12. Kodali, N., Abernethy, J., Hays, J., Kira, Z.: On convergence and stability of GANs (2017). arXiv preprint arXiv:1705.07215
13. Mao, X., Li, Q., Xie, H., Lau, R.Y., Wang, Z., Smolley, S.P.: Least squares generative adversarial networks. In: ICCV, pp. 2794–2802 (2017)
14. McCoy, J.T., Kroon, S., Auret, L.: Variational autoencoders for missing data imputation with application to a simulated milling circuit. IFAC **51**(21), 141–146 (2018)
15. Miyato, T., Kataoka, T., Koyama, M., Yoshida, Y.: Spectral normalization for generative adversarial networks. In: ICLR (2018)
16. Nazabal, A., Olmos, P.M., Ghahramani, Z., Valera, I.: Handling incomplete heterogeneous data using VAEs. Pattern Recogn. **107**, 107501 (2020)
17. Neves, D.T., Naik, M.G., Proença, A.: SGAIN, WSGAIN-CP and WSGAIN-GP: novel GAN methods for missing data imputation. In: ICCS, pp. 98–113 (2021)
18. Oh, E., Kim, T., Ji, Y., Khyalia, S.: STING: self-attention based time-series imputation networks using GAN. In: ICDM, pp. 1264–1269 (2021)
19. Qiu, W., Huang, Y., Li, Q.: IFGAN: missing value imputation using feature-specific generative adversarial networks. In: BigData, pp. 4715–4723 (2020)

20. Rekatsinas, T., Chu, X., Ilyas, I.F., Ré, C.: Holoclean: holistic data repairs with probabilistic inference. PVLDB **10**(11), 1190–1201 (2017)
21. Ryu, S., Kim, M., Kim, H.: Denoising autoencoder-based missing value imputation for smart meters. IEEE Access **8**, 40656–40666 (2020)
22. Song, S., Sun, Y., Zhang, A., Chen, L., Wang, J.: Enriching data imputation under similarity rule constraints. IEEE Trans. Knowl. Data Eng. **32**(2), 275–287 (2020)
23. Stekhoven, D.J., Bühlmann, P.: MissForest - non-parametric missing value imputation for mixed-type data. Bioinformatics **28**(1), 112–118 (2012)
24. Tihon, S., Javaid, M.U., Fourure, D., Posocco, N., Peel, T.: DAEMA: denoising autoencoder with mask attention. In: ICANN, pp. 229–240 (2021)
25. Van Buuren, S., Groothuis-Oudshoorn, K.: mice: multivariate imputation by chained equations in R. J. Stat. Softw. **45**, 1–67 (2011)
26. Vaswani, A., et al.: Attention is all you need. In: NIPS, pp. 5998–6008 (2017)
27. Wu, R., Zhang, A., Ilyas, I., Rekatsinas, T.: Attention-based learning for missing data imputation in HoloClean. MLSys **2**, 307–325 (2020)
28. Yoon, J., Jordon, J., Schaar, M.: GAIN: missing data imputation using generative adversarial nets. In: ICML, pp. 5689–5698 (2018)

CF-SAFF: Collaborative Filtering Based on Self-attention Mechanism and Feature Fusion

Weixin Kong, Xiaoye Wang[(✉)], and Yingyuan Xiao

School of Computer Science and Engineering, Tianjin University of Technology,
Tianjin 300384, China
jiaoliu456@163.com, yyxiao@tjut.edu.cn

Abstract. User features and item features are important information for recommendation systems, and their interaction significantly improves the accuracy of recommendations. Existing work classifies types of interactions as internal and cross interactions. However, it does not take into consideration the different importance of information in cross interactions, nor captures more complex dependencies between features with higher-order interactions. To solve this problem we propose Collaborative Filtering based on Self-Attention Mechanism and Feature Fusion (CF-SAFF). It explicitly uses internal interactions for user and item feature learning, and assigns different weights to cross interactions based on the Self-Attention mechanism to represent the different importance of interaction nodes, thereby performing preference matching on recommendations. At the same time, the fusion operation of internal interaction information and cross interaction information is used to discover high-order feature combination information, which improves the prediction accuracy and generalization ability of the model. The model has conducted extensive experiments on public standard datasets, and the results show that the model has achieved better results than previous mainstream models in all performance indicators.

Keywords: recommendation systems · Self-Attention · high-order interactions · Collaborative Filtering

1 Introduction

Research on recommendation systems dates back to the 1990s, and many content based and collaborative filtering-based approaches have been developed. Collaborative filtering (CF) is one of the earliest proposed recommendation techniques [11,22], and it is also the most researched and most practically applied recommendation system based on matrix factorization technology. It learns the embedding of users and items in user-item interactions, and shows the interaction between users and items [18]. Considering attribute interactions for each data sample when learning attribute embedding [16,17] provides useful information for accurate predictions. Typically, attribute interaction-aware CF models

C. Strauss et al. (Eds.): DEXA 2023, LNCS 14147, pp. 274–288, 2023.
https://doi.org/10.1007/978-3-031-39821-6_22

consider attribute interactions to determine the final predictor Factorization Machine (FM) [16] models each attribute interaction as a dot product of two embedding vectors and linearly aggregates all modeling results. However, limited by the linear nature of factorization models, they lack performance when dealing with large-scale and complex data, such as complex attribute interactions and may contain complex semantic information. With the application of Graph Neural Network (GNN) in recommendation systems, Li et al. [13] and Su et al. [19] used the relational modeling capability of GNN to capture more complex attribute interaction information. These models treat all attribute interactions equally, modeling and aggregating them in the same way, which may not give satisfactory results since different attribute interactions should have different effects. Attribute interactions are classified into two categories [20] in Neural Graph Matching based Collaborative Filtering model (GMCF). The interactions between user attributes (or item attributes) are described as internal interactions. The interaction between user attributes and item attributes are called cross interactions. If the application scenario is to predict the possibility of male users liking to watch sci-fi movies, the cross feature <male, science> is more important than <male, comedy>. It is unreasonable for the above model to perform indiscriminate sum pooling for all cross interactions. At the same time, high-level interaction information is not used. We know that there are certain dependencies between features, and they are all "and" relationships. For recommendation model, it is crucial to extract high-order interactions between features.

To solve the above problem, we propose Collaborative Filtering based on Self-Attention Mechanism and Feature Fusion (CF-SAFF). CF-SAFF uses the Self-Attention mechanism to assign different weights to cross interactions to indicate the different importance of interaction nodes, thus matching preferences for recommendations. CF-SAFF attempts to perform information fusion operations to model internal and cross interaction information, capture meaningful high-order interactions between users and items, give important features greater weight, and consider interactions between multiple features, while they are not simply considered as independent factors. High-order interactions can capture more complex dependencies between features, thereby improving the predictive accuracy of the model. To the best of our knowledge, this is the first attempt to use the Self-Attention mechanism to assign weight to cross interactions and to fuse internal and cross interaction information to obtain high-order feature combination information. Our contributions can be summarized in three aspects:

- We use the Self-Attention mechanism to improve the cross interaction, and assign different weights to the cross interaction to represent the different importance of the interaction nodes, so as to perform preference matching on the recommendation.
- We perform fusion operations on internal and cross-interaction information to capture meaningful high-order interactions between users and items, give important features greater weight, and improve the prediction accuracy and generalization ability of the model.

- We conduct extensive experiments on three public standard datasets. The Self-Attention mechanism and high-order interaction modules play an important role, and the experiments confirm the superiority of our proposed model.

2   Related Work

2.1   CF Model

"Collaborative filtering" is a process that filters massive amounts of information in collaboration with everyone's feedback, evaluations, and opinions, and screens out information that target users may be interested in. Decomposing the co-occurrence matrix in a collaborative filtering algorithm into a user matrix and an item matrix, and using the inner product of the user hidden vector and the item hidden vector to rank and recommend, is the basic idea of the classical model of collaborative filtering, FM [16]. The feature field [10] is introduced to increase the concept of field, and the same feature uses different latent vectors for different fields, so that the model modeling is more accurate, or add a gate [14], to remove useless features. However, limited by the natural linear nature of the factorization model, its performance in dealing with large-scale and complex data is lacking, such as complex attribute interactions, and the item side may contain complex semantic information. Adding more complex nonlinear structures such as Multi-Layer Perceptron (MLP) to capture implicit information [2,4,7,9] has not been proven effective [1]. GMCF [20] aggregates attribute interactions in an explicit way to obtain more effective attribute interaction modeling and structural information. On this basis, we perform fusion operations on internal and cross interaction information to capture meaningful high-order interactions between users and items, and improve the prediction accuracy and generalization ability of the model.

2.2   Self-attention Mechanism

Attention mechanisms have been widely used in various fields of deep learning in recent years, whether it is image processing, speech recognition, or natural language processing, attention models can be clearly seen in the processing of various tasks in recommendation systems [15,24]. Self-attention mechanism can reduce the dependence on external information, and is good at mining the internal correlation of features from the data itself. Items in the input sequence are first encoded in parallel into multiple representations, called keys, queries, and values. Self-Attention mechanism has played an important role in various tasks such as reading comprehension, textual implicit, sentence representation, abstract summaries, and recommendation systems [6,12,21]. Our proposed model uses the Self-Attention mechanism to assign different weights to user-item cross interactions to represent the different importance of interaction nodes and optimize preference matching.

2.3   Graph Neural Networks

Graph Neural Networks (GNNs) are powerful tools for processing graph struc-tured data [23]. The molecular data are the data of the graph structure, with the atoms acting as nodes in the graph and the bonds connecting them as the edges of the graph. GNNs can improve the quality of node embedding by consid-ering neighborhood information extracted from the underlying graph topology. They have been widely used in a variety of applications, including knowledge graphs, drug discovery, neuroscience, social networks, recommendation systems [3,5,8,25]. GNN is essentially suitable for node interaction modeling on graph structure features. Li et al., [13] and Su et al., [19] utilize GNNs for attribute interaction modeling and aggregation as a graph learning process. But these models do not differentiate attribute interactions, which is inefficient in captur-ing useful structural information. Our data is modeled with a graph structure, and the final mixed features are generated by fusion of internal interaction infor-mation, cross interaction information, basic graph structure information, and high-order interaction information, and then aggregated in a graph matching structure to make recommendation more efficient.

3   The Proposed Model: CF-SAFF

3.1   Symbolic Description

The user's attribute space is represented by $\mathcal{J}^U$ and the item's attribute space is represented by $\mathcal{J}^I$. Attribute-value pair is represented as (att, val), where att represents the attribute and val represents the attribute value. For exam-ple, $(female, 1)$ means that the user's gender is $female$, where $female \in \mathcal{J}^U$ means that $female$ is a user attribute. Let D be N training pairs with $D = (x_n, y_n)_{1 \leq n \leq N}$, where x_n is called the data sample consisting of $C_n^U = c^U = (att, val)_{att \in \mathcal{J}^U}$ and $C_n^I = c^I = (att, val)_{att \in \mathcal{J}^I}$. Respectively, C_n^U and C_n^U are called the feature representation of the user and the item. The user's implicit feedback on the item (e.g., liked, watched) is denoted as $y_n \in \mathbb{R}$. Our goal is to devise a prediction function $F_{saff}(x_n)$ that gives the correct feedback y_n for the input x_n.

Each attribute $att \in \mathcal{J}^U \cup \mathcal{J}^I$ is embedded in d-dimensional space $\mathbb{R}^d$ as a vector v. The same attribute att in all data samples shares the same embed-ding vector v_{att}, but their corresponding attribute values val are different. The attribute-value pairs are represented as u_{att}, $u_{att} = val \cdot v_{att}$. For simplicity, the subscripts of v and u are omitted hereinafter. The co-occurrence of two attributes att_1 and att_2 is defined as an attribute interaction. The function $f(u_1, u_2) : \mathbb{R}^{2 \times d} \longrightarrow \mathbb{R}^\ell$ is modeled as an attribute interaction. Where ℓ represents the dimension of the output, and the collaborative information of interactions in different data samples will further help to discover the interactions between attributes that have never happened simultaneously. Therefore, $f(\cdot, \cdot)$ actually learns attribute embedding that capture the collaborative information between attributes (i.e. similar attributes will have similar embedding) [16].

$G = <V, E>$ is represented as a graph, where $v = v_{i1<i<n}$ is the set of n nodes, each represented by a vector v_i. Neighborhood information between nodes is denoted as E. Graph neural networks aggregate neighborhood information through message passing. Specifically, node's message passing first aggregates the vector representations of its neighbors. Then, aggregate vectors by fusing nodes and neighbor nodes. The updated nodes are represented as:

$$v'_i = f_{fuse}(v_i, Aggr(v_j))_{j \in N(i)} \tag{1}$$

where the $f_{fuse}(\cdot, \cdot)$ fusion function, $Aggr(.)$ is an aggregation function that aggregates neighborhood information into a fixed dimensional representation (e.g., element sum), and $N(i)$ is the set of neighborhoods of node i.

3.2   Attention Mechanism Attempts

In order to prove that the Self-Attention mechanism is more suitable for our model, we tried different attention mechanisms.

Attention Net. An attention network is added to the interaction of second-order cross features. Figure 1 illustrates the basic structure of Attention net. Attention is given in the form of a net, an MLP is used to parameterize the attention score, and the Attention net is defined as:

$$a'_{ij} = h^T Relu(W(u_i \odot \hat{u}_j) + b) \tag{2}$$

$$a_{ij} = \frac{exp(a'_{ij})}{\sum_{j \in N_i} exp(a'_{ij})} \tag{3}$$

The structure of Attention net is a simple single fully connected layer with a softmax output layer. Where, $W \in \mathbb{R}^{N \times d}$, $b \in \mathbb{R}^N$, $h \in \mathbb{R}^N$, $j \in N_i$ is the set of all neighbors of the node i. The learned model parameters are the weight matrix W and bias vector b from the feature intersection layer to the fully connected layer of the Attention network and the weight vector h from the fully connected layer to the softmax output layer. u_i is the embedding of a node i in one graph, $\hat{u}_j$ is the embedding of node j in another graph. a_{ij} is the importance of each interaction feature. Attention net participates in the back-propagation process with the whole model to obtain the final weight parameters, and the model with Attention net is called CF-AFF.

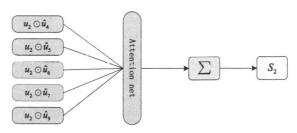

Fig. 1. Structural Model of Attention net.

Self-attention Mechanism. Self-Attention mechanism is a special case of attention mechanism that captures the relationship between each user and each item in the sequence, Self-Attention mechanism as shown in Fig. 2. Use the element-level product of u_i and $\hat{u}_j$ as input to the Self-Attention layer, $X \in \mathbb{R}^{N \times d}$:

$$X = u_i \odot \hat{u}_j \tag{4}$$

The initial values of query, key and value are all equal to input matrix X, query and key are projected into the same space by a linear transformation and they have common parameters.

$$Q = XW_q \tag{5}$$
$$K = XW_k \tag{6}$$

where $W_k = W_q \in \mathbb{R}^{d \times d}$ is the weight matrix of query and key respectively, the scaled dot product is used as the attention score function, and then the attention score mapping matrix can be obtained by calculation.

$$a_{ij} = softmax(\frac{K^T Q}{\sqrt{d}}) \tag{7}$$

The output attention score mapping matrix a_{ij} is a matrix of $N \times N$, representing the similarity of N items. Finally, the attention score mapping matrix is multiplied by value.

$$s_{ij} = a_{ij}X \tag{8}$$

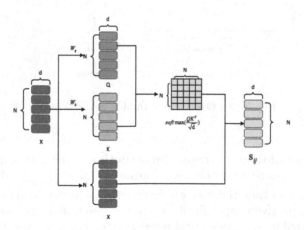

Fig. 2. A structural Model of the Self-Attention Mechanism.

3.3 CF-SAFF Model

Our model is composed of three main modules: (1) Feature embedding and graph construction modules. (2) Internal interaction based on message passing and cross interaction module based on Self-Attention mechanism. (3) Information fusion module, an overview of the CF-SAFF model is shown in Fig. 3.

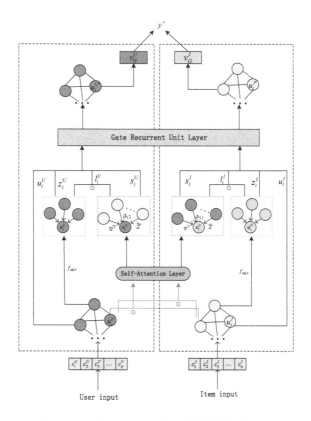

Fig. 3. An Overview of the CF-SAFF Model.

Feature Embedding and Graph Construction. Each data sample $x = c_i^U \cup c_j^I$ is feature-embedded to obtain user feature vector $c_i^U{}_{1<i<p}$ and item feature vector $c_i^I{}_{1<i<q}$, and then the user attribute graph and item attribute graph are constructed respectively. Specifically, each user (each item) is constructed as a graph, with attributes as nodes and internal interactions between attributes as edges. For each input sample x, we use the attribute representation u_i^U as the node representation for each node i. Thus, the set of nodes in user attribute graph is denoted as $V^U = u_i^U{}_{i \in \mathcal{J}^U}$. Pairs of nodes in a node set are connected by an edge to represent the interaction between two attributes, so the user attribute graph is a complete graph. Each user property graph is represented as

$G^U = \langle V^U, E^U \rangle$, where E^U is the set of edges formed by the interaction of all internal properties. The item graphs are constructed in the same way.

Internal Interaction Based on Message Passing and Cross Interaction Based on Self-attention Mechanism. In this model, message passing is used to model internal interactions, which are used for the user (item) feature learning. Specifically, we used a MLP function to model each internal interaction. The feature embeddings of two nodes are taken as input and the modelling results are output:

$$z_{ij} = f_{MLP}(u_i, u_j) \tag{9}$$

where $z_{ij} \in \mathbb{R}^d$ is the result of modelling the interaction of node pair (i, j). f_{MLP} explicitly modeling the interaction information between two attributes, combining interaction modeling and message passing in graph learning. Then, all interaction modelling results corresponding to node i are aggregated into message passing information. We use element-level sums to aggregate the interaction modelling results:

$$z_i = \sum_{j \in N_i} z_{ij} \tag{10}$$

where $z_i \in \mathbb{R}^d$ is the result of a message passing information and N_i denotes the set of neighbors of node i.

Intersection modeling is used to match nodes in the two graphs. Intuitively, if user attribute c_i^U shows a higher preference for item attribute c_j^I, then c_i^U and c_j^I have a higher matching score. We use Self-Attention mechanism to assign different weights to each cross feature to increase the weight of valuable information. Specifically, we model the cross interactions as:

$$s_{ij} = a_{ij}(u_i \odot \hat{u}_j) \tag{11}$$

where a_{ij} is the attention score calculated using the Self-Attention mechanism, s_{ij} denotes the result of matching two nodes. Similar to message passing, we use element-level sums to aggregate node matches between one node in a graph and all nodes in another graph, and aggregate the node matching results as:

$$s_i = \sum_{j \in \hat{v}_i} s_{ij} \tag{12}$$

Information Fusion Module. We use the cross interaction node s_i and the messaging of internal interactions z_i to capture meaningful high-order interactions. Modelling with element-level products as:

$$l_i = z_i \odot s_i \tag{13}$$

where l_i denotes the element-level product of high-order interactions result of node i.

Using the initial node representation u_i, the internal interaction information z_i, the cross interaction information s_i, and high-order interactions information

l_i as input sequences, the recurrent neural network model GRU as the fusion function, the final output hidden layer is the fused node representation as:

$$u_i' = f_{GRU}(u_i \oplus z_i \oplus s_i \oplus l_i) \tag{14}$$

where u_i' denotes the fused representation of node i and, $f_{GRU} \in \mathbb{R}^{4 \times d} \to \mathbb{R}^d$, $\oplus$ the connection operation.

We aggregate the fused node representations into a graph structure, and then element-level sums are used to aggregate the fused node representations, resulting in vector representations of the user attribute graph and item attribute graph:

$$v_G^U = \sum_{i \in V^I} u_i^{U'}, v_G^I = \sum_{i \in V^U} u_i^{I'} \tag{15}$$

where V^I denotes the set of nodes of the project attribute graph and V^U denotes the set of nodes of the user attribute graph.

Finally we match the two graphs using the dot product to get the final prediction function as:

$$y' = v_G^{U^T} v_G^I \tag{16}$$

Model Training. We use the L_2 norm to regularize all parameters of the CF-SAFF model, the empirical risk minimization function consisting of a cross-entropy loss function and an L_2 norm:

$$\mathcal{R}(\theta) = \frac{1}{N} \sum_{n=1}^{N} y_n log y' + (1 - y_n) \log(1 - y') + \lambda(||\theta||_2) \tag{17}$$

$$\theta^* = argmin_\theta \mathcal{R}(\theta) \tag{18}$$

All parameters in the CF-SAFF model are denoted as θ, θ^* is the optimal parameter obtained.

4   Experiments

4.1   Datasets and Data Preparation

The proposed model was evaluated on three datasets and the statistical information is presented in Table 1. User ratings of film are included in Movielens 1M. A user and a movie with the corresponding attributes are included in each data example. Other attributes are added to the dataset, such as directors and actors. Explicit ratings were transformed into implicit feedback, with ratings greater than 3 being treated as positive for Movielens 1M. Invisible user ratings of books are included in Book-crossing. A user and a book with the corresponding attributes are included in each data instance. The Taobao dataset collects records of users clicking on advertisements, and each record contains a category of users and advertisements.

Table 1. Experimental Dataset Statistics

Dataset	Data	User	Item	User attr.	Item attr.
MovieLens 1M	1,149,238	5,950	3514	30	6944
Book-Crossing	1,050,834	4,873	53168	87	43157
Taobao	2,599,463	4,532	371,760	36	434,254

4.2 Evaluation Indicators

We use the Area Under Curve (AUC), Logloss, Normalized Discounted Cumulative Gain ($NDCG$) to evaluate the performance of the proposed model and the baseline model. *Logloss* is an evaluation metric for binary classification models that measures the distance between the predicted score and the true label of each instance, $NDCG@k$ is a common metric for evaluating top k recommendations. It is an evaluation index that considers the return order. We set the values of k to 5 and 10.

4.3 Baselines

FM [16] is established that second-order feature interactions from vector inner products and all modelling results are aggregated into a final prediction. **NFM** [9] combines the linearity of FM in second-order feature interaction modelling with the non-linearity of neural networks in higher-order feature interaction modelling. **W&D** [4] can efficiently memorise sparse feature interactions and generalize to previously unseen feature interactions. **DeepFM** [7] consists of two components, the FM component and the depth component, which share the same input. **AutoInt** [17] uses a Self-Attention mechanism with residual networks to explicitly model feature interactions in low-dimensional space. **Fi-GNN** [13] models complex interactions between feature fields into modeling node interactions on feature graphs. **L$_0$-SIGN** [19] detects the most beneficial feature interactions and uses only the beneficial feature interactions for recommendations, treating each data sample as a graph, features as nodes and feature interactions as edges. **GMCF** [20] recognizes two different types of attribute interactions, aggregated into a graph matching structure, and both types of attribute interactions are effectively captured for recommendation. **CAN** [2] uses the MLP structure to mine the relationship between the user's historical behavior and the target item, which approximates the empirical feature interaction.

4.4 Experimental Setup

The training set, validation set and test set are randomly divided in the ratio of 6:2:2. The embedding dimension of the node is set to 64 (i.e. $d = 64$). The hidden layer of the MLP is set to 1 and the number of units is set to $4d$. For all baselines, the MLP structure of the interaction modeling is set to be the same

Table 2. Summary of the Performance in Comparison with Baselines

model	MovieLens 1M				Book-Crossing				Taobao			
	AUC	Logloss	NDCG@5	NDCG@10	AUC	Logloss	NDCG@5	NDCG@10	AUC	Logloss	NDCG@5	NDCG@10
FM	0.8761	0.4409	0.8143	0.8431	0.7417	0.5771	0.7616	0.8029	0.6171	0.2375	0.0812	0.1120
NFM	0.8985	0.3996	0.8486	0.8832	0.7988	0.5432	0.7989	0.8326	0.6550	0.2122	0.0997	0.1251
W&D	0.9043	0.3878	0.8538	0.8869	0.8105	0.5366	0.8048	0.8381	0.6531	0.2124	0.0959	0.1242
DeepFM	0.9049	0.3856	0.8510	1.8848	0.8127	0.5379	0.8088	0.8400	0.6550	0.2115	0.0974	0.1243
AutoInt	0.9034	0.3883	0.8619	0.8931	0.8130	0.5355	0.8127	0.8472	0.6434	0.2146	0.0924	0.1206
Fi-GNN	0.9049	0.3871	0.8705	0.9029	0.8136	0.5338	0.8094	0.8522	0.6462	0.2131	0.0986	0.1241
L0-SIGN	0.9072	0.3846	0.8849	0.9094	0.8163	0.5274	0.8148	0.8629	0.6547	0.2124	0.1006	0.1259
GMCF	0.9127	0.3789	0.9374	0.9436	0.8228	0.5233	0.8671	0.8951	0.6679	0.1960	0.1112	0.1467
CAN	0.9133	0.3773	0.9396	0.9442	0.8235	0.5143	0.8722	0.8996	0.6776	0.1919	0.1130	0.1494
CF-AFF	0.9194	0.3702	0.9412	0.9472	0.8326	0.5139	0.8859	0.9068	0.6803	0.1911	0.1145	0.1539
CF-SAFF	0.9287	0.3623	0.9477	0.9538	0.8379	0.5017	0.8965	0.9224	0.6892	0.1874	0.1210	0.1566
Improv	1.68%	3.97%	0.86%	1.02%	1.75%	2.45%	2.79%	2.53%	1.71%	2.34%	7.08%	4.82%

as the MLP in our model to allow for a fair comparison. The parameter λ for regularisation is set to 1×10^{-5}; the learning rate α is set to 1×10^{-3}; and Adam is used as the optimization algorithm for the model.

4.5   Performance Comparison

The experimental results of the nine baseline models with the proposed model on the three datasets are presented in Table 2. CF-AFF and CF-SAFF outperformed the baseline model for all metrics on all datasets. Our model achieved the best performance, which shows that this model is able to analyse attribute information effectively. FM and AFM are the worst performers, using only dot products for modelling attribute interactions and are unable to model complex interactions effectively. GMCF, AutoInt, Fi-GNN, L0-SIGN, CAN, CF-AFF, and CF-SAFF are models that explicitly model attribute interactions. They obtain better prediction accuracy than other models, which says that explicit modelling is more helpful in extracting useful attribute information. The GNN-based models (such as Fi-GNN, GMCF, CF-AFF, CF-SAFF) have better predictions than other models, which illustrate the role of GNN in attribute interaction modelling. This shows that the Self-Attention mechanism assigns different weights to cross interactions and increases high-order interaction information, effectively improving the performance of the model.

4.6   Model Analysis and Discussion

Ablation Study. To explore whether the Self-Attention mechanism and high-order interactions have a positive effect on prediction. We removed the Self-Attention mechanism module and the high-order interactions module. The Self-Attention module is denoted as -A, and the high-order interactions module is denoted as -L. Removing the high-order interactions module gives the model no-L and removing the Self-Attention mechanism module gives the model no-A. If both modules are present, our model is obtained. As shown in Fig. 4 (the results of *Logloss* and *NDCG*@5 are omitted because they have the same trend as *AUC* and *NDCG*@10), we found that both the Self-Attention module and the high-order interactions module improved the performance of the model to a certain extent, and the feasibility of the two modules was proved.

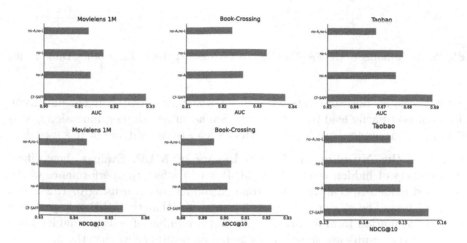

Fig. 4. Impact of Using Different Modules on The Model.

Fusion Algorithm Research. To demonstrate the validity of GRU, we further used SUM and MLP as aggregation functions, and the experimental results of these three algorithms as fusion methods are in Table 3. GRU was used as a fusion function to efficiently represent the fused nodes for accurate predictions. The worst results were obtained when SUM was used as a fusion function. This shows that information fusion is a complex process that requires powerful algorithms.

Table 3. Performance Using Different Fusion Algorithms

	MovieLens 1M		Book-Crossing		Taobao	
	AUC	NDCG@10	AUC	NDCG@10	AUC	NDCG@10
SUM	0.9143	0.9488	0.8312	0.9198	0.6729	0.1489
MLP	0.9131	0.9494	0.8304	0.9211	0.6807	0.1548
GRU	**0.9287**	**0.9538**	**0.8379**	**0.9224**	**0.6892**	**0.1566**

Dimensional Studies. The results of the proposed model on different node embedding dimensions (d) are presented in Fig. 5. Our model achieved the best

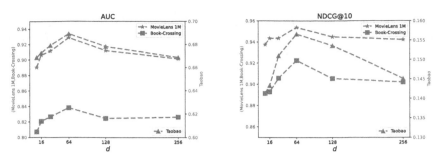

Fig. 5. Performance of CF-SAFF in Different Node (attribute) Embedding Dimensions.

performance on both datasets when $d = 64$, suggesting that higher dimension does not necessarily lead to better prediction accuracy. Higher dimension means that more parameters are fitted, which can lead to over-fitting of this model.

Effect of the Number of Hidden Layers of MLP. Evaluate how different numbers of hidden layers in the MLP would affect the performance of the proposed model. Specifically, the proposed model was run using 0, 1, 2 and 3 hidden layers. The absence of hidden layers means that the MLP only performs the transformation from linear, using the same number of cells per hidden layer $(4d)$, and the results are shown in Table 4. The results show that the process of extracting useful information is complex and requires a non-linear approach. The best results for the proposed model were obtained when the number of layers of hidden layers was 1. This shows that increasing the number of hidden layers does not make the model perform better, as higher numbers of hidden layers can lead to overfitting.

Table 4. The Performance of CF-SAFF on Using Different MLP Depths

	MovieLens 1M		Book-Crossing		Taobao	
	AUC	NDCG@10	AUC	NDCG@10	AUC	NDCG@10
0	0.9196	0.9477	0.8289	0.9112	0.6782	0.1512
1	**0.9287**	**0.9538**	**0.8379**	**0.9224**	**0.6892**	**0.1566**
2	0.9264	0.9501	0.8344	0.9176	0.6843	0.1537
3	0.9233	0.9489	0.8311	0.9155	0.6831	0.1526

Effect of Self-attention on Different Datasets. We find that the proposed model on the Taobao dataset improves the accuracy more because the Taobao dataset is more sparse compared to others. We know that a dense dataset contains more effective user information, while a sparse dataset has more dispersed effective information, and the Self-Attention mechanism can compensate for the performance impact caused by sparse data by assigning higher weights to more important information.

5 Conclusion

In this paper, we propose a new recommendation model based on Self-Attention mechanism and high-order interactions. We use a Self-Attention mechanism to assign different weights to cross interactions for preference matching. We perform fusion operations on internal and cross interaction information to capture meaningful high-order interactions between users and items, give important features greater weight, and improve the prediction accuracy and generalization ability of the model. Experimental results show that the proposed model is an effective and experiments were conducted on three standard datasets widely used by recommendation systems, the results outperformed all the baselines.

Acknowledgements. This work is supported by "Tianjin Project + Team" Key Training Project under Grant No. XC202022.

References

1. Beutel, A., et al.: Latent cross: making use of context in recurrent recommender systems. In: Proceedings of the Eleventh ACM International Conference on Web Search and Data Mining, pp. 46–54 (2018)
2. Bian, W., et al.: CAN: feature co-action network for click-through rate prediction. In: Proceedings of the Fifteenth ACM International Conference on Web Search and Data Mining (2022)
3. Chen, Y., Wu, L., Zaki, M.J.: Toward subgraph guided knowledge graph question generation with graph neural networks. arXiv preprint arXiv:2004.06015 (2020)
4. Cheng, H.T., et al.: Wide & deep learning for recommender systems. In: Proceedings of the 1st Workshop on Deep Learning for Recommender Systems, pp. 7–10 (2016)
5. Di, W.: LightFIG: simplifying and powering feature interactions via graph for recommendation. PeerJ Comput. Sci. **8**, e1019 (2022)
6. Fu, Y., Liu, Y.: CGSPN: cascading gated self-attention and phrase-attention network for sentence modeling. J. Intell. Inf. Syst. **56**, 147–168 (2020). https://doi.org/10.1007/s10844-020-00610-z
7. Guo, H., Tang, R., Ye, Y., Li, Z., He, X.: DeepFM: a factorization-machine based neural network for CTR prediction. arXiv preprint arXiv:1703.04247 (2017)
8. Hamilton, W., Ying, Z., Leskovec, J.: Inductive representation learning on large graphs. In: Advances in Neural Information Processing Systems, vol. 30 (2017)
9. He, X., Chua, T.S.: Neural factorization machines for sparse predictive analytics. In: Proceedings of the 40th International ACM SIGIR Conference on Research and Development in Information Retrieval, pp. 355–364 (2017)
10. Juan, Y., Zhuang, Y., Chin, W.S., Lin, C.J.: Field-aware factorization machines for CTR prediction. In: Proceedings of the 10th ACM Conference on Recommender Systems, pp. 43–50 (2016)

11. Koren, Y., Bell, R., Volinsky, C.: Matrix factorization techniques for recommender systems. Computer **42**(8), 30–37 (2009)
12. Li, J., Wu, M.: Deep attention factorization machine network for distributed recommendation system. In: 2022 International Conference on Machine Learning, Cloud Computing and Intelligent Mining (MLCCIM), pp. 511–517 (2022)
13. Li, Z., Cui, Z., Wu, S., Zhang, X., Wang, L.: Fi-GNN: modeling feature interactions via graph neural networks for CTR prediction. In: Proceedings of the 28th ACM International Conference on Information and Knowledge Management, pp. 539–548 (2019)
14. Liu, B., et al.: AutoFIS: automatic feature interaction selection in factorization models for click-through rate prediction. In: Proceedings of the 26th ACM SIGKDD International Conference on Knowledge Discovery & Data Mining, pp. 2636–2645 (2020)
15. Qiu, S., Wu, Y., Anwar, S., Li, C.: Investigating attention mechanism in 3D point cloud object detection. In: 2021 International Conference on 3D Vision (3DV), pp. 403–412. IEEE (2021)
16. Rendle, S.: Factorization machines. In: 2010 IEEE International Conference on Data Mining, pp. 995–1000. IEEE (2010)
17. Song, W., et al.: AutoInt: automatic feature interaction learning via self-attentive neural networks. In: Proceedings of the 28th ACM International Conference on Information and Knowledge Management, pp. 1161–1170 (2019)
18. Su, Y., Erfani, S.M., Zhang, R.: MMF: attribute interpretable collaborative filtering. In: 2019 International Joint Conference on Neural Networks (IJCNN), pp. 1–8. IEEE (2019)
19. Su, Y., Zhang, R., Erfani, S., Xu, Z.: Detecting beneficial feature interactions for recommender systems. In: Proceedings of the AAAI Conference on Artificial Intelligence, vol. 35, pp. 4357–4365 (2021)
20. Su, Y., Zhang, R., Erfani, S.M., Gan, J.: Neural graph matching based collaborative filtering. In: Proceedings of the 44th International ACM SIGIR Conference on Research and Development in Information Retrieval, pp. 849–858 (2021)
21. Sun, G., Wang, Z., Zhao, J.: Automatic text summarization using deep reinforcement learning and beyond. Inf. Technol. Control **50**, 458–469 (2021)
22. Wang, X., Zhang, R., Sun, Y., Qi, J.: Combating selection biases in recommender systems with a few unbiased ratings. In: Proceedings of the 14th ACM International Conference on Web Search and Data Mining, pp. 427–435 (2021)
23. Wu, Z., Pan, S., Chen, F., Long, G., Zhang, C., Philip, S.Y.: A comprehensive survey on graph neural networks. IEEE Trans. Neural Netw. Learn. Syst. **32**(1), 4–24 (2020)
24. Yin, Y., Huang, C., Sun, J., Huang, F.: Multi-head self-attention recommendation model based on feature interaction enhancement. In: ICC 2022 - IEEE International Conference on Communications, pp. 1740–1745 (2022)
25. Ying, R., He, R., Chen, K., Eksombatchai, P., Hamilton, W.L., Leskovec, J.: Graph convolutional neural networks for web-scale recommender systems. In: Proceedings of the 24th ACM SIGKDD International Conference on Knowledge Discovery & Data Mining, pp. 974–983 (2018)

Except-Condition Generative Adversarial Network for Generating Trajectory Data

Yeji Song[1], Jihwan Shin[1], Jinhyun Ahn[2], Taewhi Lee[3], and Dong-Hyuk Im[1]($\boxtimes$)

[1] Kwangwoon University, Seoul, South Korea
dhim@kw.ac.kr
[2] Jeju National University, Jeju, South Korea
[3] Electronics and Telecommunications Research Institute, Daejeon, South Korea

Abstract. Location data shared on social media is collected and processed as trajectory data, which exposes individuals to leakage of sensitive information, such as sensitive geographic areas. A typical countermeasure is a generative adversarial network (GAN) model that ensures data anonymity. However, generating data selectively by identifying only sensitive areas is difficult. In this study, we propose an except-condition GAN (exGAN) model that generates synthetic data while maintaining the original's utility. This model ensures the anonymity of sensitive areas and maintains the distribution of data in relatively less sensitive areas. It uses a method that assigns the remaining labels except for specific selected labels as a condition. The selected labels represent points of high sensitivity, and the trajectory data generated by the model contains only points corresponding to the labels, except for the selected labels. Furthermore, in our comparative evaluation of the exGAN model, it outperformed the original GAN model.

Keywords: Generative Adversarial Network · Privacy Protection · Trajectory Data

1 Introduction

Personal information uploaded onto social media may occasionally be shared in a manner that could violate a user's right to privacy. Therefore, significant effort has been put into protecting information that can identify individuals. Generally, individual users are largely unaware of the risks they are exposed to when posting location information on social media. Therefore, they readily share information about their visited locations [1]. Such location information can be treated as trajectory data showing the user's movement, which can be used to expose the user's sensitive information. For example, if a record of visiting a place, such as a police station or hospital, is shared, it can be combined with other information to expose sensitive personal data. Several methods of protecting user trajectory data have been studied. However, research on methods for ensuring information privacy while sharing information is limited. Existing approaches mostly protect only specific points.

Anonymization, which is different from encryption, includes methods such as k-anonymity and l-diversity [2–4]. Anonymization allows a user to distribute data to be

© The Author(s), under exclusive license to Springer Nature Switzerland AG 2023
C. Strauss et al. (Eds.): DEXA 2023, LNCS 14147, pp. 289–294, 2023.
https://doi.org/10.1007/978-3-031-39821-6_23

read by anyone while simultaneously protecting the user's personal information from being identified. Generally, even if the individual's identification information is deleted, certain quasi-identifiers can be combined to identify the individual. Thus, such attributes must be anonymized.

Trajectory data generation is one of the ways to ensure anonymity while maintaining the utility of data and is being studied of various ways. One relatively recent approach is to use GANs. Unlike in fields such as image generation, it is not a very active area of research. Therefore, various approaches are possible to the GAN model and its derivatives, which are generative models. One approach to using the GAN model is to generate approximate data. Approximate data is designed to be similar to the input data by balancing anonymity and utility using the objective and loss functions. The generated synthetic trajectory data ensures the anonymity of individuals while preserving the utility value of the data, because of its similarity to the original data. However, due to the nature of the GAN model, protecting only certain sensitive points is difficult. Accordingly, derivative models, such as conditional GAN (cGAN) have been studied [5]. Therefore, research into generating trajectory data using a derivative GAN model can contribute to the safe publication of data while maintaining its utility value.

In this study, we propose a novel GAN model that focuses on and protects certain sensitive points and retains information about relatively less sensitive points. The model is trained to generate synthetic trajectory data effectively by applying the Long Short-Term Memory (LSTM) and Attention techniques to fit the characteristics of trajectory data.

2 Background

GANs perform extremely well as image-generation models and have received a lot of attention [6]. A GAN model can be divided into two sub-models: the Generator, which executes the image generation function, and the Discriminator, which determines whether the image is correct. The two models are trained by keeping each other in check, with the goal of maximizing their performance. This can be expressed as the GAN model's overall objective function, as shown in Eq. (1):

$$\min_{G} \max_{D} V(D, G) = E_{x \sim p_{data}(x)}[\log D(x)] + E_{z \sim p_z(z)}[\log(1 - D(G(z)))] \quad (1)$$

This equation minimizes G in the objective function and maximizes D. x represents the real data and $x \sim p_{data}(x)$ represents the data sampled from the probability distribution of the entire real data. $D(x)$ uses the real data as the input to the Discriminator to distinguish the real data. $z \sim p_z(z)$ represents the data sampled from random noise using a Gaussian distribution. $G(z)$ generates data by using the noise as the input to the Generator. $D(G(z))$ takes the data generated by the Generator as input to the Discriminator and classifies it as fake data.

One of the limitations of the GAN model is that it cannot generate only the specific image that one wants. One of its improved models is the cGAN, which utilizes a condition [7]. The structure of the objective function of cGAN is nearly identical to that of the original GAN. It can be modified and written as Eq. (2) because the input data's label is given along with the input data:

$$\min_{G} \max_{D} V(D, G) = E_{x \sim p_{data}(x)}[\log D(x|y)] + E_{z \sim p_z(z)}[\log(1 - D(G(z|y)))] \quad (2)$$

In contrast to the original GAN, the value of y is given in $D(x)$. y represents the ground truth label corresponding to the real data x. Therefore, in $D(x|y)$ and $D(G(z|y))$, the label is assigned to the input data and taken as an input.

3   Except-Condition Generative Adversarial Network (ExGAN)

The exGAN model proposed in this study generates data, excluding those satisfying given conditions. In the case of cGAN, if a label is given, as a condition on the data to be generated, only data that fits the label are generated. However, to generate data by excluding certain labels among different selectable labels, the given condition must be changed. Equation (3) shows the exGAN objective function.

$$\min_{G} \max_{D} V(D, G) = E_{x \sim p_{data}(x)}[\log D(x|y_{exc}) + E_{z \sim p_z(z)}[\log(1 - D(G(z|y_{exc})))] \quad (3)$$

The object function of exGAN has the same structure as that of cGAN. y_{exc} replaces y, which corresponds to the label condition in cGAN. Therefore, it gives the exception condition in the same way that a label is given as a condition in cGAN.

3.1   Condition and Exception

For the value given as a condition, we select a label to be excluded from a set of selectable labels. Setting a single label value for trajectory data that have location, time, and category attributes is difficult. Thus, the median value of the entire distribution is used. For example, if the category attribute of the trajectory data ranges from 0 to 9, the label to be hidden is selected. If one point with a category attribute of 9 must be hidden, the value given as the input label will be the median value of those points, which reflects the overall distribution. For the location attribute, the median value is the middle value of the densest area. For the time attribute, the median value is the middle value of the most frequently visited time. Therefore, if one category attribute value must be hidden, the median value of the other category attribute for it to be replaced with is given as an except-condition.

3.2   Trajectory Data Model

Figure 1 shows the exGAN's overall structure. The input data are preprocessed to facilitate training because they comprise real number and integer values corresponding to spatial and temporal attributes. Subsequently, the values obtained by adding noise to the original data are used as inputs for effective training of the Generator. For the noise-added input data, a corresponding label value is assigned to each record and appended. Here, if the value corresponds to a pre-specified label, it is replaced with the median value of other labels. The Generator embeds the label value-connected input data, learns them through the LSTM layer, and finally decodes them to restore the original trajectory data. The data generated as output become synthetic trajectory data, which are used as distributable data after training.

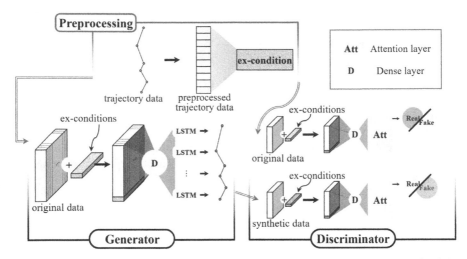

Fig. 1. exGAN's structure, consisting of Generator on the left and Discriminator on the right.

The Discriminator takes the real and the synthetic data generated by the Generator as inputs. Again, each data record is paired with a label value that meets the condition. In contrast to the Generator, the Discriminator uses an Attention layer instead of an LSTM layer.

4 Experimental Results

The data used for our experiments were collected from Foursquare [8]. We used only the records of those who had checked in from New York City during the 10-month period starting in April 2012. The check-in records comprise userID, trajectoryID, Latitude, Longitude, Temporal (Weekday, Hour), and Category attributes. The number of trajectories was 2,052, and the training and test data were split in a 2:1 ratio. In the experiment, we evaluated whether the points in the generated data were similar to those in the original data in terms of spatial data using Latitude and Longitude. In addition, Trajectory User Linking (TUL) was used in the comparison to evaluate the anonymity of the trajectory data. The models compared were the trajectory data generation model using exGAN proposed in this study and an LSTM-TrajGAN model [9]. The LSTM-TrajGAN model uses the original GAN and generates trajectory data effectively by configuring an LSTM layer internally.

By representing a single trajectory on a graph, we can evaluate the graph similarity between the original and generated data using the Hausdorff distance. Table 1 shows

Table 1. Hausdorff distance-based utility assessment results for spatial attribute data

Model	MIN	MAX	AVG	MEAN
LSTM-TrajGAN	0.006838	0.062886	0.020647	0.019752
exGAN	0.005475	0.066636	0.019705	0.018124

the results of the Hausdorff distance similarity assessment. A value approaching 0 indicates that the generated trajectory data has a higher degree of similarity to the original. Although both models show values close to 0, the trajectory data generated by exGAN shows the highest similarity, except for the MAX value. The longest trajectory is evaluated as having relatively low utility. Thus, the MAX score is lower than that of the comparative model. Because longer trajectories mean higher probabilities of including the sensitive point, more changes will occur. The anonymity score will be higher accordingly. In addition, the difference between the MAX scores is not large, and the proposed model fairs better in the average score. This implies the exGAN model's performance is similar or better than that of the comparative model.

Table 2. Anonymity evaluation results through TUL test

Model	ACC@1	ACC@5	Macro-F1	Macro-P	Macro-R
LSTM-TrajGAN	0.406037	0.651412	0.345144	0.390910	0.374779
exGAN	0.335930	0.636806	0.279527	0.334072	0.328148

The TUL evaluation examines if the user who distributed the trajectory data can be connected to the trajectory without an identifier. Table 2 shows the TUL evaluation results. The TUL evaluation result is a value between 0 and 1. A value closer to 1 indicates that trajectory data can be associated with the user, whereas a value closer to 0 indicates that the user cannot be identified based on the trajectory data alone. As shown in the results, the trajectory data generated using exGAN has values closer to 0, indicating greater anonymity.

5   Conclusion

This study proposes exGAN for trajectory data generation by changing the structure of the original GAN to overcome its limitations and facilitate effective learning of trajectory data. By evaluating data utility and comparing the anonymity provided by the various models, we demonstrated that the proposed exGAN model outperforms the conventional LSTM-TrajGAN model. Rather than modifying the data in its entirety, exGAN offers greater anonymity by selecting and replacing only certain highly sensitive points. By generating data while maintaining as much of the original data as possible for relatively less sensitive points, our model maintains the data's usefulness even if the original data at sensitive points are impaired. We expect that the proposed model will find application in various fields. For example, it can be used to generate synthetic data, excluding certain selected conditions.

However, this study has a limitation; it does not propose an alternative method for replacing a label to be given as a condition when the labels do not exist. In a follow-up study, we aim to overcome this limitation and strengthen the model by considering unlabeled data as well as trajectory data.

Acknowledgments. This work was supported by the Institute of Information & communications Technology Planning & Evaluation (IITP) grant funded by the Korean government (MSIT) (No.2021–0-00231, Development of Approximate DBMS Query Technology to Facilitate Fast Query Processing for Exploratory Data Analysis, 50%). This work was also supported by the National Research Foundation of Korea (NRF) granted by the Korean government (MSIT) (No. NRF-2021R1F1A1054739, 30%) and the Ministry of Science and ICT (MSIT), Korea, under the Information Technology Research Center (ITRC) support program (IITP-2023–2018-08–01417, 20%) supervised by the Institute for Information & Communications Technology Promotion (IITP).

References

1. Jin, F., Hua, W., Francia, M., Chao, P., Orlowska, M., Zhou, X.: A survey and experimental study on privacy-preserving trajectory data publishing. IEEE Trans. Knowl. Data Eng. **35**, 5577–5596 (2022)
2. Sweeney, L.: k-anonymity: a model for protecting privacy. Int. J. Uncertain. Fuzziness Knowl. Based Syst. **10**(5), 557–570 (2002)
3. Machanavajjhala, A., Kifer, D., Gehrke, J., Venkitasubramaniam, M.: l-diversity: privacy beyond k-anonymity. ACM Trans. Knowl. Discov. Data (TKDD) **1**(1), 3-es. (2007)
4. Temuujin, O., Ahn, J., Im, D.H.: Efficient L-diversity algorithm for preserving privacy of dynamically published datasets. IEEE Access **7**, 122878–122888 (2019)
5. Shin, J.H., Song, Y.J., Ahn, J.H., Lee, T.W., Im, D.H.: TCAC-GAN: synthetic trajectory generation model using auxiliary classifier generative adversarial networks for improved protection of trajectory data. In: 2023 IEEE International Conference on Big Data and Smart Computing (BigComp), Jeju, Republic of Korea (2023)
6. Goodfellow, I., et al.: Generative adversarial nets. In: 27th International Conference on Neural Information Processing Systems, vol. 2, pp. 2672–2680. Montreal, Canada (2014)
7. Mirza, M., Osindero, S.: Conditional generative adversarial nets (2014). https://doi.org/10.48550/arXiv.1411.1784
8. Yang, D., Zhang, D., Zheng, V.W., Yu, Z.: Modeling user activity preference by leveraging user spatial temporal characteristics in LBSNs. IEEE Trans Syst Man, Cybern. Syst. **45**(1), 129–142 (2014)
9. Rao, J., Gao, S., Kang, Y., Huang, Q.: LSTM-TrajGAN: a deep learning approach to trajectory privacy protection. In: 11th International Conference on Geographic Information Science (GIScience), pp. 12:1–12:17 (2021)

Next POIs Prediction for Group Recommendations: Influence-Based Deep Learning Model

Sayda Elmi[1,2(✉)] and Kian-Lee Tan[3]

[1] School of Medicine, Yale University, New Haven, USA
saida.elmi@yale.edu
[2] University of New Haven, West Haven, USA
[3] School of Computing, National University of Singapore, Singapore, Singapore
tankl@comp.nus.edu.sg

Abstract. A significant amount of data are significantly available due to the dramatic proliferation of location-based social networks. This has led to the development of location-based recommendation tools that assist users in discovering attractive Points-of-Interest (POIs). Next POIs recommendation is of great importance for not only individuals but also group of users since group activities have become an integral part of our daily life. However, most existing methods make recommendations through aggregating individual predictive results rather than considering the collective features that govern user preferences made within a group. This insufficiency can directly affect the completeness and semantic accuracy of group features. For this reason, we propose a novel approach which accommodates both individual preferences and group decisions in a joint model. More specifically, based on influencing users in a group, we devise a hybrid deep architecture model built with graph convolution networks and attention mechanism to extract connections between group and personal preferences and then capture the impact of each user on the group decision-making, respectively. We conduct extensive experiments to evaluate the performance of our model on two well-known real large-scale datasets, namely, Gowalla and Foursquare.

1 Introduction

Points-Of-Interest (POIs) recommendation is useful for both service providers and users by (i) estimating the number of users that may visit the POI and (ii) helping users discover various POIs for social activities occurring in the near future close to their current locations, respectively. As a natural extension of general POIs recommendation, next POIs recommendation aims at predicting the POIs that are most likely to be visited next by analyzing the users' behaviors and mobility given their check-in history. Location-based social networks (LBSNs) such as Yelp, Foursquare, Gowalla, Facebook Place and GeoLife have enabled users to share their experiences and locations via check-ins for Points-Of-Interest (POIs), e.g., restaurants, stores, tourists spots, etc. With the exponential increase in the amount of geo-tagged data collected from the LBSNs, POIs recommendation has attracted wide attention from both academia and industry [13].

Compared with the general POIs recommendation, next POIs recommendation focuses more on exploiting user movement patterns hidden in the historical check-in data for individuals. However, individuals tend to have several group activities reflecting their social life, e.g. friends often dine out and see a movie and families often watch TV programs, attend parties or travel together. For this reason, it is highly critical to develop group recommender systems to suggest relevant POIs for a group of users, known as group recommendation.

Group recommendation is much more challenging than making recommendations to individual users, as the group members may have different preferences [4]. A straightforward way to predict the next POIs for a group is to consider every group as a virtual individual and apply existing individual recommendation methods. However, it may lead to poor recommendation quality due to mainly two reasons: (i) the data of group-POIs interactions is usually very sparse, which makes it difficult to accurately learn group preferences on POIs and (ii) the influence of each member preferences is completely ignored on the group decisions and thus, recommended POIs are often not highly desired by members. Therefore, learning users' personal preferences is important to reach a consensus among group members. Authors in [1] generated the most favorable items for groups by modeling the groups' decision-making processes while authors in [10] aggregated the members' preferences but without considering interactions among members in a group. This insufficiency may lead to unsatisfactory recommendation of items for a group as a whole.

In this paper, inspired by the success of the attention mechanism architecture in learning group members interactions, we investigate how to apply a hybrid deep learning model by proposing a representation of the different factors and integrate them smoothly in a shared latent space. We propose a high-quality representation for POIs, groups and its members. We then propose our hybrid deep learning model KYD to predict the next POIs for a set of groups exploring and analysing the influence of their members.

2   Problem Definition

In this section, we first introduce the key data structures used in this paper and formally define the next POIs prediction problem for a set of groups. Following the convention, we use bold capital letters (e.g., X) to represent both matrices and graphs, and use squiggle capital letters (e.g., $\mathcal{X}$) to denote sets. We use lowercase letters with superscript $\rightarrow$ (e.g., $\vec{x}$) to denote vectors. We employ normal lowercase letters (e.g., x) to denote scalars. Assume a set of POIs $\mathcal{P} = \{p_1, p_2, \ldots, p_{|\mathcal{P}|}\}$ and a set of users $\mathcal{U} = \{u_1, u_2, \ldots, u_{|\mathcal{U}|}\}$ belonging to a LBSN $\mathcal{N}$ and a set of groups $\mathcal{G} = \{g_1, g_2, \ldots, g_{|\mathcal{G}|}\}$. The k-th group $g_k \in \mathcal{G}$ is a set of

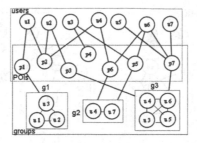

Fig. 1. Illustration of the input data: A set of users visiting a set of POIs and a set of groups interacting with the POIs set.

users and we use $\mathcal{U}_{g_k}$ to denote this set. We make use of the following graphs that reflect the interactions between the different sets $\mathcal{U}$, $\mathcal{P}$ and $\mathcal{G}$. We use bipartite graphs G_{UP} and G_{GP} as shown in Fig. 1, to represent three kinds of interactions: **(i) User-POIs Interaction:** called also a user check-in activity, is a visit of the user u to the POI p at a time t denoted by $x_{u,t}$, i.e., $x_{u,t}$ is the POI visited by the user u at time t. A sequence of historical user check-ins over time (from $t-$ n to t) is called a user check-in profile and represented by the vector $\vec{x}_u$. A given POI p is characterised by the geographical coordinates in terms of longitude and latitude coordinates denoted by l_p and its content (e.g., food, shop, service, etc.) denoted by c_p, **(ii) Group-POIs Interaction:** it is the group check-in activity related to the group g and the POI p at a time t denoted by $x_{g,t}$. The vector $\vec{x}_g$ represents the group profile over time, i.e., a set of POIs visited by the group g over a sequence of historical time-slots and **(iii) User-User Interaction:** it is the relation related to a set of users visiting a given POI at the same time forming an active group $g_k \in \mathcal{G}$.

In the following, we formally define the problem of next POIs group prediction, which consists in predicting the next POIs for all users forming a group.

Problem 1 (Next POIs Prediction). Given a set of groups $\mathcal{G}$, each comprises a set of users, our task is to predict the next POIs of all groups $\mathcal{G}$ in a given LBSN $\mathcal{N}$ that they would be interested in at a future time T where $T = t + z$ and z is the number of the future time-slots to be predicted. More formally, given $\vec{x}_{g_k}$ and $\vec{x}_u/u \in g_k$, we want to predict $x_{g_k,t+i}$ for all $g_k \in \mathcal{G}$ where $1 \leq i \leq z$ and $1 \leq k \leq |\mathcal{G}|$.

3   Methodology

In our framework, the bipartite graphs G_{UP} and G_{GP} and interactions between users in a given LBSN $\mathcal{N}$ are the data containing information about all interactions. A baseline method for predicting the next POIs for a set of users forming a group is to learn the preferences of the group by aggregating the personal preferences of its members. This method may have a high computational cost

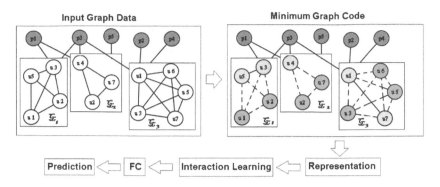

Fig. 2. KYD Framework: it consists of mainly three components: (i) Minimum Graph Code Process, (ii) Representation Process and (iii) Interaction Learning Process.

with (i) larger graphs $\mathcal{N}$ and thus a large number of groups K where $K = |\mathcal{G}|$ and (ii) larger number of users forming the groups $g_k \in \mathcal{G}, \forall k \in [1, K]$. However, when analysing the group behaviour, we have observed that the group decision is related to the preferences of some group members, that we call **influencers**, but not all. Thus, introducing the preferences of the non influencers into the aggregation process may bias the results in terms of accuracy. As shown in Fig. 2, our framework proposes the concept of "the minimum graph code" which replaces the original input graph data considering the preferences of the influencers and neglecting the non acting users. Then, we introduce a representation process which consists of modeling the different interactions between the entities users, groups and POIs sets (as shown in Sects. 3.1 and 3.2). To make the problem learnable, the graph interactions should be either represented as a sequence of visited POIs, or a series of adjacent matrices and vertex features. For sequence inputs we can use Transformers [12] to extract high-level features and learn interactions between the different entities. If the inputs are modeled as series of adjacent matrices and vertex features, we can use convolution GNN to learn vertex representations with message passing from neighborhoods. After obtaining the representations of our entities, we feed them into an interaction learning module to extract the correlated features from each side. Then we feed the output into fully-connected layers to make predictions. In the following, we give the representations of both sets Users and Groups. We then model the interactions between these two entities and the POIs set.

3.1   User Representation

The User-POI interaction at a given time denoted by $x_{u,t}$ is the POI visited by the user u at time t. For each user, we create a profile which is a set of User-POIs interactions. A user Profile is a sequence of check-in activities in chronological order from $t - n$ to t. Formally, the check-in sequence of a user u up to time-slot t is represented by the vector $\vec{x}_u$ as follows: $\vec{x}_u = [x_{u,t-n}, x_{u,t-(n-1)}, \ldots, x_{u,t-2}, x_{u,t-1}, x_{u,t}]$.

In a given social network $\mathcal{N}$, we model the check-ins of all users in $\mathcal{U}$ by designing the user check-in matrix $X_{\mathcal{U}}$ recording the check-ins sequences of all users in $\mathcal{U}$ with n historical time-slots. Formally, the user check-in matrix $X_{\mathcal{U}}$ is represented as follows.

$$X_{\mathcal{U}} = \begin{bmatrix} x_{u_1,t-n} & \cdots & x_{u_1,t-1} & x_{u_1,t} \\ x_{u_2,t-n} & \cdots & x_{u_2,t-1} & x_{u_2,t} \\ \vdots & \ddots & \vdots & \vdots \\ x_{u_{|\mathcal{U}|-1},t-n} & \cdots & x_{u_{|\mathcal{U}|-1},t-1} & x_{u_{|\mathcal{U}|-1},t} \\ x_{u_{|\mathcal{U}|},t-n} & \cdots & x_{u_{|\mathcal{U}|},t-1} & x_{u_{|\mathcal{U}|},t} \end{bmatrix} \tag{1}$$

Each row vector in $X_{\mathcal{U}}$ contains check-in data for the same user from continuous time slots from $t - n$ up to t, while each column contains check-in data for all users at the same time slot.

3.2 Group Representation

A group is a set of users visiting together one POI at the same time. The Group-POI interaction at a given time denoted by $x_{g,t}$ is the POI visited by the group g at time t. A group may have a check-in profile which is a set of Group-POIs interactions over time from $t - m$ to t. Formally, the check-in profile of the group g up to time-slot t is represented by the vector $\vec{x}_g$ such as $\vec{x}_g = [x_{g,t-m}, x_{g,t-(m-1)}, \ldots, x_{g,t-2}, x_{g,t-1}, x_{g,t}]$.

In a given social network $\mathcal{N}$, we model the check-ins of all groups g_k in $\mathcal{G}$, $\forall k \in [1, K]$ where $K = |\mathcal{G}|$ by designing the group check-in matrix $X_{\mathcal{G}}$ recording the check-ins sequences of all groups in $\mathcal{G}$ with m historical time-slots. Formally, the group check-in matrix $X_{\mathcal{G}}$ is represented such as $X_{\mathcal{G}} = \begin{bmatrix} \vec{x}_{g_1} \\ \vec{x}_{g_2} \\ \vdots \\ \vec{x}_{g_K} \end{bmatrix}$

In location-based group recommender, we can consider only $X_{\mathcal{G}}$ as an input to learn the mobility behaviour from the historical check-ins of the group as one individual. However, the group preferences may be affected by the personal preferences of its members. Thus, integrating the preferences related to the group members is of great importance for more accurate prediction. However, some of the group members may not be active in the group decision making and some of them, may have more influence than others, i.e., they affect more the group preferences. For this reason, we measure in the following the influence degree of each group member. We then propose the minimum graph code by discarding the non influencing users from the learning module.

The User Influence Modeling. The personal preferences of a user u are represented by the vector $\vec{x}_u$, extracted from the matrix $X_{\mathcal{U}}$ and reflecting the user check-in profile over time, i.e., all visited POIs, alone or as a group member. Similarly, the vector $\vec{x}_g$ records the group check-ins over time. Let $\phi(\vec{x}_u)$ and

$\phi(\vec{x}_g)$ be the sets of distinct check-ins visited by the user u and the group g, respectively.

Definition 1 (Influence Degree). *Given a group g_k in $\mathcal{G}$ where $k \in [1, |\mathcal{G}|]$. A set of users forming the group g_k is called $\mathcal{U}_{g_k}$. The influence degree for each user u in $\mathcal{U}_{g_k}$, denoted by $\varphi(u)_{u \in \mathcal{U}_{g_k}}$, is defined as follows*

$$\varphi(u)_{u \in \mathcal{U}_{g_k}} = \frac{\phi(\vec{x}_u)_{u \in \mathcal{U}_{g_k}} \cap \phi(\vec{x}_{g_k})}{|\phi(\vec{x}_{g_k})|} \tag{2}$$

The intuition behind the influence degree is to measure the impact of the personal preferences of the user u in $\mathcal{U}_{g_k}$ on the preferences of the group g_k. The more the user u has similar preferences to those of the group g_k, the more the user influenced the group behaviour.

In the following, we model the group representation by proposing the minimum group code.

The Minimum Group Code. Given a group g_k in $\mathcal{G}$ where $k \in [1, |\mathcal{G}|]$. We assume that the users u in $\mathcal{U}_{g_k}$ are either influencers or followers. A group influencers are the group members having the power to affect the decisions of others which are the followers. The set of influencing users i in the group $g_k \in \mathcal{G}$ according to an influence threshold α, is called $\mathcal{I}_{g_k}$ and defined as $\mathcal{I}_{g_k} = \{u \in \mathcal{U}_{g_k} / \varphi(u) \geq \alpha\}$ where $\alpha \in [0, 1]$.

The code of the group g_k contains the check-in profiles of all users u in $\mathcal{U}_{g_k}$. Since personal preferences can affect the group decision and thus improving the prediction accuracy, we integrate into the learning module, the check-in profiles of the users, but not all. The minimum group code is the representation of only

its influencing members i in $\mathcal{I}_{g_k}$ such as $|\mathcal{I}_{g_k}| \leq |\mathcal{U}_{g_k}|$ and $X_{\mathcal{I}_{g_k}} = \begin{bmatrix} \vec{x}_{i_1} \\ \vec{x}_{i_2} \\ \vdots \\ \vec{x}_{|\mathcal{I}_{g_k}|} \end{bmatrix}$

where $\vec{x}_{i_m} / m \in [1, |\mathcal{I}_{g_k}|]$ is the check-in profile of the influencer i_m in $\mathcal{I}_{g_k}$. The minimum group code is the code with the minimum group members with the highest influence according to an influence threshold α. Finally, each group can be represented by the corresponding minimum code, and vice versa. The input of our learning module is the group check-in profile as well as the check-in profiles of all its influencers. More formally, we design the matrix X such as

$$X = X_{\mathcal{G}} \oplus X_{\mathcal{I}_{g_k} / k \in [1, K]} = \begin{bmatrix} \vec{x}_{g_1}, X_{\mathcal{I}_{g_1}} \\ \vec{x}_{g_2}, X_{\mathcal{I}_{g_2}} \\ \vdots \\ \vec{x}_{g_K}, X_{\mathcal{I}_{g_K}} \end{bmatrix} \tag{3}$$

where $K = |\mathcal{G}|$. This representation allows the integration of the group check-in profiles as well as the influencing check-in profiles to be analysed. In the following, given the matrix X, we propose a learning module to extract the hidden features behind the group behaviour.

3.3   The Interaction Learning Module

After obtaining the representations of both users and groups interactions, inspired by existing hybrid deep learning models [3], we propose a hybrid deep learning model which comprises three major components: (1) Learning the group mobility from $X_{\mathcal{G}}$ and then by extracting the hidden patterns of the embedded influencing users using a convolution GNN, (2) a time-series learning using the attention mechanism to model the temporal dependency and (3) the POIs content extraction layer modeling the impact of POIs content.

The learning module proceeds as follows: First, we extract the hidden patterns from the group behaviour over time, i.e., features from historical group check-ins. Then, the influencing users are extracted and embedded into the matrix X. To learn the patterns hidden in the influencing preferences as well as the group preferences, KYD employs GCN as shown in Fig. 3 since it has a very deep structure that can effectively capture the dynamics behind neighbor nodes, i.e., influencing users. Then, we reshape these features to be suitable for time-series learning. Since the group mobility is directly affected by periodicity features, i.e., the mobility behavior during the week-ends may be similar on consecutive week-ends, we feed the group check-in vectors of periodic time intervals into an attention layer, a deep variant sequence modeling, to capture such temporal dynamics. We then employ a fully connected (FC) layers to extract the semantic features that describe the POIs content.

Group Encoding. In graph models, each node has a feature vector, and each edge is used to pass information from its source to its target. GNNs do not need node ids and edge ids explicitly because the adjacency information is included in an adjacent matrix. i.e., interactions between users, groups and POIs. Given that a group is represented as a set of influencing users and thus defined by a minimum graph code, the next encoding step is to transform each influencing user into a vector. As explained in Sect. 3.2, we can vectorize node labels into multi-hot vectors represented by the input matrix X as shown in the Eq. 3. In such representation, the adjacent information of the group influencers can be stored in a sparse matrix ($X_{\mathcal{I}_{g_k}/k \in [1,|\mathcal{G}|]}$) to reduce the memory usage and improve the computation speed.

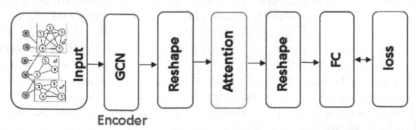

Fig. 3. Interaction Learning Module Architecture is mainly composed by two components: (1) Learning the group behaviour by embedding the influencing users and (2) Learning the sequential patterns.

Graph Convolutional Network. After obtaining the input representations, we feed them as inputs of interaction layers to extract the correlated behaviour between the group preferences and those of the group members. Graph Convolutional Network (GCN) [11] is developed specifically to handle multi-relational data in realistic knowledge bases. GCN is acting as an embedding operation or a look-up operation (LK) to integrate the influencing users. A stack of convolutions (CNN) is used to understand the connections between group preferences and user preferences. We employ the Batch Normalization (BN) after the convolution layer for faster training speed.

The input of the convolution layer is the check-in profiles of groups as well as the embedded influencing user check-in profiles $\mathcal{X}^{(l)}$. The output layer is to generate the prediction result. At an arbitrary l-th layer, we use $f^{(l)}$ filters to convolve and concatenate all matrices to get $\mathcal{X}^{(l+1)}$. The f-th matrix convolved by the f-th filter can be formulated as follows: $\mathcal{X}^{l,f} = [x_1^{l,f}, \ldots, x_k^{l,f}, \ldots, x_{|\mathcal{G}|}^{l,f}]$ where $x_k^{l,f} = F\big(LK(X_{\mathcal{G}}^{(l-1)}, I) * W^{l,f} + b^{l,f}\big)$.

Here, $*$ denotes the convolution operation which uses the f-th filter $W^{l,f}$, F is an activation function, e.g. the rectifier ReLU $F(x) = max(0, x)$ which has achieved a training effectiveness in reducing the problem of gradient vanishing, $W^{l,f}$ and $b^{l,f}$ are the learnable parameters in the l-th layer with the f-th filter. This filter aims at emphasizing the correlated features of influencing users.

Attention Mechanism. In addition to the impact of the influencing users on the group preferences, the group check-in movement may have temporal and sequential patterns, i.e., a visited POI at a given time interval (12:00 pm–02:00 pm) on Monday can be similar to the next following weekdays for a given group. For learning such patterns, our experimental results show that the attention mechanism outperforms LSTM, GRU and RNN in the benchmark datasets. We thus employ the attention mechanism [12] that has shown great success in capturing such sequential patterns. After C convolution units, we use the attention model to learn the long-term temporal patterns considering the influencing group members. On top of the C convolution units, we get the output tensor $X^C \in \mathbb{R}^{pd \times |\mathcal{G}| \times f_C}$ where f_C is the number of the convolution filters at the last C-th Conv layer and pd is the period (the number of days/weeks). We reshape X^C in the way of time sequence to feed into the attention layer. We get a tensor $X' \in \mathbb{R}^{|\mathcal{G}| \times n \times f_C}$ representing the group check-in profile vectors for all n time-slots, where $X'_{k,t} = X^C[t, k, :]$. We train the group profile vector $X'_{k,t}$ which records the check-ins for group $g_k \in \mathcal{G}$ for n time-slots. We use the transformers architecture by applying the attention mechanism based on the original implementation described in [12]. However, we have made some changes to the original architecture since we are not working with sequences of words. Additionally, we are doing a regression and not a classification of words or characters. We also need to keep the SoftMax layer from the output of the Transformer because our output nodes are probabilities of visiting a given POI x at a future time-slot. The used transformer block is a parameterized function class $F_\theta : \mathbb{R}^{|\mathcal{G}| \times n}$ with $|\mathcal{G}|$ and n are the number of groups and time-slots, respectively and $F_\theta(X'_{k,t}) = Y_k$

where F_θ is attention mechanism [12] and the parameters θ consist of the entries of the weight matrices. The input $X'_{k,t} \in \mathbb{R}^{|\mathcal{G}| \times n}$ is the check-in behavior of $|\mathcal{G}|$ groups, each with n features. The final output of the transformer layers can be represented by the sequential vector Y_S in which the last element is the predicted check-in for the next time iteration such as $Y_S = [Y_{k,t-n}, \ldots, Y_{k,t}, Y_{k,t+1}]$.

Content Learning. Groups at lunchtime tend to go to restaurants for food rather than going to a cinema or any other entertainment points. We thus extract POIs content features by a fully connected layer (FC) as shown in Fig. 3. The output is denoted by Y_C. The predicted POI at the t-th time interval, denoted by $\widetilde{Y}_T$, is defined as $\widetilde{Y}_T = tanh(W_S \circ Y_S + W_C \circ Y_C)$ where $\circ$ is Hadamard product (i.e., element-wise multiplication), W_S and W_C are the learnable parameters in the sequential and content learning component, respectively. The model output is a probability distribution on all POIs calculated by $\widetilde{Y}_T$. And then we take a gradient step to optimize the loss based on the output and one-hot representations of POIs.

4 Experimental Evaluation

Data Description: We describe in the following the data-sets used for our experimental evaluation. We used two data-sets to test our model. Details are given as follows: **(i) Foursquare Dataset:** a public data-set, on which we extract social relationships between users forming groups and content features. This data-set contains more than 450,000 check-in records from 2009 to 2011 generated by 2114 users living in USA. For each user, we acquired her POI attendance list and social friend list. For each POI, its visiting time and its content were also collected. Each check-in contains a user, a timestamp and a POI, indicating the user visited the POI at that time; users having relationships visiting the same POI at the same time are considered as a group and **(2) Gowalla Dataset:** This data-set contains 6,442,892 check-in historical records from 2009 to 2010 which is much more than the Foursquare data-set. However, it does not contain the content information about POIs. Therefore, each check-in record has the same format with the above Foursquare data-set except for POI-content. The two real data-sets are publicly available[1].

Pre-processing: Given a group profile $\vec{x}_g$ in terms of a collection of group check-in records, we first sort them according to their check-in time-slot order. We then use the 80th percentile as the cut-off point so that check-ins before this point will be used for training and the rest are for testing. A group is considered as an active group if it records 80% of the studied n time-slots. We then eliminate inactive groups from the training data-set since their mobility behavior cannot be captured with their sparse check-ins. Meanwhile, we figure out a strategy to deal with missing check-in data for active groups. We fill these

---

[1] https://sites.google.com/site/dbhongzhi/.

missing check-ins by interpolating similar check-ins at the same time-slots in the previous days/weeks. To find out the connections between group and its members, we represent the social network as a graph where nodes are users and edges are relations between users. Each edge is a link connecting two users. To learn the short-term periodicity hidden in check-in data, we divide a day into different time-slots. For POIs content information, a binary vector is given by one-hot coding to transform and represent the content for the visited POIs. We use Tanh in the output of the KYD model as our final activation. We train our network with the following hyper-parameters setting: mini-batch size (48), learning rate (0.01) with adam optimizer, a variant of Stochastic Gradient Descent (SGD). Afterwards, we continue to train the model on the full training data for a fixed number of epochs (e.g., 20, 50, 100, 200 epochs). A transformer includes L transformer blocks. The hyper-parameters of the transformer are d, k, m, H, and L. The settings of these hyper-parameters are $d = 512$, $k = 64$, $m = 2048$, $H = 8$ and $L = 6$.

Benchmarks: Several prevailing algorithms are chosen for comparisons with our proposed model KYD. **(1) PRME-G** [6]: It uses the metric embedding method to embed users and POIs into the same latent space to capture the user transition patterns; **(2) STGCN** [14]: Used to predict next POIs, it is a variant of recurrent neural networks that consider the temporal intervals between neighbor check-ins; **(3) RNN** [7]: This method leverages the temporal dependency in user's behavior sequence for next POIs problem through a standard recurrent structure; **(4) FPMC-LR** [2]: It uses the Markov chains to model the user movement in a given region. **(5) ST-RNN** [9]: Based on the standard RNN model, ST-RNN replaces the single transition matrix in RNN with time-specific transition matrices and distance specific transition matrices to model spatial and temporal contexts of next POIs problem; **(6) LSTM** [15]: This is a variant of RNN model, **(7) HST-LSTM** [8]: It introduces Spatio-Temporal preferences for location prediction into gate mechanism in LSTM and **(8) GRU** [5]: This is a variant of RNN model. Some of the baselines were used only to predict the next POIs for individuals. To test these methods on groups, we consider a group as a virtual individual.

Evaluation Metrics: The performance of our KYD model is evaluated by using two metrics Accuracy@K (**Acc@K**) and Mean Average Precision (**MAP**). These two metrics are popularly used for evaluating location based prediction results, such as [14]. These metrics are used to compare our proposed model with the eight baselines described above. Note that for an instance in testing set, Accuracy@K is 1 if the visited POI appears in the set of the top-K predicted POIs, and 0 otherwise. The overall Accuracy@K is calculated as the average value of all testing instances. In this paper, to illustrate different results of Acc@K, we only show the performance for K ∈ 1, 5, 10, 15, 20, since a greater value of K is usually ignored for the top-K recommendation tasks. **Effectiveness of the Minimum Graph Code:** In addition to KYD, we design two other algorithms KYD'' and KYD' to show the effectiveness of the minimum graph code approach: (i) KYD'' considers the input $X'' = X_{\mathcal{G}}$ to learn the mobility behaviour from the historical check-ins of the group as one individual, (ii) KYD' considers the input X' representing the group check-ins as well as all its members check-ins where $X' = X_{\mathcal{G}} \oplus X_{\mathcal{U}_{g_k}/k \in [1,|\mathcal{G}|]}$ since the group preferences may be affected by the personal preferences of its members. However, KYD employs the minimum graph code and has the input X where $X = X_{\mathcal{G}} \oplus X_{\mathcal{I}_{g_k}/k \in [1,|\mathcal{G}|]}$ as described in the Sect. 3.2. As shown in the Fig. 4, KYD' performs better than KYD'' which proves that including the preferences of the group members captures more patterns and thus improves the prediction accuracy. However, KYD outperforms KYD' since non influencing group members can bias the prediction results.

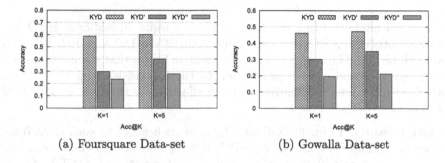

(a) Foursquare Data-set   (b) Gowalla Data-set

Fig. 4. Minimum Graph Code Evaluation.

Table 1. Evaluation of next POIs Prediction in terms of Accuracy@K and MAP on both datasets

Method	Ref	Foursquare				Gowalla			
		Acc@1	Acc@5	Acc@10	MAP	Acc@1	Acc@5	Acc@10	MAP
FPMC-LR [2]	IJCAI, 2013	0.216	0.228	0.285	0.251	0.163	0.247	0.276	0.271
PRME-G [6]	IJCAI, 2015	0.181	0.253	0.321	0.301	0.211	0.222	0.251	0.251
RNN [7]	SIGIR, 2018	0.241	0.345	0.395	0.352	0.301	0.342	0.359	0.365
STGCN [14]	AAAI, 2019	0.481	0.501	0.512	0.502	0.391	0.421	0.439	0.442
LSTM [15]	IJCAI, 2017	0.382	0.395	0.421	0.430	0.256	0.278	0.314	0.321
ST-RNN [9]	AAAI, 2016	0.452	0.473	0.491	0.486	0.332	0.351	0.392	0.398
GRU [5]	EMNLP, 2014	0.296	0.325	0.391	0.404	0.218	0.251	0.289	0.312
HST-LSTM [8]	IJCAI, 2018	0.471	0.492	0.519	0.498	0.365	0.398	0.412	0.419
ST-KYD		0.508	0.521	0.541	0.539	0.441	0.453	0.468	0.459
KYD		0.589	0.603	0.657	0.695	0.461	0.473	0.498	0.490

Method Comparison. Table 1 shows the experimental results for both Foursquare and Gowalla data-sets. Our proposed KYD, including spatial, temporal and content learning, is significantly better than existing state-of-the-art methods considering all metrics. Specifically, KYD outperforms the Markov

Table 2. Accuracy of KYD Variants on Foursquare

Method	Acc@K				
	K = 1	K = 5	K = 10	K = 15	K = 20
KYD-S	0.235	0.279	0.339	0.394	0.475
KYD-T	0.212	0.267	0.364	0.376	0.445
KYD-C	0.519	0.529	0.551	0.582	0.601
KYD	**0.589**	**0.603**	**0.657**	**0.681**	**0.691**

chain based methods considerably by a large margin. In addition, KYD consistently outperforms five RNN-based methods: RNN, LSTM, GRU, ST-RNN, and HST-LSTM. The significant improvement indicates that the mechanism to model temporal and spatial, in addition to the content learning in KYD can better catch the user's behaviors and are effective for the task of next POIs prediction.

Ablation Study. To further validate the benefits brought by each learned factor, we design four variants of KYD. KYD-S is the first variant where we remove the spatial learning and we only keep learning other factors. KYD-T is the second simplified version of the KYD model where the temporal feature extraction is eliminated. As the third simplified version of KYD, KYD-C does not consider the content effect by removing the category feature extraction. To explore the benefits of incorporating the spatial impact, temporal effect and content effect into KYD model respectively, we compare our KYD model with KYD-S, KYD-T and KYD-C. Since Gowalla dataset has no content information, we only show the results on Foursquare dataset in Table 2. From the result, we first observe that KYD consistently outperforms all variants. That is because KYD benefits from considering the simultaneously learning of three factors in a joint way. Second, as shown in Table 2, the contribution of each learning block to improve the prediction accuracy is different. Specifically, according to the importance of every extracted feature, they can be ranked as follows: Spatial Feature, Temporal Feature and then the Content Feature. However, we find that there is a slight difference between the spatial and temporal learning blocks, which shows that they are similarly contributing to KYD model. In other words, the performance gap between ⟨KYD and KYD-S⟩ and ⟨KYD and KYD-T⟩ are the most significant, indicating that spatial and temporal effects play an important role in next POIs prediction. This shows that spatial and temporal related factors matter a lot in group users' daily routines (Fig. 5).

Impact of the Parameter p: Since the input check-in data record user POIs over three time-slots per day, i.e., morning check-in, afternoon check-in and evening-night check-in, the number of predicted check-ins per day denoted by p can vary from 1 up to 3, i.e., p can be 1, 2 or 3 to denote morning, afternoon or evening check-in, respectively. In the following, we first show the impact of varying this parameter on the performance of KYD as well as other methods. We

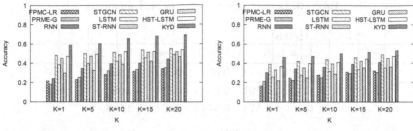

(a) Evaluation on Foursquare Data-set   (b) Evaluation on Gowalla Data-set

Fig. 5. The impact of the Parameter K

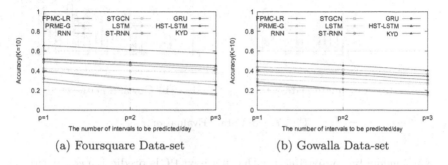

(a) Foursquare Data-set   (b) Gowalla Data-set

Fig. 6. the impact of the Parameter p.

then show the KYD performance varying p under different values of K. Figure 6 shows the comparative performances varying the number of predicted check-ins for all the state-of-art methods. As shown in the figure, KYD achieves the best performance when varying p. Note that the performance becomes worse with larger p since the correlation between check-in in time intervals being predicted and check-ins in current moment decreases. However, this fact does not have much impact on our KYD model, which reveals the robustness of our model. In the following, Table 3 shows the performance of the KYD model varying the parameter p under different values of K on Foursquare data-sets. Similar observations can be made on both Foursquare and Gowalla datasets. The results show that the prediction accuracy of KYD is not highly sensitive to the dimension p. For a given value of K, the prediction accuracy does not change much when increasing the number of predicted check-ins per day p. This fact proves that our proposed model KYD has a strong capability of capturing long-term features.

Evaluation of Cold Start: Since the group check-in data are very sparse, there are many groups with few check-in records in the Foursquare as well as the Gowalla datasets. If a group of users just visits a few POIs in the datasets, which means we can hardly learn group preference on POIs, we suppose that the group is a cold case. Specifically, we consider the groups having only one check-in per day as cold groups in our experiments. We then evaluate the performance of

Table 3. Impact of time interval on Foursquare

p	Acc@K				
	K = 1	K = 5	K = 10	K = 15	K = 20
1	0.589	0.603	0.657	0.681	0.691
2	0.569	0.591	0.613	0.669	0.698
3	0.541	0.578	0.578	0.671	0.689

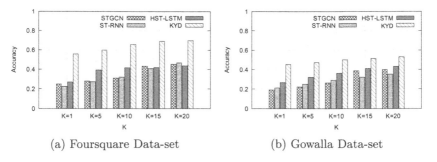

(a) Foursquare Data-set   (b) Gowalla Data-set

Fig. 7. Cold Start Evaluation.

the KYD model by comparing it with other next POIs prediction competitors for cold-start groups. Figure 7 shows the experiments conducted on Foursquare and Gowalla datasets. We show the results of different methods using Accuracy@K with varying K where K = $\{1, 5, 10, 15, 20\}$. As shown in the figure, we can observe that the KYD model performs much better than the other methods under cold start scenario. The reason is that our proposed KYD model figures out a strategy to deal with missing check-ins by interpolating values of similar time-slots in the previous day, which proves that our method can work out well with sparse data.

5   Conclusion

In this paper, GCN-attention based network named KYD, was proposed for group next POIs prediction. In KYD model, a new representation of the group preferences is introduced and given by the minimum graph code. A hybrid architecture is employed to capture the hidden patterns considering the social relations connecting users in a group. An attention mechanism is then employed to extract the sequential features. We evaluate our model on two real location-based networks, Foursquare and Gowalla, achieving performances which are significantly beyond eight existing methods.

References

1. Cao, D., He, X., Miao, L., An, Y., Yang, C., Hong, R.: Attentive group recommendation. In: The 41st ACM SIGIR, pp. 645–654 (2018)

2. Cheng, C., Yang, H., Lyu, M.R., King, I.: Where you like to go next: successive point-of-interest recommendation. In: The 23rd IJCAI (2013)
3. Elmi, S., Tan, K.-L.: DeepFEC: energy consumption prediction under real-world driving conditions for smart cities. In: WWW, pp. 1880–1890. ACM/IW3C2 (2021)
4. Elmi, S., Tan, K.L.: Influence-based deep network for next POIs prediction. In: Hagen, M., et al. (eds.) ECIR 2022. LNCS, vol. 13185, pp. 170–183. Springer, Cham (2022). https://doi.org/10.1007/978-3-030-99736-6_12
5. Cho, K., et al.: Learning phrase representations using RNN encoder-decoder for statistical machine translation. In: EMNLP, pp. 1724–1734 (2014)
6. Feng, S., et al.: Personalized ranking metric embedding for next new POI recommendation. In: Proceedings of the 24th IJCAI, pp. 2069–2075 (2015)
7. Huang, J., Zhao, W.X., Dou, H., Wen, J.-R., Chang, E.Y.: Improving sequential recommendation with knowledge-enhanced memory networks. In: The 41st ACM SIGIR Conference, pp. 505–514 (2018)
8. Kong, D., Wu, F.: HST-LSTM: a hierarchical spatial-temporal long-short term memory network for location prediction. In: The 27th IJCAI (2018)
9. Liu, Q., Wu, S., Wang, L., Tan, T.: Predicting the next location: a recurrent model with spatial and temporal contexts. In: AAAI 2016, pp. 194–200 (2016)
10. McCarthy, J.F., Anagnost, T.D.: MusicFX: an arbiter of group preferences for computer supported collaborative workouts. In: CSCW 1998 (1998)
11. Schlichtkrull, M., Kipf, T.N., Bloem, P., van den Berg, R., Titov, I., Welling, M.: Modeling relational data with graph convolutional networks. In: Gangemi, A., et al. (eds.) ESWC 2018. LNCS, vol. 10843, pp. 593–607. Springer, Cham (2018). https://doi.org/10.1007/978-3-319-93417-4_38
12. Vaswani, A., Shazeer, N., Parmar, N., Uszkoreit, J., Jones, L., Gomez, A.N.: Attention is all you need. In: NIPS, pp. 5998–6008 (2017)
13. Wang, Q., et al.: Next point-of-interest recommendation on resource-constrained mobile devices. In: WWW, pp. 906–916 (2020)
14. Zhao, P., Zhu, H., Liu, Y., Xu, J.: Where to go next: a spatio-temporal gated network for next POI recommendation. In: AAAI (2019)
15. Zhu, Y., et al.: What to do next: modeling user behaviors by time-LSTM. In: IJCAI (2017)

Interpreting Deep Text Quantification Models

YunQi Bang$^{(\boxtimes)}$, Mohammed Khaleel, and Wallapak Tavanapong

Department of Computer Science, Iowa State University, Ames, IA 50011, USA
{vickys,mkhaleel,tavanapo}@iastate.edu

Abstract. Quantification learning is a relatively new deep learning task. Differing from a classic classification problem where the class of a single instance is predicted, a quantification model predicts the distribution of classes within a given set of instances. Quantification learning has applications in various domains. For example, in designing political campaign ads, it is important to know the proportion of different aspects voters care about. QuaNet is a recent deep learning quantification model that was shown to achieve good quantification performance. Like many deep learning models, there is no explanation about the contributions of different inputs QuaNet uses to predict a class distribution. In this study, we propose a method to provide such an explanation, which is important to increase users' trust in the model. Our method is the first work on interpreting deep learning quantification models.

Keywords: Deep learning · Interpretation · Quantification

1 Introduction

Quantification learning is formally defined as "given a labeled training set, induce a quantifier that takes an unlabeled test set as input and returns its best estimate of the class distribution" [9]. For instance, given all votes for two presidential candidates, a quantifier predicts the class distribution of the votes for the two candidates as 1:2. Predicting the class distribution or class ratios is important for numerous domains such as social sciences, epidemiology, healthcare, and market research [20]. For example, in healthcare, the proportion of patients in different age groups affected by a specific disease is important for developing appropriate treatment. In e-commerce, a company should know customers' ratings on different aspects of its product.

Different methods were proposed for quantification learning [7]: Classify and Count (CC) [6,10], an ensemble-based method [21], distribution matching [15,25], and direct quantification [12,16,17]. CC methods use a classifier to classify each document and tally the number of documents predicted in each class. CC suffers when the distribution of the test dataset is significantly different from

---

Y. Bang and M. Khaleel—Equal contribution.

C. Strauss et al. (Eds.): DEXA 2023, LNCS 14147, pp. 310–324, 2023.
https://doi.org/10.1007/978-3-031-39821-6_25

that of the training dataset. However, effective quantification learning methods are expected to handle different class distributions. Deep learning quantifiers were recently proposed [8, 23]. Deep learning models have performed excellently in various tasks, such as image captioning, object detection, and time series forecasting. There are plenty of deep learning models for classification tasks, but there exist only two deep learning models for quantification tasks: QuaNet [8] and Deep Quantification Network (DQN) [23]. Both approaches showed better performance than non-deep learning models for quantification tasks. However, both models do not provide insights into the extracted features and the influence of these features on the final class distribution prediction. There has been active research to interpret/explain what a deep-learning model learns and uses to make a decision given an input (local interpretation) or the inner working of the model (global interpretation) for classification tasks [13]. However, none exists for quantification tasks. This motivates our study to design and evaluate an interpretation method for deep quantification models.

QuaNet [8] and DQN [23] are very different in their architecture design, making it difficult to design one interpretation method that works for both models. QuaNet relies on a separate classifier to predict the class of an individual document, whereas DQN is an end-to-end deep-learning model. In this work, we focus on the interpretation of QuaNet since it was the top-performing method among thirteen different quantification methods in one of the datasets. QuaNet ranked fourth in performance based on the average absolute error on eleven different datasets in a recent study [20]. Furthermore, a public Python library, QuaPy [19] created by the authors of QuaNet is available.

Our interpretation method is built on Layer-wise Relevance Propagation (LRP) [4]. LRP is a well-known interpretation method that has been applied to interpret text and image classification models [2, 4]. LRP was also adapted to interpret Long Short-Term Memory (LSTM) recurrent networks [3]. LSTM is a component of QuaNet, which makes LRP suitable for interpreting QuaNet. Additionally, recent studies [1, 26] showed that LRP outperformed several well-established explanation methods. Our unique contributions described in this paper are as follows.

- The first interpretation of a deep quantification model, QuaNet, using LRP. Our interpretation method provides insights into which input component of QuaNet plays an important role in the predicted class distribution; the code for the interpretation will be available on a public github site upon acceptance of the paper.
- Explanation of the QuaNet models on three public datasets: IMDB [14], YELP [27], and AG-NEWS [27], and two political science datasets [22].

The remainder of this paper is organized as follows. Section 2 provides the necessary background. Section 3 introduces our proposed interpretation method. In Sect. 4, we show the experiments, results, and findings. Finally, Sect. 5 concludes our study and states some future directions.

2 Background and Related Work

We present an overview of QuaNet and LRP as they are most related to our work. Other quantification methods [5,6,10,11,17,21,25] are not discussed due to limited space. Readers interested in the overview of these methods are referred to [13].

QuaNet [8] is a deep-learning quantification model built on top of a classifier. QuaNet utilizes a probabilistic classifier to obtain a document representation and a posterior probability for each document in the input set of documents. Furthermore, to predict the class distribution, QuaNet also uses four class distributions predicted by Classification and Count (CC) [10], Adjusted Classification and Count (ACC) [10], Probabilistic Classification and Count (PCC) [6], and Probabilistic Adjusted Classification and Count (PACC) [6], respectively. CC counts the ratio of the number of documents predicted in each class to the total number of documents [10]. ACC adjusts the predicted class distribution by CC using true positive and false positive rates. PCC and PACC are similar to CC and ACC, respectively, except that each document is probabilistically assigned to one or more classes.

Figure 1 shows the architecture of QuaNet given $|C| = 2$. C denotes the set of classes and $|C|$ denotes the cardinality of the set. The bi-LSTM recurrent network produces a quantification embedding given a list of document representation embedding for each document x and posterior probabilities $(Pr(c|x))$ from the trained classifier. This list is later referred to as L. Let $\mathbf{x}$ denote an embedding vector of x. The fully connected layers with ReLU activation process the concatenation of the quantification embedding and the class distribution predictions from CC, ACC, PCC, and PACC. The final layer with $|C|$ neurons and softmax activation produces the final prediction.

Training Process: Let D^L be the training dataset where each document has an associated class label. Given an already trained classifier f on D^L, the training procedure of QuaNet is as follows. Run the classifier on each document x in D^L to obtain the document representation and posterior probabilities. The true positive rate and the false positive rate for each class is calculated by applying the trained classifier f on a separate validation dataset. Run CC and PCC on the training dataset. Run ACC and PACC quantifiers using the calculated true and false positive rates to output the predicted class distribution. The sorted list L (input component 1) is then created and passed through the bidirectional LSTM layer (bi-LSTM) to obtain the corresponding quantification embedding. The quantification embedding is a vector of $2h$ elements where h is the size of each LSTM cell in bi-LSTM. The quantification embedding is then concatenated with the four quantifiers' predicted class distributions, where each is a vector of $|C|$ elements. These are the last four input components of QuaNet. The resulting vector of size $2h + 4|C|$ is given to the fully connected layers with ReLU activation. The final layer with $|C|$ neurons and softmax activation outputs the final prediction. The loss function is the Mean Square Error of the prediction compared to the ground truth. The loss function is minimized via stochastic gradient descent.

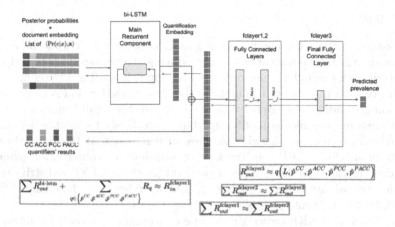

Fig. 1. QuaNet's architecture given $|C| = 2$; **black forward arrows** show the inference flow of QuaNet. For each document x, two purple boxes represent the two posterior probabilities, one for each class, and orange boxes represent the document embedding. The predicted class distribution by each quantifier has two values (blue boxes), one for each class. **Our proposed work: The green backward arrows** show the interpretation flow of our Algorithm 1. The mathematical expressions in the green boxes show the conservation principle of LRP. (Color figure online)

Testing Process: QuaNet uses the same process as its training procedure to produce its input from the testing dataset, D^U. The list L and the four quantifiers' predictions then pass through the trained QuaNet to arrive at a final prediction.

2.1 Performance Metrics for Quantification Models

Unlike classification problems, the performance metrics for quantification models are different. The commonly used metrics are Mean Absolute Error (MAE), Relative Mean Absolute Error (RMAE), and Kullback-Leibler Divergence (KLD).

$$MAE(P, \hat{P}) = \frac{1}{|C|} \sum_{c \in C} |P(c) - \hat{P}(c)| \tag{1}$$

$$RMAE(P, \hat{P}) = \frac{1}{|C|} \sum_{c \in C} \frac{|P(c) - \hat{P}(c)|}{P(c)} \tag{2}$$

$$KLD(P, \hat{P}) = \sum_{c \in C} P(c) \log \frac{P(c)}{\hat{P}(c)} \tag{3}$$

where P is the true class distribution; $\hat{P}$ is the predicted class distribution; $|C|$ is the total number of classes; and $|P(c) - \hat{P}(c)|$ is the absolute value of the difference between the true and the predicted ratios of class c. A small value of error or divergence is desirable.

2.2   LRP

LRP computes the relevance flow in a neural network [4]. It uses the network weights and the neural activations created during the forward pass to propagate the output back through the network until the input layer. Then the contribution of each pixel (with an image as input) to the classifier's decision is visualized for users' understanding. A large relevance score means the input is more important. A positive score means the input contributes positively to the prediction, and a negative score means otherwise. A score of 0 means that the input does not contribute anything. LRP uses the layer-wise relevance conservation principle. For a deep learning model f, input x, and target class c, LRP redistributes the posterior probability of class c for x, $f_c(x)$, layer by layer, following different rules [3,18] for different types of layers.

The benefits of LRP outweigh its time-consuming drawback for interpretation. (1) LRP is not dependent on the input type. Thus, the input can be a mix of two or more forms, for example, a list of words and numerical scores. (2) The application of LRP is straightforward and does not require the training of an external model to derive the explanation. (3) LRP is applicable to different model architectures [2,4]. (4) Finally, the output relevance scores are signed values, which differentiate whether a highly relevant feature contributes positively or negatively to the final prediction.

3   Proposed Work: Interpreting QuaNet

We focus on two research questions in our attempt to gain insights into the inner-working of QuaNet.

- RQ1: What input components are important to the class distribution prediction by QuaNet?
- RQ2: Does sorting L by $Pr(c|x)$ increase the performance of QuaNet?

Section 3.1 presents Algorithm 1 that adapts LRP for interpreting QuaNet by deriving a relevance score for each input of the input components of the QuaNet model to the final prediction. Section 3.2 describes how to utilize the relevance scores to calculate the contribution amount of each of the five input components where each component has several input elements.

3.1   LRP-Based Algorithm to Calculate Relevance Scores

Algorithm 1 is based on the QuaNet implementation by the authors of QuaNet [8] as summarized in Sect. 2. Algorithm 1 takes as its input the trained QuaNet model to be interpreted and S^U, the QuaNet's input components based on the test dataset U. Recall that QuaNet has five input components. Thus, S^U has five different components: (1) a sorted list L of pairs $\langle Pr(c|x), \mathbf{x} \rangle$ for all document $x \in D^U$, where $\mathbf{x}$ is the document embedding of x and L is sorted by the values of $Pr(c|x)$, (2) predicted class distribution using CC ($\hat{p}^{CC}(D^U)$), (3) predicted class

---

Algorithm 1. Interpreting QuaNet

Input: Trained QuaNet model and S^U
Output: Relevance score of each input component of QuaNet in S^U
Notation: See Table 1.

1: Run QuaNet on S^U to get intermediate input and output of each layer
2: $R_{out}^{\text{fclayer3}} \leftarrow$ List of neuron activations of fclayer3
3: **for** $\ell \in$ [fclayer3, fclayer2, fclayer1] **do**
4:     $R_{in}^{\ell} \leftarrow \text{LRP}(\ell, I_\ell, O_\ell, R_{out}^{\ell})$
5:     $R_{out}^{\ell'} \leftarrow R_{in}^{\ell}$ ▷ $\ell' = \ell$ of the next iteration; when $\ell = fclayer1$, ℓ' is null.
6: **end for**
7: $R_{out}^{\text{bi-LSTM}} \leftarrow$ first $|O_{\text{bi-LSTM}}|$ elements of R_{in}^{fclayer1}
8: $R_L \leftarrow \text{LRP}(\text{bi-LSTM}, I_{\text{bi-LSTM}}, O_{\text{bi-LSTM}}, R_{out}^{\text{bi-LSTM}})$
9: $R_{CC}, R_{ACC}, R_{PCC}, R_{PACC} \leftarrow$ all elements after the $|O_{\text{bi-LSTM}}|$-th elements in R_{in}^{fclayer1}
10: **return** $R_{CC}, R_{ACC}, R_{PCC}, R_{PACC}, R_L$

---

distribution using ACC ($\hat{p}^{ACC}(D^U)$), (4) predicted class distribution using PCC ($\hat{p}^{PCC}(D^U)$), and (5) predicted class distribution using PACC ($\hat{p}^{PACC}(D^U)$). L is input to the bi-LSTM layer to obtain a quantification embedding which is then concatenated with the input components 2–5.

The green arrows in Fig. 1 show the overall interpretation process of Algorithm 1. In this algorithm, Step 1 runs the QuaNet model on the input to obtain the intermediate input and output of each layer. In Step 2, the relevance scores of the last layer, $fclayer3$, are set as its neuron activations. Steps 3–6 iterate over the three QuaNet's fully connected layers, from the last layer, $fclayer3$, to the first fully connected layer, $fclayer1$. The relevance scores of each layer are backpropagated to the layer before it. In Step 4, LRP is used to calculate the relevance scores to be propagated to the previous layer, R_{in}^{ℓ}, by using the layer information of layer ℓ, the input I_ℓ, the output O_ℓ, and the relevance scores of the layer ℓ, R_{out}^{ℓ}. Step 5 then assigns the relevance scores to be propagated as the relevance scores of the layer ℓ in the next iteration. At the last iteration, the relevance scores to be propagated are for the bi-LSTM layer and the four quantifiers' results.

Recall that h is the hidden size of an LSTM cell, and the quantification embedding is a vector of $2h$ elements. The purpose of Step 7 and Step 9 is to allocate elements in R_{in}^{fclayer1} to $R_{out}^{\text{bi-LSTM}}$, R_{CC}, R_{ACC}, R_{PCC}, R_{PACC}. Step 7 assigns the first $2h$ relevance scores as $R_{out}^{\text{bi-LSTM}}$ for Step 8. In Step 8, the relevance scores of the bi-LSTM layer are backpropagated to list L. To interpret the bi-LSTM layer of QuaNet in Step 8, we followed the implementation of [3][1]. Finally, Step 9 assigns the rest of R_{in}^{fclayer1} as the relevance scores for the four quantifiers' results.

---

[1] https://github.com/alewarne/Layerwise-Relevance-Propagation-for-LSTMs.

Table 1. Notation used in Fig. 1 and Algorithm 1

C	Set of classes with the cardinality denoted as $|C|$
I_ℓ and O_ℓ	Input and output of layer ℓ, respectively
R_{in}^ℓ	List of relevance scores to be propagated to the layer before ℓ, $|R_{in}^\ell| = |I_\ell|$
R_{out}^ℓ	List of relevance scores for all the neurons in layer ℓ, $|R_{out}^\ell| = |O_\ell|$
$\text{LRP}(\ell, I_\ell, O_\ell, R_{out}^\ell)$	LRP denotes the function that implements the LRP algorithm. It returns the relevance score of each element of the input of this layer ℓ
q	Trained quantifier (QuaNet)
$Pr(c|x)$	Class posterior probability of document x for class c, which is obtained from classifier f
L	Sorted list of pairs of document representation ($\mathbf{x}$) and $Pr(c|x)$ for a chosen class c for all input documents
bi-LSTM, fclayer1, fclayer2, fclayer3	BiLSTM and the three fully connected layers of QuaNet
$\hat{p}^{CC}(D)$	CC quantifier prediction on D with given classifier, $|\hat{p}^{CC}(D)| = |C|$
S^U	List of five elements given unlabelled documents/training set: L, $\hat{p}^{CC}(D^U)$, $\hat{p}^{ACC}(D^U)$, $\hat{p}^{PCC}(D^U)$, $\hat{p}^{PACC}(D^U)$
$I_{\text{bi-LSTM}}$	Same as L
I_{fclayer1}	$O_{\text{bi-LSTM}} \oplus \hat{p}^{CC}(D^U) \oplus \hat{p}^{ACC}(D^U) \oplus \hat{p}^{PCC}(D^U) \oplus \hat{p}^{PACC}(D^U)$ where $\oplus$ denotes a concatenation

3.2 Method to Calculate the Amount of the Contribution of Each Input to QuaNet

We propose Eq. 4 using the relevance scores output by Algorithm 1. Let R_X denote the sum of the absolute relevance scores of each element of the input component X. For example, the CC input component is a list of $|C|$ numbers, where each number represents the class ratio of the corresponding class. In Algorithm 1, LRP returns a relevance score for each number. Thus, the total contribution of the CC component is the sum of the absolute relevance scores of all $|C|$ predictions by CC. The contribution of the input component is calculated as a percentage for a better comparison and visualization between results from different test datasets.

$$\text{contribution of input component } X \; (\%) = \frac{\sum R_X}{\sum_{y \in S^U} R_y} \times 100 \qquad (4)$$

Given a test dataset, we consider two methods to apply Eq. 4 to calculate the contribution amount for each input component.

- Naive method: We input the entire test dataset. The output is the amount of contribution for each input component. However, this method does not provide any statistics, such as a mean, median, or variance of the contribution amount.
- Sampling method: We design different sampling methods to obtain multiple subsets of the entire test dataset. Since the design of our sampling methods considers the number of classes and class imbalance ratios, we describe them in more detail in the next section.

4 Experiment Design and Results

This section describes the datasets, the experimental settings, and the results of the interpretation of QuaNet. Our two research questions guided the design of the experiments.

4.1 Datasets

We used five different classification datasets as summarized in Table 2. Three of them are publicly available and commonly used for quantification tasks. They are also balanced datasets, i.e., having the same number of documents for each class. The Standford Large Movie Review dataset (IMDB) has movie reviews with binary sentiment classification [14]. Yelp Polarity Reviews (YELP) [27] dataset was constructed for binary sentiment classification by grouping reviews with one or two stars in a negative class and reviews with three or four stars in a positive class. AG-NEWS [27] is a balanced multi-class classification dataset. For this study, only the titles in AG-NEWS were kept, while the article and description of each document were removed.

Table 2. Statistics of the five datasets used for experiments.

Dataset	#classes	# training documents	#testing documents	Average #words
IMDB	2	25,000	25,000	231
YELP	2	560,000	76,000	133
AG-NEWS	4	120,000	7,600	8
PAPT-Iowa	20	10,124	1,114	16
PAPT-Nebraska	20	5,132	559	17

The PAPT dataset [22] is a highly imbalanced dataset constructed by collecting tweets from state legislators in the US. These tweets were manually classified into 20 topics by Policy Agenda Project [24]. PAPT-Iowa and PAPT-Nebraska represent the PAPT dataset, specifically from Iowa and Nebraska legislators.

4.2   Experimental Settings

The training of both the underlying classifier and QuaNet was done by using the default implementation in the `QuaPy` library [19] available online[2]. This implementation uses a bi-LSTM network for the classifier model with all the same hyperparameters as in [8]. The classifier was trained to minimize the Cross-Entropy loss for a maximum of 200 epochs with an early stopping criterion. The training loop was ended after no improvement in the validation accuracy was observed for ten consecutive epochs. No data augmentation was used for any of the datasets. QuaNet was trained with stochastic gradient descent to minimize the Mean Square Error of the predictions. A batch size of 100 was used with a maximum number of iterations of 100. The training data was split into three parts, where the first two parts remained as the training data, and the third part was used as the validation data. The same early stopping criterion was also applied in training QuaNet. In [19], the list of document representation and posterior probabilities, L, was sorted based on a given target class for binary quantification tasks. L was not sorted for multi-class quantification tasks. We kept this default configuration in obtaining the results in Sect. 4.3, as [8] only tackled binary quantification and did not mention sorting L for multi-class quantification.

4.3   RQ1: What Input Features are Important to the Final Class Distribution Prediction of QuaNet?

We show the results of the two methods discussed in Sect. 3.2.

Interpretation Results Using the Naive Method: This method inputs the entire test dataset to QuaNet. Hence, Algorithm 1 produces only one contribution score for each of the five input components to QuaNet. Recall that the contribution is calculated using Eq. 4. Figure 2 shows the pie charts of the contributions.

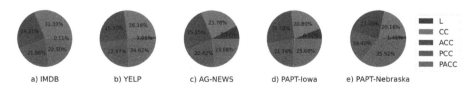

Fig. 2. Percentage of contributions of each input component towards the final prediction of QuaNet on different test datasets.

Figure 2 shows that the sorted list L has the lowest contributions across all the datasets by a large margin. For IMDB, the contribution of L is almost

---

[2] https://github.com/HLT-ISTI/QuaPy.

negligible (0.11%), but not so for AG-NEWS and PAPT-Iowa. For the balanced test datasets, IMDB, YELP, and AG-NEWS, the predicted class distribution by CC contributes the most, but not too far ahead from the other predicted class distributions by ACC, PCC, and ACC. For PAPT-Nebraska, we see the most contributions from PACC. However, for PAPT-Iowa, there is a tie in the contributions by PACC and ACC, with the sorted list L contributing more. The results on the two imbalanced PAPT datasets indicate that adjusting the predicted class distribution using false positive and true positive rates via PACC or ACC is important for dealing with class imbalance.

Interpretation Results Using the Sampling Method: We created multiple equal-sized subsets from a given test dataset. Each subset (sample) had 100 documents. We designed different sampling strategies to reflect the difference in the number of classes and the class imbalance ratios for different datasets.

For binary quantification tasks (IMDB and YELP datasets), we made the prevalence of a positive class for different samples take a value in $\{0, 0.01, 0.02, ..., 0.99, 1\}$. We ensured that the class ratios of the positive and the negative classes sum to one, e.g., if the positive class ratio is 0.99, then the negative class ratio is 0.01. This strategy created 101 samples with different distributions.

For four-class quantification tasks on AG-NEWS, the prevalence of each class took a value in $\{0, 0.1, 0.2, ..., 0.9, 1\}$. We ensured that the sum of the four class ratios was one. This strategy resulted in 282 different samples. When assigning documents to a given sample, sampling without replacement was used if the specified prevalence for the class was larger than the actual prevalence of the class. Otherwise, sampling with replacement was used. For twenty-class quantification on both the PAPT datasets, we randomly selected 100 documents per sample instead of restricting each class to a specified class ratio for the following reasons. 1) To avoid too many combinations of class distributions. 2) To prevent the creation of samples with many duplicate documents caused by sampling with replacement since the PAPT datasets are small, and the class distribution is highly imbalanced.

Figure 3 and 4 show the boxplot of contributions of each of the five input components. Recall that the contributions were calculated from the relevance scores given by Algorithm 1 using Eq. 4. There are five data points per sample in Fig. 3 where each data point for the sample is the contribution of the corresponding input component shown on the x-axis towards the final prediction of QuaNet. For each input component on the x-axis, the box in the boxplot indicates the center of the distribution of the contributions for that component. The solid line across the box shows the median value. The bottom and top of the box are drawn at the first quartile and the third quartile, while the whiskers represent 1.5 times an interquartile range away from the first and third quartile, respectively. Any data point outside of this range is considered an outlier.

Figures 3 and 4 show that the sorted list L consistently has the lowest contribution towards the final prediction. This component has a median contribution of less than 5% in all except AG-NEWS. Furthermore, there are relatively less

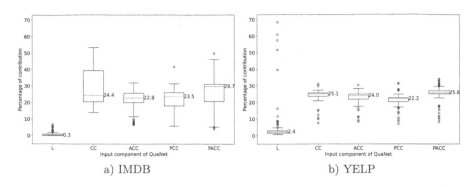

a) IMDB   b) YELP

Fig. 3. Percentage of contributions of each input component towards the final prediction of QuaNet on the testing sets for binary quantification tasks.

outliers on the contributions of L for the PAPT datasets. We found that all the large outliers in YELP and AG-NEWS resulted from samples with highly skewed class prevalence. This shows that QuaNet relies on L when the training and test data class distributions are very different. This also explains why the data points from PAPT datasets are highly clustered and have fewer outliers, as the samples were produced randomly, leading to similar class distributions between different samples. Interestingly, the contributions were spread out more for IMDB compared to YELP and AG-NEWS.

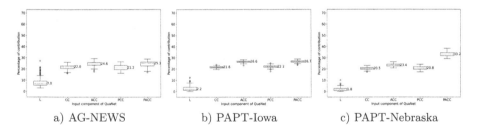

a) AG-NEWS   b) PAPT-Iowa   c) PAPT-Nebraska

Fig. 4. Percentage of contributions of each input component towards the final prediction of QuaNet on the testing sets for multiclass quantification tasks.

The predictions from the four quantifiers, CC, ACC, PCC, and PACC, contribute on average more than 20% respectively, totaling over 90% of the contribution towards QuaNet's decisions. PACC is consistently the highest contributing component, meaning that PACC's prediction is considered the most valuable information in QuaNet's decision. This aligns with the finding in [20], where PACC was the top-performing quantifier among these four quantifiers. Having a total of over 90% contribution from the quantifiers reflects that QuaNet learns to rely heavily on the results from these quantifiers more than the quantification embedding. We trained a simple neural network called Small-NN to predict the class prevalence with only these four quantifiers' predicted class prevalence as input.

Table 3. Performance of QuaNet vs Small-NN. Small-NN is a simple fully connected network that takes the four quantifiers' predicted class prevalence as input on the same test data for each dataset. MAE, KLD, and RMAE are performance metrics mentioned in Sect. 2.1. Bold numbers indicate better performances.

Model	QuaNet			Small-NN		
Dataset	MAE	KLD	RMAE	MAE	KLD	RMAE
IMDB	0.0907	0.0613	0.2278	**0.0560**	**0.0292**	**0.1686**
YELP	0.0216	**0.0026**	**0.0800**	**0.0179**	0.0273	0.0801
AG-NEWS	0.0370	0.0749	0.3130	**0.0177**	**0.0122**	**0.1224**
PAPT-Iowa	0.0612	1.1222	1.5903	**0.0117**	**0.0626**	**0.3489**
PAPT-Nebraska	0.0514	0.9164	1.1115	**0.0112**	**0.0570**	**0.3361**

Small-NN has one fully connected layer with 64 neurons with ReLU activation, followed by a final layer of $|C|$ neurons with softmax activation. The training and test data for Small-NN were produced by discarding the input component L to QuaNet during the training and testing. 30% of the training data was used as the validation data for training purposes. Small-NN was trained to optimize MAE with Adam optimization at a learning rate of 10^{-3}. It was trained for 50 epochs with the stopping criterion to end the training after no improvement in validation loss for 2 epochs. Table 3 shows that Small-NN, a much less sophisticated model, outperforms QuaNet on all the datasets based on MAE. This outcome aligns with our interpretation that the predicted class distributions of the four quantifiers are more valuable as quantification features.

4.4   RQ2: Does Sorting L by $Pr(c|x)$ Increase the Performance of QuaNet?

Esuli et al. [8] proposed to sort the list L because of the intuition that RNN can learn to count by observing the ordered sequence of $Pr(c|x)$ values, thus recognizing the switch point between classes. This RQ is framed to examine if bi-LSTM can recognize the switch point.

We performed an experiment to compare the performance of QuaNet on the IMDB dataset by sorting L on the test data with two other sorting methods. Method 1 sorts the posterior probabilities of a class different from the class used in training. Method 2 does not sort the posterior probabilities. Algorithm 1 was used in this experiment with its inputs S^U produced by the different sorting methods. The MAE provided by QuaNet on the IMDB test dataset (Table 2) is as follows (best to worst): 0.0253 for sorting based on the default class used in QuaNet, 0.0706 for no sorting (Method 2), and 0.4835 for Method 1. This result shows that QuaNet achieved the best performance with sorted L as its input.

One hypothesis is that the trained bi-LSTM layer observes the sorted order in L, i.e., sorted by $Pr(c|x)$. Once the posterior probabilities of the test data are not sorted according to the order it learned (sorted by $Pr(c'|x)$ where $c' \neq c$ or not sorted at all), L has a different pattern that the quantifier has not seen before, thus affecting its prediction. To confirm this hypothesis, we trained different versions of QuaNet by applying different sorting methods on L from the training data and compared the performance. With each dataset, the same classifier is used to train the different versions of QuaNet. We tested each QuaNet with samples created as described in Sect. 4.3.

Table 4 shows the average performance using the common metrics for quantification defined in Sect. 2.1. QuaNet with or without sorting L performs similarly. We conclude that sorting L does not necessarily increase the average performance of QuaNet.

Table 4. Performance of QuaNet models trained with different sorting methods. The average performance on samples created as described in Sect. 4.3 is shown. QuaNet denotes the model that was trained with default sorting. QuaNet-unsorted denotes the model trained with unsorted L. QuaNet-revsort denotes the model trained by sorting L on the posterior probabilities of a different class than the default.

Dataset	IMDB			YELP		
Model	MAE	KLD	RMAE	MAE	KLD	RMAE
QuaNet	0.0907	0.0613	0.2278	0.0216	0.0026	0.0800
QuaNet-unsorted	0.0837	0.0509	0.2168	0.0195	0.0026	0.0828
QuaNet-revsort	0.0972	0.0553	0.2442	0.0216	0.0033	0.0792

(a) Binary quantification tasks

Dataset	AG-NEWS			PAPT-Iowa			PAPT-Nebraska		
Model	MAE	KLD	RMAE	MAE	KLD	RMAE	MAE	KLD	RMAE
QuaNet-sorted	0.0377	0.0651	0.3214	0.0510	1.0872	1.2434	0.0516	0.8162	0.9433
QuaNet-unsorted	0.0370	0.0749	0.3130	0.0612	1.1222	1.5903	0.0514	0.9164	1.1115

(b) Multi-class quantification tasks

5   Conclusion and Future Work

We present the first attempt at interpreting a deep text quantifier. By using LRP, we calculate the relevance of each input component to a quantifier's predicted class distribution and identify important quantification features. This work presents an interpretation of QuaNet, a recent deep-learning quantifier for text quantification tasks. Our findings are as follows. 1) QuaNet relies the most on one of its inputs, the PACC prediction. QuaNet learns that PACC is the most reliable feature to guide its prediction. This finding is consistent with the study by [20], that PACC is the top performing quantifier among CC, ACC, PCC, and PACC. 2) The quantification embedding used in QuaNet [8] is an important quantification feature when the class prevalence of a given test dataset is significantly different from that of the training dataset. 3) On the five test datasets used in this study, a small fully connected network with CC, ACC, PCC, and PACC as input gives a better average quantification performance than QuaNet.

As QuaNet relies on a separate classifier to classify each document, our method cannot provide relevance scores for individual words in the input documents. Achieving such interpretation requires an interpretation method on the classifier model and integration with the contributions calculated from Algorithm 1.

In our future work, we plan to interpret the other deep text quantifier, DQN. This is where the advantage of LRP on returning signed relevance scores should stand out. DQN is a direct quantification model which does not need any classifier. Modifying LRP to interpret DQN can trace the relevance back to each word in a document in the input. This allows word-level interpretation, where we can identify both the positive and negatively contributing words to the quantification prediction.

Acknowledgement. This work is partially supported in part by the NSF SBE Grant No. 1729775. Any findings and opinions expressed in the paper are of the authors and do not necessarily reflect the view of the funding agency.

References

1. Ali, A., Schnake, T., Eberle, O., Montavon, G., Müller, K.R., Wolf, L.: XAI for transformers: better explanations through conservative propagation. In: ICML (2022)
2. Arras, L., Horn, F., Montavon, G., Müller, K.R., Samek, W.: "What is relevant in a text document?": an interpretable machine learning approach. PloS One **12**(8), e0181142 (2017)
3. Arras, L., Montavon, G., Müller, K.R., Samek, W.: Explaining recurrent neural network predictions in sentiment analysis. In: Proceedings of the Workshop on Computational Approaches to Subjectivity, Sentiment and Social Media Analysis, pp. 159–168. Association for Computational Linguistics (2017)
4. Bach, S., Binder, A., Montavon, G., Klauschen, F., Müller, K.R., Samek, W.: On pixel-wise explanations for non-linear classifier decisions by layer-wise relevance propagation. PloS One **10**(7), e0130140 (2015)
5. Barranquero, J., Díez, J., del Coz, J.J.: Quantification-oriented learning based on reliable classifiers. Pattern Recognit. **48**(2), 591–604 (2015)
6. Bella, A., Ferri, C., Hernández-Orallo, J., Ramírez-Quintana, M.J.: Quantification via probability estimators, pp. 737–742. IEEE (2010)
7. Esuli, A., Fabris, A., Moreo, A., Sebastiani, F.: Methods for learning to quantify. In: Esuli, A., Fabris, A., Moreo, A., Sebastiani, F. (eds.) Learning to Quantify. The Information Retrieval Series, vol. 47, pp. 55–85. Springer, Cham (2022). https://doi.org/10.1007/978-3-031-20467-8_4
8. Esuli, A., Moreo Fernández, A., Sebastiani, F.: A recurrent neural network for sentiment quantification. In: Proceedings of ACM International Conference on Information and Knowledge Management, Torino, Italy. Association for Computing Machinery (2018)
9. Forman, G.: Counting positives accurately despite inaccurate classification. In: Gama, J., Camacho, R., Brazdil, P.B., Jorge, A.M., Torgo, L. (eds.) ECML 2005. LNCS (LNAI), vol. 3720, pp. 564–575. Springer, Heidelberg (2005). https://doi.org/10.1007/11564096_55

10. Forman, G.: Quantifying trends accurately despite classifier error and class imbalance. In: Proceedings of ACM SIGKDD, pp. 157–166 (2006)
11. González-Castro, V., Alaiz-Rodríguez, R., Alegre, E.: Class distribution estimation based on the Hellinger distance. Inf. Sci. **218**, 146–164 (2013)
12. Jerzak, C.T., King, G., Strezhnev, A.: An improved method of automated non-parametric content analysis for social science. Polit. Anal. **1**(17) (2019)
13. Khaleel, M., Qi, L., Tavanapong, W., Wong, J., Sukul, A., Peterson, D.A.M.: IDC: quantitative evaluation benchmark of interpretation methods for deep text classification models. J. Big Data **9**(1), 1–14 (2022)
14. Maas, A., Daly, R.E., Pham, P.T., Huang, D., Ng, A.Y., Potts, C.: Learning word vectors for sentiment analysis. In: Proceedings of the 49th Annual Meeting of the Association for Computational Linguistics: Human Language Technologies, pp. 142–150 (2011)
15. Maletzke, A., Reis, D.D., Hassan, W., Batista, G.: Accurately quantifying under score variability. In: 2021 IEEE International Conference on Data Mining (ICDM), pp. 1228–1233 (2021). https://doi.org/10.1109/ICDM51629.2021.00149
16. Martino, G., Gao, W., Sebastiani, F.: Ordinal text quantification. In: Proceedings of International ACM SIGIR Conference on Research and Development in Information Retrieval, pp. 937–940 (2016)
17. Milli, L., Monreale, A., Rossetti, G., Giannotti, F., Pedreschi, D., Sebastiani, F.: Quantification trees. In: IEEE International Conference on Data Mining, pp. 528–536 (2013)
18. Montavon, G., Binder, A., Lapuschkin, S., Samek, W., Müller, K.-R.: Layer-wise relevance propagation: an overview. In: Samek, W., Montavon, G., Vedaldi, A., Hansen, L.K., Müller, K.-R. (eds.) Explainable AI: Interpreting, Explaining and Visualizing Deep Learning. LNCS (LNAI), vol. 11700, pp. 193–209. Springer, Cham (2019). https://doi.org/10.1007/978-3-030-28954-6_10
19. Moreo, A., Esuli, A., Sebastiani, F.: QuaPy: a python-based framework for quantification. In: Proceedings of ACM International Conference on Information & Knowledge Management, pp. 4534–4543 (2021)
20. Moreo, A., Sebastiani, F.: Tweet sentiment quantification: an experimental re-evaluation. PLoS One **17**(9), e0263449 (2022)
21. Pérez-Gállego, P., Castaño, A., Ramón Quevedo, J., del Coz, J.J.: Dynamic ensemble selection for quantification tasks. Inf. Fusion **45**, 1–15 (2019)
22. Qi, L.: Quantification learning with deep neural networks (2021)
23. Qi, L., Khaleel, M., Tavanapong, W., Sukul, A., Peterson, D.: A framework for deep quantification learning. In: Hutter, F., Kersting, K., Lijffijt, J., Valera, I. (eds.) ECML PKDD 2020. LNCS (LNAI), vol. 12457, pp. 232–248. Springer, Cham (2021). https://doi.org/10.1007/978-3-030-67658-2_14
24. Qi, L., Li, R., Wong, J., Tavanapong, W., Peterson, D.A.: Social media in state politics: mining policy agendas topics. In: Proceedings of the 2017 IEEE/ACM International Conference on Advances in Social Networks Analysis and Mining 2017, pp. 274–277 (2017)
25. Saerens, M., Latinne, P., Decaestecker, C.: Adjusting the outputs of a classifier to new a priori probabilities: a simple procedure. Neural Comput. **14**(1), 21–41 (2002)
26. Ullah, I., Rios, A., Gala, V., Mckeever, S.: Explaining deep learning models for tabular data using layer-wise relevance propagation. Appl. Sci. **12**(1), 136 (2021)
27. Zhang, X., Zhao, J., LeCun, Y.: Character-level convolutional networks for text classification. In: Advances in Neural Information Processing Systems, vol. 28 (2015)

NExtGCN: Modeling Node Importance of Graph Convolution Network by Neighbor Excitation for Recommendation

Jingxue Zhang[1] , Ning Wu[2] , and Changchun Yang[1(✉)]

[1] Changzhou University, Changzhou, Jiangsu Province, China
`s21150812001@smail.cczu.edu.cn`, `ycc@cczu.edu.cn`
[2] Beihang University, Beijing, China
`wuning@buaa.edu.cn`

Abstract. To alleviate information overload, the recommender system is pushing personalized contents to users and improving the efficiency of information distribution. Graph Convolution Networks (GCNs), which can better gather structured information and becomes a new state-of-the-art for collaborative filtering. Many current works on GCNs tend to be easier to train and have better generalization ability like LightGCN. However, they care less about the importance of nodes. In this work, we propose a new model named *NExtGCN (Neighbor Excitation Graph Convolutional Network)*, which models the node importance of GCN by neighbor excitation. The NExtGCN can learn the importance of nodes via the global and local excitation layer which is inspired by the Squeeze-Excitation network. Furthermore, we propose a neighbor excitation layer that can fully utilize graph structure and make this model practical to large-scale datasets. Extensive experimental results on four real-world datasets have shown the effectiveness and robustness of the proposed model. Especially on the Amazon-Books dataset, our NExtGCN has improved by 10.95%, 49.36%, and 26.8% in Recall@20, MRR@20, and NDCG@20 compared to LightGCN. We also provide source code (https://github.com/clemaph/NExtGCN.git) to reproduce the experimental results (This job is supported by Postgraduate Research & Practice Innovation Program of Jiangsu Province, the item number is KYCX22_3071).

Keywords: Recommendation · Graph Convolution Networks · Collaborative Filtering

1 Introduction

The recommender system has become the infrastructure of information services, providing users with personalized contents. LightGCN [1] simplifies GCNs by only keeping the neighborhood aggregation module and achieves better empirical performance than other sota methods. UltraGCN [4] uses several constraint losses. SGL [7] explores self-supervised learning on useritem graph. The above models ignore the weights of neighbor nodes, so it is difficult to effectively distinguish the influence of important nodes from non-important nodes. There has been some existed works [2,5,6] that

© The Author(s), under exclusive license to Springer Nature Switzerland AG 2023
C. Strauss et al. (Eds.): DEXA 2023, LNCS 14147, pp. 325–330, 2023.
https://doi.org/10.1007/978-3-031-39821-6_26

can model node or item importance by attention mechanism, typical attention mechanism is not suitable for user-item bipartite graph. As shown in Fig. 1(a), the learned attention score in GAT has a high variance and it makes the model hard to optimize. Besides, we observe that k3 and k7 always have a relatively high attention score for any query node. Intuitively, k3 and k7 are representative works in romance and science fiction movies. We make an assumption that there is an intrinsic importance score for each node in the user-item graph and we need to learn a static importance score for each user and item node. It should be much easier to optimize and requires fewer parameters than a dynamic attention mechanism. The importance scores learned by our NExtGCN are also presented in Fig. 1(b). We propose to apply Squeeze-Excitation network [3] on node importance modeling and design global and local modules to capture the node-wise and dimension-wise importance weights of neighbor node embeddings by Squeeze-Excitation network. We propose the neighbor excitation layer to adapt the large node number and high-order connectivity in graph node importance modeling.

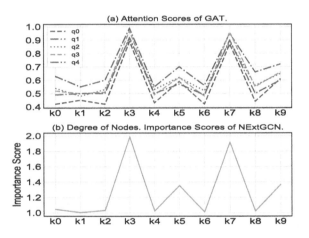

Fig. 1. The importance of ten nodes in the Movielens-1M. k0 to k9 represent 10 different movies: k0 (*Digimon*), k1 (*Contender*), k2 (*Tigerland*), k3 (*Roman Holiday*), k4 (*Bikini Beach*), k5 (*Meet the Parents*), k6 (*Urban Legends*), k7 (*2001: A Space Odyssey*), k8 (*Beach Party*), and k9 (*Kronos*). q0 to q4 denotes 5 user nodes.

2   Method

2.1   Input Layer

At the beginning, we formulate the recommendation task to be addressed in this paper: **Input:** user-item bipartite graph $\mathcal{G}$. **Output:** A function that predicts the likelihood $\hat{y}_{ui}$ of user u choosing item i. D denotes the embedding size, $e_u \in \mathbb{R}^D$ denotes user embeddings while $e_i \in \mathbb{R}^D$ denotes item embeddings. And the parameter matrix after splicing the two is as follows:

$$E = [e_{u_1}, ..., e_{u_N}, e_{i_1}, ..., e_{i_M}]. \tag{1}$$

2.2 NExtGCN Layer

The importance weights s_u and s_i of user u and item i are employed in aggregation, and our graph convolution operation is defined as follows:

$$e_u^{(l+1)} = \sum_{i \in N_u} \frac{s_u}{\sqrt{|N_u|}\sqrt{|N_i|}} e_i^{(l)}, e_i^{(l+1)} = \sum_{u \in N_i} \frac{s_i}{\sqrt{|N_i|}\sqrt{|N_u|}} e_u^{(l)}. \qquad (2)$$

where N_u denotes neighbors of user u, N_i denotes neighbors of item i. As mentioned above, not all neighbor nodes are equally important when aggregating extended neighbors and we need to give them different scores to capture different effects via the two-level excitation layer.

Modeling Global and Local Node Importance. The i-th feature embeddings are squeezed into a scalar value z_i which represents the global information about i-th feature embedding. All features are squeezed into scalar values and two linear layers are followed to model the final importance of all features. We consider squeezing node embedding in Eq. 3, and calculate global importance by Eq. 4:

$$z^0 = \text{MaxPooling}(\boldsymbol{E}^{(0)}) \qquad (3)$$

$$z^{k+1} = \text{Dropout}(\sigma(\boldsymbol{W}_k z^k)), z_u^{k+1} = \sum_{u \in N_i} \boldsymbol{W}_{i,u}^k z_u^k \qquad (4)$$

where we use MaxPooling denotes the max pooling mechanism. $\boldsymbol{W}_k$ denotes a learnable parameter matrix in the k-th excitation layer, Dropout denotes dropout and σ denotes ReLU activation function. A typical excitation layer consists of linear layer, dropout operation and ReLU activation function. To capture the overall importance score of a node, it is necessary to first pass the embedding through a layer of global maximum pooling to obtain an initial global importance score with a shape of $N \times 1$. Then several excitation layers will be applied to the global importance scores. We set the shape of all parameter matrix $\boldsymbol{W}$ as $N \times N$, which will keep the input and output dimension of all excitation layers' as N.

Except the global excitation layer, the local excitation mechanism layer is to input an embedding of shape $N \times D$ and generates a dimension-wise weight matrix. The local modules are computed via a similar structure without pooling as follows:

$$\boldsymbol{X}^{k+1} = \text{Dropout}(\sigma(\boldsymbol{W}_k \boldsymbol{X}^k)), \boldsymbol{X}_u^{k+1} = \sum_{u \in N_i} \boldsymbol{W}_{i,u}^k \boldsymbol{X}_u^k \qquad (5)$$

with the local module, we can compute fine-grain importance score $\boldsymbol{X}^{k+1}$ for each dimension of all nodes. The final importance score $\boldsymbol{X}^{k+1}$ should still keep the $N \times D$ shape. Given the global importance score z^K and local importance score $\boldsymbol{X}^K$, we obtain the final importance scores of user and item embeddings as follows:

$$s = \text{MeanPooling}(\lambda * \sigma(z^K \oplus \boldsymbol{X}^K)) \qquad (6)$$

where $\oplus$ denotes the broadcasting addition and σ denotes Sigmoid activation function. The Sigmoid activation function is used to predict the importance score. Since the range of Sigmoid is $[0, 1]$, we set $\lambda = 2$ to scale up its range. Finally, we add a mean-pooling mechanism on the addition embedding which has a shape of $N \times D$ to obtain an importance score for each node.

2.3 Prediction Layer

After neighbor aggregation layer, we get weighted $e_u^{(0)}$ for all users and $e_i^{(0)}$ for all items. We get a final representation of the user and item via L-layers normalized sum:

$$e_u = \sum_{l=0}^{L} a_l e_u^{(l)}; e_i = \sum_{l=0}^{L} a_l e_i^{(l)} \tag{7}$$

where α_l means the importance of l-th layer, which keeps the same setting with Light-GCN [1]. We use the inner product of user and item final representations to define the model prediction:

$$\hat{y}_{ui} = e_u^T e_i \tag{8}$$

3   Experiments

3.1   Experimental Settings

We conduct experiments on four benchmark datasets: MovieLens-1M, Gowalla, Yelp2018, and Amazon-Books, randomly rearrange the data, set the dataset to be divided into 8:1:1 training, test, and evaluation sets, and use full sort. We mainly compare three indicators: *Recall@20*, *NDCG@20*, and *MRR@20*. For the embedding layer and optimization method, we use $\mathcal{L}_2$ regularization with 10^{-4} weight, and the learning rate is set to 10^{-3}, the batch size of training to 4096 and batch size of evaluating to 204800, the negative sampling radio $\mathcal{R}$ to 300, we fix the embedding size to 64.

3.2   Performance Comparison

We present the results of the performance comparison in Table 1. First, GCNs are generally better than embedding based methods and matrix factorization based methods, because GCNs can utilize high-order connectivity in user-item bipartite. Second, NExtGCN has the best performance on all four datasets, especially on the Amazon-Books dataset. The results in the table illustrate our considerable improvement over state-of-the-art GCN models. Finally, GAT doesn't achieve higher results compared with other GCN-based recommendation methods. It's possible that the dynamic attention mechanism in GAT is not suitable to a user-item bipartite graph, because it's too complicated to be optimized on a simple graph structure.

Table 1. Overall performance comparison. NExtGCN denotes our model which consists of global and local modules. R, M, and N denote Recall, MRR, and NDCG, respectively. Improv. denotes the relative improvements over the best GNN-based baselines. The results of significance testing indicate that our improvements over the current strong GCN-based baseline are statistically significant(p-value< 0.05).

Metrics		LINE	BPRMF	NeuMF	GCMC	GAT	NGCF	DGCF	LightGCN	NExtGCN	Improv.
Amazon-Books	R@20	0.0358	0.0968	0.0655	0.0908	0.0921	0.0937	0.1068	0.1123	**0.1246**	10.95%
	M@20	0.0241	0.0678	0.0389	0.5860	0.6171	0.0629	0.0693	0.0778	**0.1162**	49.36%
	N@20	0.0191	0.0389	0.0336	0.0557	0.0574	0.0580	0.0588	0.0638	**0.0809**	26.8%
Gowalla	R@20	0.1047	0.1517	0.1447	0.1497	0.1514	0.1572	0.1811	0.1898	**0.2070**	9.06%
	M@20	0.0750	0.1114	0.0901	0.0904	0.1538	0.1074	0.1257	0.1307	**0.1421**	8.72%
	N@20	0.0610	0.1001	0.7950	0.0807	0.0866	0.0904	0.1058	0.1106	**0.1211**	9.49%
Yelp2018	R@20	0.0491	0.1000	0.7610	0.0924	0.0939	0.0963	0.1132	0.1134	**0.1206**	6.35%
	M@20	0.0341	0.0690	0.0466	0.0601	0.0625	0.0644	0.0757	0.0772	**0.0835**	8.16%
	N@20	0.0266	0.0557	0.0396	0.0495	0.0514	0.0523	0.0604	0.0628	**0.0674**	7.32%
Movielens-1M	R@20	0.2419	0.2665	0.2483	0.2630	0.2691	0.2725	0.2799	0.2677	**0.3000**	7.18%
	M@20	0.3994	0.4248	0.3906	0.4232	0.4211	0.4169	0.4386	0.4333	**0.4614**	5.20%
	N@20	0.2277	0.2495	0.2296	0.2493	0.2501	0.2513	0.2608	0.2540	**0.2795**	7.17%

3.3 Ablation of NExtGCN

Effect of Neighbor Excitation Layer Numbers. Figure 2 shows how the neighbor excitation layer benefits from high-order connectivity on the graph. We can conclude that NExtGCN outperforms LightGCN, means that it can always learn the importance of neighbor nodes w/ or w/o high-order information.

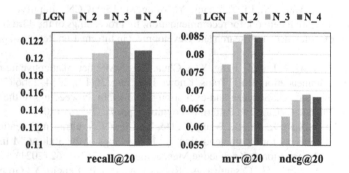

Fig. 2. Performance comparison of setting the different number of the neighbor excitation layers on Yelp2018. N_2, N_3, N_4 denotes the model with 2, 3, 4 neighbor excitation layer numbers.

Effect of Global and Local Modules. Figure 3 shows the effect of the global module, the local module, and the combination of the two. Whether the two modules are used alone or in combination, they perform better than LightGCN and the combination of the two modules is better than using one alone module.

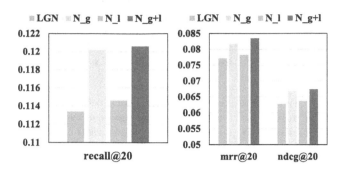

Fig. 3. Performance comparison of different modules on Yelp2018. N_g, N_l, N_g+l denotes the model with the global module, local module, and both modules.

4   Conclusions

In this work, we explore the influence of static node importance and dynamic node importance on recommendation results. Due to the drawbacks of the state-of-the-art models, we propose NExtGCN to model nodes' static importance of GCN by global and local excitation for the recommendation. Furthermore, the NExtGCN can capture graph structure to enhance importance modeling, and makes node importance modeling to be more applicable when facing dozens of thousands of nodes.

References

1. He, X., Deng, K., Wang, X., Li, Y., Zhang, Y., Wang, M.: LightGCN: simplifying and powering graph convolution network for recommendation. In: Proceedings of the 43rd International ACM SIGIR Conference on Research and Development in Information Retrieval, pp. 639–648 (2020)
2. He, X., He, Z., Song, J., Liu, Z., Jiang, Y.G., Chua, T.S.: NAIS: neural attentive item similarity model for recommendation. IEEE Trans. Knowl. Data Eng. **30**(12), 2354–2366 (2018)
3. Hu, J., Shen, L., Sun, G.: Squeeze-and-excitation networks. In: Proceedings of the IEEE Conference on Computer Vision and Pattern Recognition, pp. 7132–7141 (2018)
4. Mao, K., Zhu, J., Xiao, X., Lu, B., Wang, Z., He, X.: UltraGCN: ultra simplification of graph convolutional networks for recommendation. In: Proceedings of the 30th ACM International Conference on Information & Knowledge Management, pp. 1253–1262 (2021)
5. Veličković, P., Cucurull, G., Casanova, A., Romero, A., Lio, P., Bengio, Y.: Graph attention networks. arXiv preprint arXiv:1710.10903 (2017)
6. Wang, X., He, X., Cao, Y., Liu, M., Chua, T.S.: KGAT: knowledge graph attention network for recommendation. In: Proceedings of the 25th ACM SIGKDD International Conference on Knowledge Discovery & Data Mining, pp. 950–958 (2019)
7. Wu, J., et al.: Self-supervised graph learning for recommendation. In: Proceedings of the 44th International ACM SIGIR Conference on Research and Development in Information Retrieval, pp. 726–735 (2021)

Dual Congestion-Aware Route Planning for Tourists by Multi-agent Reinforcement Learning

Kong Yuntao[1,2(✉)] , Peng Chen[2] , Nguyen Minh Le[2] , and Ma Qiang[3]

[1] Department of Social Informatics, Kyoto University, Kyoto, Japan
[2] School of Information Science, Japan Advanced Institute of Science and Technology, Ishikawa, Japan
oldkong007@gmail.com
[3] Department of Information Science, Kyoto Institute of Technology, Kyoto, Japan

Abstract. Tourist route planning has been extensively investigated due to the rapid development of the tourism industry. Existing methods often recommend popular and optimized (short or quick) routes based on interest of tourists. Consequently, the sightseeing area becomes more crowded, and the overtourism issue becomes more serious. As a solution, we propose a multi-agent reinforcement learning (RL) framework for congestion-aware routes planning. Our work contains two parts: first, we introduce a novel tourism RL environment that can interact with multiple tourists. Then we propose a dual-congestion awareness model. Specifically, both the local congestion of visited sightseeing spot and the global tourists' distribution in the target city are considered to model the reward in our multi-agent RL framework. We conduct experiments by using real datasets collected from Kyoto, one of the most popular sightseeing cities in the world. The experimental results demonstrate that our model is superior to previous methods in terms of tourists' distribution in the target city.

Keywords: Multi-agent Reinforcement Learning · Sustainable Tourism Development · Multiple Tourists Route Planning

1 Introduction

Tourism route planning models [1–3] have been developed from various aspects, such as popular route optimization and route personalization. Existing studies have primarily considered interest of tourists, resulting in that visits are mainly planned on popular points-of-interest (POIs) and some minor POIs remain unvisited at the end of the trip. We term this problem as popularity-biased route planning which not only poses a burden on the popular POIs but also discourages the development of sustainable tourism.

To address this issue, we propose a route planning model with consideration of tourists distribution (RPMTD). We first design a novel tourism multi-agent reinforcement learning (RL) environment which can interact with multiple tourists. Subsequently, based on this environment, we propose dual-congestion awareness model. Specifically, local and global congestion rewards are considered. The local congestion reward evaluates the crowdness of visited POI and the global congestion reward evaluates the evenness of tourists' distribution across all POIs.

We implement our method on Kyoto real human-mobility data with scalable tourists settings. The model is evaluated in terms of the fairness of POIs' visits and the evenness of tourists' distribution. Result indicates that our model generally outperforms other models and shows robustness in various tourists scenarios.

2   Related Work

Numerous works have developed single tourist route planning schemes considering different aspects. Some studies [5–7] aim to maximize trip reward. Some other works focus on personalized route planning [2,8–10]. Conversely, few studies focus on multiple tourists. Sylejmani et al. [11] plans a group of tourists with considerations of individual preferences and mutual social relationship. Sarkar et al. [12] applies Subgame-Perfect Nash equilibrium of game theory to recommend various routes for a group of tourists. Kong et al. [4] applies single-agent RL framework to diversify multiple tourists routes. Different from previous methods, our model is the first to generate routes considering the POI's interest.

3   Multi-agent Reinforcement Learning Environment

Our proposed environment consists of the mobility matrix, tourists and POIs. The $U \times V$ mobility matrix M presents the POIs' population change over one day, where U is the number of POIs; V is the number of time slots representing one day. Cell$_{uv}$ of M indicates the number of people in POI u at the specific time slot v. There is a current time indicator $time^{cur}$ for M, indicating the current time slot and moving to next time slot after one interaction. For each tourist t_j, there is an indicator of activate time $time_j^{act}$. Before the activate time, t_j does not interact with the environment. Specifically, t_j is either in sightseeing or unstarted before the activate time. For each POI p_i, the popularity is set according to current time slot of the mobility matrix in each interaction.

At the beginning, the current time indicator $time^{cur}$ is initialized by the tourists' starting time. The POIs' popularity is set according to current time slot from the mobility matrix M. Environment generates the observation O_j for the tourist t_j, which is stacking of each POI p_i's observation O_j^i. In O_j^i, there are 8 features: capacity, time cost, score, visit count, popularity, tourist remaining time, tourist time budget and tourist moving time. Thus, the size of O_j is $n \times 8$, where n represents the number of POIs. Given observation O_j, agent$_j$ outputs the action A_j which is the next visiting POI. Then the mobility matrix M is updated, and the reward is given. $time_j^{act}$ moves to the time slot where is the end of this POI visiting.

4   Dual-Congestion Aware Routes Planning Model

4.1   Multi-agent Reinforcement Learning Implementation

Our work is based on fully decentralized approach to accommodate our model to scalable tourists scenarios. The actor-critic method is adopted and independent PPO (IPPO) [15] is utilized for training. Since all the agents are homogeneous, namely, they having the same state space, action space and optimization objective, parameter sharing is conducted.

Formally, in the actor-critic method, the observation $\mathbf{O}$ is given for an agent from the environment in each interaction. Specifically, $\mathbf{O} = \{O^1, ..., O^n\} \in \mathbb{R}^{n \times d}$, where n and d denotes the number of POIs and the POI's feature dimension, respectively. The actor firstly maps $\mathbf{O}$ into the hidden space with a feed-forward network (FNN) to obtain the hidden representation: $\mathbf{O}' = \mathrm{FFN}_1(\mathbf{O}) \in \mathbb{R}^{n \times h}$, where h denotes the hidden dimension.

Then the self-attention is applied to update hidden representation $\mathbf{O}'$, which is motivated by its effectiveness in combinatorial optimization studies [16]. Specifically, it learns the relations between each pair of POIs, and updates each POI's representation from all POIs. We use transformer encoder [17] as self-attention module, and stack 2 layers with 8 heads: $\mathbf{H} = \mathrm{TransformerEncoder}(\mathbf{O}') \in \mathbb{R}^{n \times h}$.

The updated representation $\mathbf{H}$ is further fed into another FNN to calculate the logits of each POI, and then is normalized by softmax. The final action is obtained by random sampling. The critic has the similar structure with actor except that it finally calculates a scalar value.

4.2   Dual-Congestion Mechanism

For the unbiased routes planning, we propose a dual-congestion mechanism to construct reward function. Specifically, the local congestion considers visited POI congestion and the global congestion considers evenness of tourists overall distribution. The total reward is the sum of local and global congestion reward:

$$Reward_{total} = Reward_{global} + Reward_{local} \tag{1}$$

Reward Based on Local Congestion: $Reward_{local}$ evaluates the visited POI's crowdness. The intuition is that less tourists poses less burden on the POI and makes tourists more satisfied, which results in higher reward. We use a linear function to define the local congestion reward without considering negative reward as follows:

$$Reward_{local} = \max(score_{sc} \cdot (1 - \frac{num^t}{c}), 0) \tag{2}$$

where $score_{sc}$ is the scaled POI score, which depends on each POI's visit number and overall trip process, aiming to improve the minor POIs' visiting. The intuition is that POIs with less visit will have higher score.

$$score_{sc} = \max(I(x) \cdot Sc(num^t), 1) \cdot score$$

$$I(x) = \frac{1}{1 + e^{-20(x - \lambda_1)}}$$

$$Sc(num^t) = 1 + \frac{20}{(num^t + 1)^4}$$

(3)

where $I(x)$ is the trigger function. x is the percentage of total tour process. $Sc(num^t)$ is the score scaling function. $I(x)$ triggers score scaling function after a certain time. λ_1 is a hyperparameter and set to 0.6 as default, indicating that the score scaling function is triggered from the 60% of total touring process.

Reward Based on Global Congestion: To more fairly distribute the tourists over all POIs, we propose the global congestion reward $Reward_{global}$, which is inversely proportional to the variance of all POIs' attendance percentage:

$$Reward_{global} = \frac{\lambda_2}{Var(ap_1, ..., ap_n)}, \ where \ ap_i = \frac{num_i^t}{c_i}$$

(4)

λ_2 is the hyperparameter and set to 0.1 in our experiment.

5   Experimental Evaluation

5.1   Environment Setting

Mobility Setting. We conduct experiments on Kyoto real data settings: FYP-Data. Details are given as following:

- **F-Data**: We count the number of Flicker users who post photos on Flicker in each POI and different time slots. It represents the Kyoto tourists' mobility and the maximum mobility is about 4000.
- **Y-Data**: This data set is provided by Yahoo Japan Corporation. It includes 6667031 trajectories of 757878 mobile users using Yahoo! JAPAN services in 16 days. The data is pre-anonymized and random noise is added. It represents the Kyoto local residents' mobility and the maximum mobility is about 25000.
- **P-Data**: Based on Y-data, we use a density model [14] to generate trajectory data of 500 pseudo-users to simulate the mobility of random tourists.
- **FYP-Data**: This data is the combination of **Y-Data**, **F-Data** and **P-Data**, which is closer to reality.

Tourists Setting. We generate 100, 200 tourists as small-scale tourists and 500, 1000 tourists as large-scale tourists.

5.2   Baseline Setting

- MARLRR [4] diversifies tourists' routes by dynamic reward function and applies single-agent RL framework.
- Pointer-NN [7] is one state-of-the-art model for single tourist route planning. It constructs model with pointer networks and dynamic graph self-attention.

Table 1. Result based on FYP-Data.

		Gini_{POI}	$\text{Gini}_{tourist}$	Var_{ave}	Var_{max}	ED_r
FYP-Data 100 tourists	RPMTD	**0.612**	**0.168**	**0.036**	**0.067**	**1315.286**
	MARLRR	0.713	0.176	0.038	0.071	1287.936
	Pointer-NN	0.821	0.196	0.275	0.356	69.328
FYP-Data 200 tourists	RPMTD	**0.625**	0.189	**0.032**	**0.071**	2265.351
	MARLRR	0.736	0.191	0.107	0.181	**2313.025**
	Pointer-NN	0.868	**0.187**	0.292	0.478	93.836
FYP-Data 500 tourists	RPMTD	**0.651**	0.193	**0.043**	**0.081**	**3619.283**
	MARLRR	0.692	0.269	0.151	1.498	3592.875
	Pointer-NN	0.921	**0.186**	0.412	1.754	151.802
FYP-Data 1000 tourists	RPMTD	**0.674**	0.216	**0.045**	**0.096**	**5812.231**
	MARLRR	0.851	0.659	0.398	0.967	5762.922
	Pointer-NN	0.948	**0.201**	1.104	3.257	184.064

5.3 Evaluation Metrics

Performance is evaluated by following metrics: Maximum and average variance of all POIs' percentage of attendance (Var_{max} and Var_{ave}) throughout the whole trip; Gini coefficient [13] of POIs (Gini_{POI}) and tourists ($\text{Gini}_{tourist}$); Average of edit distance of planned routes (ED_r).

5.4 Experimental Results

Table 1 shows the results of FYP-Data. RPMTD shows the smallest Gini_{POI}, while MARLRR and Pointer-NN show much larger Gini_{POI}, indicating stronger bias of POI. Pointer-NN shows the smallest $\text{Gini}_{tourist}$ and ED_r, which means that tourists' rewards are similar and routes are homogeneous. RPMTD and MARLRR show comparable $\text{Gini}_{tourist}$ in small-scale tourists scenarios. However, $\text{Gini}_{tourist}$ of MARLRR increases significantly in large-scale tourists scenarios. In terms of Var_{ave} and Var_{max}, RPMTD shows the best performance. Pointer-NN performs much worse than other two models because tourists are planned in similar POIs.

Three models show same tendency in Gini_{POI} that bias always grows in large-scale tourists. MARLRR outperforms us in small-scale tourists for $\text{Gini}_{tourist}$, however, it increases significantly in large-scale tourists. For the Var_{ave} and Var_{max}, MARLRR shows the same unstable performance with large-scalar tourists. Pointer-NN performance indicates that existing single tourist planning method cannot be applied to multiple tourists planning.

6 Conclusion

We introduce the popularity-biased routes planning problem. We solve this problem using our multi-agent RL framework, which includes an environment and a dual-congestion aware model. In our experiments, we consider novel mobility data to make it closer to reality. Experiments with baseline models are conducted. The result reveals the superior performance and robustness of our model. Additionally, our work is the first to plan multiple tourists' routes from the interest of POIs, which could be incorporated with existing models to meet the sustainable development goals for tourism.

Acknowledgements. This work was partly supported by JSPS KAKENHI (23H03404) and MIC SCOPE(201607008).

References

1. Chen, L., Zhang, L., Cao, S., Wu, Z., Cao, J.: Personalized itinerary recommendation: deep and collaborative learning with textual information. Expert Syst. Appl. **144**, 113070 (2020)
2. Padia, P., Lim, K., Cha, J., Harwood, A.: Sentiment-aware and personalized tour recommendation. In: IEEE Big Data, pp. 900–909 (2019)
3. Lim, K., Chan, J., Karunasekera, S., Leckie, C.: Personalized itinerary recommendation with queuing time awareness. In: ACM SIGIR, pp. 325–334 (2017)
4. Kong, W., Zheng, S., Nguyen, M., Ma, Q.: Diversity-oriented route planning for tourists. In: Strauss, C., Cuzzocrea, A., Kotsis, G., Tjoa, A.M., Khalil, I. (eds.) DEXA 2022. LNCS, pp. 243–255. Springer, Cham (2022). https://doi.org/10.1007/978-3-031-12426-6_20
5. Hsieh, H., Li, C., Lin, S.: TripRec: recommending trip routes from large scale check-in data. In: WWW, pp. 529–530 (2012)
6. Gunawan, A., Yuan, Z., Lau, H.: A mathematical model and metaheuristics for time dependent orienteering problem. In: PATAT, pp. 202–217 (2014)
7. Gama, R., Fernandes, H.: A RL approach to the orienteering problem with time windows. Comput. Oper. Res. **133**, 105357 (2021)
8. Gionis, A., Lappas, T., Pelechrinis, K., Terzi, E.: Customized tour recommendations in urban areas. In: ACM WSDM, pp. 313–322 (2014)
9. Taylor, K., Lim, K., Chan, J.: Travel itinerary recommendations with must-see points-of-interest. In: WWW, pp. 1198–1205 (2018)
10. Xu, G., Fu, B., Gu, Y.: Point-of-interest recommendations via a supervised random walk algorithm. IEEE Intell. Syst. **31**, 15–23 (2016)
11. Sylejmani, K., Dorn, J., Musliu, N.: Planning the trip itinerary for tourist groups. Inf. Technol. Tourism. **17**, 275–314 (2017)
12. Sarkar, J., Majumder, A.: gtour: multiple itinerary recommendation engine for group of tourists. Expert Syst. Appl. **191**, 116–190 (2022)
13. Ceriani, L., Verme, P.: The origins of the gini index: extracts from Variabilità e Mutabilità (1912) by Corrado Gini. J. Econ. Inequal. **10**, 421–443 (2012)
14. Pappalardo, L., Simini, F., Barlacchi, G., Pellungrini, R.: scikit-mobility: a python library for the analysis, generation and risk assessment of mobility data. ArXiv Preprint ArXiv:1907.07062 (2019)
15. Schulman, J., Wolski, F., Dhariwal, P., Radford, A., Klimov, O.: Proximal policy optimization algorithms. ArXiv Preprint ArXiv:1707.06347 (2017)
16. Mazyavkina, N., Sviridov, S., Ivanov, S., Burnaev, E.: Reinforcement learning for combinatorial optimization: a survey. Comput. Oper. Res. **134**, 105400 (2021)
17. Vaswani, A., et al.: Attention is all you need. In: NIPS, vol. 30 (2017)

Subspace Clustering Technique Using Multi-objective Functions for Multi-class Categorical Data

Rahmah Brnawy[✉] and Nematollaah Shiri

Department of Computer Science and Software Engineering,
Concordia University, Montreal, Canada
{r_brnawy,shiri}@encs.concordia.ca

Abstract. Clustering categorical data has been a challenging task. Most known clustering techniques for such data often achieve excellent performance over binary data, however, their performance on multi-class data has not been as good. We believe this could be improved by taking into account the overlapping and imbalance issues in the clustering process. This would allow detecting small and informative clusters as well, and hence improved cluster quality. In this paper, we introduce a novel subspace technique for multi-class data. For this, we use the notion of entropy from information theory in conjunction with Coa's density function to identify and combine potential sub-clusters from the same classes, revealing smaller clusters. We compare our results to those of well-known subspace clustering approaches, density-based algorithms, and ensemble clustering. The results of our numerous experiments indicate that our proposed technique outperforms the majority of existing solutions.

Keywords: Clustering · Categorical data · Density function · Entropy · Overlapping

1 Introduction

Categorical data is conceptual and inherits layer structure, so it makes sense to analyze it based on its attributes to find clusters in the subspace, which resulted in the development of subspace-based clustering algorithms. These algorithms can be divided into two groups: weighted-based and standalone clustering algorithms. The algorithms in the first group embed the attribute weight in the clustering process. The assigned weight denotes the importance of each categorical attribute and is used to control the contribution of attribute-cluster similarity to object-cluster similarity. In this technique, weight is calculated as the weighted distance between the object and the mode categories of the cluster on each attribute. We note that most weighted-based algorithms were successful when applied to binary data. However, their applications to multi-class data were not as successful. This weakness, we believe, is rooted in two problems. Firstly, the

C. Strauss et al. (Eds.): DEXA 2023, LNCS 14147, pp. 337–343, 2023.
https://doi.org/10.1007/978-3-031-39821-6_28

initial clustering used in those algorithms is mainly based on either k-modes or fuzzy k-modes. The underlying assumption of such algorithms is that clusters are "cohesive" and mapped around a real or virtual center. However, the cohesion degree in multi-cluster data could vary. Obviously, such distance computation methods can easily lead to loss of non-mode categories in small and scattered clusters [4]. In contrast to the algorithms in the first group, standalone-based algorithms avoid forming clusters around the modes. Instead, they sequentially scan the entire space and search for dense clusters in subspaces. After identifying all potential dense clusters, an integration process is used to generate the desired number of clusters. Examples of popular such algorithms include ROCK, COOLCAT, and ROCAT [5], to name a few. Those algorithms have the following drawbacks. They need a number of parameters to define dense regions, and setting these parameters poses challenges [5]. Another drawback is that small clusters are lost during the merging phase. This is because the heuristic search used generates a large number of redundant sub-clusters [4,5]. Furthermore, these algorithms are affected by the order of data objects or attributes [5].

In this paper, we propose a new subspace algorithm to address the above problems. For this, we study the application of entropy and a density function at every iteration of the clustering process. First, we present an iterative technique for detecting sub-clusters and overlapping clusters. We then merge sub-clusters (dense clusters) and re-cluster the overlapping clusters. This process is repeated until the necessary number of clusters is obtained. Following that, using expected entropy, non-clustered objects are reassigned to the closest cluster.

The rest of this paper is organized as follows. In Sect. 2, we introduce our proposed techniques. The experiments and results are reported in Sect. 3. Concluding remarks and future work are presented in Sect. 4.

2   The Proposed Multi-objective Technique

Our proposed clustering technique is built based on two assumptions. First, categorical data is complex and has a layer structure. Therefore, hidden patterns may be difficult to find using a single clustering objective function [4]. Second, the detection of small clusters is less likely to happen during the early steps of the process as they can be easily hidden inside larger clusters. Based on these assumptions, Cao's algorithm [1] and the entropy concept are used, as they employ functions that are both adequate and complementary. Cao's algorithm creates clusters based on the frequency of attribute values and matching similarity. The ROCAT [5] algorithm used the Shannon entropy to find overlapping clusters based on attribute homogeneity. Both algorithms use categorical data properties to define clusters; thus, we believe their combined application is appropriate and beneficial. Moreover, they complement each other and avoid their weaknesses. That is, Cao's algorithm outputs sub-clusters of a golden cluster as primary clusters [4], whereas ROCAT [5], on the other hand, sequentially searches the entire space for such clusters. Therefore, instead of searching the entire space, we localize the search within the clustering solution found.

Before going into the details, let's define some notations. Let $D = \{x_1, x_2, \cdots, x_n\}$ be a categorical dataset of n objects describes by m features. $F = \{f_1, f_2, \cdots, f_m\}$. V_j is domain of attribute $f_j \in F$; $V_j = \{v_{j1}, v_{j2}, \cdots, v_{jr}\}$, where r is the number of categories of attribute f_j.

2.1   Cao Clustering Algorithm

The Cao algorithm [1] is an extension of the k-modes clustering algorithm. The algorithm seeks the cluster center of categorical data by calculating the average density of the objects. Given a dataset D, Cao's algorithm partitions the data into K clusters as follows: it selects the first center which has the highest density in the dataset, computed as the following

$$Dens(x) = \sum_{f \in F} Dens_f(x)/|F| \tag{1}$$

where $Dens_f(x)$ is the density of objects x in D with respect to f, given by

$$Dens_f(x) = \frac{|\{y \in D | Val(x, f) = Val(y, f)\}|}{|D|} \tag{2}$$

For the rest of $k - 1$ centers, both density and the distance from the selected centers x_m are applied.

$$center(c) = d(x_i, x_{cen}) \times Dens(x_i) \tag{3}$$

where $d(x_i, x_{cen})$ is the simple matching method.

2.2   Step One: Finding and Analyzing the Clusters

Identifying Sub-clusters. This is the first step of the proposed technique. Figure 1 illustrates the idea. Suppose we apply the Cao algorithm on the dataset D and obtain the clusters $\Pi = \{C_1, C_2, C_3, C_4\}$. For each cluster $C_i \in \Pi$, we identify the subset of features that contributed to the discovery of the corresponding cluster [5]. We refer to these features as relevant features of C_i, denoted as $Relv(C_i)$, which are selected based on their level of variability in the clusters. To measure the variability of a categorical feature, we follow $ROCT$ and use Shannon entropy. Then,

$$Relv(C_i) = \{f_j \mid H(f_j \mid C_i) < \Delta, \forall f_j \in F\}$$

where $H(f_j \mid C_i)$ is the entropy of feature f_j in cluster C_i computed as :

$$H(f_j \mid C_i) = -\sum_{q=1}^{|V_j|} P(v_{jq} \mid C_i) \log P(v_{jq} \mid C_i) \tag{4}$$

and $P(v_{jq} \mid C_i)$ is the probability of objects in the cluster C_i obtaining the same attribute values

$$P(v_{jq} \mid C_i) = \frac{|\{x_i \in C_i \mid x_{ij} = v_{jq}\}|}{|C_i|} \tag{5}$$

As shown in Fig. 1, $Relv(C_1) = \{f_1, f_2, f_3, f_4, f_7, f_9\}$ is the set of relevant features of C_1 because these features have fewer categories, indicated by different colors. We do the same to clusters C_2 and C_3. Depending on the entropy threshold, relative features divided the clusters into equivalence classes[1]. Large equivalence classes are considered dense sub-clusters; equivalence classes with only one item are deleted and relocated in the last step, and the remaining equivalence classes proceed to the second iteration. Consider the value C_2 in Fig. 1. The first two items form a dense sub-cluster, whereas the latter two are redistributed in the final step. The proposed technique is based on a density function, and hence we expect some clusters to be spatially sparse. We refer to such clusters as overlapped clusters. These clusters are described by a small number of relative features or the full subspace F. In the literature, such clusters are considered uninteresting [6]. However, our view is to consider such clusters to be composed of smaller clusters that are scattered throughout the space. Consequently, a detailed analysis may be further applied to identify them. C_4 in Fig. 1 is an illustration of overlapping clusters.

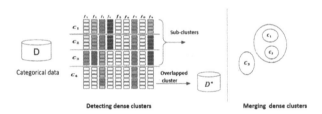

Fig. 1. Overview of the proposed technique

Merging Dens-Clusters. Following each iteration, the similarity between the current dens clusters is computed. To achieve this, we used the merge function proposed in [2], which is given as follows:

$$sim(C_x, C_y) = \prod_{i=1}^{|F_{xy}|} \left[\sum_{j=1}^{|V_i|} \min\{P(v_{ij} \mid C_x), P(v_{ij} \mid C_y)\} + \epsilon \right] \tag{6}$$

where $|F_{xy}|$ is size of the union of the relative features, $P(v_{ij}|C_x)$ denotes the probability of the jth value for the ith feature in cluster C_x, $|f_i|$ the size of the domain value of the ith feature, and ϵ is a very small value. A higher similarity

---

[1] A set of objects share strong associations defined by their relative features.

score means that the distribution of attribute values in two clusters is similar, and should be merged. Note that the transitive concept to group the dense clusters is applied. In Fig. 1, C_1 and C_2 can be merged since they are similar; however, C_3 has low similarity with both clusters.

2.3   Step Two: Reassigning Non-clustered Objects

Some objects may not be assigned to any cluster. Those objects are assigned to a cluster to which they have the highest similarity. Suppose that from step one, we obtained K clusters $C^K = \{C_1, \ldots, C_k\}$ where each C_i has n_i objects. Let $X = \bigcup C^K$ and d be the union of the relative feature of C^K. One of the solutions to find the best cluster is expected entropy, computed as the following [3]

$$O(C^K) = \frac{1}{d}\left(\hat{H}(X) - \frac{1}{n}\sum_{i=1}^{K} n_i \hat{H}(C_i)\right) \qquad (7)$$

where

$$\hat{H}(X) = -\sum_{i=1}^{d}\sum_{j=1}^{|V_i|} p(v_{ij} \mid C) \log p(v_{ij} \mid C) \qquad (8)$$

Using the above, objects are assigned to the cluster, resulting in minimal changes to the overall clustering structure C^K.

3   Experiments and Results

In this section, we present the experiments conducted to study the performance of the proposed multi-objective techniques and report the results obtained using well-known datasets from the UCI machine learning repository. Table 1 provides a brief description of these datasets. To evaluate the overall performance of the proposed technique, we followed the literature and used three well-known evaluation measures: clustering accuracy (AC), and normalized mutual information (NMI) measures. In Tables 2 and 3, we compare the performance of the proposed technique against well-known clustering techniques. The results showed that the proposed techniques yield high AC and NMI in all datasets except Mushrooms, where KALM provides better results. As the proposed technique

Table 1. Descriptions of the datasets

Datasets	Zoo	Dermatology	lymphoma	Mushroom
No.Objects	101	366	148	8123
No.Attributes	16	35	18	23
Classes	7	6	3	2

Table 2. Comparison of the proposed techniques based on AC

Dataset	Proposed	Al-Razgan	KALM	FKMAWCW	Khan	COOLCAT
Dermatology	**0.9261**	0.6180	0.7889	0.7926	0.6175	*
Zoo	**0.8996**	0.7907	0.8476	0.8218	0.8911	0.785
Lymphoma	**0.7260**	0.0061	0.4624	0.6139	0.5068	0.574
Mushroom	0.8132	0.7474	**0.8496**	0.8182	0.8288	0.531

requires three parameters, we tested the sensitivity of the suggested technique to changes in parameters. Figure 2 shows the effect of entropy over the clustering results. For the minRelf parameter, we observed that $minRelf = 5$ works for all three dataset datasets except the Zoo, for which we set the value to 10. Finally, for the merger threshold, we followed the author's suggestion [2] and used 0.9 as the default value.

We also examine the proposed technique to identify smaller clusters. The confusion matrix in Fig. 4, shows that the proposed technique effectively detected the entire cluster C_0 in dermatology. In the case of the Zoo dataset, we can see that the proposed solution accurately discovered all four objects in cluster C_5. Although they are merged with the objects in cluster C_3, this is much more effective than merging with huge clusters, as in the case of Cao's technique. The same applies to the Lymphoma dataset where our technique output the smaller cluster C_0 with very high purity, unlike Cao's algorithm (Fig. 3).

Table 3. Comparison of the proposed techniques based on NMI

Dataset	Proposed	Al-Razgan	KALM	FKMAWCW	Khan	COOLCAT
Dermatology	**0.8811**	0.7769	0.7418	0.7466	0.6175	*
Zoo	**0.9027**	0.8150	0.8485	0.7806	0.8911	0.6615
Lymphoma	**0.4793**	0.0061	0.2020	0.4221	0.4226	0.05
Mushroom	0.3812	0.4323	**0.4766**	0.4053	0.19155	0.2812

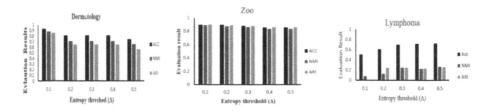

Fig. 2. Proposed sensitivity to the entropy value

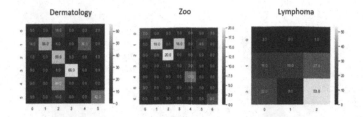

Fig. 3. The confusion matrix of clustering results by Cao's algorithm

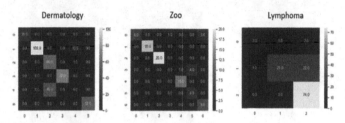

Fig. 4. The confusion matrix of clustering results by the proposed technique

4   Conclusion and Future Work

We introduced a new multi-objective function for clustering categorical data. The proposed technique is intuitive, straightforward, and yet powerful. The results of our experiments and analyses indicated the efficiency of the proposed technique in handling multi-class data. Currently, we are working on more effective ways to solve the entropy threshold problem. We also look into employing classifiers to handle imbalance issues and study the impact this could have on clustering performance.

References

1. Cao, F., Liang, J., Bai, L.: A new initialization method for categorical data clustering. Expert Syst. Appl. **36**(7), 10223–10228 (2009)
2. Chang, C.H., Ding, Z.K.: Categorical data visualization and clustering using subjective factors. Data Knowl. Eng. **53**(3), 243–262 (2005)
3. Chen, K., Liu, L.: The "best k" for entropy-based categorical data clustering (2005)
4. Chen, L., Wang, S., Wang, K., Zhu, J.: Soft subspace clustering of categorical data with probabilistic distance. Pattern Recogn. **51**, 322–332 (2016)
5. He, X., Feng, J., Konte, B., Mai, S.T., Plant, C.: Relevant overlapping subspace clusters on categorical data. In: Proceedings of the 20th ACM SIGKDD International Conference on Knowledge Discovery and Data Mining, pp. 213–222 (2014)
6. Müller, E., Assent, I., Günnemann, S., Krieger, R., Seidl, T.: Relevant subspace clustering: mining the most interesting non-redundant concepts in high dimensional data. In: 2009 Ninth IEEE International Conference on Data Mining, pp. 377–386. IEEE (2009)

Neural Networks

Multi-task Graph Neural Network for Optimizing the Structure Fairness

Jiahui Wang, Meng Li, Fangshu Chen$^{(\boxtimes)}$, Xiankai Meng, and Chengcheng Yu

School of Computer and Information Engineering, Shanghai Polytechnic University, Shanghai, China
{wangjh,20221513096,fschen,xkmeng,ccyu}@sspu.edu.cn

Abstract. Graph neural networks, the mainstream paradigm of graph data mining, optimize the traditional feature-based node classification models with supplementing spatial topology. However, those isolated nodes not well connected to the whole graph are difficult to capture effective information through structural aggregation and sometimes even bring the negative local over-smoothing phenomenon, which is called structure fairness problem. To the best of our knowledge, current methods mainly focus on amending the network structure to improve the expressiveness with absence of the influence of the isolated parts. To facilitate this line of research, we innovatively propose a Multi-task Graph Neural Network for Optimizing the Structure Fairness (GNN-OSF). In GNN-OSF, nodes set is divided into diverse positions with a comprehensive investigation of the correlation between node position and accuracy in global topology. Besides, the link matrix is constructed to express the consistency of node labels, which expects isolated nodes to learn the same embedding and label when nodes share similar features. Afterward, the GNN-OSF network structure is explored by introducing the auxiliary link prediction task, where the task-shared and task-specific layer of diverse tasks are integrated with the auto-encoder architecture. Our comprehensive experiments demonstrate that GNN-OSF achieves superior node classification performance on both public benchmark and real-world industrial datasets, which effectively alleviates the structure unfairness of the isolated parts and leverages off-the -shelf models with the interaction of auxiliary tasks.

Keywords: Graph Neural Network · Structure fairness · Over-smoothing · Isolated parts · Multi-task Learning

1 Introduction

Driven by various social needs, Internet application scenarios are showing an explosive growth trend [1]. The data entities behind not only include their

Supported by the National Natural Science Foundation of China (Grant No. 62002216), the Shanghai Sailing Program (Grant No. 20YF1414400), the Shanghai Polytechnic University Research Projects (Grant No. EGD23DS05).

C. Strauss et al. (Eds.): DEXA 2023, LNCS 14147, pp. 347–362, 2023.
https://doi.org/10.1007/978-3-031-39821-6_29

internal feature information but also contain a large number of complicated topological associations [2]. Therefore, the direction of academic research is migrating to the non-Euclidean graph structure [3,4]. While current data mining theoretical systems are mainly established on the structured Euclidean data, the effective integration of topological structure is still in the exploratory stage.

Recently, Graph Neural Networks (GNN) break the limitations of conventional models without graph structure by neighborhood aggregation with a tremendous capability of node representation. However, there are significant differences in the prediction effect between the isolated part(small sub-graph or single nodes) and the main part(tightly connected nodes). For example, GNN got 70% accuracy (60.8% of isolated part, 74.7% of main part) on the widely discussed benchmark Citeseer dataset according to statistics [5]. Thus, there is strong evidence that the isolated part significantly restricts the capability of the GNN model with 13.9% lower on accuracy metric.

In-depth exploration of the reason, GNN cannot capture effective information for isolated parts through conventional structure aggregation, which is called structure fairness problem. Also, the isolated parts may cause the negative over-smoothing problem that the node representation will converge to the local sub-graph representation [6]. While, most models focus on optimizing the network structure from the global perspective, especially the neighborhood convolutional aggregation strategies, which could obtain more accurate latent feature representation for classification. To our best knowledge, very limited work pays attention to the isolated part that significantly restricts the performance of GNN.

Motivated by the above observation, this paper innovatively proposes the multi-task Graph Neural Network for Optimizing the Structure Fairness (GNN-OSF), which is dedicated to alleviating the negative influence of the isolated part in the classical node classification task on the homogeneous graph.

Considering that the information captured is very limited to the isolated part, GNN-OSF intends to introduce an auxiliary link prediction task to describe the consistency of the label predicted with the high-level latent node embedding, which enhances the feature representation of the isolated parts. Through collaborative training of multiple tasks, GNN-OSF is expected to learn a more accurate and extensive model with information sharing with different tasks.

With the link prediction task, those nodes pairwise containing the isolated nodes and their similar feature nodes expect to be connected by some particular links, indicating that they are highly likely to share the same label. Besides, pairwise composed of the training nodes are also utilized to describe the consistency of labels in the link matrix, where the positive and negative ones mean consistency and inconsistency, respectively. By introducing the multi-task mechanism, the original N node-level comparison items have been expanded with N^2 edge-level comparison items, which exponentially expands the number of comparison items in the loss function and enables the model to learn more information. On the other hand, the interaction of different tasks might be conducive to the node representation. The experiment also gives strong evidence that GNN-OSP achieves relatively high performance compared with other off-the-shelf models. In summary, this work makes the following contributions:

1. Investigate the correlation between node position and classification accuracy in global topology, and deliver a graph partition algorithm to divide the nodes into diverse positions.
2. Construct a link matrix to express the consistency of labels as the reference standard for link prediction tasks, which expects isolated nodes to learn the same embedding and label as nodes with similar features.
3. Propose a novel multi-task graph neural network for optimizing the structure fairness, where the task-shared and task-specific layer of diverse tasks are integrated with the auto-encoder architecture.
4. Establish clear superiority of GNN-OSF through comprehensive experimental results that link matrix and multi-task mechanism can learn more accurate node embedding and enhance the expressive capability.

2 Related Work

Compared with traditional structured Euclidean data, graph data supplements the spatial topological relationship based on the inherent node features. Current data mining theoretical systems are mainly established on the structured data directly extracted from relational databases, but the effective integration of topological structure in previous models is still in the exploratory stage.

Motivated by some previous exploration work on convolution definition, Henaff [7] established the general definition of the convolution operation. The first-generation convolution kernel was defined as a diagonal matrix of eigenvalues to simplify the sophisticated process of spectral decomposition process. Afterward, Defferrard [8] utilized the Tylor and Chebyshev polynomials to propose the second and third-generation convolution kernels, which improved the aggregation capability. As a milestone, Kipf [9] introduced a novel approach for semi-supervised classification on graph-structured data, which injected new vitality into the field of graph deep learning.

Although vanilla GCN has the advantage of low computational complexity, the aggregation process is only simply to superimpose neighbor information, which causes the pitfall of node representation and restricts the performance. To solve the problem, researchers focused on optimizing the subprocesses of the initial model. The GraphHeat [10] and PPNP [11] models utilized the Heated kernel and PageRank function to revise the original linearly connected convolution layer. GCNII [12] added the self-mapping and residual connections mechanisms. GWNN [13] replaced the original Fourier transform with wavelet transform, which solved the defect of Fourier transform in locality representation. SGC [14] combined multiple convolutional layers into one layer, which could greatly reduce the number of parameters and achieve similar effects.

Considering that the essence of graph convolution is the aggregation of neighbor nodes, Hamilton [15] pioneered the inductive framework GraphSAGE, which can naturally generalize to unseen nodes. Different from the spectral method, the eigendecomposition process of the Laplacian matrix has been completely abandoned, which can be regarded as the spatial graph model. GAT [16] and UniMP

[17] took into account the different contributions of neighbors, and then integrated the attention mechanisms. PNA [18] and CPAN [19] integrate potential information of nodes for signal scaling and Attention aggregation.

Since the advent of GNN, node classification with GNN as a basic task has attracted worldwide attention. A large number of scholars have perfected related theories from different perspectives. While, compared with the nodes in the main body, those nodes in the isolated marginal groups are more likely to converge to the subgraph representation. For those nodes in the marginal part, the structure might be meaningless and even cause obstruction. So, the model will unavoidably face the challenge of the structure fairness problem.

To overcome this pitfall, DropEdge [20] utilized the strategy of drop edge and JK-NET [21] concatenated the convolution layers, which considered the difference between the k-order neighborhood of the marginal and core edges. PairNorm [22] designed a regular term to keep the features intra-community concertation and inter-community scatter. P-reg [23] explored a new node-based graph regularization, instead of the traditional edge-based Laplacian regularization. MADGAP [24] introduced MAD metrics to quantify the smoothness.

However, to our best knowledge, current models only consider the global over-smoothing problem, most of them still have little effect when dealing with the local over-smoothing phenomenon of structure fairness issue. Under this situation, it is difficult for GNN models to distinguish the nodes with different labels in the same isolated group. The position will extremely affect the reliability of the feature and structure. Therefore, it is important to optimize the structure fairness of GNN, especially the isolated parts.

Besides, most approaches merely focus on the classification task itself with the absence of other relevant tasks to enhance the capability of models. Therefore, this paper intends to leverage the performance of the isolated parts with the multi-task learning mechanism.

3   Methodology

3.1   Overview of Architecture

Generally, the neighbor aggregation with graph convolution is beneficial for extracting high-level feature representation of nodes. While, this method is not always effective and sometimes even brings negative impact for those isolated nodes, which shows the phenomenon of structure unfairness of GNN models.

To deal with this issue, we innovatively propose a novel Multi-task Graph Neural Network for Optimizing the Structure Fairness (GNN-OSF), which leverages the feature expression of isolated nodes. GNN-OSF breaks the limitation of traditional models that ignore the isolated parts such as [9–12], which significantly improves the representation and learning ability.

GNN-OSF consists of position partition, link matrix construction, and multi-task graph networks module as shown in Fig. 1. Specifically, GNN-OSF first divides the original graph into several diverse connected groups with position information, and then partitions the nodes into main parts and isolated parts.

Afterward, the link matrix is constructed with the similarity to express the consistency of labels, which expects isolated nodes to learn the same embedding and label as nodes with similar features. The link matrix is also utilized as the ground truth of the link prediction auxiliary task, which is trained cooperatively with the conventional node classification task. Under this circumstance, the isolated part can learn more effective embedding representation for achieving more accurate classification, since the interaction and collaborative training of different tasks enriches the expressive ability of the model.

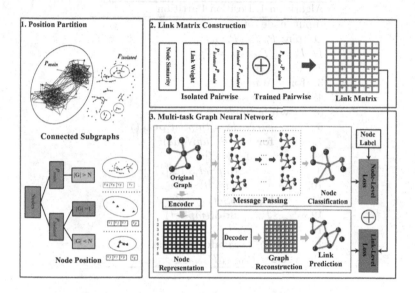

Fig. 1. GCN-OSF architecture framework

3.2   Position Partition

Typically, those nodes in the main body are closely connected to other parts. After the graph convolution procedure, better node embedding representations can be obtained with the supplement of topological structure information. However, for the nodes in the isolated part, the node representation will easily converge to the subgraph representation. Therefore, the diverse positions will significantly affect the predictive performance.

The given graph is defined as $G(V,E)$, where $V = \{v_1, v_2, ..., v_n\}$ and $E = [(v_{s1}, v_{e1}), (v_{s2}, v_{e2}) ... (v_{sm}, v_{em})]$ represent the vertex and edge set, respectively. Besides, each node in the graph has a corresponding d-dimensional feature vector $x_i \in \mathbb{R}^d$ and $X = [x_1, x_2, ..., x_n]$ denotes the entire feature matrix. Then, the binary adjacency matrix $A \in \mathbb{R}^{n \times n}$ and the degree matrix $D \in \mathbb{R}^{n \times n}$ could be easily calculated by the edge information.

In response to the aforementioned problems, GNN-OSF divides the nodes into the main part P_{main} and isolated parts $P_{isolated}$ as Algorithm 1. First, the deep-first-search strategy is exploited to find all connected subgraphs of the graph. Specifically, the nodes are traversed according to the adjacency matrix $A \in \mathbb{R}^{n \times n}$ of the graph. When the current node v is traversed, its neighbor nodes $R(v)$ are visited and added to the subgraph according to the deep-first-search order until the maximal connected subgraph is constructed. If the number of nodes in the subgraph is greater than a specific threshold N, the connected subgraph is added to the main part of the graph $P_{main} \in \mathbb{R}^n$. The remaining connected subgraphs are then divided into the isolated part.

Algorithm 1: Position Partition

Input: node set V

Output: P_{main}, $P_{isolated}$

1. $P_{main}, P_{isolated} = \{\}, \{\}$
2. $visit = [1] * N$
3. **for** v **in** $nodeSet$:
4. $tmpSet.\mathbf{add}(v)$
5.   **while** $tmpSet$:
6. $curNode = tmpSet.\mathbf{pop}.()$
7.   **for** t **in** $R(curNode)$:
8.   **if** $visit[t]$:
9. $tmpSet.\mathbf{add}.(t)$
10. $visit[t] = 0$
11.   **if** $\mathbf{len}(tmpSet) \geq N$:
12. $P_{main}.\mathbf{add}(tmpSet)$
13.   **else**:
14. $P_{isolated}.\mathbf{add}(tmpSet)$
15. **return** P_{main}, $P_{isolated}$

3.3   Link Matrix Construction

GNN-OSF intends to enhance the feature embedding and weaken the invalid structural information of the isolated parts. Therefore, the link matrix L is constructed to describe the consistency of labels, which expects isolated nodes to learn the same embedding and label as nodes with similar features. Input as the reference standard of link prediction task, the link matrix is helpful in capturing more effective embedding representation for alleviating the structural unfairness and reducing the negative impact of the local over-smoothing phenomenon. According to the node position, the construction of link matrix L is divided into two components, where L_{iso} includes $P_{isolated} - P_{main}$, $P_{isolated} - P_{isolated}$ node pair-wises, and L_{lab} includes $P_{train} - P_{train}$ node pair-wises.

For the L_{iso} component, GNN-OSF calculates the similarity $sim(x_i, x_j)$ between nodes as Eq. (1), where x_i, x_j are the node features corresponding to v_i, v_j in the Euclidean feature space, each component is expressed as x_{ik}, x_{jk}. According to the value range, feature variables can be divided into two categories: continuous variables and discrete variables. The discrete data with one-hot encoding will utilize Jaccard similarity to calculate the distance between nodes v_i, v_j, which measures the number of the same attributes. Besides, cosine similarity is utilized to measure the similarity of continuous data. Afterward, the link matrix L_{iso} can be set as ρ with the comparison with the preset threshold t as Eq. (2).

$$sim(x_i, x_j) = \begin{cases} \dfrac{|x_i \cap x_j|}{|x_i \cup x_j|} & x_i, x_j \in \{0,1\} \\ \dfrac{x_i \cdot x_j}{||x_i|| \times ||x_j||} & otherwise \end{cases} \tag{1}$$

$$L_{iso}[i][j] = \begin{cases} 0 & if\ sim(x_i, x_j) < t \\ \rho & if\ sim(x_i, x_j) \geq t \end{cases} \tag{2}$$

For the L_{lab} component, the link weight matrix is calculated according to the labels' consistency of the data node of the training set. For those nodes that share the same labels, the corresponding value of the link matrix $L_{lab}[i][j]$ is 1, which is equivalent to strengthening the structural information between the two nodes. If the labels are opposite, the value is -1, which is regarded as weakening the structural information between nodes. The formulation is shown in Eq. (3)

$$L_{lab}[i][j] = \begin{cases} +1 & if\ lab((i) == lab(j)\ and\ i,j \in P_{train} \\ 0 & otherwise \\ -1 & if\ lab(i)! = lab(j)\ and\ i,j \in P_{train} \end{cases} \tag{3}$$

$$L[i][j] = L_{lab}[i][j] + L_{iso}[i][j] \tag{4}$$

3.4 Multi-task Graph Neural Network

GNN-OSF expects those nodes with a high probability of sharing the same label can learn similar representations before classification layer. So, we construct a novel network structure to optimize the high-level latent embedding with the integration of the link matrix. GNN-OSF consists of three key components: node embedding, message passing, and graph reconstruction. The pseudocode is shown in Algorithm 2, and the details will be elaborated on below.

First, the node embedding $f_e(x)$ is extracted with the strategy of prediction then propagation [11]. As regular, the ReLU activation function is added for non-linear representation and the dropout mechanism is designed for preventing the overfitting phenomenon in Line 1–6.

Afterward, the transmission of node information on the graph is simulated in the message-passing module. In GNN-OSF, the propagation progress has been defined with the PageRank matrix. The propagation progress could be expressed by Eq. (5), where $A \in \mathbb{R}^{n \times n}$ and $D \in \mathbb{R}^{n \times n}$ denote the adjacency and degree matrix of the graph, respectively.

$$Z^{(k)} = (1 - \alpha)AD^{-1}Z^{(k-1)} + \alpha f_e(x) \tag{5}$$

Furthermore, the node vectors have been decoded into the graph structure in the graph reconstruction module, which attempts to predict whether there is

---
Algorithm 2:GNN-OSF

Input: Graph Data

Output: $L_{nodeCls}, L_{edge}$

1. //**Node Embedding**

2.$X^{(1)} = \textbf{ReLU}(XW_1)$

3.**dropout**$(X^{(1)})$

4.$X^{(2)} = \textbf{ReLU}(X^{(1)}W_2)$

5.$Z^{(0)} = f_e(X) = \textbf{dropout}(X^{(2)})$

6. //**Message Passing**

7.$for\ k\ in\ range(2, K):$

8. $Z^{(K)} = (1 - \alpha)AD^{-1}Z^{(k-1)} + \alpha Z^{(0)}$

9.$\hat{C}(x_i) = \textbf{softmax}((1 - \alpha)AD^{-1}Z^{(k)} + \alpha Z^{(0)})$

10.$L_{nodeCls} = \sum_i loss(\hat{C}(x_i), C(x_i)),\ i \in S_{train}$

11. //**Graph Reconstruction**

12.$\hat{L}(i, j) = \textbf{tanh}(f_e(x), f_e(x)^T)$

13.$L_{edge} = \sum_i \sum_j loss(\hat{L}(i, j), L(i, j)),\ i, j \in S_{train}$

14. $return\ L_{nodeCls}, L_{edge}$

---

an edge between two nodes. In the network structure, the dot decoder approach is conducted to transform the node representation into the link matrix $\hat{L}$ as Eq. (6). Considering the scope of $L(i, j)$, the activated function is utilized to perform non-linear representation and scale the value between -1 and 1.

$$\hat{L}(i, j) = \textbf{tanh}(f_e(x)f_e(x)^T) \tag{6}$$

For classification, the node vector output from the message passing is then fed into a softmax classifier to predict its label, and the cross-entropy loss function is utilized to minimize the error of node multi-classification as Eq. (7), where N_t denote to the number of training set.

$$Loss_{nodeCls} = -\frac{1}{N_t} \sum_{i=1}^{N_t} C(i) \log \hat{C}(i) \tag{7}$$

Besides, GNN-OSF introduces the link matrix to express the labels' consistency with link prediction task. Thus, the edge-level loss is defined as Eq. (8) that minimizes the difference between the generated graph structure and link matrix, where N_{link} denotes the number of virtual links in the link matrix L.

$$Loss_{edge} = \frac{1}{N_{link}(N_{link} - 1)} \sum_{i=1}^{N_{link}} \sum_{j=1, j \neq i}^{N_{link}} (\hat{L}(i, j) - L(i, j))^2 \tag{8}$$

Essentially, $Loss_{nodeCls}$ pays attention to the feature information, which minimizes the prediction label and ground truth of the training set. From another point of view, $Loss_{edge}$ takes into account the information of the virtual edge formed by the training node, which expects the nodes with the same label to obtain a similar representation. Each task is treated as equal in GNN-OSF and the final loss is defined as Eq. (9) to train the model cooperatively.

$$Loss = Loss_{nodeCls} + Loss_{edge} \tag{9}$$

4   Experimental Analysis and Discussion

4.1   Data Description

To evaluate the effectiveness, we evaluate GNN-OSF on both experimental and industrial datasets in Table 1. For the public datasets, we select three different citation networks: Cora, Citeseer, and Pubmed, which have been widely compared in the literature on graph convolutional networks. Furthermore, we implement our model in the industrial financial scenario of fraud detection. The Fraud dataset consists of 20 consumption and operation features extracted from the real business data provided by the famous e-commerce payment platform.

As illustrated in Table 1, a very limited amount of labeled data is available for training in the citation network. Therefore, the model is confronted with the problem of lacking sufficient annotated data. Besides, the model is expected to detect more highly suspicious hidden users from very limited fraudulent users that have been discovered in the fraud detection task. So, we partition the Fraud dataset with different proportions for testing the robustness, especially the small training scale.

Table 1. Data Description

Domain	Dataset	Nodes	Features	Class	Edges	Train/Test	Labeled
Citation Network	Cora	2708	1433	7	5278	140/1000	5.15%
	Citeseer	3327	3703	6	4552	120/1000	3.61%
	Pubmed	19717	500	3	44324	60/1000	0.30%
Fraud Detection	Fraud	12917	20	2	588699	/	/

4.2   Performance of GNN-OSF

For evaluation, we compare GNN-OSF with some current frontier models on the benchmark datasets in Table 2. Node2vec [25], Planetoid [26], and GPNN [27] are three classic methods. PKGCN-JS [28], HLHG [29], and MGCN [30] are very recent algorithms published in higher impact factor journals and GCN [8], SGC [14], GAT [16], and PPNP [11] are four well-recognized methods proposed in the latest influential conferences. The models with asterisks are reproduced by ourselves, and the others are directly quoted from the original papers.

Table 2 reveals that GNN-OSF achieves relatively high accuracy than current models. The accuracy has been improved by 1.14 and 1.09 percent on Cora and Citeseer datasets, respectively. Under the multi-task mechanism, 19,460 edge-level comparison items of high reliability ($P_{train} - P_{train}$ pairwise) have been added. While in the Pubmed dataset, only 3,540 edge-level comparison items have been introduced, where the proportion of the supplementary information is too low compared with the original 19,717 nodes and 44,324 edges. Thus, the improvement on the Pubmed dataset is not as significant as that on Cora and Citeseer datasets.

It is worth mentioning that GNN-SAT [31] also transfers the multi-task learning mechanism to the fundamental graph data mining tasks. While, the experiments also indicate GNN-OSF has achieved better performance than GNN-SAT.

Table 2. Performance on Citation Network

Methods	Datasets					
	Cora		Citeseer		PubMed	
	Fixed	Random	Fixed	Random	Fixed	Random
Node2vec	74.9	72.9	54.7	47.3	75.3	72.4
Planetoid	75.7	–	64.7	–	77.2	–
GPNN	81.8	79.9 ± 2.4	69.7	68.6 ± 1.7	79.3	76.1 ± 2.0
PKGCN-JS	82.94	–	70.61	–	77.44	–
HLHG	82.7 ± 0.29	–	71.5 ± 0.39	–	79.3 ± 0.39	–
MGCN	82.6 ± 0.53	–	72.1 ± 0.44	–	79.4 ± 0.20	–
GNN-SAT	82.17 ± 0.09	–	71.14 ± 0.12	–	79.33 ± 0.07	–
GAT*	83.1 ± 0.4	81.0 ± 1.4	70.8 ± 0.5	69.2 ± 1.9	78.5 ± 0.3	78.3 ± 2.3
GCN*	80.86 ± 0.62	78.78 ± 1.93	70.26 ± 1.00	67.04 ± 1.52	79.15 ± 0.36	76.89 ± 2.27
PPNP*	83.30 ± 0.97	81.27 ± 1.99	71.70 ± 0.78	68.88 ± 1.24	79.28 ± 0.87	**78.98 ± 2.57**
GNN-OSF*	**84.44 ± 0.50**	**81.69 ± 2.00**	**72.61 ± 0.93**	**70.41 ± 1.65**	**79.45 ± 0.43**	78.79 ± 2.59

4.3 Ablation Experiments

Moreover, we further demonstrate the effectiveness of introducing the multi-task learning mechanism and optimizing the structure fairness. From the perspective of model structure, GNN-OSF is regarded as an optimized version of conventional single-task GNN. Hence, we also conduct the ablation experiments on the benchmark datasets, where GNN-ML introduce the multi-task mechanism with the link matrix L_{lab} that only considers the $P_{train} - P_{train}$ pairwise and GNN-OSF further expand the link matrix as $L_{lab} + L_{iso}$ that supplement the $P_{isolated} - P_{isolated}$, and $P_{isoalted} - P_{main}$ pairwise. The structure of the single-task and multi-task versions are the same, only the link prediction is added in the multi-task version of GNN-ML, and GNN-OSF.

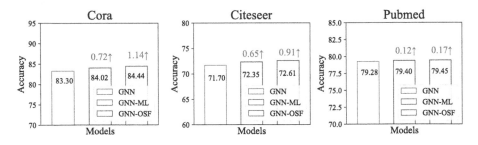

Fig. 2. Comparison of single-task and multi-task learning

Figure 2 compares the performance of GNN, GNN-ML, and GNN-OSF with bar charts. As shown, GNN-ML achieves higher accuracy on all the datasets, which constructs the link matrix to express the labels' consistency of the nodes in P_{train} and cooperatively trains the model with multi-task learning. Meanwhile, GNN-OSF optimizes the link matrix with the supplement of the isolated pairwise and gains 84.44, 72.61, 79.45 accuracy, which is absolutely 1.14, 0.91,0.17 higher than the original single version GNN, which significantly breaks the limitation with the optimizing the structure fairness. Therefore, GNN-OSF strengthens the feature-level information and eliminates the negative impact of invalid structure information of the isolated parts. Under this situation, more effective node representation and richer information can be learned with our proposal.

4.4 Effect of Different Similarity Distances Threshold t

When constructing the link matrix, the similarity distances threshold t is an important parameter, which significantly affects the number of virtual links. Figure 3 displays the accuracy and virtual link numbers under different similarity distance thresholds t. Since Cora and Citeseer datasets utilize one-hot encoding to represent features, we simplify the Jaccard similarity as the intersection of the number of features between two nodes as $|x_i \cap x_j|$.

When the threshold is set too strictly, the information learned from the link matrix is very limited. On the other hand, too many noisy edges will be introduced with a too-relaxed threshold t, where the performance will be significantly reduced. According to the curve, it can be found that the accuracy increases rapidly at first, and then tends to decline after the peak. GNN-OSF utilizes the link matrix to ensure the isolated parts can learn the similar representation in the embedding space with similar nodes on the original graph, which optimize the structure fairness to some extent. When the threshold is set as 4 and 8 on the Cora and Citeseer datasets, GNN-OSF achieves 84.44% and 72.61% accuracy, respectively. Carefully observing the bar chart in the figure, we can find that the peak value can be reached when the increment of the edge in the link matrix L_{iso} is about 20%, which can guide the parameter setting.

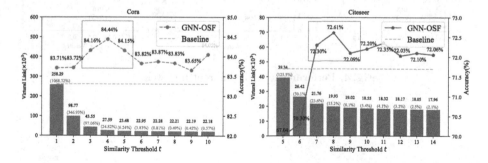

Fig. 3. Performance with different similarity threshold t

4.5   Effect of Different Link Weight ρ

The link weight is another important parameter during the link matrix construction. Hence, we discuss the influence of the link weight ρ in Fig. 4. Specifically, We fix the threshold near the optimal value and then discuss the accuracy with different link weights ρ. All the curves show the tendency of increasing first and then decreasing. With in-depth analysis, the links with a higher similarity threshold are of high confidence, so the link weight should be set larger to achieve higher accuracy.

4.6   Impact of Multi-task Learning on Node Representation

GNN-OSF intends to optimize the structure fairness with link matrix and multi-task learning. The multi-task mechanism exponentially increases the edge-level comparison items, which enhances the node representation of the isolated parts through the interaction of different tasks. Therefore, we design further experiments to explore whether multi-task is helpful to obtain a more accurate node representation.

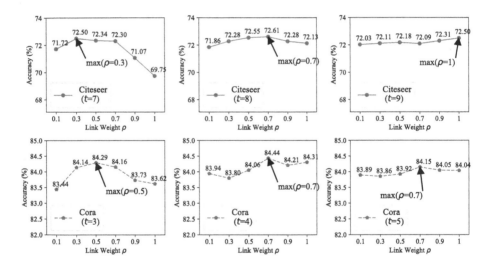

Fig. 4. Performance with different link weight ρ

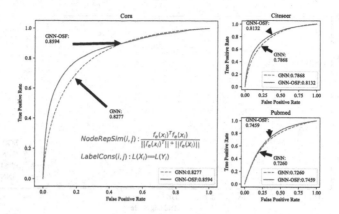

Fig. 5. Performance of node representation with multi-task learning

Obviously, we expect the embedding representation of nodes with the same label to be as close as possible, and that of the nodes with different labels to be as far as possible in the new representation space outputted from the encoder layer. Figure 5 records the AUC score between the similarity of node representation $NodeRepSim(i,j) = \frac{f_e(x_i)^T f_e(x_j)}{||f_e(x_i)^T||\times||f_e(x_j)||}$ and the consistency of the taxonomy $LabelCons(i,j)$. Compared with the single-task-based GNN network, GNN-OSF achieves a relatively higher AUC score, which means that multi-task learning has the capability of obtaining more effective node embedding for the later classification task.

4.7 Performance of Fraud Detection

In addition to the public datasets, we also conduct experiments on e-commerce fraud detection tasks. Fraudulent users often maliciously steal marketing budgets with some illegal methods. Due to the lack of annotated data, semi-supervised learning methods are usually exploited to find more "escape fish" from the blacklist database. So, we treat the fraud detection as the semi-supervised node classification problem with reference to [9]. Besides, the model is expected to equip with the capability to detect more fraudulent users based on rare labeled data. So, we testified the performance with different training scales with AUC metric, especially in the case of small annotation data as shown in Fig. 6.

According to the tendency, the accuracy gradually increases as the scale of labeled data increases in all methods. Besides, GNN-OSF sharply reaches 0.9 when only 5% of the labeled data are available and then shows a gentle growth trend. But, the other methods exhibit a linear increasing trend. Thus, GNN-OSF is more advantageous under the condition of scarcity of annotations. Besides, there is no obvious difference in the effects of each model when the annotation data is extremely large or small. Therefore, comprehensive experimental results indicate that GNN-OSF can effectively deal with different training scales and achieve an impressive superiority over other state-of-the-art models in most situations.

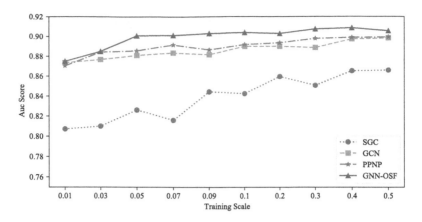

Fig. 6. Performance on fraud detection with different training scale

5   Conclusion

Graph neural networks show structural unfairness to the isolated parts and sometimes even bring the negative local over-smoothing phenomenon. To facilitate
this line of research, this paper proposes a Multi-task Graph Neural Network for
Optimizing the Structure Fairness (GNN-OSF), which is dedicated to alleviating
the negative influence of the isolated part in the classical node classification task
on the homogeneous graph.

GNN-OSF consists of position partition, link matrix construction, and multi-
task graph networks. Firstly, all the nodes are divided into P_{train}, P_{main},
$P_{isolated}$. Then, the link matrix is exploited to describe the label's consistency,
where those nodes pairwise $P_{train} - P_{train}$, $P_{main} - P_{isolated}$, $P_{isolated} - P_{isolated}$
of high probability to share the same label are expected to be positive in the
link matrix. Afterward, the multi-task learning network structure is exploited
to enhance the feature representation of the nodes in the $P_{isolated}$ and weaken
the meaningless structure. Our exhaustive experiments also establish the clear
superiority of GNN-OSF, which enhances expressive capability and leverages
off-the-shelf models with the optimization of structure fairness.

References

1. Besta, M., Iff, P., Scheidl, F., et al.: Neural graph databases. In: Learning on Graphs
   Conference, vol. 198, p. 31 (2022)
2. Ren, H., Galkin, M., Cochez, M., et al.: Neural graph reasoning: complex logical
   query answering meets graph databases, CoRR abs/2303.14617 (2023)
3. Zhang, Z., Cui, P., Zhu, W.: Deep learning on graphs: a survey. IEEE Trans. Knowl.
   Data Eng. **34**(1), 249–270 (2022)
4. Wu, Z., Pan, S., Chen, F., et al.: A comprehensive survey on graph neural networks.
   IEEE Trans. Neural Netw. Learn. Syst. **32**(1), 4–24 (2021)

5. Wang, J., Guo, Y., Wang, Z., et al.: Graph neural network with feature enhancement of isolated marginal groups. Appl. Intell. **52**(14), 16962–16974 (2022)
6. Kang, J., Zhu, Y., Xia, Y., et al.: RawlsGCN: towards Rawlsian difference principle on graph convolutional network. In: ACM Web Conference, pp. 1214–1225 (2022)
7. Henaff, M., Bruna, J., LeCun, Y.: Deep convolutional networks on graph-structured data. arXiv preprint arXiv:1506.05163 (2015)
8. Defferrard, M., Bresson, X., Vandergheynst, P.: Convolutional neural networks on graphs with fast localized spectral filtering. In: Advances in Neural Information Processing Systems, vol. 29, pp. 3837–3845 (2016)
9. Kipf, T.N., Welling, M.: Semi-supervised classification with graph convolutional networks. In: International Conference on Learning Representations, ICLR (2017)
10. Xu, B., Shen, H., Cao, Q., et al.: Graph convolutional networks using heat kernel for semi-supervised learning. In: International Joint Conference on Artificial Intelligence, IJCAI, pp. 1928–1934 (2019)
11. Klicpera, J., Bojchevski, A., Günnemann, S.: Predict then propagate: graph neural networks meet personalized pagerank. In: International Conference on Learning Representations, ICLR (2019)
12. Chen, M., Wei, Z., Huang, Z., et al.: Simple and deep graph convolutional networks. In: International Conference on Machine Learning, vol. 119, pp. 1725–1735 (2020)
13. Xu, B., Shen, H., Cao, Q., et al.: Graph wavelet neural network. In: International Conference on Learning Representations, ICLR (2019)
14. Wu, F., Souza, A., Zhang, T., et al.: Simplifying graph convolutional networks. In: International Conference on Machine Learning. Proceedings of Machine Learning Research, vol. 97, pp. 6861–6871 (2019)
15. Hamilton, W.L., Ying, Z., Leskovec, J.: Inductive representation learning on large graphs. In: Advances in Neural Information Processing Systems, vol. 30, pp. 1024–1034 (2017)
16. Velickovic, P., Cucurull, G., Casanova, A., et al.: Graph attention networks. In: International Conference on Learning Representations, ICLR (2018)
17. Shi, Y., Huang, Z., Feng, S., et al.: Masked label prediction: unified message passing model for semi-supervised classification. In: International Joint Conference on Artificial Intelligence, pp. 1548–1554. ijcai.org (2021)
18. Corso, G., Cavalleri, L., Beaini, D., et al.: Principal neighbourhood aggregation for graph nets. In: Advances in Neural Information Processing Systems (2020)
19. Zhang, S., Xie, L.: Improving attention mechanism in graph neural networks via cardinality preservation. In: International Joint Conference on Artificial Intelligence, IJCAI, pp. 1395–1402 (2020)
20. Rong, Y., Huang, W., Xu, T., et al.: DropEdge: towards deep graph convolutional networks on node classification. In: International Conference on Learning Representations, ICLR (2020)
21. Xu, K., Li, C., Tian, Y., et al.: Representation learning on graphs with jumping knowledge networks. In: International Conference on Machine Learning, ICML, vol. 80, pp. 5449–5458 (2018)
22. Zhao, L., Akoglu, L.: PairNorm: tackling oversmoothing in GNNs. In: International Conference on Learning Representations. OpenReview.net (2020)
23. Yang, H., Ma, K., Cheng, J.: Rethinking graph regularization for graph neural networks. In: AAAI Conference on Artificial Intelligence, pp. 4573–4581 (2021)
24. Chen, D., Lin, Y., Li, W., et al.: Measuring and relieving the over-smoothing problem for graph neural networks from the topological view. In: AAAI Conference on Artificial Intelligence, pp. 3438–3445. AAAI Press (2020)

25. Grover, A., Leskovec, J.: Node2vec: scalable feature learning for networks. In: ACM SIGKDD International Conference, pp. 855–864 (2016)
26. Yang, Z., Cohen, W.W., Salakhutdinov, R.: Revisiting semi-supervised learning with graph embeddings. In: International Conference on Machine Learning, ICML, vol. 48, pp. 40–48 (2016)
27. Liao, R., Brockschmidt, M., Tarlow, D., et al.: Graph partition neural networks for semi-supervised classification. In: International Conference on Learning Representations, ICLR (2018)
28. Yu, S., Yang, X., Zhang, W.: PKGCN: prior knowledge enhanced graph convolutional network for graph-based semi-supervised learning. Int. J. Mach. Learn. Cybern. **10**, 3115–3127 (2019)
29. Lei, F., Liu, X., Dai, Q., et al.: Hybrid low-order and higher-order graph convolutional networks. Comput. Intell. Neurosci. **2020**, 3283890:1–3283890:9 (2020)
30. Chen, S., Tian, X., Ding, C.H.Q., et al.: Graph convolutional network based on manifold similarity learning. Cogn. Comput. **12**(6), 1144–1153 (2020)
31. Manessi, F., Rozza, A.: Graph-based neural network models with multiple self-supervised auxiliary tasks. Pattern Recogn. Lett. **148**, 15–21 (2021)

Few-Shot Multi-label Aspect Category Detection Utilizing Prototypical Network with Sentence-Level Weighting and Label Augmentation

Zeyu Wang[ID] and Mizuho Iwaihara[✉][ID]

Graduate School of Information, Production, and Systems, Waseda University,
Kitakyushu 808-0135, Japan
wangzeyu@akane.waseda.jp, iwaihara@waseda.jp

Abstract. Multi-label aspect category detection is intended to detect multiple aspect categories occurring in a given sentence. Since aspect category detection often suffers from limited datasets and data sparsity, the prototypical network with attention mechanisms has been applied for few-shot aspect category detection. Nevertheless, most of the prototypical networks used so far calculate the prototypes by taking the mean value of all the instances in the support set. This seems to ignore the variations between instances in multi-label aspect category detection. Also, several related works utilize label text information to enhance the attention mechanism. However, the label text information is often short and limited, and not specific enough to discern categories. In this paper, we first introduce support set attention along with the augmented label information to mitigate the noise at word-level for each support set instance. Moreover, we use a sentence-level attention mechanism that gives different weights to each instance in the support set in order to compute prototypes by weighted averaging. Finally, the calculated prototypes are further used in conjunction with query instances to compute query attention and thereby eliminate noises from the query set. Experimental results on the Yelp dataset show that our proposed method is useful and outperforms all baselines in four different scenarios.

Keywords: Aspect category detection · Few-shot learning · Meta-learning · Prototypical network · Label augmentation

1 Introduction

Aspect category detection (ACD) [13,14] is a sub-task of aspect-based sentiment analysis (ABSA) [9]. ACD is to categorize user reviews on products and services such as hotels and restaurants into a pre-defined set of aspect categories. Examples of aspect categories for hotels are location, price, room, while those of restaurants are food, service, interior, etc. ACD will facilitate access to viewpoint information for users and provide assistance for making decisions. As in

© The Author(s), under exclusive license to Springer Nature Switzerland AG 2023
C. Strauss et al. (Eds.): DEXA 2023, LNCS 14147, pp. 363–377, 2023.
https://doi.org/10.1007/978-3-031-39821-6_30

Table 1. Example of a 3-way 2-shot meta-task. The bolded parts with gray background represent the target aspects, while the square marked parts indicate the noise aspects.

	Support Set
(A) experience	(1) Perhaps we'll try one more time and hope our **experience** is better.
	(2) The **experience** and ⸤service⸥ is very great!
(B) drinks	(1) It was happy hour so the **drinks** were a little ⸤less expensive⸥.
	(2) Just an hour in the afternoon and only ⸤50 cents⸥ or so off the **drinks** with no ⸤food⸥ specials.
(C) food	(1) They also have rotating **dining** specials.
	(2) The **food** was good and ⸤price⸥ was reasonable.
	Query Set
(A) and (C)	My **experience** as far as ⸤service⸥ and the **food** are the same.
(B)	**Drinks** were tasty and quick, and the ⸤atmosphere⸥ was cool.

practical scenarios, user reviews are generally diversified and contain more than one aspect, the task of multi-label aspect category detection becomes essential. This task can also be perceived as a special case of multi-label text classification tasks.

Few-shot learning (FSL) [3,4] enables a quick adaption to novel classes with a limited number of samples after learning a large amount of data, being an effective solution to the issues of finite data and data sparsity. FSL problems can be dealt with by a meta-learning [7] strategy, which is also known as "learning to learn." In the meta-training phase, the dataset is divided into separate meta-tasks to learn the generalization capability of the model in the case of category changes. The meta-task adopts the N-way K-shot setting, as demonstrated in Table 1, which is an example of a 3-way 2-shot meta-task, meaning that there are altogether three classes (aspect categories) in the support set and in each class there are two samples (sentences). The prototypical network [18] utilized in this paper follows exactly the meta-paradigm described above.

A prototypical network aims to extract a prototype for each class by averaging all the instances in one class to measure the distance with the instance of the query set. However, as shown in Table 1, it is evident that the number of noise aspects contained in the sentences of each support set class is different. In terms of that, simply averaging all instances in a class neglects the differences between sentences and treats samples from the same classes equally. Related work of multi-label few-shot learning for ACD [8] merely uses attention mechanism to denoise the sentence at word-level, but just denoising over words is not enough, since variance of noise between sentences still exists.

Moreover, in the context of few-shot text classification tasks, there are several papers that incorporate label text information and have obtained promising results. In the work exemplified by [25], although a higher boost in word-level attention using label embedding was obtained, it still has some deficiencies. We point out that there are many semantically similar or poorly expressed labels in the Yelp dataset we are using. For instance, the labels of *food_food_meat* and *food_food_chicken* are semantically similar in their label texts, which may lead to confusion in the classification. Furthermore, there are labels whose meanings are rather obscure or ambiguous. Take the label *drinks_ alcohol_ hard* as an example, it is known that the word "*hard*" is a polysemous word. The word "*hard*" in this class name is related to hardness, difficulty, etc. It is obvious that "*hard*" modifies "*alcohol*", but the word order is reversed, which may confuse the classier. In this case, augmenting the label name with a word related to the label, such as "*vodka*", can give more specific meaning to the label name and assist separation from other aspects.

For the purpose of tackling all the issues mentioned above, we propose a novel model named **S**entence-**L**evel **W**eighted prototypical network with **L**abel **A**ugmentation (Proto-SLWLA) that can well solve the current multi-label few-shot ACD task. Our model mainly consists of two parts, LA and SLW. Specifically, in the LA part, we concatenate the synonyms obtained from the original label text with the label itself and incorporate it into the existing word-level attention. The augmented label words will add certain auxiliary information to the label, which will make the label information become more adequate. For the SLW part, we propose assigning corresponding weights to different sentences in a class inspired by [10], which likewise treats the samples in one class as differentiated individuals. After mitigating noises at word-level, we implement our idea as giving lower weights to sentences with more noise and higher weights to sentences with less noise, by means of a sentence-level attention mechanism. Then prototypes by giving weighted averages to the instances are obtained. With these two methods, the prototype can be more representative of the current class. Our experiments conducted on Yelp_review [1] dataset shows that our method Proto-SLWLA outperforms the baselines in nearly all conditions, which demonstrates the effectiveness of our method.

The rest of this paper is organized as follows: Sect. 2 covers related work. Section 3 describes the proposed method of Proto-SLWLA. In Sect. 4, performance evaluation of Proto-SLWLA and comparision with baseline methods are shown. Section 5 presents concluding remarks and future work.

2   Related Work

Aspect Category Detection. Previous research on ACD has concentrated on a single aspect, which includes unsupervised and supervised methods. Unsupervised methods use semantic association analysis based on pointwise mutual information [19] or co-occurrence frequency [6,17] to extract aspects. However, these approaches require large corpus resource and the performance is hardly satisfactory. Supervised methods exploit representation learning [26] or topic-attention network [12] to identity different aspect categories. In practice, these methods have shown to be effective, yet they heavily rely on a massive amount of labeled data for each aspect to train discriminative classifiers. In addition, a review sentence often encompasses multiple aspects due to the diversity and arbitrariness of human expression, which motivates the multi-label aspect category detection.

Few-Shot Learning. FSL is a paradigm to solve the problem of scarcity of data. Meta-learning for solving FSL problems has been widely adopted, notably in model-based approaches, optimization-based approaches, and metric-based approaches. In our paper, we concentrate on the metric-based method, whose representative models are matching network [22], relation network [20], prototypical network [18], and so forth. An essential element of their idea is to learn a feature mapping function in which support and query samples are projected into an embedding space, and to classify queries by learning some metrics in that space.

Multi-label Few-Shot Learning. Compared to single-label FSL, the potential of multi-label FSL is yet to be stimulated. In the NLP domain, Proto-HATT [5] is proposed for an intent classification task, while designing a meta calibrated threshold mechanism with logits adaption and kernel regression. Proto-AWATT [8] focuses on multi-label few-shot aspect category detection and it is also the first work to focus on this task. It utilizes attention mechanisms to alleviate noise aspects, achieving remarkable results. However, its prototypical network assigns an equal weight to all samples, even if certain samples contain abundance of noises and multiple aspects. This is relatively disadvantageous for a multi-label few-shot learning task. In our work, we introduce a sentence-level attention module to give different weights to different instances.

Using Label Information for Text Classification. Label embedding is currently widely used in NLP for text classification tasks [23] to enhance generalization ability, and it is also very common in zero-shot and few-shot settings. In the context of zero-shot learning, prompt-based strategies [15,16] to match text against class names in an implicit way have been developed. For few-shot learning, [11] extracts semantics of class names and simply appends class names to the input support set and query set sentences to guide the feature representation. A work close to our task is [25] which takes label embeddings as a supplementary information in its LAS and LCL part. In our work, we extend its LAS part by applying a label augmentation method to expand the label information.

3   Methodology

3.1   Overview

A meta-task consists of a support set and a query set. We assume that in an N-way K-shot meta-task, the support set is denoted as $S = \{(x_1^n, x_2^n, ..., x_K^n), y^n\}_{n=1}^N$, where x_k^n represents the k-th sentence in n-th class and y^n is the common aspect that all x^n sentences contain. The query set is denoted as $Q = \{(x_m, y_m)\}_{m=1}^M$, where x_m indicates a query instance and y_m is its corresponding N-bit binary label from the support classes.

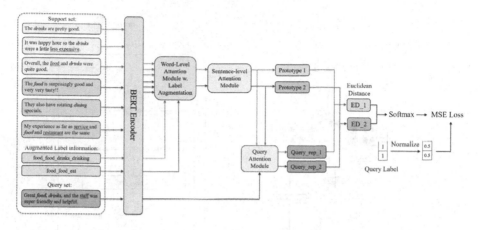

Fig. 1. General architecture of our proposed model.

Our proposed model mainly consists of three components, which are support-set attention (word-level attention), sentence-level attention and query-set attention modules, as illustrated in Fig. 1(a). Given a sentence $x = [w_1, w_2, ..., w_l]$ with length l, we utilize the BERT [2] pre-trained model as the encoder and obtain an embedding matrix $H = [h_1, h_2, ..., h_l]$, where $H \in \mathbb{R}^{d \times L}$ and L is the maximum length of BERT input.

3.2 Word-Level Attention with Label Augmentation

Support Set Attention. Following the work of [8], we primarily alleviate the noises in the support set by using support set attention (word-level attention). We extract the common aspect vector $v^n \in \mathbb{R}^d$ out of the K-shot instances by mean pooling each instance and then perform a word-level average on the K instances.

$$v^n = avg(H_1^n, H_2^n, ..., H_K^n) \tag{1}$$

In order to further remove the noises, we adopt the approach of [24] following [8] to train a dynamic attention matrix by feeding the repeated common aspect vector [21] into a linear layer. This approach is possible to learn to accommodate the common aspect and pick up on its different perspectives.

$$W^n = W(v^n \otimes e_M) + b, \tag{2}$$

where $(v^n \otimes e_M) \in \mathbb{R}^{e_M \times d}$ denotes repeating v^n for e_M times and $W^n \in \mathbb{R}^{d \times d}$. The linear layer has parameter matrix $W \in \mathbb{R}^{d \times e_M}$ and bias $b \in \mathbb{R}^d$. As different classes are trained, the parameters of this linear layer are constantly updated to accommodate the new classes. Then we use the common aspect vector to

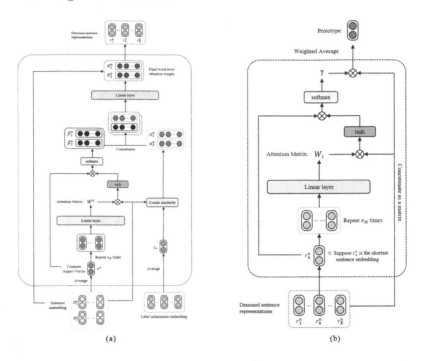

Fig. 2. Specific framework of our proposed model. (a) Structure of the word-level attention module. (b) Structure of the sentence-level attention module.

calculate the attention with each instance and multiply the obtained word-level weights on each sentence.

$$\beta_k^n = \text{softmax}(v^n \tanh(W^n H_k^n)), \tag{3}$$

where $n \in [1, N]$ and $k \in [1, K]$. So far, we have achieved a preliminary word-level attention weight to enhance the focus on the target aspect, for reducing the effect of noise aspects to some extent.

Label-Guided Attention Enhanced by Label Augmentation. As previously stated, since there are semantically similar and ambiguous labels in the dataset, we attempt to augment label texts with supplementary words to enrich label information. In particular, in order to dig words that are relevant to the sentence as well as the label name, we design a template whose format is: "[X]. It is about [Label], and its synonym is [MASK]." In this template, [X] represents a sentence of a given aspect category in the dataset. [Label] stands for its aspect category label, and [MASK] denotes the mask token. We then supply the embedding vector $d \in \mathbb{R}^l$ into the BERT pre-trained masked language model (MLM) to predict the word that should appear at the [MASK] position, as shown in Fig. 3. The MLM head will output a probability distribution which

Table 2. Part of the finally obtained words relevant to a label name

Label Name	Predicted Words
food_food	eat, delicious, dining, cooking, meal, eating, foods, ...
parking	cars, space, traffic, parking, cars, driving, bike, ...
restaurant_entertainment_music	song, pop, jazz, opera, melody, rock, blues, folk, ...
drinks_alcochol_hard	vodka, tequila, rum, gin, bitter, bourbon, ...
restaurant_location	place, destination, locality, spot, geography, ...
entertainment_casino	gambling, vegas, gaming, poker, casinos, game, ...
building_hall	hallway, lobby, corridor, wall, halls, library,

indicates the likelihood of each word w appearing at the [MASK] position over all the vocabulary V.

$$p(w|d) = \text{softmax}(W_2\sigma(W_1 d + b')), \qquad (4)$$

where $W_1 \in \mathbb{R}^{l \times l}$, $W_2 \in \mathbb{R}^{|V| \times l}$ and $b' \in \mathbb{R}^l$ are learnable parameters that have been pre-trained with the MLM objective of BERT, and $\sigma(\cdot)$ is the activation function.

After we have obtained a list of candidate label name related words, we filter the predicted words from each sentence by removing stop words, punctuations, words identical to the class name, etc. The final predicted augmenting words for each class name are shown in Table 2. Then we take the top words of all the predicted words in each category, and find the top m words with the highest frequency among the total number of sentences multiplied by m words.

Once we obtained these words, we append them to the original label name with an underline to form a new label. For instance, take $m = 1$ as an example, the original label name *drinks_alcohol_hard* will be transformed into *drinks_alcohol_hard_vodka*, which nicely emphasizes the meaning of "*liquor*", thus eliminating its interference with other meanings in the synonym.

So far, we have accomplished the process of label augmentation. Now we need to integrate augmented label information into the word-level attention to give more guidance to support sentences on label information as [25]. Specifically, we enter the augmented label information into the BERT model to obtain its embedding, and compute the cosine similarity between the label information embedding and sentence embedding:

$$\alpha_k^n = \cos(L_n, H_k^n), \qquad (5)$$

where $L_n \in \mathbb{R}^d$ is the label information embedding of class n in the support set, which is calculated by averaging in terms of the length of label information. H_k^n is the word embedding matrix of the k-th sentence in the n-th class. We then combine the calculated cosine similarity α_k^n with the previously obtained

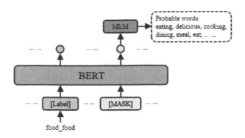

Fig. 3. Using BERT masked language model (MLM) to predict words relevant to a label name.

word-level attention β_k^n in (3). We concatenate the two vectors and enter them into the linear layer to obtain the final attention weight.

$$\theta_k^n = W_g[\alpha_k^n; \beta_k^n] + b_g, \tag{6}$$

where $\theta_k^n \in \mathbb{R}^l$, W_g and b_g are trainable parameters of the linear layer. $[\cdot; \cdot]$ denotes the concatenation operation. We then renormalize the attention score by the softmax function to make the weight more reliable.

$$\tilde{\theta}_k^n = \text{softmax}(\theta_k^n) \tag{7}$$

Eventually, we assign the final word-level attention weight to each sentence in the support set.

$$r_k^n = \tilde{\theta}_k^n H_k^n, \tag{8}$$

where $n \in [1, N]$ and $k \in [1, K]$. So far, we have constructed a collection of $R_n = [r_1^n, r_2^n, ..., r_K^n]$ which consists of denoised support set representations. The whole process of this word-level attention module is illustrated in Fig. 2(a).

3.3   Sentence-Level Attention

In this section, we describe calculation of sentence-level weights. The architecture is depicted in Fig. 2(b). As previously mentioned, we would like to adjust weights on different sentences in the classes depending on the amount of estimated noises involved in the sentence. Hence, we introduce a method to compute the attention by centering on the shortest sentence in the class. As illustrated in Fig. 4, we can observe that as the length of the sentence increases, the number of aspects therein increases as well. That is, the noises of the sentence are also extending. Consequently, it is reasonable to surmise that the shorter the sentence, the fewer number of aspects the sentence contains, which will then have a larger possibility of having a single aspect. Since each sentence always contains the target aspect, the chance of having noisy aspects becomes smaller than longer sentences.

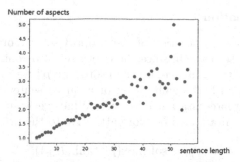

Fig. 4. Distribution of sentence length and number of aspects in the Yelp dataset. We randomly selected 8000 samples in the dataset and averaged the number of aspects for all sentences of the same length.

Thus, we introduce a mechanism that emphasizes aspects learned from the shortest sentence and apply this aspect weighting to all the sentences in the support set. We first locate the denoised support representation r_{min}^n of the shortest sentence and then repeating it for e_M times, which aims to learn different perspectives of the shortest sentence. Then we feed it to a linear layer to obtain the attention matrix W_s^n, where $W_s^n \in \mathbb{R}^{d \times d}$ and W_s, and b_s are trainable parameters.

$$W_s^n = W_s(r_{min}^n \otimes e_M) + b_s \tag{9}$$

Similarly, we follow the preceding method to directly use the shortest sentence embedding r_{min}^n to compute sentence-level attention on all the denoised sentence representations. Particularly, R^n is multiplied with the attention matrix W'^n to exploit the relationships between the shortest sentence and other sentences from different perspectives. The weight is calculated as follows:

$$\gamma^n = \text{softmax}(r_{min}^n \tanh(W_s^n R^n)), \tag{10}$$

where $\gamma^n \in \mathbb{R}^k$, and $R^n \in \mathbb{R}^{d \times k}$ represents the concatenation of all the denoised representations in one class as shown in (8). By doing this, longer sentences which are dissimilar to the shortest sentence and meanwhile contain more noise will obtain lower weights. Finally, we perform a weighted average to the representations to derive the final prototype $p^n \in \mathbb{R}^d$.

$$p^n = \gamma^n R^n \tag{11}$$

In this way, we obtain N prototypes $[p^1, p^2, ..., p^N]$ after processing all the classes in the support set.

3.4   Query Attention

Following the denoising operation of the support set, we ought to mitigate the noises in the query set as well, since the query set also contains some irrelevant aspects. To achieve this goal, we use the prototype p^n we just obtained to compute the attention with the embedding of a query sentence $H_m \in \mathbb{R}^{d \times L}$ and acquire the query representation $r_m \in \mathbb{R}^d$. What we want to accomplish is to enable query representation to be more attentive to the prototype aspect.

$$r_m = \text{softmax}(p^n \tanh(H_m)) \tag{12}$$

Up to this point, we have completed introducing all the modules of the model and finished construction of representations of a given support set and query set.

3.5   Training Objective

In this paper, we use the Euclidean Distance (ED) to measure the distances between prototypes and query representations. The final prediction of a query instance is the negative distances and then we use a softmax to normalize the result.

$$\hat{y} = \text{softmax}(-\text{ED}(p^n, r_m)), \tag{13}$$

where $n \in [0, N]$, and $m \in [0, M]$. Lastly, we use the mean square error (MSE) loss to be our final training objective.

$$L = \sum (\hat{y} - y_m)^2, \tag{14}$$

where y_m is the N-bit golden label for query instance x_m. Note that since $\hat{y}$ is softmaxed, our golden label y_m should be normalized as well. During the training process we allow the predicted values to be as close as possible to the golden label.

4   Experiments

In this section, we primarily introduce the dataset used in our work, together with baselines, evaluation metrics, and implement details. Thereafter, we present and analyze the experimental results on our dataset in four different settings.

4.1   Dataset

Since the Yelp dataset having review aspects and used in [8] is not publicly available, we construct the dataset by combining the Yelp_dataset_round8 and Yelp_review_aspect [1] which are datasets consisting of extensive user reviews. After processing the raw data into sentences and their corresponding aspects, we collected the sentences for each aspect and selected 100 aspects from all the 135 aspects and remove 35 of them. The selected aspects are split without

Table 3. Statistics of the Yelp dataset. #**cls.** indicates the number of classes. #**inst./cls.** indicates the number of instances per class. #**inst.** indicates the total number of instances.

Dataset	#cls.	#inst./cls.	#inst.
FewAsp	100	630	63000

intersection into 64 aspects for training, 16 aspects for validation, and 20 aspects for testing. We randomly sample 800 meta-tasks from the 64 gathered aspect sentences for training, 600 meta-tasks from the 16 gathered aspects for validation and 600 meta-tasks from the 20 gathered aspects for testing, following [8]. The statistics of our dataset is shown in Table 3. Note that at each epoch of training, the 800 meta-tasks are resampled.

4.2 Baseline Models

Our method is compared with the following methods: Matching Network [22], Relation Network [20], Prototypical Network [18] and Proto-AWATT w/o DT [8] and Proto-SLW. Note that we use BERT as the encoder for all baseline models for the sake of fairness.

Matching Network [22]. It learns an embedding mapping function first, combines the samples of support set and query set samples and enters them into Bi-LSTM, and finally adopts the cosine similarity as the distance measure to obtain the classification result.

Relation Network [20]. Instead of a fixed distance metric, it uses a deep neural network with multiple layers of convolution to compute the relationship between query samples and support samples.

Prototypical Network [18]. By averaging the corresponding support samples, it computes a prototype for each class and uses the negative Euclidean distance between the query samples and the prototype for the few-shot classification task.

Proto-AWATT [8]. It is the first approach for multi-label aspect category detection tasks. It mitigates the adverse effects caused by noisy aspects using support set and query set attention mechanisms.

Proto-SLW. As an ablation setting, this model is removed of the LA part from our proposing model and only utilizing the sentence-level attention to assign different weights to different sentences in one class in the support set.

Proto-SLW+LAS. We add the LAS part from [25] to our SLW model to take the label name itself as a complementary information of the attention weights in the support set. Note that this case is actually equivalent to the case of $m=0$ in Proto-SLWLA.

Table 4. Experimental results of our model with AUC and macro-F1(%) evaluated on FewAsp. m represents for the number of words augmented by each label. The symbol † indicates p-value < 0.05 of the T-test comparing with Proto-AWATT, while symbol ‡ indicates that p-value < 0.05 of the T-test comparing with Proto-SLW.

Models	5-way 5-shot		5-way 10-shot		10-way 5-shot		10-way 10-shot	
	AUC	F1	AUC	F1	AUC	F1	AUC	F1
Matching Network	0.9025	66.59	0.9230	70.97	0.8834	51.54	0.9085	53.84
Prototypical Network	0.9017	65.71	0.9318	71.55	0.8991	53.82	0.9063	55.44
Relation Network	0.8463	54.72	0.8473	55.54	0.8428	42.92	0.8325	45.86
Proto-AWATT w/o DT	0.9061	66.32	0.9319	71.67	0.8999	53.86	0.9125	57.75
Proto-SLW	0.9116	67.27	0.9387	72.83	**0.9062**	54.74	0.9156	57.33
Proto-SLW+LAS ($m = 0$)	0.9123	67.94	0.9374	72.90	0.9037	54.98	0.9119	57.09
Proto-SLWLA ($m = 1$)	0.9156	68.11	**0.9391†**	**73.20†‡**	0.9024	54.79	**0.9179†‡**	**58.69†‡**
Proto-SLWLA ($m = 2$)	**0.9157†‡**	**68.30†‡**	0.9377	72.87	0.9026	55.18	0.9156	58.20
Proto-SLWLA ($m = 3$)	0.9117	67.57	0.9380	72.97	0.9038	**55.63†‡**	0.9154	57.94

4.3   Evaluation Metrics

Traditional single-label FSL tasks in the past typically used accuracy to measure the performance of the model. In the multi-label task, we follow Proto-AWATT and choose the AUC (Area Under Curve) score which is used to select model and macro-F1 score as evaluation metrics.

4.4   Experimental Settings

For parameter settings, we set $m = 1$, $d = 768$, $L = 50$, $e_M = 4$, $Q = 5 \times N$. We train our model on GeForce RTX 3090 GPU and set our learning rate as 1e-5, batch size as 4 when $N = 5$, as 2 when $N = 10$. When performing label augmentation, we randomly select 2000 sentences in the query set of each class for subsequent operations.

Regarding the threshold value, we set $\tau = 0.3$ in all the conditions. We adopt an early stop strategy when the AUC score is no longer increased in 3 epochs. Then we will select the epoch which has the best result of the AUC score in the validation phase for testing.

4.5   Experimental Results and Discussions

The experimental results are shown in Table 4. The results demonstrates that both our word-level attention with label augmentation module and sentence-level attention module are effective. It is worth mentioning that in the 10-way 10-shot scenario, Proto-SLWLA(m = 1) improves F1 score by 1.36% from Proto-SLW, which is considered a significant improvement in our experimental results.

When Proto-SLW is compared with Proto-AWATT, the results of Proto-SLW are better than those of Proto-AWATT in all four scenarios, which fully illustrates the effectiveness of our sentence-level attention module. Furthermore, the result on 10-shot has slightly more improvement than that of the 5-shot result. This suggests that a noise-less short sentence is more likely to be included in more shots (sentences). However, the 10-way boost is a little less pronounced than the 5-way, since 10-way (10-classes) causes more chance for short sentences to contain noise aspects than 5-way (5-classes).

Compared with Proto-SLW+LAS, the results of Proto-SLWLA outperform in all these four scenarios. This indicates that our LA part is effective, and suggests that it is reliable to use label-related words to enhance the label itself. In addition, for Proto-SLWLA we evaluated with three different values of m, which represents the number of words augmented by each label. We can observe that our model can achieve the best results when $m = 1$ or 2, whereas for $m = 3$, the performance of the model is decreasing and sometimes even lower than Proto-SLW+LAS. This demonstrates that the first or the second of the augmenting words are highly relevant to the label. However as the number of augmenting words increases, lower-ranked words seem to be causing drifts on label semantics.

5   Conclusion

In this paper, we proposed Proto-SLWLA, which is based on prototypical network with sentence-level weighting and label augmentation to tackle the multi-label few-shot aspect category detection task. Existing methods in this domain often utilize prototypical network, but they perform denoising merely at the word-level and do not focus on the variations between instances. Since the concentrations of noises in the target aspects are varying between instances, we introduced sentence-level attention to assign specific weights to the instances after using word-level attention. Also, another existing approach incorporates label text information into the word-level attention module to improve the performance, but the label name texts are not sufficient, because label names are often semantically similar and ambiguous each other, causing separation hard in the representation space. To improve separation by label names, we introduced label name augmentation in which a template is designed and masked language model prediction is utilized to generate words related to each label name, and append them to the label information, which is then used for a guidance of the word-level weights. Our experimental evaluations by AUC and macro-F1 score demonstrate that our design is feasible and effective, outperforming nearly all the baseline models.

References

1. Bauman, K., Liu, B., Tuzhilin, A.: Aspect based recommendations: recommending items with the most valuable aspects based on user reviews. In: Proceedings of the 23rd ACM SIGKDD International Conference on Knowledge Discovery and Data Mining, pp. 717–725 (2017)
2. Devlin, J., Chang, M.W., Lee, K., Toutanova, K.: BERT: pre-training of deep bidirectional transformers for language understanding. arXiv preprint arXiv:1810.04805 (2018)

3. Fei-Fei, L., Fergus, R., Perona, P.: One-shot learning of object categories. IEEE Trans. Pattern Anal. Mach. Intell. **28**(4), 594–611 (2006)
4. Fei-Fei, L., et al.: A Bayesian approach to unsupervised one-shot learning of object categories. In: Proceedings Ninth IEEE International Conference on Computer Vision, pp. 1134–1141. IEEE (2003)
5. Gao, T., Han, X., Liu, Z., Sun, M.: Hybrid attention-based prototypical networks for noisy few-shot relation classification. In: Proceedings of the AAAI Conference on Artificial Intelligence, vol. 33, pp. 6407–6414 (2019)
6. Hai, Z., Chang, K., Kim, J.: Implicit feature identification via co-occurrence association rule mining. In: Gelbukh, A.F. (ed.) CICLing 2011. LNCS, vol. 6608, pp. 393–404. Springer, Heidelberg (2011). https://doi.org/10.1007/978-3-642-19400-9_31
7. Hochreiter, S., Younger, A.S., Conwell, P.R.: Learning to learn using gradient descent. In: Dorffner, G., Bischof, H., Hornik, K. (eds.) ICANN 2001. LNCS, vol. 2130, pp. 87–94. Springer, Heidelberg (2001). https://doi.org/10.1007/3-540-44668-0_13
8. Hu, M., et al.: Multi-label few-shot learning for aspect category detection. arXiv preprint arXiv:2105.14174 (2021)
9. Hu, M., Liu, B.: Mining and summarizing customer reviews. In: Proceedings of the Tenth ACM SIGKDD International Conference on Knowledge Discovery and Data Mining, pp. 168–177 (2004)
10. Ji, Z., Chai, X., Yu, Y., Pang, Y., Zhang, Z.: Improved prototypical networks for few-shot learning. Pattern Recogn. Lett. **140**, 81–87 (2020)
11. Luo, Q., Liu, L., Lin, Y., Zhang, W.: Don't miss the labels: label-semantic augmented meta-learner for few-shot text classification. In: Findings of the Association for Computational Linguistics: ACL-IJCNLP 2021, pp. 2773–2782 (2021)
12. Movahedi, S., Ghadery, E., Faili, H., Shakery, A.: Aspect category detection via topic-attention network. arXiv preprint arXiv:1901.01183 (2019)
13. Pontiki, M., Galanis, D., Papageorgiou, H., Manandhar, S., Androutsopoulos, I.: SemEval-2015 task 12: aspect based sentiment analysis. In: Proceedings of the 9th International Workshop on Semantic Evaluation (SemEval 2015), pp. 486–495 (2015)
14. Pontiki, M., et al.: SemEval-2016 task 5: aspect based sentiment analysis. In: ProWorkshop on Semantic Evaluation (SemEval-2016), pp. 19–30. Association for Computational Linguistics (2016)
15. Puri, R., Catanzaro, B.: Zero-shot text classification with generative language models. arXiv preprint arXiv:1912.10165 (2019)
16. Schick, T., Schütze, H.: Exploiting cloze questions for few shot text classification and natural language inference. arXiv preprint arXiv:2001.07676 (2020)
17. Schouten, K., Van Der Weijde, O., Frasincar, F., Dekker, R.: Supervised and unsupervised aspect category detection for sentiment analysis with co-occurrence data. IEEE Trans. Cybern. **48**(4), 1263–1275 (2017)
18. Snell, J., Swersky, K., Zemel, R.: Prototypical networks for few-shot learning. In: Advances in Neural Information Processing Systems, vol. 30 (2017)
19. Su, Q., Xiang, K., Wang, H., Sun, B., Yu, S.: Using pointwise mutual information to identify implicit features in customer reviews. In: Matsumoto, Y., Sproat, R.W., Wong, K.-F., Zhang, M. (eds.) ICCPOL 2006. LNCS (LNAI), vol. 4285, pp. 22–30. Springer, Heidelberg (2006). https://doi.org/10.1007/11940098_3
20. Sung, F., Yang, Y., Zhang, L., Xiang, T., Torr, P.H., Hospedales, T.M.: Learning to compare: relation network for few-shot learning. In: Proceedings of the IEEE Conference on Computer Vision and Pattern Recognition, pp. 1199–1208 (2018)

21. Vaswani, A., et al.: Attention is all you need. In: Advances in Neural Information Processing Systems, vol. 30 (2017)
22. Vinyals, O., Blundell, C., Lillicrap, T., Wierstra, D., et al.: Matching networks for one shot learning. In: Advances in Neural Information Processing Systems, vol. 29 (2016)
23. Wang, G., et al.: Joint embedding of words and labels for text classification. arXiv preprint arXiv:1805.04174 (2018)
24. Zhao, F., Zhao, J., Yan, S., Feng, J.: Dynamic conditional networks for few-shot learning. In: Proceedings of the European Conference on Computer Vision (ECCV), pp. 19–35 (2018)
25. Zhao, F., Shen, Y., Wu, Z., Dai, X.: Label-driven denoising framework for multi-label few-shot aspect category detection. arXiv preprint arXiv:2210.04220 (2022)
26. Zhou, X., Wan, X., Xiao, J.: Representation learning for aspect category detection in online reviews. In: Proceedings of the AAAI Conference on Artificial Intelligence, vol. 29 (2015)

Toward Healthy Aging: Temporal Regression for Disability Prediction and Warning Decision-Making

Jianfei Zhang[1](✉), Lifei Chen[2], and Shengrui Wang[1,2]

[1] Université de Sherbrooke, Sherbrooke, Québec, Canada
jianfei.zhang@usherbrooke.ca
[2] Fujian Normal University, Fuzhou, Fujian, China

Abstract. Virtually all countries in the world are experiencing growth in the number and proportion of seniors in their population. Almost half of these seniors live with one or more disabling conditions. This highlights the concern about when, and how probably, a disability is likely to occur in aging people. In this paper, we mathematicize this concern as a prediction and a warning of the onset of disability. As such, we propose to start by transforming the fitting problem into a series of independent survival learning and prediction problems. Our approach can use all repeated measures of disability-specific factors and, more importantly, effectively quantify their different impacts on the onset of disability. We also present a new approach for estimating the time of onset and determining the better-timed warning of disability onset. To evaluate our time predictions and warning decisions, we develop four evaluation metrics based on the criteria we explore for the aging study. The results of comparative experiments and ablation studies on the elderly cohorts across Canada demonstrate the effectiveness of our approach.

1 Introduction

Each of us is aging. In every country in the world, the population is aging faster than ever before. As of November 2022, the global population aged 65 or over numbers 800 million. Over the next three decades, this number is projected to more than double, reaching over 1.5 billion in 2050 [24]. Statistics Canada said that over 861,000 Canadian people (which is 2.3% of the population) aged 85 and older were counted in the 2021 Census, more than twice the number observed in the 2001 Census [19]. Over the next 25 years (by 2046), the population aged 85 and older could triple to almost 2.5 million people in Canada [19]. Age has a significant impact on health – *the older you get, the more likely you are to become disabled.* The number of people aging with disabilities is on the rise all around the world. According to the 2017 Canadian Survey on Disability [18], roughly 37.8% of Canadians aged 65 and above had been identified as having some form of disability regarding flexibility, dexterity, mobility, cognition, vision, hearing, pain-related, etc. The disability was severe enough to limit them to some extent in their daily activities, especially for those aged 85 and older.

C. Strauss et al. (Eds.): DEXA 2023, LNCS 14147, pp. 378–392, 2023.
https://doi.org/10.1007/978-3-031-39821-6_31

The concern of this work is a generation of better-timed warnings of the onset of disability. The warning in advance induces healthcare providers to take early actions toward healthy aging. For example, they can make early informed decisions regarding healthcare and social services, which will enable the elderly to delay the onset and reduce the severity of disabilities, thereby maintaining optimal health and quality of life. What is needed for this concern is, from a computational perspective, to predict *'when and how probably a disability is likely to occur in aging people'*, and determine *'when a warning of disability onset should be issued'*. Hence, we have two tasks: *time-to-onset prediction* – i.e., predicting the time elapsing from the beginning of the follow-up until the onset of disability, and *better-timed warning decision* – i.e., determining the time when the lowest onset-free probability below which the warning should be issued. There is as yet no documented work to make such a time-to-onset prediction and the warning decision based on that prediction. For example, one may be able to estimate that a 65-year-old woman has a 70% probability of living without a disability for one year and only a 10% probability for two years but is incapable of determining the exact time when this woman will be disabled and nailing down the time of warning. Making such a prediction and decision is arduous. Late warnings might lead to conditions' worsening, while too early warnings may increase unnecessary nursing and sometimes make the actions unfeasible. This is the motivation for the work reported here, whose aim is to accurately predict the *onset-free probability over time*, estimate the *time-to-onset*, and determine the *better-timed warning*.

To make better predictions and decisions, we shall examine the link between the onset of disability and various social, demographic, geographic, and economic factors that impact healthy aging, such as general health and well-being, physical activity, and the use of healthcare services. For example, among older adults, injuries due to falls threaten independent living, mobility, and functional ability (such as the ability to engage in regular exercise), and increase the risk for future disabilities [16]. Note, however, that these factors may be repeatedly measured and therefore the measures may vary over time (e.g., taking measurements with an interval of a particular time frame). For example, the seniors may be asked, *'Have you had a fall in the past 6, 12, and 18 months?'* Here, the factor 'fall' is measured repeatedly at 6, 12, and 18 months, producing three measures. Handling these repeated measures has been challenging [6] because of the difficulty in using these measures together to train the existing time-to-event data analysis methods [27]. Besides this, we suffer from *censoring*: most of the older people in our study had not experienced a disability by the close of the study period (e.g., the past 6, 12, or 18 months), and therefore the time of the onset is unknown for a subset of the seniors. *How to build the link between these unknown onset times and repeated measures is the key issue of this work.*

In this paper, we first formalize aging data in two parts: factors' repeated measures and encoded onset times, and then develop temporal survival regression (TR) to explore the link between these two time-dependent parts. It is trained by learning the measures of various factors and the onset of disability over time.

With the learned model, we predict the onset-free probabilities at different times, estimate the time of onset in the future, and determine the better-timed warning of disability. We conduct experiments on real-life aging data and compare the performance of our approach with state-of-the-art models. An extensive ablation study is also performed to investigate the effectiveness of each elaboration of our approach. The experimental results demonstrate that our approach performs better than other baselines and TR's variants, yielding more accurate predictions of onset-free probability and estimates of onset time. We summarize the contributions as follows:

- Our approach is computationally simple: it requires only a sequence of logistic regression fits and the operations are easily understood in terms of regression modeling. We transform the fitting problem into a series of independent learning problems by predicting the encoded responses that we redefined based on the observations. The prediction tasks at different times are independent of each other, but the predicted results are highly related.
- Our approach is conceptually interpretable: it performs predictions and makes prediction-based healthcare decisions on a reasoning basis so that people readily understand how factors are jointly related to form the final predictions and decisions. We impose the impact of the factors and their repeated measures on the onset of disability, where the impact values explored by the model identify the risk or protective effect of both time-varying and time-invariant factors, thereby guiding healthcare actions.
- Our approach is empirically effective: it performs a retrospective study on a Canadian cohort including a bunch of non-disabled people, where the results demonstrate not just the high accuracy achieved by the model but its predictive confidence in dealing with censoring contexts. We develop our own assessment criteria based on the consensuses in life science and mathematicize the criteria as evaluation metrics in the absence of ground truth.

2   Related Work

We will review the work related to time-to-event data analysis. Broadly speaking, time-to-event data analysis can be classified into two main categories: statistical methods and machine-learning-based methods, which share the common goal of predicting the time of the event. Statistical approaches can be grouped into nonparametric (e.g., Nelson-Aalen estimator [1], Kaplan-Meier estimator [9]), parametric (e.g., accelerated failure time model [26]), and semi-parametric estimators (e.g., semi-parametric Cox proportional-hazards model [4]), developed primarily for retrospective cohort studies, each has its inherent disadvantages. Nonparametric approaches are intended to generate unbiased descriptive statistics, but generally cannot be used to assess the effect of multiple factors on the response variable (e.g., time-to-event probability), and the parametric approaches suffer from a critical weakness, relying as they do on the assumption that the underlying failure distribution (i.e., how the probability of failure changes over time) has been correctly specified. For semi-parametric approaches, the assumption about how the factors influence the risk of failure is often violated in practical use. The increasing availability of a wide variety of data (e.g., time-varying factors) poses more challenges to the statistical approaches and is stimulating numerous research efforts that use machine learning methods in conjunction with time-to-event modeling. For example, feed-forward neural networks have been used for time-to-event data analysis [28]. Although the feed-forward network can preserve most of the advantages of a typical Cox proportional-hazards hypothesis, it was still not the optimal way to model the baseline variations. This was the rationale for deep learning studies [11]. Additionally, recurrent neural networks [29] proposed to compute the survival function by considering a series of binary classification problems, each leading to the estimation of the survival probability in a given interval of time. Besides neural networks, typical examples include multi-task learning [13], Gaussian process [7], active learning [25], transfer learning [12], etc.

In practice, the factors' effects (e.g., the effect of a treatment) may change over time with longer follow-up [8]. To accommodate such situations, there has been a surge of interest in learning time-varying coefficients instead of time-invariant ones. The varying coefficient models are a very important tool to explore the dynamic pattern. The association between repeated measures and the outcome has been modeled in various ways. The work related to repeated measures focuses mainly on the analysis of time-varying risk factors. For the Cox model, [23] estimated time-varying coefficients by maximizing a kernel-weighted partial likelihood, while [22] employed a local empirical partial likelihood smoothing. Time-varying coefficients were also used in [14] to describe the potential time-varying effects of factors on breast cancer. The proportionality assumption may not hold in practice when factor effects change over time.

3   Problem Statement

In our study, each senior is identified by two response variables – i.e., 'censor' $C \in \{0, 1\}$ and 'stamp' time $S \in \mathbb{R}^+$. The response 'censor' $C = 0$ indicates that the time of disability onset, say T, is uncensored, where the disability occurred right at the 'stamp' time S (i.e., the last recorded time), that is, $T = S$. If $C = 1$, the onset time T is censored, and the stamp time S underestimates the true but unknown T, i.e., $T > S$. As an example for three old Canadians aged over 65 shown in Table 1 (e.g., $T = 7$ months for the 77-year-old Québecois), and otherwise $C = 1$ means T is censored (i.e., unknown) because of dropout (e.g., the 69-year-old Ontarian dropped out of the study at 11 months) or early end of study (e.g., the 86-year-old Albertan has been free of disability throughout the 18-month study period). Each senior is described by D factors, such as 'province' and 'age' that are time-invariant and 'depression', 'sleep', and 'fall' that are time-varying (i.e., repeatedly measured). All these factors' measures can be expressed as $\mathbf{X} = (\mathbf{x}_1, \ldots, \mathbf{x}_V) \in \mathbb{R}^{D \times V}$ at V different times $\tau_1 < \cdots < \tau_V$, where $\mathbf{x}_v = (x_{v1}, \ldots, x_{vD})^\top$. For time-invariant factor d, we have $x_{1d} = x_{2d} = \cdots = x_{Vd}$.

Table 1. Factors and responses for three aging Canadians.

Risk Factors						Response	
time-invariant			time-varying			Censor	Stamp
Province	Age	··	Depression	··	Fall	Smoking	(C)	(S)
Québec	77	··	mild	··	no	sometimes	0	7
			mild	··	yes	sometimes		
			··	··	··	··		
			no	··	no	seldom		
Ontario	69	··	severe	··	no	often	1	11
			moderate	··	no	seldom		
			··	··	··	··		
			moderate	··	no	sometimes		
Alberta	86	··	no	··	yes	never	1	18
			no	··	no	never		
			··	··	··	··		
			mild	··	no	never		

To say that someone remains onset-free means that they are still at risk of the onset of disability, that is, the disability has not yet occurred in them. Hence, the onset-free probability at time t means the probability of remaining onset-free up to time t, i.e., the probability that the disability will not occur earlier than time t, that is, $\Pr(T \geq t)$. Our learning task is to find a mapping function $f_{\mathbf{W}}$ ($\mathbf{W}$ is the learnable parameters) for predicting the probabilities at $t_1 < t_2 \cdots < t_K$. Then, based on these predictions, we shall determine the onset time T at which the newly designed error E is lowest and the warning policy p minimizing the newly defined warning cost Q, as follows:

$$
\left.
\begin{array}{l}
\mathbf{x}_1, \ldots, \mathbf{x}_{v:\tau_v \leq t_1} \xrightarrow{f_{\mathbf{W}}} \Pr(T \geq t_1) \\
\vdots \\
\mathbf{x}_1, \ldots, \mathbf{x}_{v:\tau_v \leq t_K} \xrightarrow{f_{\mathbf{W}}} \Pr(T \geq t_K)
\end{array}
\right\}
\begin{array}{l}
\xrightarrow{\arg\min E} T \\
\xrightarrow{\arg\min Q} p
\end{array}
$$

4   Our Approach

4.1   Encoding

The first thing we need to do is encoding the dualistic response, C and S, as the encoded responses Y, where its value at time t can be given by

$$
y[t] = (-\mathbb{1}(C = 1))^{\mathbb{1}(S < t)},
$$

which takes the value 1 if the disability happened before or at time t, i.e., $S \geq t$, and otherwise either 0 if $C = 0$ or –1 if $C = 1$. Here, $\mathbb{1}(judgment)$ is an indicator function taking a value of 1 if the *judgment* is true and 0 otherwise. Table 2 shows the encoded response for the three Canadians. Each encoded response $y[t_k]$ indicates whether the disability has occurred by time t_k, being 1 if it has occurred, 0 if it has not occurred, and –1 if unknown. Once $y[t_k]$ becomes "0" it will not turn over to "1". There are thus $K + 1$ legally possible sequences of the form $(1, 1, \ldots, 0, 0, \ldots)$, including the sequences composed of all "1"s and all "0"s. For the 67-year-old Québecois, the encoded response remains "1" until $S = 7$ and "0" thereafter; for the other two, whose onset time is censored, the encoded response is "1" until the stamp time and "–1" thereafter.

Table 2. An example of the encoded responses for three Canadian seniors aged over 60.

Encoded Temporal Responses											Response	
$y[1]$	$\cdots$	$y[7]$	$y[8]$	$\cdots$	$y[11]$	$y[12]$	$\cdots$	$y[18]$	$y[19]$	$\cdots$	C	S
1	1	1	0	0	0	0	0	0	0	0	0	7
1	1	1	1	1	1	–1	–1	–1	–1	–1	1	11
1	1	1	1	1	1	1	1	1	–1	–1	1	18

4.2 Training

For senior $i \in \mathcal{G}_0 = \{i | \forall i : C_i = 0\}$ with known encoded responses $Y = (y[t_1], \ldots, y[t_K])$ at times $t_1 < \cdots < t_K$ and measures $\mathbf{X}$ ($\tau_V \leq t_K$), we can estimate the probability of Y via the generalized logistic regression

$$\Pr(Y \mid \mathbf{X}; \mathbf{W})_0 = \frac{\exp(\mathbf{W} * \mathbf{X} \cdot \mathbf{\Delta} \cdot \mathbb{1}(\mathbf{y} \leq 0))}{\exp(\mathbf{W} * \mathbf{X} \cdot \mathbf{\Delta} * \mathbf{A}) \cdot \mathbf{1}} \tag{1}$$

$$\mathbf{W} = (\mathbf{w}_1, \ldots, \mathbf{w}_V) \in \mathbb{R}^{D \times V}$$

$$\mathbf{w}_v = (w_{v1}, \ldots, w_{vD})^{\top} \in \mathbb{R}^{D \times 1}, \forall v = 1, 2, \ldots, V$$

$$\mathbf{W} * \mathbf{X} = (\mathbf{w}_1 \cdot \mathbf{x}_1, \ldots, \mathbf{w}_V \cdot \mathbf{x}_V) \in \mathbb{R}^{1 \times V}.$$

The regression coefficient $\mathbf{W}$ quantifies how the factors and their repeated measures affect the chance of the senior remaining free of disability, where the coefficients $\mathbf{w}_v$ shows the contribution of D measures at time τ_v. The sum of transformed measures across V time points is given by the column-wise Hadamard product [15]. To quantify the different impacts of the D factors and their V repeated measures on the appearance of Y, we develop the time-decay ratio, which is determined by the elapsed time, as follows,

$$\mathbf{\Delta} = \exp(\delta(k, v)) \in \mathbb{R}^{K \times V} \tag{2}$$

$$\delta(k, v) = -(t_k - \tau_v) \times \mathbb{1}(t_k \geq \tau_v). \tag{3}$$

Here, **exp** means the element-wise exponential matrix. The decay ratio $\exp(\sigma(k,v))$ is the exponential of the time difference between t_k and τ_v, indicating that the impact of measure at time τ_v on the probability at time t_k decreases as time elapses. The more time elapses, the more the impact is reduced. For instance, a fall-caused injury poses an effect on the onset and this effect will go down as time goes on. We use the lower triangular identity matrix $\mathbf{A} = (\alpha_1, \ldots, \alpha_K) \in \mathbb{R}^{K \times K}$, where $\alpha_{ij} = 1$ if $i \geq j$ and 0 otherwise, to explore the onset-free probabilities. The denominator means the total score (where $\mathbf{1} = (1, \ldots, 1) \in \mathbb{R}^{1 \times K}$) of the probability during the period $[t_k, t_{k+1})$.

For senior $i \in \mathcal{G}_1 = \{i | \forall i : C_i = 1\}$ with unknown onset time, the encoded responses before the stamp time are consistent. Hence, we have

$$\Pr\left(Y \mid \mathbf{X}; \mathbf{W}\right)_1 = \frac{\exp(\mathbf{W} * \mathbf{X} \cdot \boldsymbol{\Delta} * \mathbf{A} \cdot \mathbb{1}(\mathbf{y} \leq 0)) \cdot \mathbf{1}}{\exp\left(\mathbf{W} * \mathbf{X} \cdot \boldsymbol{\Delta} * \mathbf{A}\right) \cdot \mathbf{1}}. \tag{4}$$

The numerator is the sum of the risks of the target responses appearing. We learn $\mathbf{W}$ by minimizing the negative logarithm of the above likelihood across all seniors via expectation-maximization [5] with suitable initialization, as follows,

$$\min_{\mathbf{W}} \; P(\mathbf{W}) - \sum_{i \in \mathcal{G}_0} \log(\Pr(Y_i | \mathbf{X}_i; \mathbf{W})_0) - \sum_{i \in \mathcal{G}_1} \log(\Pr(Y_i | \mathbf{X}_i; \mathbf{W})_1), \tag{5}$$

where the elastic-net penalty $P(\mathbf{W}) = \lambda_1 \|\mathbf{W}\|_1 + \lambda_2 \|\mathbf{W}\|_2^2$ makes the loss function strongly convex and leads to a unique minimum. We choose the hyperparameters λ_1 and λ_2 on an independent validation set.

4.3   Prediction

Given the known measures $\mathbf{X}'$ for a new senior at times $\tau_1 < \cdots < \tau_{V'}$ before time t_k and the learned parameters $\widehat{\mathbf{W}} \in \mathbb{R}^{D \times V}$, predicting the onset-free probability at time t_k is actually predicting the encoded responses $Y'[t_k] = (1, 1, \ldots, 1) \in \mathbb{R}^{1 \times k}$. This probability, $\Pr\left(Y'[t_k] \mid \mathbf{X}'; \widehat{\mathbf{W}}\right)$, can be estimated by the temporal regression shown in Eq. 1. (In what follows, the notation with a "'" indicates it is redefined on the test set.)

4.4   Decision-Making

When our model yields the predicted onset-free probabilities at different times for the new senior, we will make two decisions: the time of onset for this senior and the time of warning.

- To estimate the time of onset, we define the relative error at time t as follows:

$$E(t) = \sum_{k=1}^{K} (|\log t - \log t_k|) \Pr\left(Y'[t_k] \mid \mathbf{X}'; \widehat{\mathbf{W}}\right),$$

thereby seeking the time at which E is the lowest, i.e., the time of onset is

$$\widehat{T} = \operatorname{argmin}_{t \in \{t_1, \ldots, t_K\}} E(t).$$

- To determine the better-timed warning, we define p_L as the threshold for warning generation, so that a warning is put in place as soon as the predicted probability goes down below this threshold, where L is the ideal lead time between the time of warning and the time of onset T. The better-timed warning can be issued at the time when either this senior becomes disabled or his/her predicted probability becomes lower than the threshold, whichever occurs first, i.e.,

$$\mathcal{T}(p_L) = \min\{T, \inf\{t| \Pr(Y[t]|\mathbf{X}; \widehat{\mathbf{W}}) < p_L\}\}, t \in \{t_1, \ldots, t_K\}.$$

For any decision-maker-determined L, the optimal policy $\widehat{p}_L$ should minimize the total cost of the warning generation, i.e.,

$$\widehat{p}_L = \operatorname{argmin} \sum_{i \in \mathcal{G}_0} Q_\alpha(L, T_i - \mathcal{T}_i(p_L)),$$

where $Q_\alpha(L, T_i - \mathcal{T}_i(p_L))$ is the cost of a warning at $T_i - \mathcal{T}_i(p_L)$ days before the actual onset time T_i for senior i. Particularly, $Q_\alpha(L, L)$ (which equals 0) is the cost of the warning at L days early and $Q_\alpha(L, 0)$ at the onset time (i.e., no early warning). We use the pinball loss [21] as the cost in this paper.

5   Experiments

5.1   Data

The Canadian Community Health Survey provided a survey between 2015 and 2017 that focused on the health of Canadians living in the ten provinces by examining various factors that affect healthy aging. In this study, we extracted data from the survey respondents from 3,604 Canadians aged 65 and over. Table 3 (top) presents the statistics of the data. For the factors, we tested for pairwise correlations between independent factors using a correlation matrix; if the Pearson correlation coefficient [10] was greater than 0.75 for two factors, we removed the one we deemed to be of lesser importance. In doing so, we obtained 24 factors, of which 12 are time-invariant and the other 12 are time-varying, see Table 3 (bottom). The numeralization of categorical factors (except the two numerical factors: age and BMI) was implemented through Softmax normalization.

5.2   Baselines

Since there are no methods particularly able to handle repeated measures, we selected as our baselines seven survival models that can deal with survival data a bit like our aging data.

- The nonparametric Kaplan-Meier estimator [9] considers the number of disabled people, m_k, at t_k and non-disabled people, n_k, by time t_k, producing the onset-free probability $\Pr(T \geq t) = \prod_{k:t_k \leq t}(1 - m_k/n_k)$.
- The parametric approach assumes $T \sim \text{Weibull}(\xi, q) = q\xi^q t^{q-1}$, where the scale of the distribution is determined by ξ and the shape by q, and predicts the probability at time t as $\Pr(T \geq t) = \exp(-(\xi t)^q)$.

Table 3. Data statistics (top) and 24 factors (bottom).

Population by	number of people
gender: M / F	2,045 (56.7%) / 1,559 (43.3%)
age group: 65–74 / 75–84 / 85+	2,089 (58.0%) / 1,188 (33.0%) / 327 (9.0%)
health group: disabled / censored	711 (19.7%) / 2,893 (80.3%)
12 time-invariant factors	**12 time-varying factors**
age, gender, BMI, marital,	caregiving, medication use, oral health,
province, education, cognition,	depression, transportation, fall, smoking,
dietary supplement, sleep, pain,	social participation, physical activity,
chronic conditions, alcohol use	care receiving, lifestyle changes, pension

- The semi-parametric Cox regression [4] estimates the probability $\Pr(T \geq t \mid \mathbf{X}) = \exp(-H_0(t)\exp(\beta\mathbf{x}))$. The baseline hazard H_0 is the cumulative hazard with no consideration of the factors and determined by the Breslow's estimator [3] and the coefficient β describes the relationship between outcome and factors. We adopted this model to include a single measure $\mathbf{x}$ at time t.
- MTLR [13] builds multiple independent logistic regressors for prediction. Given the measures $\mathbf{x}_k$ at t_k, it predicts the probability at t_k, i.e., $\Pr(T \geq t_k|\mathbf{X}) = \exp\sum_{k=1}^{K}\mathbf{w}_k\mathbf{x}_k / \sum_{k=0}^{K}\exp(\sum_{l=k+1}^{K}(\mathbf{w}_l\mathbf{x}_l))$.
- DeepHit [11] uses a neural network to predict the probability, by adding all outcomes up, i.e., $\Pr(T \geq t_k|\mathbf{X};\omega) = \sum_{k=1}^{K}\Pr(Y[t_k]|\mathbf{x}_k;\omega)$, where ω is the link between the factors and the onset. We used the single-hit setting here.
- SNN [29] uses neural networks to estimate binary classification scores of remaining onset-free at fixed time intervals, where the network's outcomes are considered as the onset-free probability at different time intervals and scaled by a sigmoid function σ: $\Pr(\tau_k|\mathbf{X};\mathbf{W}) = \sigma(\mathbf{w}_k^{\text{out}} \cdot \sigma(\mathbf{W}^{\text{hide}}\mathbf{X}^{\text{hide}}))$, where the networks' parameters $\mathbf{w}_k^{\text{out}}$ and $\mathbf{W}^{\text{hide}}$ are the weights for the output layer and the hidden layers, respectively.

5.3   Variants

To analyze the significance of our model's settings, we conducted an extensive ablation study, in which we developed our model variants by respectively reducing the time-dependent impact (i.e., $\boldsymbol{\Delta}$ in Eq. 1), the repeated measures (i.e., $\mathbf{X}$ in Eq. 1), the estimate for censoring seniors (i.e., $\Pr(Y \mid \mathbf{X};\mathbf{W})$ in Eq. 4), and the elastic-net penalty (i.e., $P(\mathbf{W})$ in Eq. 5). The four variants are thus:

- TR-censor, which ignores the seniors with no confirmed onset and does not include Eq. 4 in the objective function when learning model parameters.
- TR-impact, which ignores the time-dependent impact of repeated measures on the outcomes and causes each element of $\boldsymbol{\Delta}$ to be $\exp 0 = 1$.
- TR-static, which discards repeated measures and uses $\mathbf{W} = (0,\ldots,0,\mathbf{w}_V) \in \mathbb{R}^{D \times V}$ and $\mathbf{X} = (0,\ldots,0,\mathbf{x}_V) \in \mathbb{R}^{D \times V}$ to predict the onset-free probability.

- TR-penalty, which excludes the elastic-net penalty $P(\mathbf{W})$ from Eq. 5 and therefore learns the regression coefficients $\mathbf{W}$ without regularization.

5.4   Evaluation

To evaluate the models' predictive power, we consider three questions that a good model should be able to answer well:

- **Is the senior would be likely to be disabled in 18 months?**
- **Which one of the two seniors is more likely to be disabled?**
- **How accurate is the prediction that a senior will be disabled?**

These questions can be answered based on the following facts:

- Seniors who have an onset confirmed during the study period should have a lower onset-free probability at the end than those without an onset.
- Every senior who has an onset should have a smaller onset-free probability at his/her onset time than all those who remain onset-free.
- Seniors who have an onset of a disability at a certain time should have an onset-free probability as close as to 0%, while, for those without the onset, the probability should be approaching 100%.

We formalize these facts to develop three evaluation metrics that can be used to quantify a model's ability to address the questions. The restricted concordance index (rC-index) measures the models' predictive ability at T^* (i.e., at the end of the study). The unrestricted concordance index (uC-index) is a generalization of rC-index [20] across all comparable pairs $\mathcal{P} = \{(i,j)|\forall i,j : T_i \leq S_j\}$. The censoring mean squared error (C-MSE) is an overall error of predicted probability across all stamp times. They are defined as follows:

$$r\text{C-index} = \frac{1}{|\mathcal{G}_0'| \times |\mathcal{G}_1'|} \sum_{i \in \mathcal{G}_0'} \sum_{j \in \mathcal{G}_1'} \mathbb{1}\left(\Pr(Y_i'[T^*]|\mathbf{X}_i) < \Pr(Y_j'[T^*]|\mathbf{X}_j)\right)$$

$$u\text{C-index} = \frac{1}{|\mathcal{P}|} \sum_{i,j \in \mathcal{P}} \mathbb{1}\left(\Pr(Y_i'[T_i]|\mathbf{X}_i) \leq \Pr(Y_j'[S_j]|\mathbf{X}_j)\right)$$

$$\text{C-MSE} = \frac{1}{|\mathcal{G}_0' \cup \mathcal{G}_1'|} \sum_{i \in \mathcal{G}_0' \cup \mathcal{G}_1'} \left(C_i - \Pr(Y_i'[S_i]|\mathbf{X}_i)\right)^2 .$$

The three metrics are highly independent of each other. This means that a model which performs very well on one may not do well on the other two. A sophisticated prediction model should achieve a high rC-index and uC-index with a low C-MSE. Additionally, to evaluate the predicted onset time, we define the relative error (RE) that measures the difference between the estimated time $\widehat{T}$ and ground truth T for all seniors with a disability in the test set:

$$\text{RE} = \frac{1}{|\mathcal{G}_0'|} \sum_{\forall i \in \mathcal{G}_0'} \min\left\{|(\widehat{T}_i - T_i)/T_i|, 1\right\} .$$

Table 4. The 10 cross-validation rC-index, uC-index, C-MSE, and RE test results, expressed as mean (standard deviation). The best results are in **bold**. There is no definite answer to the ideal lead time; 60-days is deemed reasonable, according to the health care system.

		rC-index	uC-index	C-MSE	RE	avg. Q_α ($L = 60$ days)	
						$\alpha = 0.2$	$\alpha = 0.5$
Baselines	KM	.673(.031)	.669(.028)	.372(.029)	.583(.059)	20.6(2.2)	37.9(3.4)
	Weibull	.687(.027)	.682(.034)	.274(.036)	.410(.094)	32.9(6.5)	24.9(3.3)
	Cox	.693(.024)	.673(.033)	.313(.035)	.542(.073)	28.3(5.4)	26.9(2.5)
	MTLR	.744(.035)	.739(.030)	.197(.021)	.311(.053)	17.9(1.5)	31.7(3.8)
	DeepHit	.735(.032)	.701(.029)	.234(.025)	.392(.085)	23.6(4.3)	28.9(2.2)
	TR	**.757**(.022)	**.753**(.029)	**.192**(.022)	**.308**(.068)	**12.7**(2.5)	**18.3**(3.4)
Variants	TR-censor	.718(.024)	.702(.033)	.269(.022)	.426(.072)	33.9(6.5)	24.5(3.0)
	TR-impact	.730(.037)	.717(.026)	.294(.034)	.465(.068)	17.2(6.6)	29.4(1.7)
	TR-static	.698(.020)	.685(.023)	.293(.033)	.504(.072)	25.7(2.0)	25.8(4.5)
	TR-penalty	.735(.028)	.723(.032)	.238(.019)	.533(.065)	19.5(3.6)	26.4(7.1)

5.5   Results

Table 4 shows the 10 cross-validation results on the test data. Our approach outperforms all the other baselines, achieving an rC-index improvement of 1.3% and an uC-index improvement of over 2% in comparison with the second-best model MTLR. MTLR performs better than other baselines mainly due to its particular sequencing setting [13], which is a bit similar to our approach here. Overall, the machine-learning-based models (MTLR and DeepHit) perform better than the statistical models (KM, Weibull, and Cox), revealing the strong processing capabilities and applicability of multi-task learning and neural networks for dealing with complex aging data. Weibull performs badly partly because the log-logistic distribution assumption does not fit our aging distribution well. In most cases, the accuracy in terms of the uC-index is slightly lower than the rC-index because sorting the onset time of all people accurately in the log-rank test [2] is more challenging. More importantly, our approach achieves not only accurate predictions of the onset-free probability but accurate estimates of onset time, compared to all the baselines. On the other hand, TR consistently outperforms all variants. Specifically, the superior performance of TR relative to TR-censor reveals that censors are also informative for model learning. In addition, the improvement of TR over the TR-impact demonstrates the benefit of the dynamic impact of repeated measures on the outcomes. Moreover, the comparison between TR and the TR-static reveals the importance of considering repeated measures. The superior performance of TR relative to TR-penalty demonstrates the effectiveness of the regularization. By using our TR, the warning can be issued at the lowest cost in the context of a 60-day ideal lead time.

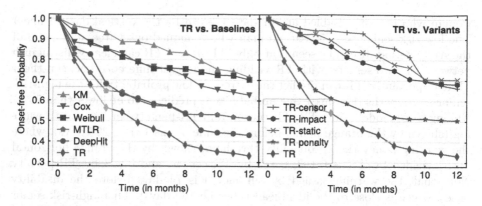

Fig. 1. Average predicted onset-free probability for all disabled seniors.

Predicted Probability. Figure 1 shows the average of the predicted onset-free probability for all disabled seniors through the 12 months. It can be seen that TR yields probabilities that differ highly from other models, where the probability produced by our model at every time interval (e.g. between 4 and 5 months) is much lower than others. The curves produced by the statistical models are close together and much higher than for the other two machine-learning-based models. Among the four variants, TR-penalty produces much lower probabilities than the other three variants.

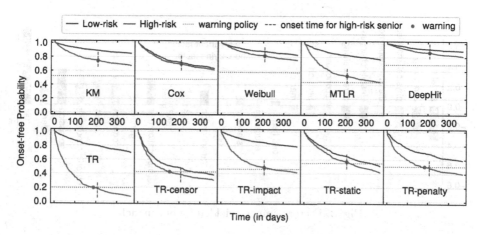

Fig. 2. Our model's predicted onset-free probability for the low-risk and high-risk seniors, estimated onset time for the high-risk senior, and warning policy.

Case Study. For the sake of investigation, Fig. 2 shows the predicted onset-free probabilities for two seniors: a high-risk versus a low-risk senior. (*N.B.:* We consider seniors who experienced a disability during the study period to be at high risk and the others at low risk.) The high-risk senior here is a 69-year-old woman who became disabled at 207 days (shown by the vertical dashed lines). The low-risk senior is a 75-year-old man who remained onset-free at the end of the study period. It can be seen that only MTLR and TR can clearly distinguish between the two seniors, with TR yielding the largest difference between them. More importantly, TR can predict the extremely low probability for the high-risk senior at 7 months (the onset-free probability is predicted to be lower than 20%). Among these models, only TR, TR-censor, and TR-penalty can generate a warning (shown by the orange point) before the time of onset (i.e., 207 days), where the warning policy (i.e., probability threshold, shown by the orange horizontal dash) yielded by TR is 0.193, 0.422 by TR-censor, and 0.503 by TR-penalty. Note that, the warning issued by our model is 74 days before the disability onset, which is close to the ideal lead time (i.e., 60 days). This high-risk senior could thus be issued a warning and offered advice on early intervention. This is crucial for aging people who are likely to have a severe disability at a specific time.

5.6   Factors' Impact

Figure 3 shows the factors' impact produced by our approach. Here positive impacts reveal a risky effect while negative for a protective effect [17]. The risk factors (e.g., fall and sleep) are associated with a higher likelihood of disability onset, while the protective factors (e.g., lifestyle changes and caregiving) have a cumulative effect on the development of disability.

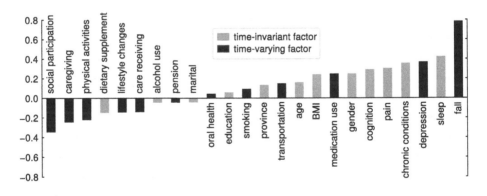

Fig. 3. Factors' impact yielded by our model.

6   Conclusions

This paper proposes a complete paradigm for modeling aging data, predicting onset-free probability, determining the time of onset and better-timed warning, and evaluating the predictions and decisions. The proposed approach successfully addresses the particularities of aging data (i.e., censored responses and repeated measures) when performing onset prediction and decision-making for a Canadian cohort. It achieves a high prediction accuracy and a low decision error and warning cost, in comparison with various baseline models and our approach's variants. It is the first attempt to develop a machine-learning-based method for predicting older people's possible disabilities, and, of course, we will further improve our approach to address other issues in aging research, e.g., the heterogeneous cohorts (e.g., interprovincial difference) and the time inconsistency (e.g., the difference between measurement times and the times of onset).

Acknowledgments. This work was supported by the Natural Sciences and Engineering Research Council of Canada (NSERC) under Discovery Grant RGPIN-2020-07110 and Discovery Accelerator Supplements Grant RGPAS-2020-00089, the National Natural Science Foundation of China (NSFC) under Grant No. U1805263.

References

1. Aalen, O.: Nonparametric estimation of partial transition probabilities in multiple decrement models. Ann. Stat. 534–545 (1978)
2. Bland, J.M., Altman, D.G.: The logrank test. BMJ **328**(7447), 1073 (2004)
3. Breslow, N.: Covariance analysis of censored survival data. Biometrics 89–99 (1974)
4. Cox, D.R.: Regression models and life tables. J. R. Stat. Soc. Ser. B Stat. Methodol. **34**, 187–220 (1972)
5. Dempster, A.P., Laird, N.M., Rubin, D.B.: Maximum likelihood from incomplete data via the EM algorithm. J. R. Stat. Soc. Ser. B Stat. Methodol. 1–38 (1977)
6. Everitt, B.S.: The analysis of repeated measures: a practical review with examples. J. R. Stat. Soc. Ser. D Stat. **44**(1), 113–135 (1995)
7. Fernández, T., Rivera, N., Teh, Y.W.: Gaussian processes for survival analysis. In: NeurIPS, pp. 5021–5029 (2016)
8. Fisher, L.D., Lin, D.Y.: Time-dependent covariates in the Cox proportional-hazards regression model. Annu. Rev. Public Health **20**(1), 145–157 (1999)
9. Kaplan, E.L., Meier, P.: Nonparametric estimation from incomplete observations. J. Amer. Stat. Assoc. **53**(282), 457–481 (1958)
10. Kirch, W. (ed.): Pearson's Correlation Coefficient, pp. 1090–1091. Springer, Netherlands, Dordrecht (2008). https://doi.org/10.1007/978-1-4020-5614-7_2569
11. Lee, C., Zame, W.R., Yoon, J., van der Schaar, M.: DeepHit: a deep learning approach to survival analysis with competing risks. In: AAAI, pp. 2314–2321 (2018)
12. Li, Y., Wang, L., Wang, J., Ye, J., Reddy, C.K.: Transfer learning for survival analysis via efficient l2,1-norm regularized Cox regression. In: ICDM, pp. 231–240 (2017)
13. Lin, H.C., Baracos, V., Greiner, R., Chun-nam, J.Y.: Learning patient-specific cancer survival distributions as a sequence of dependent regressors. In: NeurIPS, pp. 1845–1853 (2011)
14. Liu, M., Lu, W., Shore, R.E., Zeleniuch-Jacquotte, A.: Cox regression model with time-varying coefficients in nested case - control studies. Biostatistics **11**(4), 693–706 (2010)
15. Liu, S., Trenkler, G.: Hadamard, Khatri-Rao, Kronecker and other matrix products. Int. J. Inf. Syst. Sci. **4**(1), 160–177 (2008)

16. Public Health Agence of Canada: Seniors' falls in Canada (2014). https://www. phac-aspc.gc.ca/seniors-aines/publications/public/injury-blessure/seniors_falls-chutes_aines/assets/pdf/seniors_falls-chutes_aines-eng.pdf

17. Seeman, T., Chen, X.: Risk and protective factors for physical functioning in older adults with and without chronic conditions: Macarthur studies of successful aging. J. Gerontol. Ser. B **57**(3), S135–S144 (2002)

18. Statistics Canada: Canadian survey on disability reports (2017). https://www150. statcan.gc.ca/n1/pub/89-654-x/89-654-x2018002-eng.htm

19. Statistics Canada: Census of population (2021). https://www12.statcan.gc.ca/census-recensement/2021/as-sa/index-eng.cfm

20. Steck, H., Krishnapuram, B., Dehing-oberije, C., Lambin, P., Raykar, V.C.: On ranking in survival analysis: bounds on the concordance index. In: NeurIPS, pp. 1209–1216 (2008)

21. Steinwart, I., Christmann, A.: Estimating conditional quantiles with the help of the pinball loss. Bernoulli **17**(1), 211–225 (2011)

22. Sun, Y., Sundaram, R., Zhao, Y.: Empirical likelihood inference for the Cox model with time-dependent coefficients via local partial likelihood. Scand. J. Stat. **36**(3), 444–462 (2009)

23. Tian, L., Zucker, D., Wei, L.: On the Cox model with time-varying regression coefficients. J. Am. Stat. Assoc. **100**(469), 172–183 (2005)

24. United Nations Population Division: World population prospects 2022 (2022). https://www.un.org/development/desa/pd/sites/www.un.org.development.desa. pd/files/wpp2022_summary_of_results.pdf

25. Vinzamuri, B., Li, Y., Reddy, C.K.: Active learning based survival regression for censored data. In: CIKM, pp. 241–250 (2014)

26. Wei, L.J.: The accelerated failure time model: a useful alternative to the Cox regression model in survival analysis. Stat. Med. **11**(14–15), 1871–1879 (1992)

27. Wu, L., Liu, W., Yi, G.Y., Huang, Y.: Analysis of longitudinal and survival data: joint modeling, inference methods, and issues. J. Probab. Stat. **2012** (2011)

28. Zhang, J., Chen, L., Ye, Y., Guo, G., Vanasse, A., Wang, S.: Survival neural networks for time-to-event prediction in longitudinal study. Knowl. Inf. Syst. **62**, 3727–3751 (2020). https://doi.org/10.1007/s10115-020-01472-1

29. Zhang, J., Wang, S., Chen, L., Guo, G., Chen, R., Vanasse, A.: Time-dependent survival neural network for remaining useful life prediction. In: Yang, Q., Zhou, Z.-H., Gong, Z., Zhang, M.-L., Huang, S.-J. (eds.) PAKDD 2019. LNCS (LNAI), vol. 11439, pp. 441–452. Springer, Cham (2019). https://doi.org/10.1007/978-3-030-16148-4_34

A Label Embedding Method
via Conditional Covariance Maximization
for Multi-label Classification

Dan Li, Yunqian Li, Jun Li, and Jianhua Xu[✉]

School of Computer and Electronic Information, Nanjing Normal University,
Nanjing 210023, Jiangsu, China
{28190127,19190114}@stu.njnu.edu.cn, {lijuncst,xujianhua}@njnu.edu.cn

Abstract. Multi-label classification has revealed broad applications and outstanding importance in machine learning and data mining. In the multi-label domain, in addition to the problems including large instance size and high-dimensional features which need to be addressed urgently, the dimensionality of label space is increasing rapidly, leading to high computational overhead and curse of dimensionality. For tackling this issue, a typical technique called label space dimensionality reduction (LSDR) has been proposed, whose various corresponding methods are responsible for converting the high-dimensional label space into a low-dimensional one, including label embedding (LE) and label selection (LS), and then solve a low-dimensional classification or regression problem efficiently. In this paper, we focus on LE approaches, whose performance greatly depends on a preferable LE criterion. We investigate the conditional covariance operator, which originally is a nonlinear kernel-based dependence between two random vectors, formulated as a trace operation including label Gram matrix and feature Gram inverse matrix. We simplify this operator with linear label kernel matrix, which is then maximized under orthonormal constraints, resulting in an eigenvector problem. Therefore, our compression and recovery matrices correspond to those eigenvectors with several top-ranked eigenvalues. Our proposed novel LE method based on conditional covariance maximization is termed as ML-CCM for short. The experiments on four public benchmark data sets demonstrate that our proposed technique is statistically superior to three state-of-the-art LE methods on the basis of two performance evaluation metrics.

Keywords: Multi-label classification · Dimensionality reduction · Label space · Conditional covariance operator · Kernel function · Eigenvalue problem

Supported by the Natural Science Foundation of China (NSFC) under grants 62076134 and 62173186.

1 Introduction

Traditional supervised classification deals with the problem of each instance being associated with only a single class label [4,12,18], named as a single-label classification task, whose classes are mutually exclusive from one another. However, in many real-world applications, it is more common that an instance may belong to several different labels at the same time, which is referred to as a multi-label classification problem [8,27,34]. Nowadays, multi-label classification has many promising applications in a broad variety of domains, such as sound analysis [5,28], computer vision [6,11], text analysis [13,24], and recommender systems [19,41].

As much attention has been paid to multi-label classification continuously, a proliferation of methods to solve multi-label classification problems are proposed effectively. Apart from traditional large instance size and high-dimensional feature space, the scaling evolution has rendered problems in which the number of labels is extremely large [22,30,38], which results in a high computational burden and even an unsatisfactory classification performance. It is well known that dimension reduction methods can well handle multi-label high-dimensional features via feature extraction and selection [34,40]. In principle, this dimensionality reduction strategy for feature space can be also applied to multi-label high-dimensional label space [34].

Similarly to feature selection and feature extraction in feature space, the existing label space dimensionality reduction techniques can also be classified into two categories: label selection (LS) and label embedding (LE) [40]. The former method prefers to select an informative label subset directly from the original label set, and subsequently deals with a low-dimensional multi-label classification problem [23,29]. The latter method aims to convert the binary high-dimensional label space into a real or binary low-dimensional one, which then induces a low-dimensional regression or classification problem [7,34]. In this paper, we mainly focus on LE approaches.

The pioneering LE method for multi-label classification is ML-CS [20] based on compressed sensing theory (CS) [31,32]. In the training stage of ML-CS, each high-dimensional binary label vector is transformed into a low-dimensional real vector via a random Gaussian matrix, which corresponds to a multi-output regression task. In the predicting stage of ML-CS, a low-dimensional real vector is predicted via a trained regressor and then reconstructed back to its high-dimension label vector via solving an ℓ_1-regularized minimization problem. Following this framework, subsequently, various LE approaches have been proposed, which would be separated into two categories: feature-independent and feature-dependent techniques.

Among existing feature-independent techniques, only label information is exploited, as in ML-CS [20]. Several representatives include principal label space transformation (PLST) [35], cost-sensitive label embedding with multi-dimensional scaling (CLEMS) [21], binary matrix decomposition (MLC-BMaD) [39], binary linear compression (BILC) [45], minimizing local variance method (ML-mLV) [37], and etc. Particularly, PLST [35] essentially conducts principal component analysis (PCA) [1] on label space, which results in a pair of effective compression and recovery matrices with several top-ranked eigenvectors, due to the orthonormal property. This work inspires some researchers to pursue a fast recovery procedure. On the other hand, in CLEMS [21], multidimensional scaling (MDS) [33] is applied to reduce the high-dimensional label vector into a low-dimensional real one nonlinearly, which needs a nearest neighbor strategy as its independent recovery step and thus induces a kind of two-stage LE techniques, just as ML-CS [20].

In order to improve classification performance for LE techniques further, the feature information is also integrated into feature-dependent LE algorithms. The first feature-dependent LE technique is conditional principal label space transformation (CPLST) [10], which at first generalizes canonical correlation analysis (CCA) to construct its orthonormal version (OCCA), and then combines OCCA with PCA. Several typical approaches were proposed successively, such as, dependence maximization-based label space dimension reduction (DMLR) [42], Bayesian multi-label compressed sensing (BML-CS) [25], and feature-aware implicit label space encoding (FaIE) [26], label compression coding method (LCCMD) [7], and so on. In principle, these methods aim to maximize the correlations or dependency between feature space and compressed label space to design some better compression and recovery matrices.

The success of LE methods is greatly associated with their LE criteria. Therefore, how to design or search for a desirable LE criterion has become a challenging task. In this study, we pay special attention on conditional covariance operator (CCO) [2], which has been successfully applied in nonlinear feature extraction [14,15], and non-linear feature selection [9]. This operator primarily describes a nonlinear kernel dependence for two sets of random vectors, and is formulated as a trace operation consisting of label kernel matrix and inverse feature kernel matrix. To adapt to our LE task, at first, a linear label kernel matrix is applied to construct a simplified CCO (SCCO). We maximize this SCCO, subject to an orthonormal constraint, to build a constrained optimization problem, which then is converted into an eigenvalue problem. Some eigenvectors with the largest eigenvalues would be comprised of our compression matrix, whose transpose matrix becomes our recovery matrix naturally. This novel LE method is referred to as an LE method via conditional covariance maximization, or ML-CCM simply. Experiments on four public data sets illustrate that our proposed LE approach is more effective than three state-of-the-art LE techniques, including PLST [35], CPLST [10] and FaIE [26]), according to two metrics (precision@n and discounted gain@n, where $n = 1$, 3 and 5).

Summarily, there are three contributions in this study: (a) We build a new LE criterion via adapting the conditional covariance operator with linear label kernel; (b) A novel LE method (ML-CCM) is proposed as an eigenvalue problem, resulting in effective compression and recovery matrices; (c) We validate our ML-CCM experimentally via comparing with three existing LE techniques.

This paper is organized as follows. Section 2 will give some preliminaries for our work. In Sect. 3, we review conditional covariance operator, build its simplified version and propose our novel LE method based on conditional covariance maximization. Some detailed experiments are reported and analyzed in Sect. 4. Finally some conclusions and discussion will end up this paper.

2   Preliminaries

Mathematically, a multi-label training set with N instances can be formulated as follows:

$$\{(\mathbf{x}_1,\mathbf{y}_1),\dots,(\mathbf{x}_i,\mathbf{y}_i),\dots,(\mathbf{x}_N,\mathbf{y}_N)\} \tag{1}$$

where for the i-th instance its $\mathbf{x}_i \in R^D$ and $\mathbf{y}_i \in \{0,1\}^C$ refer to its D-dimensional feature column vector and C-dimensional binary label column vector, respectively.

Let $\mathbf{z}_i \in \{0,1\}^c$ denote the low-dimensional label vector of $\mathbf{y}_i$ after label compression, where c is the dimension of the compressed label vector ($c < C$). For the aforementioned training date set, we define three corresponding matrices: original feature matrix $\mathbf{X}$, original label matrix $\mathbf{Y}$ and compressed label matrix $\mathbf{Z}$, respectively, i.e.,

$$\begin{aligned}
\mathbf{X} &= [\mathbf{x}_1,\dots,\mathbf{x}_i,\dots,\mathbf{x}_N]^T \in R^{N \times D} \\
\mathbf{Y} &= [\mathbf{y}_1,\dots,\mathbf{y}_i,\dots,\mathbf{y}_N]^T = [\mathbf{y}^1,\dots,\mathbf{y}^j,\dots,\mathbf{y}^C] \in \{0,1\}^{N \times C} \\
\mathbf{Z} &= [\mathbf{z}_1,\dots,\mathbf{z}_i,\dots,\mathbf{z}_N]^T = [\mathbf{z}^1,\dots,\mathbf{z}^j,\dots,\mathbf{z}^c] \in R^{N \times c}
\end{aligned} \tag{2}$$

where the column vectors $\mathbf{y}^j$ and $\mathbf{z}^j$ indicate the j-th label in $\mathbf{Y}$ and $\mathbf{Z}$, respectively.

Here, we define a centering matrix of size $N \times N$ as follows:

$$\mathbf{H} = \mathbf{I} - \mathbf{1}\mathbf{1}^T/N \tag{3}$$

where $\mathbf{I}$ is the identity matrix of size $N \times N$, and $\mathbf{1}$ indicates the column vector with N one elements. Using $\mathbf{H}$, we could centralize feature and label matrices:

$$\begin{aligned}
\bar{\mathbf{X}} &= \mathbf{H}\mathbf{X} = [\bar{\mathbf{x}}_1,\dots,\bar{\mathbf{x}}_i,\dots,\bar{\mathbf{x}}_N]^T \\
\bar{\mathbf{Y}} &= \mathbf{H}\mathbf{Y} = [\bar{\mathbf{y}}_1,\dots,\bar{\mathbf{y}}_i,\dots,\bar{\mathbf{y}}_N]^T.
\end{aligned} \tag{4}$$

In particular, the mean of label matrix $\mathbf{Y}$ is defined as

$$\mathbf{m}_y = \frac{1}{N}\sum_{i=1}^{N}\mathbf{y}_i \tag{5}$$

which will be used to recover a high-dimension label vector from its centralized predicted label vector.

Suppose that the feature and label kernel matrices are represented as $\mathbf{K}_x$ and $\mathbf{K}_y$, respectively, which are usually centralized doubly to obtain two Gram matrices, i.e.,

$$\mathbf{G}_x = \mathbf{H}\mathbf{K}_x\mathbf{H}, \quad \mathbf{G}_y = \mathbf{H}\mathbf{K}_y\mathbf{H} \tag{6}$$

which will usually used in some kernel-based criteria [9].

The original multi-label classification is to learn a classifier $f : \mathbf{x} \in R^D \rightarrow \mathbf{y} \in \{0,1\}^C$, directly. However, for LE methods for multi-label classification, we need to design a label compression operator: $\phi : \mathbf{y} \rightarrow \mathbf{z}$, and its corresponding inverse recovery operator $\phi^{-1} : \mathbf{z} \rightarrow \mathbf{y}$.

For some linear LE methods, the aforementioned compression operator is simplified into a matrix $\phi = \mathbf{Q}$. Further, if $\mathbf{Q}$ satisfies an orthonormal property, i.e., $\mathbf{Q}^T\mathbf{Q} = \mathbf{I}$, we have $\phi^{-1} = \mathbf{Q}^T$.

When $\mathbf{z}$ is a binary vector, we still learn a multi-label or multi-class classifier $g : \mathbf{x} \rightarrow \mathbf{z} \in \{0,1\}^c$, otherwise a multi-output regressor $g : \mathbf{x} \rightarrow z \in R^c$.

In general, an LE-type method for multi-label classification consists of two more sophisticated training and testing phases [7,8], which are embedded into compression and recovery operators, respectively. During its training phase, there are three key missions: (a) designing a compression operator and its corresponding recovery operator; (b) compressing each high-dimensional label vector into a low-dimensional compressed label vector; and (c) training a multi-label/class classifiers, or a multi-output regressor, on the reduced training data set. In its testing phase, three steps are executed: (a) predicting a low-dimensional label vector for each testing instance using the trained classifier or regressor; (b) recovering a high-dimensional label vector from the predicted low-dimensional label vector by the recovery operator; and (c) evaluating some classification for an entire test data set.

In the next section, we will propose a novel LE-like method for multi-label classification.

3 A Novel Label Embedding Approach via Conditional Covariance Maximization

In this section, we review the conditional covariance operator [2], and then maximize this operator to introduce our novel label embedding algorithm for multi-label classification.

3.1 Conditional Covariance Operator and Its Simplified Form

The conditional covariance operator is primarily introduced by Baker [2], which characterizes a conditional dependence for two sets of random variables. It is well known that this operator has been applied to nonlinear feature extraction [14,15], and nonlinear feature selection [9]. In this sub-section, we will show a brief overview of this operator and then define its simplified version.

Let $(\mathbb{H}_x, k_x)$ and $(\mathbb{H}_y, k_y)$ denote the reproducing kernel Hilbert spaces (RKHSs) of the functions on spaces $\mathcal{X}$ and $\mathcal{Y}$, respectively, where $k(\cdot,\cdot)$ indicates some specific kernel function. Additionally, a pair of random vectors $(\mathbf{x}, \mathbf{y})$ is defined on $\mathcal{X} \times \mathcal{Y}$ with joint distribution $P_{x,y}$. Assuming that the kernels k_x and k_y are bounded in expectation:

$$\begin{aligned} \mathbb{E}_x\left[k_x(\mathbf{x}, \mathbf{x})\right] < \infty \\ \mathbb{E}_y\left[k_y(\mathbf{y}, \mathbf{y})\right] < \infty. \end{aligned} \tag{7}$$

Under this assumption, the cross-covariance operator associated with a pair $(\mathbf{x}, \mathbf{y})$ is the mapping $\boldsymbol{\Sigma}_{xy} : \mathbb{H}_x \to \mathbb{H}_y$, defined by the following relation for all $f \in \mathbb{H}_x$ and $g \in \mathbb{H}_y$

$$\langle g, \boldsymbol{\Sigma}_{yx} f \rangle_{\mathbb{H}_y} = \mathbb{E}_{x,y}[(f(\mathbf{x}) - \mathbb{E}_x[f(\mathbf{x})])(g(\mathbf{y}) - \mathbb{E}_y[g(\mathbf{y})])] \tag{8}$$

where $\langle \cdot, \cdot \rangle$ indicates the inner product operation.

Baker [2] showed there is a unique bounded operator $\mathbf{V}_{yx}$ such that

$$\boldsymbol{\Sigma}_{yx} = \boldsymbol{\Sigma}_{yy}^{1/2} \mathbf{V}_{yx} \boldsymbol{\Sigma}_{xx}^{1/2}. \tag{9}$$

Then the definition of the conditional covariance operator can be given as:

$$\boldsymbol{\Sigma}_{yy|x} = \boldsymbol{\Sigma}_{yy} - \boldsymbol{\Sigma}_{yy}^{1/2} \mathbf{V}_{yx} \mathbf{V}_{xy} \boldsymbol{\Sigma}_{yy}^{1/2}. \tag{10}$$

Fukumizu et al. [14,15] showed that the conditional covariance operator describes the conditional covariance of $\mathbf{y}$ given $\mathbf{x}$.

It is also indicated that the trace of the conditional covariance operator can characterize the dependence between $\mathbf{x}$ and $\mathbf{y}$ in [9].

For the training set (1), according to [15], the empirical estimate of the conditional covariance operator is defined as:

$$\begin{aligned} \text{CCO} &:= \text{trace}(\boldsymbol{\Sigma}_{yy|x}) \\ &= \text{trace}[\boldsymbol{\Sigma}_{yy} - \boldsymbol{\Sigma}_{yx}(\boldsymbol{\Sigma}_{xx} + \varepsilon I)^{-1} \boldsymbol{\Sigma}_{xy}] \\ &= \varepsilon \text{trace}[\mathbf{G}_y(\mathbf{G}_x + N\varepsilon I)^{-1}] \end{aligned} \tag{11}$$

where $\mathbf{G}_x$, $\mathbf{G}_y$ are defined in (6), and ε is a regularization constant to avoid the singularity of inverse matrix.

In the above CCO (11), there are two Gram matrices, which are associated with a nonlinear kernel function. For feature kernels, three kernel functions have been widely investigated, i.e., linear, polynomial and Gaussian kernels [17,36], which are defined as follows:

$$\begin{aligned} k(\mathbf{x}_i, \mathbf{x}_j) &= \mathbf{x}_i^T \mathbf{x}_j \quad \text{(linear kernel)} \\ k(\mathbf{x}_i, \mathbf{x}_j) &= (\mathbf{x}_i^T \mathbf{x}_j + 1)^p \quad \text{(polynomail kernel)} \\ k(\mathbf{x}_i, \mathbf{x}_j) &= \exp\left(-\frac{\|\mathbf{x}_i - \mathbf{x}_j\|^2}{2\sigma^2}\right) \quad \text{(Gaussian kernel)} \end{aligned} \tag{12}$$

where p is the degree of polynomial kernel and σ is the width of Gaussian kernel. It is worth noting that these kernels could be also calculated using the centralized feature vectors.

For the label kernel, we use the linear kernel, which corresponds to the following Gram matrix:

$$\mathbf{G}_y = \mathbf{H}\mathbf{Y}\mathbf{Y}^T\mathbf{H} = \bar{\mathbf{Y}}\bar{\mathbf{Y}}^T. \tag{13}$$

In this setting, we obtain a simplified CCO:

$$\text{SCCO} = trace\left(\bar{\mathbf{Y}}\bar{\mathbf{Y}}^T(\mathbf{G}_x + N\varepsilon\mathbf{I})\right) \tag{14}$$

which will be regarded as an LE criterion to build our novel linear LE method.

3.2   A Novel Label Embedding Method for Multi-label Classification

Now, let $\mathbf{Q}$ be a compression matrix, which reduces the centralized high-dimension label vector into a low-dimensional one, i.e.,

$$\mathbf{Z} = \bar{\mathbf{Y}}\mathbf{Q}. \tag{15}$$

The above SCCO (14) is applied to measure the dependence between the reduced label matrix $\mathbf{Z}$ and the original feature matrix $\bar{\mathbf{X}}$:

$$\begin{aligned} \text{SCCO} &= trace\left(\mathbf{ZZ}^T(\mathbf{G}_x + N\varepsilon\mathbf{I})\right) \\ &= trace\left(\mathbf{Z}^T(\mathbf{G}_x + N\varepsilon\mathbf{I})\mathbf{Z}\right) \\ &= trace\left(\mathbf{Q}^T\bar{\mathbf{Y}}^T(\mathbf{G}_x + N\varepsilon\mathbf{I})\bar{\mathbf{Y}}\mathbf{Q}\right). \end{aligned} \tag{16}$$

Further, we apply an orthonormal constraint on $\mathbf{Q}$, i.e., $\mathbf{Q}^T\mathbf{Q} = \mathbf{I}$, and maximize the aforementioned SCCO to build the following maximization problem:

$$\begin{aligned} \max \quad & trace(\mathbf{Q}^T\bar{\mathbf{Y}}^T(\mathbf{G}_x + N\varepsilon\mathbf{I})^{-1}\bar{\mathbf{Y}}\mathbf{Q}) \\ \text{s.t.} \quad & \mathbf{Q}^T\mathbf{Q} = \mathbf{I}. \end{aligned} \tag{17}$$

In order to solve this task, we apply the Lagrangian method [3], whose function could be defined as

$$J(\mathbf{Q}, \mathbf{\Lambda}) = trace\left(\mathbf{Q}^T\bar{\mathbf{Y}}^T(\mathbf{G}_x + N\varepsilon\mathbf{I})^{-1}\bar{\mathbf{Y}}\mathbf{Q} - \mathbf{\Lambda}(\mathbf{Q}^T\mathbf{Q} - \mathbf{I})\right) \tag{18}$$

where $\mathbf{\Lambda} = \text{diag}(\lambda_1, ..., \lambda_C)$ denotes a diagonal matrix composed of non-negative Lagrangian multipliers. Based on the trace differential equation in [16,43,44]: $\frac{\partial tr(\mathbf{AXX}^T\mathbf{B})}{\partial \mathbf{X}} = (\mathbf{BA} + \mathbf{A}^T\mathbf{B}^T)\mathbf{X}$, we have

$$\frac{\partial J}{\partial \mathbf{Q}} = 2(\bar{\mathbf{Y}}^T(\mathbf{G}_x + N\varepsilon_N\mathbf{I})^{-1}\bar{\mathbf{Y}})\mathbf{Q} - 2\mathbf{\Lambda}\mathbf{Q}. \tag{19}$$

Let this relation be zeros, we can obtain an eigenvalue problem:

$$\bar{\mathbf{Y}}^T(\mathbf{G}_x + \varepsilon N\mathbf{I})^{-1}\bar{\mathbf{Y}}\mathbf{Q} = \mathbf{\Lambda}\mathbf{Q} \tag{20}$$

where $\mathbf{\Lambda}$ also could be explained as a diagonal matrix with non-negative eigenvalues from the matrix $\bar{\mathbf{Y}}^T(\mathbf{G}_x + N\varepsilon\mathbf{I})^{-1}\bar{\mathbf{Y}}$, and $\mathbf{Q}$ are those corresponding eigenvectors. In this case, the problem can be rewritten as

$$J(\mathbf{Q}, \mathbf{\Lambda}) = trace(\mathbf{\Lambda}) = \sum_{i=1}^{C} \lambda_i. \tag{21}$$

In order to maximize the objective in (17), according to (21), we ought to select the largest c eigenvalues and their corresponding eigenvector to construct a compression matrix as our compression operator $\phi = \mathbf{Q} \in R^{C \times c}$. On the other hand, due to $\mathbf{Q}^T\mathbf{Q} = \mathbf{I}$, our recovery operator becomes a transpose matrix of $\mathbf{Q}$, i.e., $\phi^{-1} = \mathbf{Q}^T \in R^{c \times C}$. Therefore, our compression and recovery procedures could be formulated as

---

Algorithm 1. Pseudo-code of our ML-CCM

Training Stage:

Input:

    $\mathbf{X}$ and $\mathbf{Y}$: feature and label matrices.

    c: reduced dimensionality of label space.

    $k_x(\cdot, \cdot)$: a proper feature kernel function and its parameter.

Process:

  1: To centralize feature and label matrices to obtain $\bar{\mathbf{X}}$ and $\bar{\mathbf{Y}}$.

  2: To normalize the training data set if neccesdary.

  3: To calculate the feature kernel matrix $\mathbf{K}_x$ and its Gram matrix $\mathbf{G}_x$.

  4: To solve the eigenvalue problem (20).

  5: To determine compression matrix $\mathbf{Q}$ with the c largest eigenvalues and their eigenvectors (column vectors).

  6: To calculate compressed label matrix $\mathbf{Z} = \bar{\mathbf{Y}}\mathbf{Q}$.

  7: To learn a regressor $g(\mathbf{x}) : \mathbf{x} \rightarrow \mathbf{z}$.

Output:

    $\mathbf{Q}$: the compression matrix of size $C \times c$.

    $g(\mathbf{x})$: a trained regressor.

Testing Stage:

Input:

    $\mathbf{x}$: some testing instance.

    $\mathbf{Q}$: the recovery matrix.

    $g(\mathbf{x})$: the trained regressor.

Process:

  1: To calculate the c-dimensional real label vector $\mathbf{z} = g(\mathbf{x})$.

  2: To reconstruct the C-dimensional real label vector $\bar{\mathbf{y}} = \mathbf{Q}\mathbf{z}$.

  3: To de-centering operation to achieve a real label vector $\hat{\mathbf{y}}$ if necessary.

  4: To detect a binary label vector $\mathbf{y}$ using round operation on $\hat{\mathbf{y}}$.

Output:

    $\hat{\mathbf{y}}$ or $\mathbf{y}$: a predicted high-dimensional real or binary label vector.

$$\mathbf{z} = \mathbf{Q}^T \bar{\mathbf{y}} \text{ or } \mathbf{Z} = \bar{\mathbf{Y}}\mathbf{Q}$$
$$\bar{\mathbf{y}} = \mathbf{Q}\mathbf{z} \text{ or } \bar{\mathbf{Y}} = \mathbf{Z}\mathbf{Q}^T. \tag{22}$$

Further, we de-centralize $\bar{\mathbf{y}}$ to obtain a high-dimensional label vector $\hat{\mathbf{y}}$ using the label mean $\mathbf{m}_y$ in (5), which could directly be used to calculate the ranking-based metrics [8]. For instance-based and label-based metrics [8], we need to recover its binary label vector $\mathbf{y}$.

Based on the above description, we propose a novel label embedding algorithm in label space dimensionality reduction for multi-label classification, which is summarized as the training stage and the testing stage in **Algorithm 1**. This label embedding method via conditional covariance maximization proposed in this paper is termed as ML-CCM for short.

Table 1. Statistics of four multi-label data sets.

Dataset	Domain	Train	Test	Features	Labels	Cardinality
Birds	Audio	179	172	260	19	1.01
CAL500	Music	300	202	68	174	26.04
Human	Biology	1864	1244	440	14	1.19
Plant978	Biology	588	390	440	12	1.08

4 Experiments

In this section, we evaluate our proposed ML-CCM on four public benchmark data sets, and compare ML-CCM with three classic LE approaches including PLST [35], CPLST [10] and FaIE [26]. To begin with, we briefly introduce the four data sets and two different evaluation metrics used in our experiments. Subsequently, extensive comparative studies are also presented.

4.1 Four Data Sets and Two Evaluation Metrics

To validate the performance of our algorithm, our experiments are carried out on four benchmark multi-label data sets downloaded from Mulan[1] and Labic[2], including Birds, CAL500, Human and Plant978, which come from three application domains (audio, music and biology). Table 1 shows some detailed information with respect to these four data sets, e.g., the number of instances in split training and testing, the dimensionality of features and labels, and the label cardinality.

Two ranking-based evaluation metrics coming from [22] are used as our performance measure for various label sets: precision@n and (DisCounted Gain) DCG@n (n=1, 2, 3, ...). For a testing instance $\mathbf{x}$, its ground label vector is $\mathbf{y} = [y_1, ..., y_i, ...y_C]^T$ and predicted function values $\hat{\mathbf{y}} = [\hat{y}_1, ..., \hat{y}_i, ...\hat{y}_C]^T$, and then such two metrics can be described as follows:

$$
\begin{aligned}
\text{Precision@}n &= \frac{1}{n} \sum_{i \in rank_n(\hat{\mathbf{y}})} y_i \\
\text{DCG@}n &= \frac{1}{n} \sum_{i \in rank_n(\hat{\mathbf{y}})} \frac{y_i}{\log_2(i+1)}
\end{aligned}
\tag{23}
$$

where $rank_n(\hat{\mathbf{y}})$ returns the top n label indexes of $\hat{\mathbf{y}}$. For statistical comparison, average scores are calculated for overall performance measure across all testing instances. In addition, it is expected to achieve higher metric values for a better performance on a certain technique.

4.2 Three Compared Methods

In this subsection, we introduce three compared methods mathematically, i.e., PLST [35], CPLST [10] and FaIE [26].

---

[1] https://mulan.sourceforge.net/datasets-mlc.html.
[2] http://ceai.njnu.edu.cn/Lab/LABIC/LABIC_Software.html.

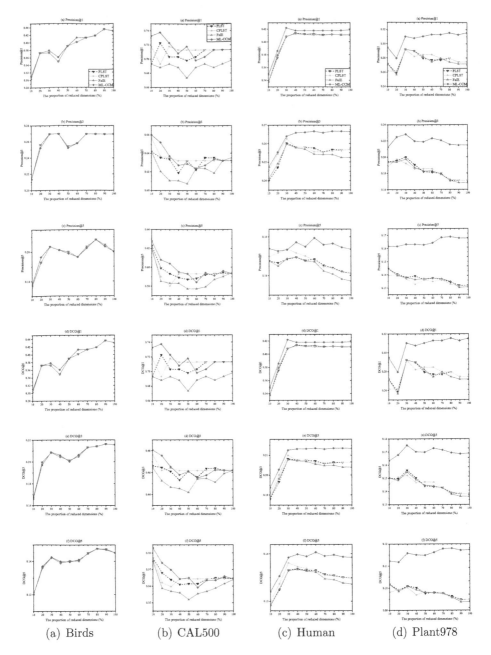

(a) Birds   (b) CAL500   (c) Human   (d) Plant978

Fig. 1. Performance of the proposed ML-CCM using two metrics on the four benchmark data sets.

Table 2. The number of tops for each method and metric across four data sets.

Metric	Precision@1	Precision@3	Precision@5	DCG@1	DCG@3	DCG@5	Total tops
PLST	11	15	9	11	10	8	64
CPLST	13	13	12	13	11	10	72
FaIE	7	10	9	7	7	6	46
ML-CCM	**36**	**33**	**33**	**36**	**32**	**33**	**203**

Table 3. Two metrics from four methods and four data sets with 30% reduced labels.

Metric	Precision@1	Precision@3	Precision@5	DCG@1	DCG@3	DCG@5
Birds						
PLST	0.37209	**0.26938**	0.20349	0.37209	0.20838	0.14229
CPLST	0.37209	**0.26938**	0.20349	0.37209	0.20838	0.14229
FaIE	0.37209	**0.26938**	0.20349	0.37209	0.20838	0.14229
ML-CCM	**0.37791**	**0.26938**	**0.20465**	**0.37791**	**0.20884**	**0.14252**
CAL500						
PLST	0.70297	0.63366	0.59307	0.70297	0.46100	0.36390
CPLST	0.69307	0.63531	0.58020	0.71782	0.46409	0.35973
FaIE	0.71782	0.61056	0.58317	0.69307	0.44672	0.35690
ML-CCM	**0.72277**	**0.63696**	**0.60396**	**0.72277**	**0.46530**	**0.36988**
Human						
PLST	0.40836	0.26018	0.18424	0.40836	0.20815	0.13654
CPLST	0.40916	0.26259	**0.18682**	0.40916	0.20995	0.13808
FaIE	0.40836	0.25965	0.18441	0.40836	0.20778	0.13651
ML-CCM	**0.42203**	**0.26393**	0.18666	**0.42203**	**0.21258**	**0.13924**
Plant978						
PLST	0.29231	0.20000	0.13795	0.29231	0.15599	0.10097
CPLST	0.29231	0.19231	0.13692	0.29231	0.15215	0.10029
FaIE	0.29231	0.19744	0.13795	0.29231	0.15471	0.10088
ML-CCM	**0.31539**	**0.22821**	**0.16256**	**0.31539**	**0.17573**	**0.11606**

Essentially, the PLST [35] conducts principal component analysis on the centralized label matrix, whose eigenvalue problem is formulated as

$$\bar{Y}^T \bar{Y} Q = \Lambda Q. \tag{24}$$

The CPLST [10] considers the feature information, whose eigenvalue problem is described as

$$\bar{Y}^T \left(\bar{X} \bar{X}^+ \right) \bar{Y} Q = \Lambda Q \tag{25}$$

where $\bar{X}^+$ is the Penrose-Moore inverse of $\bar{X}$ [44].

The FaIE [26] is formulated as the following eigenvalue problem:

$$\left(\bar{\mathbf{Y}}^T \bar{\mathbf{Y}} + \alpha \bar{\mathbf{X}}^T \left(\bar{\mathbf{X}}^T \bar{\mathbf{X}}\right)^{-1} \bar{\mathbf{X}}\right) \mathbf{Z} = \Lambda \mathbf{Z} \tag{26}$$

where α is a balanced factor to control the tradeoff between two terms. It is worth noting that the top-ranked values are used to detect several eigenvectors to construct compressed matrix $\mathbf{Z}$ directly and the recovery operator becomes $\mathbf{Z}^T \bar{\mathbf{Y}}$.

4.3   Experimental Settings

In our experiments, we validate our method ML-CCM and three existing methods: PLST [35], CPLST [10], and FaIE [26]. For this purpose, a linear ridge regressor is chosen as our baseline, in which the regularization constant is set to 0.01. In order to investigate the effect of the number of reduced labels (i.e., c) on classification performance, the dimension proposition after label reduction is predefined ranging from 10% to 100% with a step of 10%.

As for feature kernel functions in our ML-CCM, we have provided three popular functions, i.e., linear, polynomial and Gaussian kernel forms. After a primary experimental comparison, we pick up the Gaussian kernel, whose width parameter σ is set to the number of features, just as in [9]. For FaIE, we use its default and recommended setting: $\alpha = 1.0$. Additionally, we set $n = 1, 3$, and 5 for two evaluation metrics (23).

4.4   Performance Comparison and Analysis

According to the aforementioned experimental settings, two classification metrics focus on summarizing different dimension proportions after compressed label dimensionality, as shown in Fig. 1.

It can be observed from these figures that our method ML-CCM performs the best compared with the other techniques, especially when the label dimensionality is compressed to around $20 \sim 40\%$. On Human and Plant978, ML-CCM even achieves a complete victory, which benefits from the key part in our method for maximizing the correlation between feature space and reduced label space.

To provide more insight into these experimental results, we also use the top index in [7], which indicates how many times some method performs the best among four approaches. As can be seen from Table 2, the number of tops in ML-CCM is at least twice as many as, or up to four times as much as, those in the other three methods.

Moreover, given the 30% reduced dimensionality in label space, Table 3 shows two evaluation metric values. It is found out that our ML-CCM achieves 22 wins (i.e., the highest value), an incredible percentage of 91.67%, and one equivalent value as the other three methods for the best metric values. Additionally, some results have much higher values than, or even three times as many as, those from three state-of-the-art compared methods. These results substantially demonstrate the superiority of our proposed ML-CCM when compared with the other three competitors.

5   Conclusions

To copy with the issue on high-dimensional label space, the label dimensionality reduction (including label selection and label embedding) has become one important line of research in multi-label classification. In particular, the success of a label embedding approach largely depends on a desirable LE criterion. In this paper, we investigated the conditional covariance operator, originally formulated as a trace operation consisting of label Gram kernel matrix and feature Gram kernel inverse matrix. This operator is simplified via linear label kernel from scratch, and is then maximized under an orthonormal constraint to build our label embedding method based on conditional covariance maximization. Thus, our proposed technique is finally described as an eigenvalue problem, which could be solved by various existing solvers. Comprehensive evaluations and extensive comparative studies on the four diverse benchmark data sets demonstrate the efficacy and superiority of our ML-CCM compared with the other LE approaches.

In our future work, more benchmark data sets and advanced label embedding methods will be involved in our experiments for further performance evaluation.

References

1. Abdi, H., Williams, L.J.: Principal component analysis. Wiley Interdiscip. Rev. Comput. Stat. **2**(4), 433–459 (2010)
2. Baker, C.R.: Joint measures and cross-covariance operators. Trans. Am. Math. Soc. **186**, 273–289 (1973)
3. Bertsekas, D.P.: Nonlinear programming. J. Oper. Res. Soc. **48**(3), 334–334 (1997)
4. Bishop, C.M.: Pattern Recognition and Machine Learning. Springer, Singapore (2006)
5. Briggs, F., et al.: New methods for acoustic classification of multiple simultaneous bird species in a noisy environment. In: Proceedings of 2013 IEEE International Workshop on Machine Learning for Signal Processing, pp. 1–8 (2013)
6. Cabral, R., De la Torre, F., Costeira, J.P., Bernardino, A.: Matrix completion for weakly-supervised multi-label image classification. IEEE Trans. Pattern Anal. Mach. Intell. **37**(1), 121–135 (2014)
7. Cao, L., Xu, J.: A label compression coding approach through maximizing dependence between features and labels for multi-label classification. In: Proceedings of the 27th International Joint Conference on Neural Networks, pp. 1–8 (2015)
8. Herrera, F., Charte, F., Rivera, A.J., del Jesus, M.J.: Multilabel Classification: Problem Analysis, Metrics and Techniques. Springer, Cham (2016). https://doi.org/10.1007/978-3-319-41111-8
9. Chen, J., Stern, M., Wainwright, M.J., Jordan, M.I.: Kernel feature selection via conditional covariance minimization. In: Proceedings of the 31st International Conference on Neural Information Processing Systems, pp. 6949–6958 (2017)
10. Chen, Y.N., Lin, H.T.: Feature-aware label space dimension reduction for multi-label classification. In: Proceedings of the 25th International Conference on Neural Information Processing Systems, pp. 1529–1537 (2012)
11. Deng, J., Dong, W., Socher, R., Li, L.J., Li, K., Li, F.F.: ImageNet: a large-scale hierarchical image database. In: Proceedings of 2009 IEEE Conference on Computer Vision and Pattern Recognition, pp. 248–255 (2009)

12. Duda, R.O., Hart, P.E., Stork, D.G.: Pattern Classification, 2nd edn. John Wiley and Sons, New York (2001)
13. Elghazel, H., Aussem, A., Gharroudi, O., Saadaoui, W.: Ensemble multi-label text categorization based on rotation forest and latent semantic indexing. Expert Syst. Appl. **57**, 1–11 (2016)
14. Fukumizu, K., Bach, F.R., Jordan, M.I.: Dimensionality reduction for supervised learning with reproducing kernel hilbert spaces. J. Mach. Learn. Res. **5**, 73–99 (2004)
15. Fukumizu, K., Bach, F.R., Jordan, M.I.: Kernel dimension reduction in regression. Ann. Stat. **37**(4), 1871–1905 (2009)
16. Gentle, J.E.: Matrix Algebra: Theory, Computations and Applications in Statistics, 2nd edn. Springer, Switzerland (2007). https://doi.org/10.1007/978-0-387-70873-7
17. Genton, M.G.: Classes of kernels for machine learning: a statistics perspective. J. Mach. Learn. Res. **2**, 299–312 (2001)
18. Goodfellow, I., Bengio, Y., Courville, A.: Deep Learning. MIT Press, Cambridge MA, USA (2016)
19. Herlocker, J.L., Konstan, J.A., Terveen, L.G., Riedl, J.T.: Evaluating collaborative filtering recommender systems. ACM Trans. Inf. Syst. **22**(1), 5–53 (2004)
20. Hsu, D., Kakade, S.M., Langford, J., Zhang, T.: Multi-label prediction via compressed sensing. In: Proceedings of the 22nd Annual Conference on Neural Information Processing Systems, pp. 772–780 (2009)
21. Huang, K.H., Lin, H.T.: Cost-sensitive label embedding for multi-label classification. Mach. Learn. **106**(9), 1725–1746 (2017)
22. Jain, H., Prabhu, Y., Varma, M.: Extreme multi-label loss functions for recommendation, tagging, ranking & other missing label applications. In: Proceedings of the 22nd ACM SIGKDD International Conference on Knowledge Discovery and Data Mining, pp. 935–944 (2016)
23. Ji, T., Li, J., Xu, J.: Label selection algorithm based on Boolean interpolative decomposition with sequential backward selection for multi-label classification. In: Proceedings of the 16th International Conference on Document Analysis and Recognition, pp. 130–144 (2021)
24. Jiang, J.Y., Tsai, S.C., Lee, S.J.: FSKNN: multi-label text categorization based on fuzzy similarity and k nearest neighbors. Expert Syst. Appl. **39**(3), 2813–2821 (2012)
25. Kapoor, A., Jain, P., Viswanathan, R.: Multilabel classification using Bayesian compressed sensing. In: Proceedings of the 25th Annual Conference on Neural Information Processing Systems, pp. 2645–2653 (2012)
26. Lin, Z., Ding, G., Hu, M., Wang, J.: Multi-label classification via feature-aware implicit label space encoding. In: Proceedings of the 31st International Conference on Machine Learning, pp. 325–333 (2014)
27. Liu, W., Shen, X., Wang, H., Tsang, I.W.: The emerging trends of multi-label learning. IEEE Trans. Pattern Anal. Mach. Intell. **44**(11), 7955–7974 (2022)
28. Lo, H.Y., Wang, J.C., Wang, H.M., Lin, S.D.: Cost-sensitive multi-label learning for audio tag annotation and retrieval. IEEE Trans. Multimed. **13**(3), 518–529 (2011)
29. Peng, T., Li, J., Xu, J.: Label selection algorithm based on iteration column subset selection for multi-label classification. In: Proceedings of The 33th International Conference on Database and Expert Systems Applications, pp. 287–301 (2022)

30. Prabhu, Y., Varma, M.: FastXML: a fast, accurate and stable tree-classifier for extreme multi-label learning. In: Proceedings of the 20th ACM SIGKDD International Conference on Knowledge Discovery and Data Mining, pp. 263–272 (2014)
31. Qaisar, S., Bilal, R.M., Iqbal, W., Naureen, M., Lee, S.: Compressive sensing: from theory to applications, a survey. J. Commun. Netw. **15**(5), 443–456 (2013)
32. Rani, M., Dhok, S.B., Deshmukh, R.B.: A systematic review of compressive sensing: concepts, implementations and applications. IEEE Access **6**, 4875–4894 (2018)
33. Saeed, N., Nam, H., Haq, M.I.U., Bhatti, D.M.S.: A survey on multidimensional scaling. ACM Comput. Surv. **51**(1), Article-1 (25pages) (2018)
34. Siblini, W., Kuntz, P., Meyer, F.: A review on dimensionality reduction for multi-label classification. IEEE Trans. Knowl. Data Eng. **33**(3), 839–857 (2019)
35. Tai, F., Lin, H.T.: Multilabel classification with principal label space transformation. Neural Comput. **24**(9), 2508–2542 (2012)
36. Vapnik, V.N.: Statistical Learning Theory. Wiley, New York (1998)
37. Wang, X., Li, J., Xu, J.: A label embedding method for multi-label classification via exploiting local label correlations. In: Gedeon, T., Wong, K.W., Lee, M. (eds.) ICONIP 2019. CCIS, vol. 1143, pp. 168–180. Springer, Cham (2019). https://doi.org/10.1007/978-3-030-36802-9_19
38. Weston, J., Makadia, A., Yee, H.: Label partitioning for sublinear ranking. In: Proceedings of the 30th International Conference on Machine Learning, pp. 181–189 (2013)
39. Wicker, J., Pfahringer, B., Kramer, S.: Multi-label classification using Boolean matrix decomposition. In: Proceedings of the 27th Annual ACM Symposium on Applied Computing, pp. 179–186 (2012)
40. Xu, J., Mao, Z.H.: Multilabel feature extraction algorithm via maximizing approximated and symmetrized normalized cross-covariance operator. IEEE Trans. Cybern. **51**(7), 3510–3523 (2021)
41. Yang, B., Lei, Y., Liu, J., Li, W.: Social collaborative filtering by trust. IEEE Trans. Pattern Anal. Mach. Intell. **39**(8), 1633–1647 (2016)
42. Zhang, J.J., Fang, M., Wang, H., Li, X.: Dependence maximization based label space dimension reduction for multi-label classification. Eng. Appl. Artif. Intell. **45**, 453–463 (2015)
43. Zhang, X.D.: Matrix Analysis and Applications. Cambridge University Press, Cambridge, UK (2017)
44. Zhang, X.D.: A Matrix Algebra Approach to Artificial Intelligence. Springer, New York (2020). https://doi.org/10.1007/978-981-15-2770-8
45. Zhou, W.J., Yu, Y., Zhang, M.L.: Binary linear compression for multi-label classification. In: Proceedings of the 26th International Joint Conference on Artificial Intelligence, pp. 3546–3552 (2017)

Integrally Private Model Selection
for Deep Neural Networks

Ayush K. Varshney$^{(\boxtimes)}$ and Vicenç Torra

Department of Computing Sciences, Umeå University, 90740 Umeå, Sweden
{ayushkv,vtorra}@cs.umu.se

Abstract. Deep neural networks (DNNs) are one of the most widely used machine learning algorithms. In the literature, most of the privacy related work to DNNs focus on adding perturbations to avoid attacks in the output which can lead to significant utility loss. Large number of weights and biases in DNNs can result in a unique model for each set of training data. In this case, an adversary can perform model comparison attacks which lead to the disclosure of the training data. In our work, we first introduce the model comparison attack for DNNs which accounts for the permutation of nodes in a layer. To overcome this, we introduce a relaxed notion of integral privacy called ϵ-integral privacy. We further provide a methodology for recommending ϵ-Integrally private models. We use a data-centric approach to generate subsamples which have the same class-distribution as the original data. We have experimented with 6 datasets of varied sizes (10k to 7 million instances) and our experimental results show that our recommended private models achieve benchmark comparable utility. We also achieve benchmark comparable test accuracy for 4 different DNN architectures. The results from our methodology show superiority under comparison with three different levels of differential privacy.

Keywords: Data privacy · Integral privacy · Deep neural networks · Privacy-preserving ML

1 Introduction

In today's world, Artificial Intelligence (AI) plays a crucial role in our day-to-day life. AI techniques are widely used in object recognition, speech recognition, medical imaging, robotics and many other fields. AI approaches and Machine Learning (ML) in particular are very data hungry [1]. They tend to improve with the quality and quantity of data. The data often include sensitive and personal information which must be guarded to ensure security/privacy of each individual or organization. Several guidelines exists such as Europe's General Data Protection Regulation (GDPR), to regulate the use of data in ML. GDPR requires

This work was partially supported by the Wallenberg AI, Autonomous Systems and Software Program (WASP) funded by the Knut and Alice Wallenberg Foundation.

that the analysis to be made should use the minimum amount of data and must be privacy-preserving. There are several data masking and privacy-preserving models such as k-anonymity [2], differential privacy [3], integral privacy [4], etc. which try to protect privacy of individuals and organizations from any adversaries. Adversaries aim to gain sensitive information about individuals or a group of individuals making inferences from ML models.

Data masking is used to modify sensitive information so that a record can not be uniquely identified. K-anonymity is one of the most used data masking methods. A database satisfies k-anonymity if for each record there are k-1 other indistinguishable records. This can be implemented using clustering (replacing k similar records with their mean or with their generalization). In the recent years, much attention has been given to differential privacy (DP) and its variants (see [5] for more details). Differential privacy is satisfied if the outputs of a query on neighbouring datasets are similar i.e. addition or removal of one record should not affect the outcome of the query. Differential privacy depends on a parameter ϵ that establishes the level of this similarity. Theoretically, DP offers sound privacy-preserving models but it has practical limitations such as the amount of noise for small ϵ (high privacy) can be very high. Therefore, high sensitivity queries require high amount of noise. However, in case of multiple queries as the privacy budget is limited, high amount of noise is also required. High noise leads to a loss of utility for ML models. In our approach, we have considered Integral Privacy as an alternative to DP to achieve high utility privacy-preserving machine learning.

Integral Privacy models [4] are the data-driven models that appear recurrently with different training data sets. This makes inferences on sensitive information harder for an intruder. Formally, the set of integrally private models are the set of recurrent models, i.e. generated by different datasets for the same problem. This approach has practical limitations, as in general, we rarely have a huge number of different datasets. The first practical approach for Integral private model selection was given for decision trees [6], where instead of having an available set of datasets, the authors have used sampling approaches to build the model space and eventually suggesting models which are integrally private. The authors expanded the idea with integral privacy guarantees for linear regression. This is given in [7]. In [8], authors have shown how maximal c-consensus meets (see [9] for further details) can be used in the context of integral privacy to find datasets which can produce the same models. The work presented in [6] generates or approximates the model space for a given dataset. A stratified sub-sampling approach is used to approximate the model space for small datasets ($\approx$ 200 instances). The authors approximate the model space using 100k, 150k and 300k subsamples from each datasets. This can be time consuming and 100–300k subsamples may not be enough to approximate the model space for real-world big datasets. Overall, the approach is computationally expensive.

Deep Neural Networks is one of the most successful machine learning paradigms for several computer vision tasks such as image classification [10], object detection [11], video classification [12], and many other areas. However, DNNs are known to be highly dependent on the input data. In the last few years, interest in adversarial DNN examples has grown [13]. DNNs are assumed to work well with large datasets. They have large number of weights and biases which can result in very few generators (unique in many of the cases) for each model. In other words, generation or discovery of recurrent models in DNNs is difficult.

Considering these challenges in mind, we introduce a relaxed variant of integral privacy called 'ϵ-Integral Privacy' where models in the ϵ range are considered perturbated version of each other and, thus, they are considered ϵ-integrally private. We also propose a model selection strategy for choosing ϵ-integrally private models for Deep Neural Networks (DNNs). Our algorithm recommends the mean of the top recurrent models as the private model. We distribute the data in disjoint subsamples having same class-distribution as the original dataset. We find that large enough disjoint subsets having same class-distribution as the original dataset leads to the generation of the models which are utmost ϵ-different, with utility comparable to the benchmark model. This way we do not need to generate 100–300k sub samples. Our approach also supports the data-centric approach [14]. We are able to generate benchmark comparable models with samples sizes 1/100th of the original dataset. There hasn't been much work in the literature which discusses about using smaller datasets for training DNNs. The work in [15] improves the quality of data by eliminating the invalid instances, our approach is focused on maintaining the class-distribution of the data.

In this paper, we have also extended the potential model comparison attack [6] for DNNs. In this type of attack, an intruder gets access to the training data by comparing the models learned by the intruder obtained from original data and the model obtained from a modified dataset. In case of DNNs, the attack becomes tricky as any permutation of the similar set of nodes at any given layer l results in the same learning. We incorporate this to extend the model comparison attack on DNNs.

We have arbitrarily chosen a 3-hidden layered DNN for 6 datasets with varied sizes. Our experimental results show that large enough disjoint sets lead to the generation of ϵ-integral private models with benchmark comparable utility and loss. We get benchmark metrics by training and testing on our chosen DNN on 70-30 split for each data. We have also compared ϵ-integral private models with high DP (differential privacy) model, moderate DP model and low DP model; we found integrally private models have better utility in many cases and have significant improvement in terms of loss for most of the datasets.

This paper is organized as follows. In Sect. 2 we introduce the model comparison attack for DNNs; In Sect. 3 we introduce the notion of ϵ-integral privacy and present the algorithm for private model selection procedure for DNNs; In Sect. 4 we present the experimental analysis to support our claim and in Sect. 5 we present our conclusion and directions for future work.

2   Model Comparison Attack for DNNs

In this section, we describe our model comparison attack for deep neural networks. Deep neural networks are machine learning models which were created to learn like the human brain. The underlying architecture of DNNs consists of the perceptron (or commonly known as neuron) which receives an array of inputs and transform them into output signal(s). DNNs learns from data by putting together a list of layers. Each layer is responsible for learning some relationship or functionality in the input. Each layer is a collection of neurons that learns to detect patterns in the input. Each neuron in the DNNs can be considered as a logistic regression. DNNs are the extension of artificial neural networks with two or more hidden layers. In each neuron, the weighted sum of the input with a bias term is computed which is then transformed using an activation function, which is then passed on to the next layer of the DNNs. Nodes at layer l receive input from the nodes at layer $l - 1$, which means each neuron has $|l - 1| + 1$ $(+1$ for bias) number of parameters to be tuned in training. Final weights and biases of each neuron highly depends on their initialization.

2.1 Framework

In this section, we propose our framework. Let X be the training set from the original dataset D, $\mathcal{G}$ be the model generated on X. In our work, we have considered DNNs as learning algorithm. Let us denote an initial architecture and weight by $Arch$ and let A be the algorithm.

We assume the intruder has some background knowledge $S^* \subseteq D$. They are the records that are known to be used to train the model. The intruder also has access to the model. That to $\mathcal{G}$ which was learned from the training set X on the initial architecture $Arch$. That is, $\mathcal{G} = DNN(Arch, X)$. With this information, the intruder aims to gain knowledge on the training set and do membership inference attacks

The intruder essentially can perform the model comparison attack once they can generate the model space associated to S^*. The intruder can perform comparison with the models in model space and his knowledge of G. After comparison, if there is a single generator for the model, the intruder gets complete access to the training set and their inferences. If there are more than one generator for the model, an intruder can do membership inference attack for dominant records by finding the intersection between the generators.

2.2 Intruders Approach

The intruder has some background information S^*. Then, they can draw a block of subsamples $S = \{S_1, S_2, ..., S_n\}$ where $S_i \subseteq S^*$ to generate the (approximated) model space. Each subsample is a set of instances from S^* which are used to generate a DNN (see Fig. 1). Generation of the complete model space can be computationally expensive but can be approximated using sampling approaches.

Comparison of two DNNs for model comparison attack is a difficult task because we need to deal with a combinatorial problem. We need to align neurons in each layer. Observe that layers in both DNNs must contain the same neurons i.e. for two DNNs to be the same they must have equal layers; and for two layers to be equal, neurons in one layer must be some permutation of the neurons in the other layer. Given r neurons, we will have r! possible permutations.

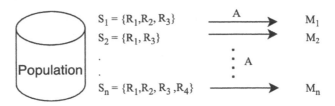

Fig. 1. Demonstration of model generation using algorithm A for subsamples $S_1, S_2, ..., S_n$.

Each model in the generated model space can be compared with the original model G. In case of DNNs, each model has one or very few generators due to the high number of parameters of the model. Therefore, after the comparison attack, the intruder may be able to uniquely identify the training set used to generate the model. When there are more than one generator for a model G, an intruder can check for membership inference by finding the dominant records from the intersection of the generators for the model.

2.3   Integral Privacy

This privacy model [4] aims to protect the disclosure of training data and inferences from a model comparison attack. Let A be an algorithm to compute model G from a given population of samples P. The model G is integrally private if it can be generated by enough number of samples from the population. Let S^* be the background information available to the intruder, then $Gen^*(G, S^*) = \{S' \setminus S^* \,|\, S^* \subseteq S' \subseteq P, A(S') = G\}$ is the possible set of generators for the model G. K-anonymous integral privacy holds when there are at least k disjoint generators in the set $Gen^*(G, S^*)$. Disjoint generators are required to avoid membership inference attacks. Formal definition for Integral privacy is as follows.

Integral Privacy. Let P be the set of samples or a dataset. For model $G \in \mathcal{G}$ generated by algorithm A on samples $S \subseteq P$, let $Gen^*(G, S^*)$ represent the set of all generators of G which are consistent with the background knowledge S^*. Then, the model G is said to be k-anonymous integrally private if $Gen^*(G, S^*)$ contains at least k sets of generators and

$$\bigcap_{S \in Gen^*(G,S^*)} S = \emptyset \tag{1}$$

3   ε-Integrally Private Model Selection for DNNs

To construct the complete model space is computationally intractable for large sets. Consider an example of a dataset with 5000 instances. Considering all possible datasets to produce all possible models of the model space (say M_c)

corresponds to producing 2^{5000} generators and the corresponding models. The alternative to M_c is to construct an approximation of the model space (M_e) using sampling. This approach was used in previous works [6,7]. Nevertheless, even in this case the number of generators and their corresponding models can be high and computationally expensive. In case of bigger datasets say with 5 million instances, the process of building an approximation of a model space will be very costly. In our approach, we have focused on reducing the huge computational requirement to recommend relaxed integrally private deep neural network models.

Let us consider the problem of finding the set of different models of the model space. First, let us recall that each neuron at layer l in DNNs receive inputs from all the neurons in layer $l-1$, which in turn require weights and bias for the neuron. The weights and biases in DNNs can take any value between -1 and $+1$. Even for a small DNN there can be a unique generator for each model or only very few models will have more than one generator. Our initial studies on DNNs confirms this even when we round-off weights to 3 digits. It is worth mentioning here that initialization of DNNs also affects the number of generators. More concretely, we may not get the same generators on differently initialized models. This makes achieving integral privacy difficult.

Because of this in our approach, we have adopted the relaxed version of integral privacy which we call 'ϵ-Integral privacy' in which models utmost ϵ different from each other are considered. In case of DNNs, two models are utmost ϵ different if and only if the difference between weights for the same connections between neurons is always less than ϵ I.e. if $G1$, $G2$ represent the weights for two DNNs then $\|G_1 - G_2\| \leq \epsilon$, where $\|G_1 - G_2\|$ represent the difference between every same connection between neurons for both DNNs. Now, let $Gen^*(G, S^*, \epsilon)$ denote the set of possible pairwise disjoint generators for the models which are utmost ϵ different than G (generators that are consistent with the background knowledge S^*), then k-anonymous ϵ-Integral privacy holds if $Gen^*(G, S^*, \epsilon)$ has at least k elements and their intersection is empty. A more formal definition follows.

ϵ-**Integral Privacy:** Let P be the set of samples or datasets. For a model $G \in \mathcal{G}$ generated by algorithm A on samples $S \subseteq P$, let $Gen^*(G, S^*, \epsilon)$ represent the set of all generators of G which are consistent with the background knowledge S^* and are utmost ϵ different. Then, the model G is said to be k-anonymous ϵ-Integrally private if $Gen^*(G, S^*, \epsilon)$ contains at least k elements and

$$\bigcap_{S \in Gen^*(G,S^*,\epsilon)} S = \emptyset \qquad (2)$$

Now, we will focus on the private model selection procedure for DNNs. Our approach to generate subsampling is data centric. We choose subsamples of size N with same class-distribution as the original dataset D. We denote these subsamples by $S_1, S_2, ..., S_n$ (here $n = \lfloor |D|/N \rfloor$). Here, we also satisfy there is no intersection between subsamples i.e. $S_1 \cap S_2 \cap ... \cap S_n = \emptyset$. This condition is

Algorithm 1. Integrally private model selection procedure for Deep Neural Networks for a given perturbed dataset D'. The algorithm returns top 5 integrally private models with their accuracies

---

Inputs: D - Perturbed Dataset
N - Size of subsamples
ϵ - Privacy parameter
A - Algorithm to generate DNNs
Output: returns a list of integrally private models with their accuracies
Algorithm:
S = Generate_subsample(D, N)   ▷ Generate n subsamples of size N
ModelList = [[]]
for S_i *in* S **do**
    $M_i \leftarrow$ A(S_i)
    present = False
    for each $m_j \in$ ModelList **do**
        if compare_model(m_j, M_i) $\leq \epsilon$ **then**
            ModelList[j].append(M_i)
            present = True
            break
        end if
    end for

    if present == False **then**
        ModelList.append(list(M_i))
    end if
end for
chosen_models = choseXModels(ModelList)
                          ▷ Chose top X recurring models
meanModels = A(mean(chosen_models))   ▷ Compute mean models
statistics = computeMetrics(meanModels)   ▷ Statistics of mean models
return meanModels, statistics

---

important to avoid membership inference attack from the intersection analysis between generators.

Now, we propose our algorithm for choosing integrally private models for DNNs. Its flowchart is given in Fig. 2. The algorithm is as follows for a given dataset D. First, we generate n subsamples each of size N having the same class-distribution as the original. Second, we compute models and cluster them so that each cluster has models that are utmost ϵ different from each other. Finally, we can choose a cluster of models which are recurring in nature and has high utility. In our methodology, we chose the mean of all the models in the cluster as our recommended model. I.e. we generate a new model whose weights are the mean of the weights of all the ϵ-integrally private models.

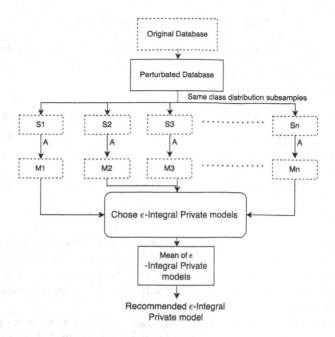

Fig. 2. Flowchart of the proposed methodology to recommend an ϵ-integral private model.

Algorithm 1 formalizes this approach. In the algorithm we have a dataset D, Algorithm A, privacy parameter ϵ and size of each subsample N as inputs. We initialize an empty list of lists and append models which are utmost ϵ distant apart from the first one. For our results we can either chose the top recurring model or X most frequent models (for more ambiguity) which is done in function choseXModels(). Our recommended model is the mean of the models in the cluster. For X ϵ−ranged models, we recommend X mean models and their statistics as the output of our proposed algorithm.

4   Experimental Results

In this section, we present our experimental results for our proposed methodology. Our approach is valid for both numerical/categorical data and for classification problems with an arbitrary number of classes. Table 1 shows the details of the datasets we have considered for our experiments namely Adult, Susy, ai4i and HepMass from UCI repository [16]; and Churn_Modelling, Diabetes [17]. Of these datasets, Churn Modelling and Adult have categorical data and Diabetes is a multi-class problem. We have considered small datasets ($\approx$10–50K instances), medium dataset ($\approx$250K instances) and large datasets ($\approx$5–7 million instances) for our experimental study. Table 1 also shows the size of the subsamples. The size is chosen so that there are enough subsamples to find integrally private models.

Table 1. Details of the used datasets

Dataset	# instances	# attribute	Data type	# classes	subsample size
Adult	48842	14	Categorical Integer	2	1000
Susy	5000000	18	Real	2	10000
ai4i	10000	14	Real	2	500
HepMass	7000000	28	Real	2	10000
Churn Modelling	10000	21	Categorical Real	2	500
Diabetes	254000	21	Real	3	5000

To compare the performance of our approach and 2 benchmark, we have used an architecture of 5-layered DNN with 3-hidden layers with 5-10-5 neurons. As we explain later, we have considered other architectures as well. Then, we have taken $\epsilon = 0.05$ for all the datasets, other values could be used depending on the application requirements.

The results of our methodology have been compared with results with a differential private solution [18] and the benchmark results. Benchmark results are obtained by training the model with 70-30 train-test split of original dataset. Now, let us look at the number of generated models from randomly chosen subsamples of the size given in Table 1. In case of the adult dataset, the total possible models which can be considered for integral privacy are 47, similarly for ai4i dataset we have 19, for susy dataset we have 498, for hepmass dataset we have 698, for churn modelling dataset we have 18, and for diabetes dataset we have 49 models to be considered for integral privacy.

Figure 3 shows the training f1 score of top 5 (for ai4i and Churn Modelling datasets there are 2 and 3 generators only) recurring models along with the training score of the benchmark model in black solid line and three level of differential privacy(DP): high privacy ($\epsilon \approx 0.1$, represented by ◆), moderate privacy ($\epsilon \approx 0.5$, represented by ·−) and low privacy ($\epsilon \approx 1.0$, represented by ●). In general, higher DP privacy (low ϵ, ◆) leads to lower training score and higher training loss. In the plots, the f1 scores of all the models are in the light shade, and the dark solid line represents the mean of the ϵ ranged integral private models. Observe from Fig. 3a and 3b, we achieve better training score than the benchmark training scores while from Fig. 3c, 3d, 3e and 3f we can observe benchmark comparable results. It can be seen from Fig. 3a, 3c and 3d, integrally private models have better training score than all three variants of differentially private models on the other hand Fig. 3b, 3e and 3f, the training utility of integrally private model is comparable with the differentially private models. We get similar results for the training loss as shown in Fig. 4. We have denoted the loss of each model in the lighter shade solid line, their mean loss in dark solid line, the benchmark model loss with solid black line and three level of differential privacy: high privacy with ◆, moderate privacy with ·− and

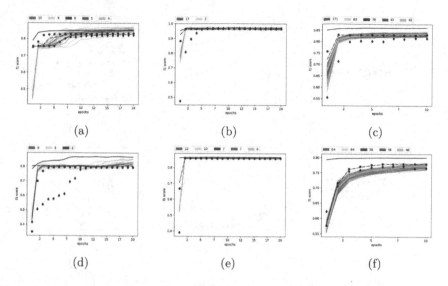

Fig. 3. f1 score of top 5 ϵ-recurring models over training data for (a) Adult (b) ai4i (c) HepMass (d) Churn Modelling (e) Diabetes (f) Susy Datasets

low privacy with ●. It can be seen that the loss for integrally private models is comparable with the benchmark model loss. We can observe from Fig. 4b, 4c and 4d, integrally private models have significant improvement in terms of training loss from DP variants while Fig 4a, 4e shows some improvement from DP variants in contrast to Fig. 4f where low, moderate DP privacy has improvement in training loss from integrally private models.

The concept of data-centric AI simply suggests that good quality of data can lead to good models. In our approach, we have only used 0.15% to 2% of the original data, but with the same class-distribution, to train our model (see Table 1 for subsample size). We got surprising result when we compared their performance on test data i.e. 30% of the original data. Figure 5 shows the result on the test data, lighter shade circles represent the test result for each model while dark solid colored circle represents their mean value. From Fig. 5, we can say that our ϵ-integrally private models achieve benchmark comparable f1 score on much bigger test datasets (15 to 200 times).

Our recommended model is the mean of all the models in the ϵ-integral private range. The result in Fig 5 motivated us to compare performance of the aggregated ϵ-integrally private models with the original training and testing datasets. Figure 6 shows the comparison of f1 score on training data (in solid color circles) and test data (in hollow circles) with benchmark training score (in solid line) and benchmark test score (in dashed line). Our recommended models have benchmark comparable f1 score on all the datasets.

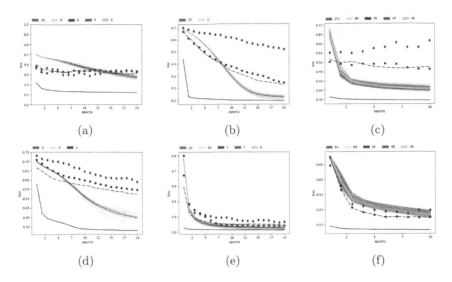

Fig. 4. Training loss of top 5 ϵ-recurring models for (a) Adult (b) ai4i (c) HepMass (d) Churn Modelling (e) Diabetes (f) Susy Datasets

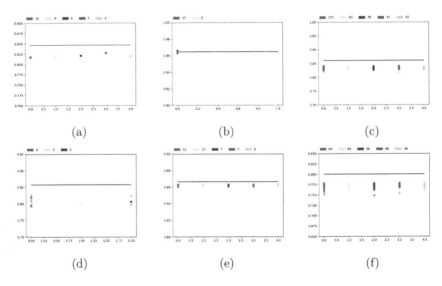

Fig. 5. f1 score of top 5 ϵ-recurring models on bigger test data for (a) Adult (b) ai4i (c) HepMass (d) Churn Modelling (e) Diabetes (f) Susy

Table 2 shows the recurrence of the recommended model with the test accuracy on much bigger test sets. We have considered 4 different architectures: DNN-1 has 3-hidden layers (with 5-10-5 neurons respectively) architecture; DNN-2 has 1- hidden layer (with 1024 neurons) architecture; DNN-3 has 3-hidden layers (with 10-20-10 neurons respective) architecture; and DNN-4 has 5-hidden

Table 2. Different architectures and their f1 score on 30% test dataset.

Dataset	DNN-1		DNN-2		DNN-3		DNN-4	
	recurrence	test_acc	recurrence	test_acc	recurrence	test_acc	recurrence	test_acc
Adult	10	0.8387	89	0.7797	16	0.8286	36	0.8284
Susy	64	0.7758	366	0.7917	8	0.7636	6	0.7882
ai4i	17	0.9647	19	0.9723	12	0.9683	10	0.9747
HepMass	171	0.8325	562	0.8344	68	0.8325	51	0.8336
Churn Modelling	9	0.8145	13	0.8520	10	0.7927	10	0.7870
Diabetes	12	0.8627	21	0.8596	13	0.8634	5	0.8596

layers (with 5-10-20-10-5 neurons respectively) architecture. Table 2 shows that the proposed methodology produces benchmark comparable results for different DNN architectures as well.

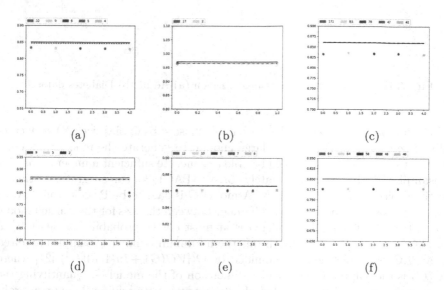

$$(a) \qquad (b) \qquad (c)$$

$$(d) \qquad (e) \qquad (f)$$

Fig. 6. f1 score on train and test data for mean of the ϵ-recurring models for (a) Adult (b) ai4i (c) HepMass (d) Churn Modelling (e) Diabetes (f) Susy

4.1 Discussion

In summary, our results with varied sized, multi-class and categorical datasets suggest that we can achieve ϵ-integral privacy with good utility (comparable to benchmark utility) from the list of the recommended models depending on the value of k (number of models in ϵ range) with no additional computational cost.

The good results of our approach can essentially be linked to the data centric AI approach where we train our model for smaller datasets with the same class-distribution as the original dataset and get good results. We further explored the impact of subsample size and compared their performance on separate 70-30 training data and testing data on moderately sized adult and diabetes datasets. Our results from Fig. 7 shows that the f1 score for both training and testing data is non-decreasing but it is neither increasing significantly with respect to the increase in subsample size. Our results are in line with [19] which highlights that one can generate arbitrarily similar model of finite floating point weights from two (or more) non-overlapping dataset. The paper [19] also suggest that we can get good results on smaller datasets as well, which aligns with the results in Fig. 7.

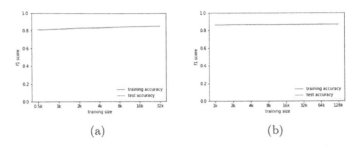

(a)   (b)

Fig. 7. f1 score of various subsample sizes on (a) Adult (b) Diabetes datasets

For our proposed methodology, we must chose subsamples size (N) very carefully. The choice for N must be large enough to generate the model with good utility at the same time it should be able to generate sufficient number of disjoint subsamples. Probably approximately correct (PAC) [20] can suggest an estimate for the choice of the parameter N. A model G is said to be PAC learnable with respect to loss l if and only if the difference between the loss for the learned model G and true (best possible) model $\bar{G}$ is at most ϵ with probability at least $1 - \delta$ i.e. $P[G_l - \bar{G}_l \leq \epsilon] \geq 1 - \delta$. With this the minimum number of samples required for a PAC learnable model is bounded by $O([VC(G) + ln(1/\delta)]/\epsilon^2)$ [21] where $VC(G)$ is the Vapnik-Chervonenkis dimension of the model G. Quantifying the VC-dimension for complex models like deep neural network is still an open problem [22]. Therefore, in the literature scientists follow the rule-of-thumbs: (1) The VC dimension of DNNs is considered equal to the number of weights in DNNs [23] and then (2) the minimum number of samples required to learn the DNN is established as 10 times the VC dimension [24]. Considering this, i.e., a sample size of 10-times the VC-dimension (number of weights) should provide a PAC learnable model. For datasets ai4i, and Churn_Modeling the number of weights are 172 and 197, respectively, and hence the minimum subsample size is estimated as 1720 and 1970 for PAC learnability. This results in very few disjoint subsamples (5 for both datasets) which may not be enough to find integrally private models. This suggests a trade-off between model complexity (number of weights) and its learning ability for integral privacy. Further study in this area is required to investigate the impact of this trade-off for integral privacy.

4.2 Limitations

Based on a critical analysis of our approach and the results obtained, we can underline the following limitations of our approach:

1. Our methodolgy may not be suitable in the presence of outliers as the outliers disturbs the distribution of the dataset.
2. Selection of private models on very small datasets with our proposed methodology is not feasible.
3. High model complexity may result in less number of models in ϵ-range.

5 Conclusion and Future Work

In this paper, we have first extended the model comparison attack to deep neural networks. We have also introduced the concept of ϵ-integral privacy which is then used to recommend integrally private models for deep neural networks. Our results show that we are able to achieve ϵ-integrally private models without any significant utility loss (improvement of utility in some cases). Our results also highlights that small data of good quality can result in a well trained model.

For our proposed methodology, we have arbitrarily chosen the size of the subsamples; the privacy parameter ϵ and the DNNs architecture. Tuning of these areas may yield interesting results. Another interesting direction is to use a data-enhancement approach to remove outliers as done in [15]. Federated Learning takes advantage of data distributed across multiple users, where learning takes place locally. Our methodology can be seen as independent and identically distributed (IID) ϵ-integral private model selection in federated learning for a single pass. Our work can further be extended into non-IID settings of federated learning.

References

1. Obermeyer, Z., Emanuel, E.J.: Predicting the future-big data, machine learning, and clinical medicine. New Engl. J. Med. **375**(13), 1216 (2016)
2. Samarati, P.: Protecting respondents identities in microdata release. IEEE Trans. Knowl. Data Eng. **13**(6), 1010–1027 (2001)
3. Dwork, C.: Differential privacy. In: Bugliesi, M., Preneel, B., Sassone, V., Wegener, I. (eds.) ICALP 2006. LNCS, vol. 4052, pp. 1–12. Springer, Heidelberg (2006). https://doi.org/10.1007/11787006_1
4. Torra, V., Navarro-Arribas, G.: Integral privacy. In: Foresti, S., Persiano, G. (eds.) CANS 2016. LNCS, vol. 10052, pp. 661–669. Springer, Cham (2016). https://doi.org/10.1007/978-3-319-48965-0_44
5. Ji, Z., Lipton, Z.C., Elkan, C.: Differential privacy and machine learning: a survey and review. arXiv preprint arXiv:1412.7584 (2014)
6. Senavirathne, N., Torra, V.: Integrally private model selection for decision trees. Comput. Secur. **83**, 167–181 (2019)

7. Senavirathne, N., Torra, V.: Approximating robust linear regression with an integral privacy guarantee. In: 2018 16th Annual Conference on Privacy, Security and Trust (PST), pp. 1–10. IEEE (2018)
8. Torra, V., Navarro-Arribas, G., Galván, E.: Explaining recurrent machine learning models: integral privacy revisited. In: Domingo-Ferrer, J., Muralidhar, K. (eds.) PSD 2020. LNCS, vol. 12276, pp. 62–73. Springer, Cham (2020). https://doi.org/10.1007/978-3-030-57521-2_5
9. Torra, V., Senavirathne, N.: Maximal C consensus meets. Inf. Fusion **51**, 58–66 (2019)
10. He, K., Zhang, X., Ren, S., Sun, J.: Deep residual learning for image recognition. In: Proceedings of the IEEE Conference on Computer Vision and Pattern Recognition, pages 770–778 (2016)
11. Ren, S., He, K., Girshick, R., Sun, J.: Faster R-CNN: towards real-time object detection with region proposal networks. In: Advances in Neural Information Processing Systems, vol. 28 (2015)
12. Jiang, Y.-G., Zuxuan, W., Wang, J., Xue, X., Chang, S.-F.: Exploiting feature and class relationships in video categorization with regularized deep neural networks. IEEE Trans. Pattern Anal. Mach. Intell. **40**(2), 352–364 (2017)
13. Oh, C., Xompero, A., Cavallaro, A.: Visual adversarial attacks and defenses. In: Advanced Methods and Deep Learning in Computer Vision, pp. 511–543. Elsevier (2022)
14. Ng, A.: MLOps: from model-centric to data-centric AI (2021). https://www.deeplearning.ai/wp-content/uploads/2021/06/MLOps-From-Model-centric-to-Data-centricAI.pdf. Accessed 09 Sept 2021
15. Motamedi, M., Sakharnykh, N., Kaldewey, T.: A data-centric approach for training deep neural networks with less data. arXiv preprint arXiv:2110.03613 (2021)
16. Dua, D., Graff, C.: UCI machine learning repository (2017). http://archive.ics.uci.edu/ml
17. Centers for Disease Control, Prevention, et al.: National diabetes statistics report, 2017. Centers for disease control and prevention, Atlanta, GA (2015, 2017)
18. Dwork, C., McSherry, F., Nissim, K., Smith, A.: Calibrating noise to sensitivity in private data analysis. In: Halevi, S., Rabin, T. (eds.) TCC 2006. LNCS, vol. 3876, pp. 265–284. Springer, Heidelberg (2006). https://doi.org/10.1007/11681878_14
19. Thudi, A., Jia, H., Shumailov, I., Papernot, N.: On the necessity of auditable algorithmic definitions for machine unlearning. In: 31st USENIX Security Symposium (USENIX Security 2022), pp. 4007–4022 (2022)
20. Vapnik, V.N., Chervonenkis, A.Y.: On the uniform convergence of relative frequencies of events to their probabilities. In: Vovk, V., Papadopoulos, H., Gammerman, A. (eds.) Measures of Complexity, pp. 11–30. Springer, Cham (2015). https://doi.org/10.1007/978-3-319-21852-6_3
21. Vershynin, R.: High-Dimensional Probability: An Introduction with Applications in Data Science, vol. 47. Cambridge University Press, Cambridge (2018)
22. Anthony, M., Bartlett, P.: Neural Network Learning: Theoretical Foundations. Cambridge University Press, Cambridge (1999)
23. Abu-Mostafa, Y.S.: Hints. Neural Comput. **7**(4), 639–671 (1995)
24. Baum, E., Haussler, D.: What size net gives valid generalization? In: Advances in Neural Information Processing Systems, vol. 1 (1988)

Gaussian Process Component Mining
with the Apriori Algorithm

Jan David Hüwel[(✉)] and Christian Beecks

Institute for Mathematics and Computer Science, University in Hagen,
Hagen, Germany
{jan.huewel,christian.beecks}@fernuni-hagen.de

Abstract. Gaussian process models are a commonly used tool for
model-based analysis of time series data. With growing database size,
the difficulty to identify the most interesting insights in order to gain
a deeper understanding of the data's underlying behavior increases. To
address this issue, we propose a novel approach for finding frequent kernel components efficiently. In this way, data scientists are empowered to
focus their investigations on the most common parts hidden in a set of
Gaussian process models. We show how to solve this task by means of
frequent item set mining methods, which are capable of analyzing large
databases efficiently. We provide evidence of our proposal with a first
series of experiments, indicating that our method is capable of detecting
frequent kernel components from Gaussian process models. Though this
short paper can be thought of as a first preliminary approach towards
analyzing Gaussian processes with conventional data mining methods, it
simultaneously opens a novel research direction of *Gaussian process mining* at the intersection between machine learning and database research.

Keywords: Frequent Item Set Mining · Gaussian Process · Time
Series

1 Introduction

Time series data is encountered in a variety of domains including finance, economics, manufacturing and marketing, to name just a few. As a complex type
of data, time series frequently comprise time-dependent measurements, such as
stock prices, inflation rates, quality control indicators or customer behavior. Due
to the wide applicability and inherent complexity of time series data, model-
based management and analysis remains a challenging issue. Many machine
learning approaches [7,8,10] have been introduced to counteract the complexity
of time series and to provide insight into underlying patterns and structures.
Gaussian Processes (GPs) [10] epitomize an adaptive, time-elastic, and probabilistic class of machine learning models for time series analysis.

This research was supported by the research training group "Dataninja" (Trustworthy
AI for Seamless Problem Solving: Next Generation Intelligence Joins Robust Data
Analysis) funded by the German federal state of North Rhine-Westphalia.

C. Strauss et al. (Eds.): DEXA 2023, LNCS 14147, pp. 423–429, 2023.
https://doi.org/10.1007/978-3-031-39821-6_34

GPs are primarily defined by means of a covariance function, also denoted as kernel, which determines the behavior of the model based on given data points. The kernel can be used to interpret the data [7], but methods to analyse similarities between different kernels have not yet been explored.

In this paper, we propose a method for transforming GP kernels into item sets, which can then be analyzed via well-established methods for frequent item set mining [5]. Our approach is simple yet efficient and it is intended to present an early step into a new area of research. This short paper spans a brief introduction into the necessary fundamentals of GPs, a description of our method and two exemplary examinations of its results.

2   Gaussian Processes

GPs are probabilistic non-parametric machine learning models [10]. Formally, a model $GP(m, k)$ depends on mean function $m : \mathbb{R} \to \mathbb{R}$ and covariance function, i.e. kernel, $k : \mathbb{R} \times \mathbb{R} \to \mathbb{R}$. Since the most information is contained within the latter, the mean function is often set to zero, leading to zero-mean GPs.

There are a number of commonly used kernels that correspond to frequently appearing behavioral patterns in the data. For example, the periodic kernel, which is defined as $k_{PER}(x, x') = \sigma^2 exp\left(-\frac{2sin^2(\pi|x-x'|/p)}{l^2}\right)$ [3], can be utilized to represent periodicity in the data. Similarly, a linear kernel and squared exponential kernel are used to indicate trends and smoothness in the data. Such base kernels can be combined via addition or multiplication to construct composite kernels, allowing GPs to adapt to more complicated data behavior [4].

For the automatic inference of a kernel befitting the training data, Duvenaud et al. [4] have introduced the Compositional Kernel Search (CKS) algorithm and have demonstrated how CKS can be used to generate textual descriptions of a given time series [7]. More recently, local kernel search approaches have been proposed, which aim to separate time series data into segments and to describe each of those segments with a suitable kernel [1,2,6].

These kernel search methods work by incrementally adding or multiplying base kernels to an existing kernel and evaluating the likelihood in the training data to find the best fit. The resulting kernel function has a structure that can be represented by means of a tree, as shown in the definition below.

Definition 1 (Tree). *A tree $T = (V, E) \in \mathbb{T}$ is a connected, acyclic directed graph. It is defined by a set of vertices V and a set of edges $E \subset V \times V$. The relevant notations are the defined as:*

1. *$r(T) \in V$, which denotes the root of tree T*
2. *$c(v) \subset V$, which denotes the set of children of node $v \in V$*
3. *$s(v) \subset \mathbb{T}$, which denotes the subtree with root $v \in V$*

Based on this data structure, the kernel can be intuitively represented as shown in the following definition.

Definition 2 (GP Kernel Representation). *Let $\mathcal{B}$ be a set of base kernels and $\mathcal{O} = \{+, \times\}$ be available operations. A GP kernel representation is a tree $T = (V, E) \in \mathbb{T}$, that fulfills the following conditions:*

1. *For every leaf $l \in V$, it holds that $l \in \mathcal{B}$*
2. *For every internal node $v \in V$, it holds that $v \in \mathcal{O}$*

To later define the transformation of a kernel into this representation, we define the evaluation $\mathcal{E}(T)$ of a tree T as the result of applying each operator in T to their respective children. If an operator only has one child, that child is treated as the result for that operator. A tree T thus represents a kernel k if $k = \mathcal{E}(T)$.

The leaves in this representation thus contain base kernels, while the inner nodes define the relationships and dependencies between these base kernels. Note that the representation of a kernel k is not necessarily unique, so $\mathcal{E}^{-1}$ is not defined. As the distributive property is valid in the function space, both $k_1 \times (k_2 + k_3)$ and $k_1 \times k_2 + k_1 \times k_3$ are the same function, but with different intuitive tree representations. This problem will be addressed in the next section.

With these definitions, we have defined a transformation of arbitrary GP kernels into trees with easily interpretable structure. In the next section, we will use this structure to extract a kernel's basic *components*, which can then be used for frequent item set mining.

3   Method

In this section, we propose a simple yet effective method for analyzing shared behavior in a set of zero-mean GP models. In order to analyze such a set, each model is decomposed into its internal components which can be processed with conventional item set mining algorithms, such as Apriori, FP-Growth or ECLAT [5].

Different operators have very distinct effects on the combination of GP kernels: A sum treats the individual kernels as independent processes, while a product defines inter-kernel dependency [7]. Accordingly, we consider products of base kernels within a kernel expression inseparable. These products will be referred to as the kernel's *components*. We use the fact that every combination of base kernels can be written as a sum of products to make the components extractable from the corresponding tree representation.

Definition 3 (Sum-of-products Representation). *A tree $T \in \mathbb{T}$ representing a kernel as defined in Definition 2 is in sum-of-products form, if the following conditions are met:*

1. *The root $r(T)$ of tree T contains the operator $+$*
2. *All nodes with depth 1 $c(r(T))$ contain the operator $\times$*
3. *The tree T has a height of 2*

The set of all sum-of-products trees is denoted as $\bar{\mathbb{T}} \subset \mathbb{T}$. The relationship between kernels and trees in this set is bijective, so we can define the function $\mathcal{R} = (\mathcal{E}|_{\bar{\mathbb{T}}})^{-1}$, which assigns each kernel its sum-of-products tree representation.

Extracting the components of a kernel k is now simplified to extracting all sub-trees starting in the multiplication nodes in $\mathcal{R}(k)$. Thus the set of components of a kernel k is defined as $\mathcal{C}(k) = \{s(x) \in \mathbb{T} | x \in c(r(\mathcal{R}(k)))\}$.

To solve the initial problem of finding frequent components, we can now extract the set of components for each given model. Each such set is then considered an item set in the classical sense, which makes (for example) the Apriori algorithm applicable to find frequently occurring components.

The effectiveness of this approach will be examined in the next section.

4   Preliminary Experiments

In this section, we investigate two different aspects of our proposal: First, we show that our treatment of components as items is viable for frequent item set mining. To do so, we will generate a set of GP models with a slight bias towards specific components and show that these components are identified as frequent items after our transformation. In the second experiment, we provide a proof of concept and demonstrate that our method generates insights from a time series. Here, we determine local GP models on the Mauna Loa [9] data set and show, that our method's generated insights comply with manual inspection. For the comparison of components in these experiments, we disregard the kernels' parameterization and thus define components as equal, if they contain the same base kernel functions.

4.1   Synthetic Data

In our first experiment, we simulate a database of GP models with varying kernels, that contain reoccurring components.

To do so, we generate a set of 1 million GP models from a set of base kernels $\mathcal{B}$ and a set of components with an increased rate of appearance $\hat{\mathcal{K}}$ via the following rules: (i) Every kernel consists of up to three components that are combined via addition, (ii) every component consists of up to three randomly selected base kernels $b \in \mathcal{B}$ that are multiplied, and (iii) every component has a 10% chance of being selected from $\hat{\mathcal{K}}$ instead.

The components are consequently randomly selected from $\mathcal{K} := \{\Pi_{i=1}^{n} b_i | n \in \{1, 2, 3\}, b_i \in \mathcal{B}\}$ with a higher occurrence rate for components in $\hat{\mathcal{K}} \subset \mathcal{K}$.

Figure 1 shows that the average support of item sets containing elements of $\hat{\mathcal{K}}$ is significantly higher than the support of other item sets. Consequently, any frequent item set mining algorithm, such as the Apriori algorithm, will return the structures that had an increased likelihood of appearing. These structures naturally correspond to repeating patterns in the underlying data and as such, provide interesting insights.

The second experiment will show a direct application of this principle (Fig. 2).

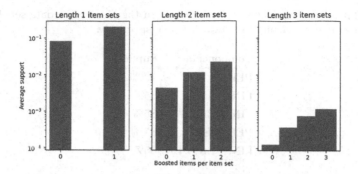

Fig. 1. The average relative support of item sets whose items consists either of kernel from $\hat{\mathcal{K}}$ or from $\mathcal{K}\backslash\hat{\mathcal{K}}$.

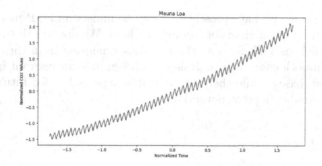

Fig. 2. The Mauna Loa time series data set. Input and output are normalized.

4.2   Real Data

In this second experiment, we use the described method to generate insights into the Mauna Loa [9] time series data set. This data set contains recordings of CO_2 levels at the Mauna Loa observatory in Hawaii. Naturally, these recordings follow a yearly periodic trend, combined with a linear increase over time.

To analyze this data, we generate GP models from a sliding window of fixed length using Event-Triggered Kernel Adjustments [6]. All generated models are then itemized as described in Sect. 3 and the items' frequency are scaled up to match the length of their respective segment. For the frequent item set mining, we use the Apriori algorithm with a minimum relative support of 0.3, meaning that all resulting components will be found in at least 30% of the data set. The frequent components together with their support are shown in Table 1. The results indicate that the most prominent components are of periodic and linear behavior, which matches our manual analysis of the time series.

Table 1. Frequent components in local models on the Mauna Loa data set. PER and LIN refer to the periodic and the linear kernel respectively.

Components	Support
PER	100%
LIN	85%
LIN, PER	85%
PER, PER	85%
LIN, PER, PER	85%

5   Conclusion

In this short paper, we have presented a way to understand GP models as item sets and apply frequent item set mining to them. We applied this approach to synthetic models as well as to a set of models generated from the Mauna Loa time series. In both cases, the analysis has proven to show relevant insights into the model database. While there are multiple aspects left for future work, all findings so far indicate great potential.

References

1. Berns, F., Hüwel, J., Beecks, C.: Automated model inference for Gaussian processes: an overview of state-of-the-art methods and algorithms. SN Comput. Sci. **3**(4), 300 (2022). https://doi.org/10.1007/s42979-022-01186-x
2. Berns, F., Schmidt, K., Bracht, I., Beecks, C.: 3CS algorithm for efficient Gaussian process model retrieval. In: 25th International Conference on Pattern Recognition (ICPR), pp. 1773–1780. IEEE (2021)
3. Duvenaud, D.: Automatic model construction with Gaussian processes. Ph.D. thesis, University of Cambridge (2014)
4. Duvenaud, D., Lloyd, J., Grosse, R., Tenenbaum, J., Zoubin, G.: Structure discovery in nonparametric regression through compositional kernel search. In: International Conference on Machine Learning, pp. 1166–1174. PMLR (2013)
5. Fournier-Viger, P., Lin, J.C.W., Vo, B., Chi, T.T., Zhang, J., Le, H.B.: A survey of itemset mining. Wiley Interdisc. Rev. Data Min. Knowl. Discov. **7**(4), e1207 (2017)
6. Hüwel, J.D., Haselbeck, F., Grimm, D.G., Beecks, C.: Dynamically self-adjusting Gaussian processes for data stream modelling. In: Bergmann, R., Malburg, L., Rodermund, S.C., Timm, I.J. (eds.) KI 2022. LNCS, vol. 13404, pp. 96–114. Springer, Cham (2022). https://doi.org/10.1007/978-3-031-15791-2_10
7. Lloyd, J.R., Duvenaud, D., Grosse, R.B., Tenenbaum, J.B., Ghahramani, Z.: Automatic construction and natural-language description of nonparametric regression models. In: AAAI, pp. 1242–1250. AAAI Press (2014)
8. Mueen, A., Keogh, E., Zhu, Q., Cash, S., Westover, B.: Exact discovery of time series motifs. In: Proceedings of the 2009 SIAM International Conference on Data Mining, pp. 473–484. SIAM (2009)

9. Thoning, K.W., Tans, P.P., Komhyr, W.D.: Atmospheric carbon dioxide at Mauna Loa Observatory: 2. Analysis of the NOAA GMCC data, 1974–1985. J. Geophys. Res. Atmos. **94**(D6), 8549–8565 (1989)
10. Williams, C.K., Rasmussen, C.E.: Gaussian Processes for Machine Learning, vol. 2. MIT Press, Cambridge (2006)

Learnable Filter Components for Social Recommendation

XianKun Zhang$^{(\boxtimes)}$ and WenJie Huang

School of Tianjin, University of Science and Technology, Tianjin, China
zhxkun@tust.edu.cn

Abstract. The recommendation system based on graph neural networks has gained increasing attention due to its superior learning ability of various additional information. Among them, the recommendation system based on social network graphs integrates social network graphs with user-item-graph interactions, aiming to extract dynamic preference features from recorded user behavior data to obtain more accurate recommendation results. However, recorded user behavior data contains noise and exhibits skewed distributions, which may lead to slightly insufficient representation performance of graph neural network models. This article proposes a new approach to address this issue by applying filtering algorithms commonly used in signal processing to noise reduction in the frequency domain. In this study, we combine learnable filter components with an MLP architecture to learn user and item latent embeddings in modeling social network graphs and user-item interaction graphs, and ultimately predict ratings using the latent embeddings of users and items. The entire MLP architecture has low time complexity, and the adaptive filter components are able to effectively suppress noise information. The model proposed in this article has been applied to real datasets and outperformed other models, achieving a 3.2% increase in the MAE evaluation index.

Keywords: Social Recommendation · Graph Neural Network · Filtering Algorithm

1 Introduction

Recommendation systems [7], which predict the likelihood of a user being interested in a particular item, are an extensively researched area. However, the recommendation performance is always unsatisfactory due to the cold-start problem [13]. To address this issue, social network-based recommendation methods have emerged [1]. In recent years, graph neural networks [15] further advance the development of social recommendation. However, user behavior data in social network graphs and user-item interaction graphs inevitably contain noise [8], including malicious forgeries [12], which can easily lead to overfitting of recommendation models [14].

Therefore, in this paper, we draw inspiration from filtering algorithms in digital signal processing and apply them in the frequency domain to eliminate noise and improve the robustness of the model to noisy data. In addition, some recent studies [10, 11]

© The Author(s), under exclusive license to Springer Nature Switzerland AG 2023
C. Strauss et al. (Eds.): DEXA 2023, LNCS 14147, pp. 430–437, 2023.
https://doi.org/10.1007/978-3-031-39821-6_35

have explored variants of the Transformer [9] architecture, mainly by modifying the multi-head attention component of the universal architecture.

Therefore, the main contribution of this article is:

i. This article proposes a novel collaborative filtering model that incorporates social networks. The model draws on the wisdom of filtering algorithms in signal processing and innovatively introduces a learnable filtering component to suppress noise and bias in user behavior data in the frequency domain.

ii. This paper replaces the traditional multi-head self-attention component in the transformer with learnable filter components and integrates MLP architecture to achieve lower time complexity. Compared with multi-head self-attention, MLP in this paper not only reduces the number of network parameters and the complexity of the model, but also optimizes the calculation speed and efficiency.

iii. This paper conducted experiments on two real-world datasets and six baselines, demonstrating the effectiveness of the proposed model.

2 Framework Overview

This paper proposes a collaborative filtering model called FMRec that combines social network graphs with recommendation systems. The model consists of three modules: user modeling, item modeling, and rating prediction. The model includes two types of relationship graphs, namely social relationship graph and user-item graph. For these two types of graphs, this paper combines learnable filter components with a full multi-layer perceptron structure to learn the latent embeddings of users and items. Finally, by aggregating the embeddings of users and items, the rating prediction result is calculated.

2.1 Problem Definition

The recommendation problem of aggregative social networks is composed of three types of entities. One is the user group $\mathcal{U} = \{u_1, u_2, \ldots, u_m\}$, and the other is the item group $\mathcal{I} = \{i_1, i_2, \ldots, i_n\}$. It is noteworthy that m and n represent the total numbers of users and items. In addition, there is also a rating matrix $R \in \mathbb{R}^{m \times n}$, where $R_{u,i}$ represents the rating of user u for item i, and the higher rating indicates a higher preference for the item. These entities contain three types of information, as shown in Fig. 1. One is the user-item graph $G_u = \{\mathcal{U}, \varepsilon_r|_\gamma^R\}$, where $\mathcal{U}$ represents the user nodes, ε_r is the weighted edge representing the user's rating for the purchased item. Another is the item-user graph $G_i = \{\mathcal{I}, \varepsilon_r|_\gamma^R\}$, where $\mathcal{I}$ represents the node of the item, and ε_r represents the ratings of all users who have purchased the item. The last one is the user social relationship graph $G_s = \{\mathcal{U}, \varepsilon_s\}$, where $\mathcal{U}$ represents the user nodes, ε_s denotes the edges of the graph, and $(u_1, u_2) \in \varepsilon_s$ represents the social connection between user u1 and user u2.

2.2 Model Architecture

The model architecture in this article is shown in Fig. 2. The model consists of three parts: user modeling, project modeling, and score prediction. In the attention module, we use a general architecture similar to the original Transformer, but replace the multi-head

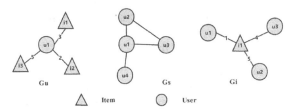

Fig. 1. Three kinds of information map

self-attention component with a learnable filter component and stack multiple neural blocks. In addition to the feature of eliminating noise effects, filters are mathematically equivalent to cyclic convolutions, which can better capture periodic features and thus have a larger perceptual range across the entire sequence. Next, we will describe in detail the role of each model component.

User Modeling

The project sequence diagram of the user not only includes information about the projects the user likes, but also includes information about the ratings the user gives to these projects.

$$a_{ir} = ReLU(W(q_i \oplus e_r) + b) \tag{1}$$

where q_i is the user's item sequence, e_r is the user's score for the item and $\oplus$ is a simple matrix stitching.

$$\tilde{x}_i^u = a_{ir} \oplus u \tag{2}$$

where u is the user to which all items belong.

$$X_i^u = Norm(\tilde{x}_i^u + Filter(\tilde{x}_i^u)) \tag{3}$$

where *Norm* is a normalized operation, *Filter* indicates a filter operation, and X_i^u is a sequence of items that carry user ratings.

In this paper, a filtering operation is performed on each feature dimension in the frequency domain as follows.

$$\tilde{X}^l = W \circ \mathcal{F}(F^l) \tag{4}$$

where F^l represents the input tensor, $\mathcal{F}$ represents the one-dimensional FFT operation and X^l is the spectral representation of F^l. $\circ$ is multiplied by elements.

$$\mathcal{F}^{-1}\left(\tilde{X}^l\right) \to \tilde{F}^l \tag{5}$$

where $\mathcal{F}^{-1}$ represents the IFFT operation.

$$n_i = softmax(W_2(ReLU\left(W_1 X_i^u + b_1\right)) + b_2) \tag{6}$$

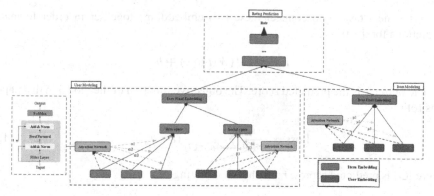

Fig. 2. The overall architecture of proposed model

$$\beta = n_i \cdot X_i^u \qquad (7)$$

where n_i is the association attention. Next is the process by which the user's friends sequence captures potential embedding.

$$m_u = softmax(W_2(ReLU\left(W_1 X_u^u + b_1\right)) + b_2) \qquad (8)$$

$$\gamma = m_u \cdot X_u^u \qquad (9)$$

where X_u^u is a sequence of user's friends after *Filter*, m_u is associated attention.

$$t = ReLU(W(\beta \oplus \gamma) + b) \qquad (10)$$

The output t represents the potential embedding of the user.

Item Modeling

Item modeling aims to learn item potential embedding by processing user sequence diagrams of items.

$$k_u = softmax(W_2(ReLU\left(W_1 X_u^i + b_1\right)) + b_2) \qquad (11)$$

$$\delta = k_u X_u^i \qquad (12)$$

where X_u^i is a sequence of all users who have purchased the item after *Filter*, k_u is associative attention.

$$y = ReLU(W\delta + b) \qquad (13)$$

where the output y is the potential embedding of the item.

Score Prediction and Optimization

Since the latent embeddings for both user and item have already been obtained in this study, the next task is to connect these two embeddings together in order to make predictions for the ratings.

$$r = ReLU(W(t \oplus y) + b) \tag{14}$$

The output r is the predicted score. The objective function of the model in this paper is as follows:

$$Loss = \frac{1}{2|\mathcal{O}|} \sum_{(u,i \in \mathcal{O})} (r - r_{ui})^2 \tag{15}$$

where $|\mathcal{O}|$ is predicted scores, r_{ui} is the actual rating of u to i.

3   Experiment

3.1   Dataset

Two representative datasets were used in this experiment, namely Ciao (http://www.ciao.co.uk) and Epinions (www.Epinions.com).

Table 1. Statistics of the datasets

Dataset	Ciao	Epinions
# of Users	7317	18088
# of Items	114975	261649
# of Ratings	283320	764352
# of Social Connections	111781	355217
# of Density (Ratings)	0.0337%	0.0162%
# of Density (Social Relations)	0.2088%	0.1086%

3.2   Evaluation Index

In this paper, MAE and RMSE were used to evaluate the quality of the recommendation models. Please note that the lower the values for both, the better the performance.

$$MAE = \frac{1}{|R_{test}|} \sum_{(u,i \in R_{test})} |r_{u,i} - r'_{ui}| \tag{16}$$

$$RMSE = \sqrt{\frac{1}{|R_{test}|} \sum_{(u,i \in R_{test})} |r_{u,i} - r'_{ui}|^2} \tag{17}$$

3.3 Baseline

In order to validate the effectiveness of the proposed algorithm in this article, we compared it with six benchmark algorithms.

i. SoRec [1]: Based on probabilistic matrix factorization, it integrates social network between users and user-item interaction matrix into the recommendation process.
ii. SoReg [2]: Based on matrix factorization, it uses user relationships as social regularization to constrain the matrix factorization objective.
iii. SocialMF [3]: The method combines trust propagation with matrix factorization structure to capture social phenomena in recommendation scenarios.
iv. DGRec [4]: Considering user interests and session-based dynamic social influence, this method uses graph attention neural network to simulate context-relevant social influence.
v. GraphRec [5]: A recommendation graph neural network model based on attention mechanism is proposed to aggregate social relationships.
vi. SR-HGNN [6]: A top-level cultural social relationship encoder is designed to capture global social dependencies between users. Based on understanding of multiple types of user-item interactions, it provides a graph structure for collaborative relationships.

3.4 Parameter Setting

For the data set, this paper uses 80% as the training set for learning parameters, 10% as the verification set for optimizing hyperparameters, and 10% as the test set for final performance evaluation. For embedded dimensions, this article tests the range {8, 16, 32, 64, 128, 256; For more, this paper test range {0.0005, 0.001, 0.005, 0.01, 0.05}; For batch sizes, the range tested in this article is {8, 16, 32, 64, 128, 256}. In order to deal with the overfitting problem, all experiments used an early stop method, that is, to stop training when the RMSE on the verification set did not improve for 5 consecutive epochs (Table 2).

3.5 Experimental Results and Analysis

Analysis of Experimental Results

According to the experimental results in Table 1, we can see that compared to SoRec, SoReg, and SocialMF, DGRec, GraphRec, and SR-HGNN have all demonstrated superior performance. This leads us to conclude that the application of graph neural networks and graph embedding layers is more effective than matrix factorization. Specifically, this method outperforms similar models in terms of the RMSE and MAE metrics, providing further evidence of its effectiveness. Moreover, our model performs even better on the Ciao dataset than the Epinions dataset. We believe that this is due to the higher data density of the Ciao dataset.

Ablation Study

In order to further evaluate the effectiveness of the proposed method, we conducted a ablation study. Specifically, we created two variants of our model, FMRec-α and FMRec-β. FMRec-α removes the attention layer, while FMRec-β removes the social connection pattern.

Table 2. Baseline algorithm comparison. The best results are in bold, Suboptimal underlined representation.

Methods	Ciao		Epinions	
	RMSE	MAE	RMSE	MAE
SoRec	1.0663	0.8279	1.1656	0.8853
SoReg	1.0876	0.8403	1.1721	0.9137
SocialMF	1.0704	0.8361	1.1682	0.8924
DGRec	1.0143	0.8031	1.0763	0.8570
GraphRec	1.0089	<u>0.7586</u>	1.0715	0.8261
SR-HGNN	<u>0.9908</u>	0.7735	<u>1.0644</u>	<u>0.8135</u>
FMRec	**0.9789**	**0.7347**	**1.0554**	**0.7983**
Improvement	1.2%	3.2%	0.8%	1.9%

Based on the results presented in Fig. 3, we can see that the FMRec model exhibits the best performance. This further confirms the importance of the learnable filter component and social connections in the model. Because of the impact of page numbers, the experiment of the model hyperparameters was omitted.

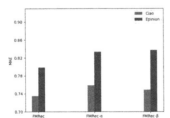 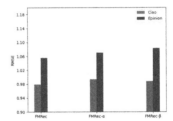

Fig. 3. MAE and RMSE of two datasets under three models

4   Conclusion and Future Work

The recommendation model is easily overfitting due to noise and malicious forgery in user behavior data. To enhance the model's robustness to noisy data, this paper adopts the filtering algorithm concept in digital signal processing and introduces a learnable filtering component. Research indicates that the learnable filterer is time-domain equivalent to circular convolution, with a broader receptive field, which can better capture periodic features. This study validates the effectiveness of the FMRec model on two real datasets. Future research can further integrate other aspects of information to solve the cold start problem and propose more effective methods to capture dynamic behavior features of users.

Acknowledgment. This work was supported by the Tianjin Graduate Research Innovation Project NO. 2021YJSS045.

References

1. Ma, H., Yang, H., Lyu, M.R., King, I.: SoRec: social recommendation using probabilistic matrix factorization. In: Proceedings of the 17th ACM Conference on Information and Knowledge Management, pp. 931–940 (2008)
2. Ma, H., Zhou, D., Liu, C., Lyu, M.R., King, I.: Recommender systems with social regularization. In: Proceedings of the fourth ACM International Conference on Web Search and Data Mining, pp. 287–296 (2011)
3. Jamali, M., Ester, M.: A matrix factorization technique with trust propagation for recommendation in social networks. In: Recsys, pp. 135–142. ACM (2010)
4. Song, W., Xiao, Z., Wang, Y., Charlin, L., Zhang, M., Tang, J.: Session-based social recommendation via dynamic graph attention networks. In: The 12th ACM International Conference on Web Search and Data Mining, pp. 555–563 (2019)
5. Fan, W., et al.: Graph neural networks for social recommendation. In: The World Wide Web Conference, WWW 2019, San Francisco, CA, USA, 13–17 May 2019, pp. 417–426. ACM (2019)
6. Xu, H., Huang, C., Xu, Y., Xia, L., Xing, H., Yin, D.: Global context enhanced social recommendation with hierarchical graph neural networks. In: 2020 IEEE Inter-national Conference on Data Mining (ICDM), pp. 701–710 (2020)
7. Sun, P., Wu, L., Wang, M.: Attentive re-current social recommendation. In: The 41st International ACM SIGIR Conference on Research & Development in Information Retrieval, SIGIR 2018, Ann Arbor, MI, USA, 08–12 July 2018, pp. 185–194. ACM (2018)
8. Chen, H., Li, J.: Data poisoning attacks on cross domain recommendation. In: Zhu, W., et al. (eds.) Proceedings of the 28th ACM International Conference on Information and Knowledge Management, CIKM 2019, Beijing, China, 3–7 November 2019, pp. 2177 2180. ACM (2019)
9. Vaswani, A., et al.: Attention is all you need. In: Advances in Neural Information Processing Systems, pp. 5998–6008 (2017)
10. Yu, W., et al.: Meta former is actually what you need for vision. arXiv:2111.11418 (2022)
11. Rao, Y., Zhao, W., Zhu, Z., Lu, J., Zhou, J.: Global filter networks for image classification. arXiv preprint arXiv:2107.00645 (2021)
12. Chen, H., Li, J.: Data poisoning attacks on cross-domain recommendation. In: CIKM 2019 (2019).https://doi.org/10.1145/3357384.3358116
13. Liu, Z., Fan, Z., Wang, Y., Yu, P.S.: Augmenting sequential recommendation with pseudo prior items via reversely pre-training transformer. In: Proceedings of the 44th International ACM SIGIR Conference on Research and Development in Information Retrieval (2021)
14. Lever, J., Krzywinski, M., Altman, N.: Points of significance: model selection and overfitting. Nat. Methods **13**(9), 703–705 (2016)
15. Hamilton, W., Ying, Z., Leskovec, J.: Inductive representation learning on large graphs. In: Advances in Neural Information Processing Systems, pp. 1024–1034 (2017)

Efficient Machine Learning-Based Prediction of CYP450 Inhibition

Gelany Aly Abdelkader[1], Soualihou Ngnamsie Njimbouom[1], Prince Delator Gidiglo[1], Tae-Jin Oh[3,4], and Jeong-Dong Kim[1,2,3(✉)]

[1] Department of Computer Science and Electronic Engineering, Sun Moon University, Asan 31460, Republic of Korea
kjd4u@sunmoon.ac.kr
[2] Division of Computer Science and Engineering, Sun Moon University, Asan 31460, Republic of Korea
[3] Genome-Based BioIT Convergence Institute, Asan 31460, Korea
[4] Department of Pharmaceutical Engineering and Biotechnology, Sun Moon University, Asan 31460, Korea

Abstract. Cytochrome P450 (CYP450) enzymes are crucial in drug metabolism and development. They are mainly involved in drug efficacy and toxicity, which are significant concerns while exploring potential pharmaceutical compounds. Machine learning-based approaches to predict the inhibitory effects of new compounds on these enzymes can help resolve the time and cost-intensive in vivo and in vitro experiments. Most proposed Machine Learning methods for predicting CYP450 isoform inhibitors often rely on complex models and numerous features that may not accurately distinguish between inhibitors and non-inhibitors. This work proposes a new approach that merges the MACCS and Morgan fingerprints with 200 other molecular descriptors to generate input features for a robust and straightforward computational model. Using 10-fold cross-validation, we trained and evaluated several classifiers using the proposed input features for each CYP450 isoform. The Support Vector Machine and LightGBM consistently showed the best performance for all five isoforms. Our models exhibited good performance proving the effectiveness of the proposed features and indicating their potential use for CYP450 inhibition prediction.

Keywords: Machine Learning · Feature Engineering · Cytochrome P450

1 Introduction

CYP450 enzymes belong to a family of enzymes essential for drug metabolism. These enzymes metabolize various endogenous and exogenous compounds, including drugs and toxins. To date, over 50 human CYP450 enzymes have been identified, but five major isozymes are responsible for the metabolism of most drugs. These major isozymes include CYP1A2, CYP3A4, CYP2D6, CYP2C9, and CYP2C19. Understanding their activity is crucial for predicting potential drug-drug interactions and optimizing drug metabolism [1].

© The Author(s), under exclusive license to Springer Nature Switzerland AG 2023
C. Strauss et al. (Eds.): DEXA 2023, LNCS 14147, pp. 438–444, 2023.
https://doi.org/10.1007/978-3-031-39821-6_36

Throughout recent years, there has been a significant surge in the utilization of Machine Learning (ML) and Deep Learning (DL) algorithms to predict the inhibition of major CYP450 isozymes, namely 1A2, 2C9, 2C19, 3A4, and 2D6. Various computational methods have been put forward to address this prediction challenge. For instance, a study by [2] utilized the random forest algorithm to predict the inhibition of the 5 major CYP450 isozymes using molecular fingerprints of various compounds collected from the PubChem Bioassay, ChEMBL, and ADME databases. MACCS keys and FP4 fingerprints were combined in Cheng et al. [3], while Support Vector Machine (SVM) with molecular signatures were used in [4]. Additionally, Grisoni et al. [5] analyzed the impact of different molecular descriptors on ML and DL models.

Most proposed computational approaches have focused on adding more data or features. While having more data and features can sometimes improve performance, there are other factors to consider, such as the quality of the data and features and the complexity of the model. It is essential to achieve a balance between having large quantity of data and features to capture the underlying patterns in the data while avoiding overfitting and keeping the model interpretable and efficient. Most of the CYP450 bioassay data available to date suffer from the unbalanced class problem, which can lead to the misinterpretation of the performance metrics of the ML and DL models. This study proposed a new approach to tackle the abovementioned challenges by integrating a large amount of data and presenting a unique combination of features without compromising the model performance. Our goal is to create a robust and straightforward computational model. To accomplish this, we aim to maintain a balance between utilizing a significant amount of data and selecting relevant features. We integrated multiple molecular descriptors that provide greater insight into the unique characteristics between the inhibitor and non-inhibitor compounds. The appropriately selected molecular descriptors include the Morgan fingerprint, MACCS fingerprint, and 200 molecular descriptors from the RDKIT Python library.

2 Proposed Model for the CYP450 Isoforms Inhibition Prediction

The steps involved in this work are shown in Fig. 1. The overall process starts with data collection from various CYP450 bioassay datasets, followed by a preprocessing step and calculation of the molecular descriptors, where we generate the input features for the ML models. Next, hyperparameter tuning and 10-fold cross-validation (10CV) of different ML models are performed for each isozyme.

2.1 CYP450 Bioassay Data Collection

Data were collected from publicly available repositories on the inhibition of human CYP450 isozymes 1A2, 2C9, 2C19, 2D6, and 3A4. Various sources were consulted, including the PubChem bioassay datasets 1851, 410, 883, 884, 899, and 891 [6]. In addition, the ChEMBL database [7] was also included. Following a prior study [2], PubChem compounds with activity class marked "active" with observed inhibition were designated as inhibitors. Conversely, compounds with an "inactive" activity class and no observed inhibition were classified as non-inhibitors.

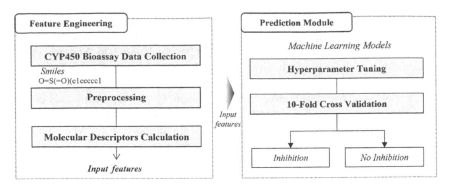

Fig. 1. Conceptual model for the CYP450 isoforms inhibition prediction

Compounds with inconclusive observations were excluded from the analysis. Following [2], ChEMBL compounds were classified as inhibitors or non-inhibitors based on their Half-maximal inhibitory concentration (IC50) values, "standard relation", and "percentage of inhibition".

2.2 Data Preprocessing

The chemical structures of each compound, mainly the smiles, were extracted. For further analysis, various steps were required to ensure the accuracy and consistency of the data. First, any syntax and semantic errors in the SMILES notation were corrected. Then, the canonical SMILES for each molecule were generated. After this step, we performed the same filtering procedure described in [2] using the RDKIT Python library version 2022.09.1 [8].

The datasets obtained from PubChem and the ChEMBL database were combined separately for each of the five CYP450 (1A2, 2D6, 2C9, 2C19, and 3A4). Compounds that shared the same smiles were considered duplicates and removed since they would produce identical molecular descriptor values. The resulting dataset was accurate and reliable by following these steps, providing a solid foundation for further analysis and research. A detailed summary of the data partition is provided in Table 1.

2.3 Molecular Descriptors Calculation

The suggested method combines molecular descriptors as input features to train ML models. These features combine the 2048 bits Morgan fingerprints [9], the 166 bits molecular access system (MACCS) fingerprints [10], and additional 200 molecular descriptors that describe diverse aspects of molecule's properties, such as its electronic structure, drug-likeness and weight. The complete list of these descriptors can be obtained from the RDKIT function MolecularDescriptorCalculator. All these descriptors were generated using the RDKIT Python library version 2022.09.1 [8].

2.4 Experimentation

To assess the efficacy and performance of our proposed input features, we applied 6 ML models for each of the five major isozymes datasets. For the ML models, we included the SVM, the LightGBM (LGBM), the Random Forest (RF), Extreme Gradient Boosting (XGB), K-NN, and the Gaussian Naive Bayes classifiers (GNB). Each isozyme set was randomly partitioned into training and testing with a ratio of 80% for training and fine-tuning and 20% for testing (Table 1). To optimize our candidate algorithms' performance, we thoroughly searched for the best hyperparameters using the GridSearchCV function from the scikit-learn library. Our search involved testing various combinations of hyperparameter values across a range of parameters for each of the corresponding models. The explored range of hyperparameters is presented in Table 2. For a more robust and accurate estimate of the performance of each model, 10CV was performed after hyperparameter tuning. This helps to ensure that the model is not just performing well on a particular subset of data.

Table 1. Distribution of training versus test set.

Partition		Isoforms				
		1A2	2C19	2C9	2D6	3A4
Training set	Inhibitors	5350	4567	3446	2410	5278
	Non-Inhibitor	6431	6625	8050	10353	9143
Testing set	Inhibitors	1348	1134	846	593	1327
	Non-Inhibitor	1598	1664	2029	2598	2279

Table 2. Optimized parameters for ML models

Models	Parameters
Random Forest	min_samples_leaf = 2, min_samples_split = 5, n_estimators = 100
Support Vector Machine	C = 10.0, Gamma = 0.01, Kernel = rbf,
LightGBM	boosting_type = dart, learning_rate = 0.1, 'max_depth = −1, n_estimator = 300,
K-Nearest Neighbors	n_neighbors = 10, Weights = distance

3 Results and Discussion

To thoroughly evaluate the performance of our proposed approach, we utilized various performance metrics. The primary metrics used were the Matthews Correlation Coefficient (MCC), Accuracy, and the Area Under the receiver operating characteristic Curve

Table 3. The average performance of each model across 10-fold cross-validation.

Isozymes	Models	Metrics		
		Accuracy	AUC	MCC
1A2	**LGBM**	**0.84**	**0.92**	**0.68**
	SVM	**0.84**	**0.91**	**0.68**
	RF	0.83	0.91	0.65
	XGB	0.82	0.90	0.63
	K-NN	0.82	0.90	0.64
	GNB	0.72	0.76	0.45
2C9	**SVM**	**0.86**	**0.92**	**0.66**
	LGBM	**0.86**	**0.92**	**0.66**
	RF	0.84	0.91	0.61
	K-NN	0.84	0.90	0.59
	XGB	0.82	0.88	0.57
	GNB	0.67	0.75	0.41
2D6	**SVM**	**0.91**	**0.92**	**0.68**
	LGBM	**0.90**	**0.92**	**0.67**
	K-NN	0.89	0.90	0.63
	RF	0.89	0.92	0.62
	XGB	0.88	0.89	0.58
	GNB	0.63	0.72	0.32
3A4	**SVM**	**0.87**	**0.94**	**0.72**
	LGBM	**0.86**	**0.93**	**0.70**
	RF	0.85	0.93	0.68
	XGB	0.83	0.90	0.62
	K-NN	0.84	0.91	0.65
	GNB	0.73	0.79	0.48
2C19	**LGBM**	**0.84**	**0.91**	**0.66**
	SVM	**0.84**	**0.91**	**0.67**
	RF	0.83	0.90	0.64
	K-NN	0.81	0.89	0.61
	XGB	0.80	0.88	0.60
	GNB	0.72	0.76	0.46

(AUC). Table 3 displays the performance of each model for the five major CYP450 isozymes using 10CV: For the 1A2 isozyme, LGBM and SVM models exhibited the best performance with 84% accuracy and 0.92 AUC on average. Likewise, the SVM and LGBM models demonstrated superior performance for other isozymes.

The LGBM and SVM models consistently achieved the highest accuracy scores for all five isozymes through 10CV, with an accuracy range of 0.84 to 0.91. These models also showed high AUC scores ranging from 0.91 to 0.94, which indicates good discrimination performance.

We also measured the performance of various models on the 20% left-out dataset. The observed performance showed that SVM and LGBM reliably performed well for all five isozymes. Both models had an overall AUC ranging from 0.91 to 0.93 and a recorded MCC between 0.66 and 0.70. These findings prove that our proposed approach can effectively predict the inhibitory effect of new molecules, with the LGBM and SVM models being the most promising for accurate and reliable predictions.

As for the future directions of our work, we plan to investigate a multimodal approach for predicting inhibitors of the five major isoforms of CYP450. To achieve this, we aim to jointly train a Convolutional Neural Network on chemical compound images and a ML model that combines the compound's features proposed in this work.

4 Conclusion

In summary, our proposed feature combination is effective for constructing a robust and scalable model capable of handling large and structurally diverse molecules without compromising performance. The proposed approach of combining the RDKIT 200 molecular descriptors to the MACCS and Morgan fingerprints has proven to be efficient in predicting the inhibition of the CYP450 isozymes.

The LGBM and SVM models consistently performed well for all five isozymes. By incorporating various molecular descriptors as features, we enhanced our models' accuracy and provided a valuable tool for researchers in the field. The proposed method can be valuable in identifying potential interactions between a drug candidate and CYP450 enzymes. By assessing the metabolic activity of a drug candidate on these enzymes, researchers can gain insights into the potential for drug-drug interactions, metabolism-mediated adverse effects, and optimization of pharmacokinetic properties.

References

1. Zanger, U.M., Schwab, M.: Cytochrome P450 enzymes in drug metabolism: regulation of gene expression, enzyme activities, and impact of genetic variation. Pharmacol. Ther. 138(1), 103–141 (2013)
2. Plonka, W., Stork, C., Šícho, M., Kirchmair, J.: CYPlebrity: machine learning models for the prediction of inhibitors of cytochrome P450 enzymes. Bioorg. Med. Chem. 46, 116388 (2021)
3. Cheng, F., et al.: Classification of cytochrome P450 inhibitors and noninhibitors using combined classifiers. J. Chem. Inf. Model. 51(5), 996–1011 (2011). https://doi.org/10.1021/ci2 00028n

4. Rostkowski, M., Spjuth, O., Rydberg, P.: WhichCyp: prediction of cytochromes P450 inhibition. Bioinformatics **29**(16), 2051–2052 (2013)
5. Grisoni, F., Consonni, V., Todeschini, R.: Impact of molecular descriptors on computational models. In: Brown, J.B. (ed.) Computational Chemogenomics. MMB, vol. 1825, pp. 171–209. Springer, New York (2018). https://doi.org/10.1007/978-1-4939-8639-2_5
6. NIH. PubChem BioAssays (2023). https://pubchem.ncbi.nlm.nih.gov/bioassay
7. Gaulton, A., et al.: The ChEMBL database in 2017. Nucleic Acids Res. **45**(D1), D945–D954 (2017). https://doi.org/10.1093/nar/gkw1074
8. Landrum, G.: RDKit: open-source cheminformatics. http://www.rdkit.org/
9. Morgan, H.L.: The generation of a unique machine description for chemical structures-a technique developed at chemical abstracts service. J. Chem. Doc. **5**(2), 107–113 (1965)
10. Durant, J.L., Leland, B.A., Henry, D.R., Nourse, J.G.: Reoptimization of MDL keys for use in drug discovery. J. Chem. Inf. Comput. Sci. **42**(6), 1273–1280 (2002)

A Machine-Learning Framework
for Supporting Content Recommendation
via User Feedback Data and Content Profiles
in Content Managements Systems

Debashish Roy[1], Chen Ding[1], Alfredo Cuzzocrea[2,3]($\boxtimes$), and Islam Belmerabet[2]

[1] Toronto Metropolitan University, Toronto, Canada
{debashish.roy,cding}@torontomu.ca
[2] iDEA Lab, University of Calabria, Rende, Italy
alfredo.cuzzocrea@unical.it
[3] Department of Computer Science, University of Paris City, Paris, France

Abstract. *Matrix Factorization (MF)* which is a *Collaborative Filtering (CF)* based model, is widely used in *Recommendation Systems (RS)*. In this research, we deal with a specific recommendation problem of recommending content to users in a *Content Management System (CMS)* utilizing users' *feedback* data.

Keywords: Matrix Factorization · TF-IDF · Collaborative Filtering · Regularization · Objective Function · Hybrid

1 Introduction

Recommender Systems (RS) are widely used to help users find items of their interest [2]. Depending on the user profile, recommender system can determine whether a particular user will like an item or not. To build a recommender system, various approaches have been developed, including Collaborative Filtering (CF), *Content-Based (CT) Filtering* and *Hybrid Filtering* [3]. In this research, a recommender system is built for a company's *Content Management System (CMS)*. All the content (posts) and users' interaction data are saved in the CMS *database*. It would be helpful to implement a *recommender system* using the saved data. The *dataset* is small in terms of number of posts, number of users, and number of interactions. Designing a recommender system for this type of dataset is a challenging task. Many existing papers on recommender systems work on some common datasets such as MovieLens, Amazon reviews, etc., which are much bigger than our dataset. The models that are working for the larger dataset might not work well for our dataset. So, we want to explore what type of information and which RS model works well for our dataset. The objectives for this research work are: 1) to study user interactions and utilize user feedback data saved in the database to generate user

This research has been made in the context of the Excellence Chair in Big Data Management and Analytics at University of Paris City, Paris, France.

profiles; 2) to use CF method for *content recommendation*, here we use feedback scores as the rating data and a MF model on the rating matrix; 3) to incorporate the similarity score as a regularization term into the MF model; 4) to analyze the effect of different categories of feedback data on the recommendation results, here we want to investigate the accuracy of recommendation if different categories of feedback data are used.

2 Related Work

Recommender systems are information filtering systems that are used to provide suggestions to users depending on user profile [1, 11]. We can apply mainly two types of approaches for content recommendation [8]: *CF-Based Approach* and a *CT-Based Approach*. CF is the most commonly used and it recommends items by identifying other users with similar interest [5, 12]. CT-Based technique recommends items based on content similarity. To get the best of the both worlds, some hybrid approaches, which use a combination of both techniques, are used. CT-based approach recommends the unseen items to the users, depending on the items already seen by the users [4]. Researchers have used different ways to build the user profiles. To model the content profile, we have used one of the main content profile modeling algorithms: *Term Frequency-Inverse Document Frequency (TF-IDF)* [13]. In our research work, to build a user profile, we have used users' explicit interactions, implicit interactions and users' social share interactions. Then, we used a weighted combination of these three types of interactions to calculate a combined feedback score to generate the user profiles. We also have used the weighted score for the individual types of feedback to generate the user profiles. We have compared CF-Based recommendation models created using different categories of feedback data. We have also compared these models with the hybrid model which considers both user feedback information and post content information.

3 Methodology

This section focuses on the proposed methods of utilizing user feedback data for content recommendations. In this work, we use two recommendation approaches: CF approach using just the feedback data and hybrid model using CF and post similarity score. To accomplish our research goal, we take the following steps: (*i*) we generate individual user profiles based on user interactions with the posts. After extracting user interactions, we classify these interactions into three major categories based on the nature of the interaction. First category is Direct Interaction, which is the user/post interaction happened in the CMS itself. Second category is Social Share, which is the interaction on user's own social network. Third category is Reading Statistics, which refers to the feedback score user provided to the content by taking a certain action, such as reading progress. All the interactions have already been assigned a weight by the CMS. We use those weights and calculate user feedback score for each user. We then compare results from each category with overall feedback score to evaluate which category reflects the user behavior the best; (*ii*) we generate content recommendations using CF algorithm; (*iii*) we find similarity score between posts; (*iv*) we evaluate the results.

3.1 System Architecture

In this paper, we mainly want to compare recommendations generated by Collaborative-Filtering approach and the hybrid approach for the CMS dataset. Furthermore, we want to find out which type of feedback data provides the most accurate recommendations. Figure 1 shows the system architecture of our recommender system. It has two major components: (*i*) a CF model and (*ii*) a model to calculate similarity between the content of the posts. At first, we extract users' implicit/explicit feedback data and contents of the posts. User feedback data is an input to the CF model and contents of the posts are input to the content profile modeling system. Post similarity score is used as an input to the hybrid model. Then, we apply MF model on user profiles to generate CF-Based recommendations. In the hybrid model, we integrate post similarity as a regularization term into the MF model.

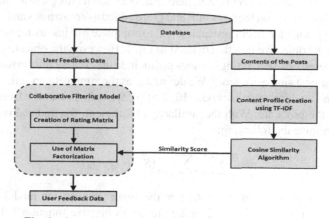

Fig. 1. Overall System Architecture for Recommendations

3.2 User Preference Matrix Generation

After examining the user's historical data and interactions, we calculate user's preference score pertaining to those categories. The interactions or user feedback data which show user/item interactions internally on the CMS such as like, comment, impression and click-through come under the category known as Direct Interaction. The interactions which show data is shared externally by the users on their social networking websites such as shares, re-shares, reply, retweets come under Social Share category. The interactions like reading duration and the number of reads which show reading progress made by a user internally on the CMS itself come under the Reading Statistics Category. Lastly, the Feedback Score is calculated by adding scores from all the categories together with the weights defined by us for each category. After calculating score for each category and the overall feedback score, we model user profile for each user. We use these calculated scores together with the content profiles of the posts that the user has interacted with to get the user profile. In this way, we created multiple user profiles based on the different user/post interaction categories and content profile developed using TF-IDF.

3.3 Matrix Factorization and Post Similarity Calculation

The process of MF [9] starts with a user-item rating matrix R. Here, the size of matrix R is: $m \times n$ where m denotes the total number of users and n denotes the total number of items. MF will decompose the rating matrix R into two low rank latent feature matrices [6] P for users and Q for items; here, the size of matrix P is: $m \times d$ and the size of matrix Q is: $n \times d$, d is the rank of the matrices and defines *dimension* for the latent features [5]. To decompose a sparse rating matrix, the following *objective function* is used by the traditional MF methods [10]:

$$L = \underset{P,Q}{min} \frac{1}{2} \sum_{(u,i)\in C} (R_{u,i} - P_u Q_i^T)^2 + \frac{\lambda}{2}(||P||_F^2 + ||Q||_F^2) \tag{1}$$

In Eq. 1, C indicates the set of (user, item) pairs of known ratings, to avoid *overfitting*, two regularization terms on the sizes of P and Q are added as *constraints* and λ is used as a regularization parameter. In our research work, to complete MF just using the rating data, we have used the objective function defined in Eq. 1. However, the objective function in Eq. 1 is only based on users' ratings, it does not include the content information. To do so, we include the post similarity score. We define S_{jn} as the similarity co-efficient between two posts j and n which satisfies: $(i) S_{jn} \in [0, 1]$; ; $(ii) S_{jn} = S_{nj}$; ; (iii) the larger S_{jn} is, the more similar the posts are. With the similarity co-efficient, the updated regularization term is to minimize the following:

$$min \frac{\alpha}{2} \sum_{j=1}^N \sum_{n=1}^N (S_{jn} - Q_j^T Q_n)^2 \tag{2}$$

To add the impact of content profiles to the basic factorization model, we add the regularization term defined in Eq. 2 to the previous objective function 1. In our work, using this updated objective function, we have designed the hybrid model and also used this model to complete MF using both rating data and content similarity score. To find out the similarity between posts, we build the content profiles. To model the post profiles, we can use bag-of-words models such as Term Frequency-Inverse Document Frequency and *Latent Dirichlet Allocation (LDA)*. Our experiment shows that TF-IDF model performs better than the LDA model. So, we have used the TF-IDF model. We have used cosine similarity to find the similarities between different posts.

4   Experiments

This section discusses experiment design and the results. We use a CMS dataset to generate recommendations for effective content suggestions to the users of the dataset. The dataset has two major entities: posts and users. The dataset consists of 250 users, 6900 posts as well as the user/post interactions. 150 users have 20868 direct interactions, 165 users have 28363 social shares, and 134 users have 10985 reading statistics. In our research work, we have used RMSE and F1-Score for evaluation [7].

4.1 Result Analysis

Evaluation of Results Using RMSE and F1-score. We have used both basic MF and hybrid MF to generate the recommendations. We calculated the RMSE scores for the generated recommendation results. For the same result sets, we calculated the F1 scores for top-10 recommendations. From Fig. 2, we see that all Interaction-Based recommendations generated the smallest RMSE, and all Interaction-Based recommendations generated the highest F1-score for both models. We find that Social Share-Based recommendation performed the best and the worst performing one is the Reading Statistics-Based recommendations. In all the cases, the hybrid model outperformed the traditional model. And social share is very close to all interaction.

Results Comparison on User Profile Types. We have in total four different categories of user profiles. The results show that All Interaction Feedback Score profile is performing the best among all four profiles. If we consider three profile categories: Social Share, Direct Interaction and Reading Statistics, the Social Share category performs much better than the other two. We get the second-best performance from Direct Interaction profile and last one is the Reading Statistics profile. The major reason behind this result is that there are more types of social shares in this category. In other words, we can say in this category, we have more types of user/post interactions as compared to the other categories and data is largely shared by users on their social networking websites. As a comparison, Reading Statistics profile has only two types of interactions. Therefore, as Social Share profile has more interaction data, it works better than Direct Interaction profile and Reading Statistics Profile. While comparing Social Share profile with all interaction Feedback Score profile, the overall Feedback Score profile works better than the social share profile. However, their differences are very small. When efficiency is a concern, we may use the social feedback only because it requires less data processing time. Finally, when we compare all the evaluation metrics on the two methods of MF: basic MF and hybrid MF, the results show that hybrid MF performs better than the basic MF for all the four types of user profiles.

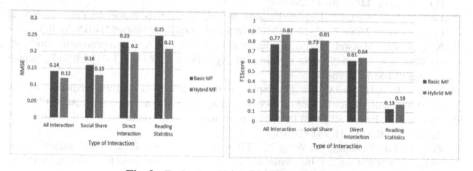

Fig. 2. Evaluation Using RMSE and F1Score

5 Conclusions and Future Work

The evaluation results show that the combination of TF-IDF content profiles with all interaction feedback, i.e., the hybrid model, generates the best recommendations for this CMS dataset. Moreover, different types of user interaction data may have different impacts on the recommendation accuracy, therefore it is necessary to explore and investigate the optimal way of using and combining them in a recommender system. As a future work, we want to compare the automatically generated user interaction weights with those that are predefined in the dataset. We will apply a regression model or by using a classification algorithm to find optimal weights based on the user interactions. Also, dealing with emerging *big data trends* (e.g., [12, 13]) is a goal of future efforts.

Acknowledgments. This research is supported by the ICSC National Research Centre for High Performance Computing, Big Data and Quantum Computing within the NextGenerationEU program (Project Code: PNRR CN00000013).

References

1. Adomavicius, G., Tuzhilin, A.: Toward the next generation of recommender systems: a survey of the state-of-the-art and possible extensions. IEEE Trans. Knowl. Data Eng. **17**(6), 734–749 (2005)
2. Aggarwal, C.C.: Recommender Systems, 1st edn. Springer, Heidelberg (2016). https://doi.org/10.1007/978-3-319-29659-3
3. Burke, R.: Hybrid recommender systems: survey and experiments. User Model. User-Adapt. Interact. **12**(4), 331–370 (2002). https://doi.org/10.1023/A:1021240730564
4. Deerwester, S., Dumais, S.T., Furnas, G.W., Landauer, T.K., Harshman, R.: Indexing by latent semantic analysis. J. Am. Soc. Inf. Sci. **41**(6), 391–407 (1990)
5. Karumur, R.P., Nguyen, T.T., Konstan, J.A.: Personality, user preferences and behavior in recommender systems. Inf. Syst. Front. **20**(6), 1241–1265 (2018). https://doi.org/10.1007/s10796-017-9800-0
6. Koren, Y., Bell, R., Volinsky, C.: Matrix factorization techniques for recommender systems. IEEE Comput. **42**(8), 30–37 (2009)
7. Li, F., Xu, G., Cao, L.: Two-level matrix factorization for recommender systems. Neural Comput. Appl. **27**(8), 2267–2278 (2016). https://doi.org/10.1007/s00521-015-2060-3
8. Nguyen, T.T., Maxwell Harper, F., Terveen, L., Konstan, J.A.: User personality and user satisfaction with recommender systems. Inf. Syst. Front. **20**(6), 1173–1189 (2017). https://doi.org/10.1007/s10796-017-9782-y
9. Tang, J., Gao, H., Hu, X., Liu, H.: Exploiting homophily effect for trust prediction. In: 6th ACM International Conference on Web Search and Data Mining, pp. 53–62. ACM (2013)
10. Turney, P.D., Pantel, P.: From frequency to meaning: vector space models of semantics. J. Artif. Intell. Res. **37**(1), 141 (2010)
11. Zhao, G., Liu, Y., Zhang, W., Wang, Y.: TFIDF based feature words extraction and topic modeling for short text. In: 2nd International Conference on Management Engineering, Software Engineering and Service Sciences, pp. 188–191 (2018)
12. Campan, A., Cuzzocrea, A., Truta, T.M.: Fighting fake news spread in online social networks: actual trends and future research directions. In: IEEE BigData 2017, pp. 4453–4457 (2017)
13. Cuzzocrea, A.: Analytics over big data: exploring the convergence of datawarehousing, OLAP and data-intensive cloud infrastructures. In: IEEE COMPSAC 2013, pp. 481–483 (2013)

Fine-Tuning Pre-Trained Model for Consumer Fraud Detection from Consumer Reviews

Xingli Tang, Keqi Li, Liting Huang, Hui Zhou(✉), and Chunyang Ye

Hainan University, Haikou, Hainan, China
{txl,kqli,litinghuang,zhouhui,cyye}@hainanu.edu.cn

Abstract. Consumer fraud is a significant problem that requires accurate and prompt detection. However, existing approaches such as periodic government inspections and consumer reports are inefficient and cumbersome. This paper proposes a novel approach named CFD-BERT, to detect consumer fraud automatically based on the group intelligence from consumer reviews. By applying the correlation between consumer reviews and official regulations to accurately mine consumer fraud patterns, and fine-tuning a pretrained model BERT to better model their semantics, which can detect fraudulent behaviors. Experimental evaluations using real-world datasets confirms the effectiveness of CFD-BERT in fraud detection. To explore its potential application and usefulness in real world scenarios, an empirical study was conducted with CFD-BERT on 143,587 reviews from the last three months. The results confirmed that CFD-BERT can serve as an auxiliary tool to provide early warnings to relevant regulators and consumers.

Keywords: Consumer fraud detection · Consumer reviews · Regulation

1 Introduction

Consumer fraud, defined as the use of false or improper means by business operators to harm the legitimate rights and interests of consumers, which poses a significant challenge for both customers and regulatory authorities [4].

Conventional approaches to consumer fraud detection rely on regulatory authorities conducting inspections via sampling, a process that is labor-intensive and often inefficient [5]. Regrettably, consumer fraud is a multifaceted problem that manifests in diverse life scenarios, and the transactions involved are frequently uncomplicated and challenging to document, thereby posing difficulties in quantifying the extent of fraud [1]. Fraud detection in the financial and credit card sectors has achieved considerable maturity, benefiting from the availability of public datasets and the development of numerous models that exhibit superior detection capabilities [2]. However, detecting consumer fraud is more challenging due to the absence of public datasets.

© The Author(s), under exclusive license to Springer Nature Switzerland AG 2023
C. Strauss et al. (Eds.): DEXA 2023, LNCS 14147, pp. 451–456, 2023.
https://doi.org/10.1007/978-3-031-39821-6_38

Some previous research has revealed that online consumer reviews can reduce the workload of manual consumer fraud detection, and scientific methods can be used to support online consumer fraud detection and prevention [7]. Lai et al. introduced a fraud detection model, BTextCAN, based on fraudulent merchant reviews [6]. However, the amount of fraud merchant data published by regulatory authorities are small, which compromises the effective extraction of fraudulent features.

Inspired by above observations, we were wondering whether the official regulations and online consumer reviews can be combined to effectively detect consumer fraud automatically. This paper presents our latest efforts on this issue and our study has yielded promising results in this regard.

This paper propose a novel approach named CFD-BERT, which aims to detect consumer fraud automatically based on the group intelligence from consumer reviews. By applying the BTM model to extract common consumer fraud patterns from consumer reviews and fine-tuning a pretrained model BERT to better model their semantics, our semantic-aware, fine-grained approach can match consumer reviews with regulations automatically to detect fraudulent behaviors. Through experimental evaluations conducted on real-world datasets, CFD-BERT demonstrates effective consumer fraud detection capabilities and successfully identifies the specific fraud categories associated with each review.

2   Methodology

Figure 1 presents an overview of CFD-BERT, which comprises three main parts. The first part is data preprocessing, which involves classifying consumer reviews. The second part is BERT fine-tuning, where a fine-tuned BERT is built and trained. The third part is consumer fraud detection, which detects consumer fraud and identifies fraud categories.

2.1   Dataset Construction

Topic Extraction. BTM directly models co-occurring words as semantic transfer units of topics, which can reveal topics better than single words. BTM labels topics with appropriate behavior by computing the probability that each document is assigned to each topic. The generated probability of each topic z is shown in (1), where z represents topic, b represents biterm and d represents document.

$$P(z \mid d) = \sum_b P(z \mid b)P(b \mid d) \tag{1}$$

$P(z \mid b)$ can be calculated via Bayes formula based on the parameters estimated in BTM as (2), BTM draws distributions of $\phi_z \sim \mathrm{Dir}(\beta)$, $\theta \sim \mathrm{Dir}(\alpha)$, $Z \sim \mathrm{Multi}(\theta)$, where α and θ are the Dirichlet priors. $P(b \mid d)$ can be estimated by the empirical distribution of biterms in the document as (3), where $n_d(b)$ is the frequency of the biterm b in the document d.

$$P(z \mid b) = \frac{\theta_z \phi_{i|z} \phi_{j|z}}{\sum_z \theta_z \phi_{i|z} \phi_{j|z}} \tag{2}$$

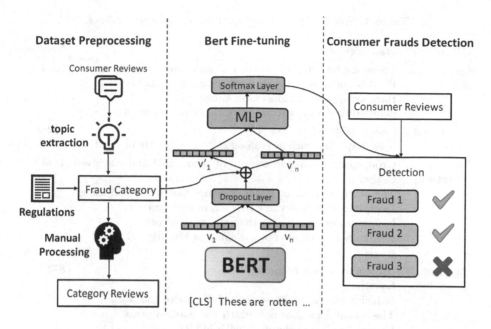

Fig. 1. Overview of CFD-BERT.

$$p(b \mid d) = \frac{n_d(b)}{\sum_b n_d(b)} \tag{3}$$

Fraud Category Collection. Based on the Law on the Protection of Consumer Rights and Interests, which defines the legitimate rights and interests of consumers, we analyze and summarize the frauds defined by two sets of regulations: "What is Fraudulent Consumer Behaviour" [4] and "Hainan Free Trade Port Anti-Consumer Fraud Regulations" [3]. We compare the similarity of the generated topic types with the regulations and ultimately classify consumer fraud.

Manual Labeling of Consumer Reviews. After identifying the fraud category, classified reviews can be collected. To do this, we employ manual methods to label and check consumer reviews. If a single review involves multiple frauds, we split the review into separate reviews, each belonging to only one consumer fraud type. These reviews are then fed into CFD-BERT for training.

2.2 Model Training

We employ fine-tuned BERT_BASE on our own data to extract and detect fraudulent features, which first maps consumer reviews to an encoded vector with contextual semantic relationships. The encoded vectors are then passed to the MLP network layer, and finally, the softmax layer is employed for fraud detection and classification.

Table 1. Fraud categories and number of consumer reviews.

Category	Fraudulent behavior	Number
Unqualified goods	There are foreign objects, such as hair, bugs, plastic, etc. Food is not fresh, odorous, spoiled, expired, moldy, etc. Have symptoms of illness after eating. Food adulteration, such as pork chops as steaks, etc.	2073
Insufficient portion	Less weight and smaller size Less in quality, such as seafood only shells without meat.	812
False advertising	Group purchase activities are not recognized and not given to use. The description of the goods does not match the real thing. Deceptive activities where the "discounted price" is equal to or even higher than the original price. Discounted goods are not the same as the original priced goods.	1115
Payment issues	No refund in case of unused fees. Overcharge. Inducing consumption, such as giving more goods. The actual price does not match the marked price. Unspecified before charging, with hidden fees.	873
Invoice issues	No invoicing. The amount of the invoice is incorrect.	195

The activation function is denoted as f, and W is a trainable parameter in the MLP, the probability vector of fraudulent category labels for each review denoted as p, v is the text vector, h_c is the fraud category vector, $p = f(W[h_c; v])$. Finally, the probability vector of the fraud category label for each word after normalization is obtained by the softmax layer. We choose the cross-entropy function as the loss function, which combined with the softmax function of the output layer can accelerate the training speed of deep learning more quickly.

3   Experiment

3.1   Data Preparation

To evaluate our model, we collected reviews from Meituan and Dianping, two popular Chinese apps that are widely used by a significant portion of consumers. We crawled public reviews from these apps between January 2021 and June 2022, resulting in a total of 640,000 negative reviews (rated below 3 stars). We randomly selected 100,000 reviews and performed topic extraction on them using the BTM topic model to obtain eight topics. After comparing the topics obtained with the regulations, we manually labeled the consumer reviews by injecting specific classification information, resulting in a total of 5,071 labeled reviews. The resulting fraud categories are as Table 1.

Table 2. Evaluation results on the consumer reviews (Best results are show in bold).

Category	Unqualified goods			Insufficient portion			False Advertising			Payment issues			Invoice issues		
	acc	pre	rec	acc	pre	rec	acc	pre	rec	acc	pre	rec	acc	pre	rec
CFD-BERT	**98%**	**98%**	**99%**	**94%**	**88%**	**96%**	**95%**	**86%**	**98%**	**94%**	**97%**	**97%**	**100%**	**97%**	**100%**
SRE	82%	96%	58%	89%	85%	37%	79%	71%	10%	85%	81%	15%	100%	97%	94%
SVM	69%	80%	27%	86%	83%	14%	76%	75%	12%	79%	90%	12%	97%	75%	26%
Text-CNN	91%	91%	94%	79%	78%	75%	85%	85%	85%	87%	88%	81%	75%	85%	100%
Text-RNN	86%	93%	93%	79%	77%	67%	78%	76%	81%	80%	88%	80%	90%	60%	100%
Text-RCNN	89%	93%	91%	75%	71%	67%	80%	68%	93%	81%	94%	62%	85%	75%	100%
Transformer	85%	95%	95%	73%	73%	73%	63%	70%	93%	87%	85%	67%	87%	67%	67%

3.2   Experimental Setup

To obtain the optimal hyper-parameters, we engaged in an iterative process of training and testing. Ultimately, a batch size of 4 and learning rate of 2e-5 yielded the best results. Throughout the training process, the model with the highest accuracy was saved for future use.

3.3   Experimental Evaluation

To establish a baseline for our study, we selected methods capable of extracting fraudulent features from consumer reviews. Based on our analysis, we categorized existing methods into three types based on their underlying techniques: (1) Semantic rule matching (SRM). (2) Machine learning with Support Vector Machines (SVM). (3) Deep learning with Text-CNN, Text-RNN, Text-RCNN, and Transformer models.

The results is shown in Table 2. Notably, the CFD-BERT model achieved superior results in all fraud category detection, owing to its capacity to construct key features based on contextual semantic relationships compared to traditional deep learning techniques. Despite this, the accuracy of all baseline models exceeded 0.7, indicating their effectiveness in detecting consumer fraud.

3.4   Empirical Study

To demonstrate the dynamic fraud detection capabilities of CFD-BERT, we crawled negative reviews from Meituan and Dianping in Haikou over the past three months, yielding a total of 9,253 reviews. The detection results as presented in Table 3. It included 5,944 reviews exhibiting consumer fraud, with the sale of unqualified goods being the most prevalent type, followed by false advertising. Invoice issues occurred less frequently than other types of fraud. Thus, CFD-BERT can effectively detect consumer fraud and between different types of fraudulent behavior. Our method can play a crucial role in maintaining consumer confidence in the marketplace.

Table 3. Evaluation results on the consumer reviews (Best results are show in bold).

Category	Comment (%)
Unqualified goods	2364 (40%)
Insufficient portion	972 (16%)
False advertising	1335 (22%)
Payment issues	1016 (17%)
Invoice issues	257 (4%)
Total	5944

4   Conclusion

This paper presents a approach CFD-BERT, which combines regulations and extracted topics to identify fraud categories and employs a novel BTM+BERT model to extract fine-grained fraudulent features. Through experimental evaluation and empirical study, the results confirming the effectiveness of our proposed approach. Our approach has the potential to help regulatory authorities better understand consumer concerns, verify fraudulent merchants, and take follow-up fraud-solving activities. We plan to further explore the impact of other features on consumer fraud detection models.

Acknowledgements. This work was supported by the Hainan Provincial Key Research and Development Program (No. ZDYF2022GXJS230, No. ZDYF2020008), the State Key Laboratory of Marine Resources Utilization in the South China Sea, and the Hainan Provincial Key Laboratory of Big Data and Intelligent Services.

References

1. Alzate, M., Arce-Urriza, M., Cebollada, J.: Mining the text of online consumer reviews to analyze brand image and brand positioning. J. Retail. Consum. Serv. **67**, 102989 (2022)
2. Bin Sulaiman, R., Schetinin, V., Sant, P.: Review of machine learning approach on credit card fraud detection. Hum.-Centric Intell. Syst. **2**(1–2), 55–68 (2022)
3. Government: Free trade port anti-consumer fraud regulations. [EB/OL]. https://www.hainan.gov.cn/hainan/dfxfg/202110/819d62e42a624a659ab0e7b37e05aa9e.shtml. Accessed 10 Jan 2023
4. Government: What is consumer fraud. [EB/OL]. https://www.gov.cn/ztzl/2009315/content_1248977.htm. Accessed 17 Jan 2023
5. Knuth, T., Ahrholdt, D.C.: Consumer fraud in online shopping: detecting risk indicators through data mining. Int. J. Electron. Commer. **26**(3), 388–411 (2022)
6. Lai, S., Wu, J., Ma, Z., Ye, C.: Btextcan: consumer fraud detection via group perception. Inf. Process. Manag. **60**(3), 103307 (2023)
7. Soldner, F., Kleinberg, B., Johnson, S.: Trends in online consumer fraud: a data science perspective. In: A Fresh Look at Fraud, pp. 167–191. Routledge (2022)

Deep Multi-interaction Hidden Interest Evolution Network for Click-Through Rate Prediction

Zhongxing Zhang, Qingbo Hao, Yingyuan Xiao[✉], and Wenguang Zheng

Tianjin University of Technology, Tianjin 300384, China
yyxiao@tjut.edu.cn

Abstract. Click-Through Rate (CTR) prediction plays a crucial role in the field of recommendation systems. Some previous works treat the user's historical behavior as a sequence to uncover the hidden interests behind it. However, these works often ignore the dependencies and dynamic interests between different user behaviors evolving over time, as well as hidden information by user representation. To solve the above problems, we propose Deep Multi-Interaction Hidden Interest Evolution Network (MIHIEN). Specifically, we first design Hidden Interest Extraction Layer (HIE) to initially mine the hidden interests of users evolving over time from it, which can better reflect the user representation. The deeper interests of users are then explored in two types of interactions in the Item-to-Item Sub-network (IISN) and the User-to-Item Sub-network (UISN), respectively. The experimental results show that our proposed MIHIEN model outperforms other previous mainstream models.

Keywords: CTR prediction · Dynamic Interests · Self-attention · Two Types of Interactions

1 Introduction

In e-commerce scenarios, the user's intent is not explicitly a single point. Therefore, it is more efficient to obtain user interest from user behavior. DIN [13] and DIEN [12] were early to focus on the historical behavior of users. However, these models only consider item-to-item interactions between historical user behavior sequences and target items. DMR [5] changes this by introducing a user-item correlation, and directly measures the user's interest preference for the target user. Nevertheless, we observe that these works often ignore the dependencies among user behaviors and the dynamic user interests evolving over time when mining the hidden interests behind users and user representations in user behavior sequences, and suffer from the problem of biased user interest mining due to insufficient mining of the hidden information of user representation.

To solve the above problems, we propose Deep Multi-Interaction Hidden Interest Evolution Network (MIHIEN). HIE treats the user's historical behavior as a sequence to uncover the hidden potential interest behind user behavior.

C. Strauss et al. (Eds.): DEXA 2023, LNCS 14147, pp. 457–462, 2023.
https://doi.org/10.1007/978-3-031-39821-6_39

IISN uses an attention mechanism to extract multiple target interests of the user. UISN introduces the correlation between users and items, where the user representation is extracted from the potential interest behavior of the users, which better reflects the evolution of user interests over time.

The main contributions are as follows: We design the MIHIEN, which introduces correlations between users and target items that evolve over time, adequately exploiting the hidden information of user representation. The HIE module is designed, which solves the problem of inadequate information mining under the evolution over time, thus better extracting the dynamic hidden interests behind user behavior sequences.

2   Related Work

In recent years, inspired by the great success of deep neural networks in various other fields [7,8], researchers have used DNN models to characterize user behavior for CTR prediction [3,4,6,11]. DHAN [10] introduces a hierarchical attention network, which introduces a multidimensional hierarchy on the first attentional level focusing on a single item. DMIN [9] further extracts multiple potential interests of the user by introducing multi-head self-attention mechanism. CAN [2] uses a new method to mine explicit feature interactions.

3   The Proposed Method

3.1   Embedding Layer

In this paper, we use four main groups of features: *User Profile, User Historical Behavior, Context* and *Target Item*. Each feature can be encoded as a high-dimensional one-hot vector [1]. These features are usually very sparse and need to be transformed into low-dimensional dense features by embedding layers. For example, the embedding matrix of the *item id* can be expressed as $E = [v_1, v_2, ..., v_k] \in \mathbb{R}^{K \times d_e}$, where K is the total number of items, d_e is the embedding size and $d_e \ll K$. With the embedding layer, *User Profile, User Historical Behavior, Context* and *Target Item* can be represented as x_p, x_b, x_c, x_i, respectively. In particular, *User Historical Behavior* consists of multiple items and can be represented as $x_b = \{e_1, e_2, ..., e_T\} \in \mathbb{R}^{T \times d_{model}}$, where T is the number of *User Historical Behavior* and d_{model} is the dimension of item embedding e_t. $p_t \in \mathbb{R}^{d_p}$ is the position code of the t-th item and d_p is the dimension of p_t.

3.2   Hidden Interest Extraction Layer

Having more hidden interest representations facilitates enriching user behavior and reflecting user interest preferences.

To better capture the dependencies with larger intervals in the temporal space, we use GRU to model the dependencies between behaviors. To be specific, the formula for GRU is as follows: $u_t = \sigma \left(W^u i_t + U^u h_{t-1} + b^u \right)$,

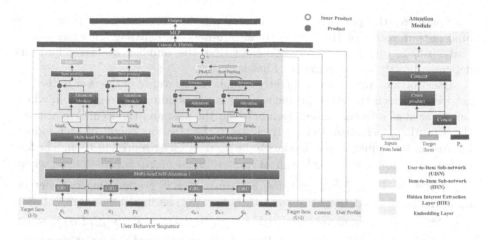

Fig. 1. The architecture of the proposed MIHIEN and Attention Module

$r_t = \sigma(W^r i_t + U^r h_{t-1} + b^r), \tilde{h}_t = tanh(W^h i_t + r_t \circ U^h h_{t-1} + b^h), h_t = (1 - u_t) \circ h_{t-1} + u_t \circ \tilde{h}_t$, where σ is the sigmoid activation function, $\circ$ is element-wise product, $W^u, W^r, W^h \in \mathbb{R}^{n_H \times d_{model}}$, $U^z, U^r, U^h \in n_H \times n_H$, n_H is the hidden size, and d_{model} is the dimension of item embedding. i_t is the input of GRU, $i_t = x_b[t]$ represents the t-th behavior of the user, h_t is the t-th hidden states. H_b is the set of hidden states h_t. This method enables the user's changing interest to be effectively captured over time.

As shown in the HIE of Fig. 1, we further pass the hidden state of the user's historical behavior sequence from the GRU into the multi-head self-attention structure. To be specific, the output of the $head_h$ is calculated as follows: $head_h = Attention(H_b W_h^Q, H_b W_h^K, H_b W_h^V) = Softmax\left(\frac{H_b W_h^Q \cdot (H_b W_h^K)^T}{\sqrt{d_h}} \cdot H_b W_h^V\right)$,

where $W_h^Q, W_h^K, W_h^V \in \mathbb{R}^{d_{model} \times d_h}$ are projection matrices of the h-th head for query, key and value respectively. Thus the latent item representation of the t-th subspace is expressed by $head_h$.

The different heads are then concatenated to obtain a completely new representation of the project, which can be defined as follows: $Z = MultiHead(H_b) = Concat(head_1, head_2, ..., head_{H_N})W^O$, where H_N is the number of heads, $W^O \in \mathbb{R}^{d_{model} \times d_{model}}$ is a linear matrix.

3.3 Item-to-Item Sub-network and User-to-Item Sub-network

In previous research, we found that user interests are not set in stone and that they change over time. To mine the weight of the correlation between each output head and the target item, we use the Attention Module, which also incorporates a position embedding containing position information. Thus the user's e-th interest can be formulated as: $int_e = \sum_{j=1}^{T} \alpha(I_{je}, x_i, p_j) I_{je} = \sum_{j=1}^{T} w_j I_{je}$, where $I_{je} \in \mathbb{R}^{d_{model}}$ represents the e-th head's vector of the j-th item. α represents the Attention Module, which represents the relevance of int_e to the target item.

Therefore, the potential target interest of the user can be obtained. The final output vector of the Item-to-Item Sub-network can now be expressed as $\{int_1, int_2, ..., int_N\} \in \mathbb{R}^{N \times d_{model}}$. Note that the number of user's interests depends directly on the number of heads N.

User-to-Item Sub-network directly models the relevance of the user to the target item through the inner product of the corresponding representation, which can be seen as a feature interaction between the user and the item. We also use the attention mechanism to adaptively learn the position weight of each hidden interest in the sequence of behaviors to be used as a user representation. The formulas are as follows: $a_t = z^T \tanh(W_p p_t + W_e I_{je} + b), \alpha_t = \frac{\exp(a_t)}{\sum_{i=1}^{T} \exp(a_i)}$, where $W_p \in \mathbb{R}^{d_h \times d_p}$, $W_e \in \mathbb{R}^{d_h \times d_{model}}$, $b \in \mathbb{R}^{d_h}$ and $z \in \mathbb{R}^{d_h}$ are learning parameters, α_t is the normalized weight of the t-th behavior. The final user representation $u \in \mathbb{R}^{d_e}$ can be formulated as: $u = V\left(\sum_{t=1}^{T}(\alpha_t I_{je})\right) = V\left(\sum_{t=1}^{T}(h_t)\right)$, where the function $V(\cdot)$ denotes the non-linear transformation of the input dimension d_{model} and the output dimension d_e, h_t is the weighted feature vector of t-th behavior. Using the user representation u and the target item representation v', we apply the inner product operation to represent the user-to-item relevance: $r = u^T v'$.

3.4 MLP and Loss Function

The results with other information are connected and passed together into the MLP for final prediction. Since the click-through prediction task is a binary classification task, the loss function is chosen to be a cross-entropy loss, usually defined as: $L_{target} = -\frac{1}{N}\left(\sum_{i=1}^{N} y \log p(x) + (1-y)\log(1-p(x))\right)$, where $x = (x_p, x_b, x_c, x_i) \in B$, B is the training set of size N. $y \in \{0, 1\}$ indicates whether the user will click on the target item. $p(x)$ is the predicted output of our network.

In addition, in the Hidden Interest Extraction Layer we use auxiliary losses, so that the global negative function is: $L = L_{target} + \beta * L_{aux}$, where β is the hyper-parameter used to balance the two subtasks.

4 Experiments

4.1 Datasets and Experimental Setup

We use the Amazon[1] Dataset, which contains product reviews and metadata from Amazon. There are 24 categories in this dataset, and we chose Electronics, CDs_and_Vinyl, Movies_and_TV, Kindle_Store, and Beauty as the datasets in our experiments.

In this experiment, we use Adam to optimize the training procedure, the dimension of d_{model} is set as 36, the embedding size of position encoding is set as 2, the dimension of user profile embedding is 18. We set the maximum value of

---

[1] http://snap.stanford.edu/data/amazon/productGraph/.

Table 1. The AUC performance and the results of ablation study on real-world datasets. All experimental results are the results of rerunning on our equipment.

Model	Ele	CDs	Movies	Kindle	Beauty
Base	0.7189	0.7905	0.7084	0.7467	0.7013
WDL	0.7345	0.8016	0.7065	0.7561	0.7086
PNN	0.7376	0.8101	0.7107	0.7602	0.7144
DIN	0.7394	0.8248	0.7136	0.7766	0.7157
DIEN	0.7509	0.8315	0.7269	0.7864	0.7169
DIEN-NO	0.7405	0.8251	0.7135	0.7726	0.7151
DHAN	0.7417	0.8361	0.7247	0.7748	0.7159
DMIN	0.7540	0.8441	0.7280	0.7944	0.7172
CAN	0.7528	0.8316	0.7240	0.7825	0.7185
MIHIEN	**0.7589**	**0.8583**	**0.7379**	**0.8036**	**0.7211**
NO HIE[a]	0.7568	0.8539	0.7341	0.7852	0.7171
NO IISN[b]	0.7540	0.8579	0.7374	0.7775	0.7144
NO UISN[c]	0.7574	0.8502	0.7361	0.7809	0.7150

[a] MIHIEN without Hidden Interest Extraction Layer
[b] MIHIEN without Item-to-Item Sub-network
[c] MIHIEN without User-to-Item Sub-network

user history behavior length to 20. All compared models have the same settings, their batch size is set to 128 and their learning rate is 0.001. The performance of the model is evaluated by using AUC.

4.2 Baseline and Result

To comprehensively evaluate our method, this paper compares the proposed method with the model baselines: **BaseModel**, **WDL** [3], **PNN** [6], **DIN** [13], **DIEN** [12], **DIEN-NO**, **DHAN** [10], **DMIN** [9], **CAN** [2].

We will compare the proposed MIHIEN model with the models in baseline methods and show the result in Table 1. From the experimental results, MIHIEN has achieved the best performance compared to the state-of-the-art methods. DIN represents the user's interest in the target product, and its results are mostly better than those of WDL and PNN. Based on DIN, DIEN further uses GRU to capture the dynamically evolving interest of users and obtains a better representation of user interest than DIN. DIEN-NO is based on DIEN and removes the auxiliary loss, which is not as effective as DIEN. DHAN uses multiple Attention Units to capture the user's hierarchical interest, which also shows the importance of capturing the evolutionary interest of users. DMIN manifests the efficacy of modeling and tracking user's multiple interests. CAN mine the information between user and item through multiple MLPs. Our proposed MIHIEN is 0.65%, 1.68%, 1.36%, 1.16% and 0.36% higher than the best AUC results of the above models in all datasets.

We also perform an ablation study on MIHIEN to verify the effectiveness of our proposed module. From the results, we can see that the AUC decreases on all five datasets regardless of which module of HIE, IISN, or USIN is removed. Therefore, we believe that HIE can filter out a large amount of useless information from users' historical behaviors, and the extracted users' hidden interests better reflect the users' own interest preferences. The combination of the two sub-networks performs better than alone, which illustrates the effectiveness of two different forms of user-item interaction, and the two forms of user-item interaction are complementary, rather than redundant.

5   Conclusion

We propose the MIHIEN, which focuses on mining the hidden interests behind user time-varying behavior sequences and mining user preferences with two forms of interactions for click-through rate prediction tasks. The results demonstrate that MIHIEN is useful for CTR prediction.

Acknowledgements. This work is supported by "Tianjin Project + Team" Key Training Project under Grant No. XC202022.

References

1. Bengio, Y., Ducharme, R., Vincent, P., Janvin, C.: A neural probabilistic language model, pp. 1137–1155 (2003)
2. Bian, W., Wu, K., Ren, L., Pi, Q., Zhang, et al.: CAN: feature co-action network for click-through rate prediction, pp. 57–65 (2022)
3. Cheng, H.T., Koc, L., Harmsen, J., Shaked, et al.: Wide & deep learning for recommender systems, pp. 7–10 (2016)
4. Han, Y., Xiao, Y., Wang, H., Zheng, W., Zhu, K.: DFILAN: domain-based feature interactions learning via attention networks for CTR prediction. In: Jensen, C.S., et al. (eds.) DASFAA 2021. LNCS, vol. 12682, pp. 500–515. Springer, Cham (2021). https://doi.org/10.1007/978-3-030-73197-7_33
5. Lyu, Z., Dong, Y., Huo, C., Ren, W.: Deep match to rank model for personalized click-through rate prediction, pp. 156–163 (2020)
6. Qu, Y., et al.: Product-based neural networks for user response prediction, pp. 1149–1154. IEEE (2016)
7. Tang, Y., et al.: An image patch is a wave: phase-aware vision MLP, pp. 10935–10944 (2022)
8. Wei, X., et al.: Learning to generalize to more: continuous semantic augmentation for neural machine translation (2022)
9. Xiao, Z., Yang, L., Jiang, W., Wei, Y., Hu, Y., Wang, H.: Deep multi-interest network for click-through rate prediction, pp. 2265–2268 (2020)
10. Xu, W., He, H., Tan, M., Li, Y., Lang, J., Guo, D.: Deep interest with hierarchical attention network for click-through rate prediction, pp. 1905–1908 (2020)
11. Yuan, Z., Xiao, Y., Yang, P., Hao, Q., Wang, H.: Deep user and item inter-matching network for CTR prediction. In: Wang, X., et al. (eds.) DASFAA 2023. LNCS, vol. 13944, pp. 195–204. Springer, Cham (2023). https://doi.org/10.1007/978-3-031-30672-3_13
12. Zhou, G., et al.: Deep interest evolution network for click-through rate prediction, pp. 5941–5948 (2019)
13. Zhou, G., et al.: Deep interest network for click-through rate prediction, pp. 1059–1068 (2018)

Temporal Semantic Attention Network for Aspect-Based Sentiment Analysis

Bin Yang[1], Xinyang Tong[2], Ying Xing[2(✉)], Qi Shen[2], Huiying Zhao[1], and Zhipu Xie[1]

[1] China Unicom Research Institute, Beijing, China
{zhaohy699,xiezp12}@chinaunicom.cn
[2] School of Artificial Intelligence, Beijing University of Posts and Telecommunications, Beijing, China
{tongxinyang,xingying,s_q}@bupt.edu.cn

Abstract. Aspect-based sentiment analysis aims to predict the polarity of sentiment towards a specific aspect in the context. In this paper, we propose the Temporal Semantic Attention Network (TSAN) model for ABSA tasks, which comprising a Global Semantic Feature Network for feature extraction and an Interact Dual Attention module to capture the dependencies of text-target interaction. Experiments on four ABSA benchmark datasets validates the effectiveness of our modules in extracting aspect-level sentiment features.

Keywords: Aspect-based sentiment analysis · attention · feature extraction

1 Introduction

Sentiment classification based on comment data has become a hot topic in natural language processing. As technology advances, this field has been refined to Aspect-Based Sentiment Analysis (ABSA), which aims to distinguish sentiment polarities of different aspect words when they appear simultaneously.

In recent years, more and more researchers have used neural networks to solve sentiment classification problems. Among them, the Long Short-Term Memory (LSTM) network has become a typical method in the ABSA field. To better capture important parts of a sentence, Tang et al. [1] proposed TD-LSTM and TC-LSTM, which further considered the position and content relationship between aspect and context information based on the LSTM approach.

In addition, attention mechanisms have been shown to be effective in enhancing the performance of neural network models in sentiment analysis, which are used to capture the interaction between the sentiment aspect and context information. For example, ATAE-LSTM was designed to focus on different parts of a sentence depending on the input aspect [2]. Ma et al. [3] designed an attention-based model called IAN, which focuses attention on the parts of the text that contribute to the semantic meaning of the sentiment word. Huang et al. [4] proposed an attention-over-attention (AOA) mechanism that generates mutual attention between sentimental aspects and their contextual information. Tang et al. [5] introduced MemNet to capture the importance of each context word of an aspect with multiple computational layers.

By using LSTM networks and attention layers, probabilities of different sentiment classifications are generated, resulting in different sentiment polarity outputs. However, LSTM models encounter challenges in capturing semantic information as sentence length increases, and are having difficulty considering semantic relationships between non-adjacent text terms. Moreover, while attention models often emphasize the significance of textual words that contribute to aspects, fewer models address the semantic reliance of aspects on context. For example, "The food is too hard to eat" may indicate eating difficulty, while "The computer case is hard" may indicate high quality. Both sentences use the word "hard", but express completely different emotions, indicating that the meaning of adjectives for each aspect may differ.

To address these limitations, we propose the Temporal Semantic Attention Network (TSAN) model. We apply convolutional structures and extract contextual features from both sequence and spatial aspects. Furthermore, in order to capture the dependency relationship between aspect words and text words, we developed an Interact Dual Attention mechanism that takes into account the semantic influences of text and aspect interaction.

Our contributions are as follows:

- We proposed a global feature extraction structure that extracts spatial and sequence semantic features, which enhanced text representation capabilities.
- We design the Interact Dual Attention (IDA) structure to extract semantic dependencies by fusing aspect-to-text attention and text-to-aspect attention.

Experiments on four benchmark datasets show that TSAN effectively alleviates both LSTM and attention limitations, and performs better than other LSTM-based attention models in ABSA tasks (Fig. 1).

2   Methodology

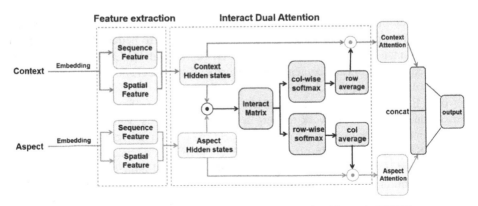

Fig. 1. Overview of the Temporal Semantic Attention Network (TSAN)

2.1 Embedding

We map each word in the sentence to a set of word vectors, using a pre-trained word embedding matrix. In our experiments, we use pre-trained GloVe and BERT word embedding methods to initialize word embeddings.

2.2 Global Semantic Feature Network

After obtaining word vectors, we extract further text features through two parallel layers. The initial layer combines LSTM and temporal convolutional networks to achieve spatial semantic extraction, while the subsequent layer employs a Bi-LSTM network for sequence semantic comprehension. Finally, we combine semantic and spatial representation from two layers by averaging hidden feature outputs.

Sequence Feature Extraction. We use Bi-LSTM to capture contextual sequence information for text and aspect. Word vectors of aspects and context enter the network to generate hidden states respectively.

Spatial Feature Extraction. Phrases can be crucial in depicting the meaning of a sentence, which contains a combination of words. Therefore, to extract semantic information from spatial order, we designed a spatial semantic feature extraction structure. It applies methods such as causal convolution, dilated convolution, and residual linking, which fuse feature information extracted from the temporal convolution network and Bi-LSTM. To augment a more comprehensive representation of semantic features, the output of the convolution module is further regenerated by a LSTM network.

2.3 Interact Dual Attention

We further apply attention mechanisms to extract the dependency relationships between text words and aspect words.

We first calculate their interaction matrix I, which is the dot product between extracted hidden context and aspect word vectors. The dot product result I_{ij} represents the semantic similarity between the i-th word in the context, while the j-th word in the aspect word reflects the semantic similarity between the context word and the aspect word. In addition, by performing softmax by row and column, we obtain the attention score correlation matrix α from aspect to context, as well as the correlation matrix β from context to aspect. We then average α by row to obtain the attention score vector at the text level, and average β by column to obtain the attention score vector at the aspect word level.

Finally, according to the attention mechanism, we apply the obtained attention score matrices to update the semantic feature vectors by using dot product. By considering the semantic relevance between aspect and context, we learn the important weights for each context word contributing to aspect, as well as the weights for each aspect word contributing to context.

2.4 Output

We concatenate two hidden vectors from context and aspect as the final classification feature. Through a fully connected layer and a softmax normalization layer, feature vectors are mapped to the aspect space of three classes, which are positive, neutral, and negative.

3 Experiment

3.1 Experiment Settings

Datasets. In this paper, we experiment on four datasets: restaurant14, restaurant15, restaurant16, and twitter [6–9], which are standard datasets for the ABSA task. Within each dataset, the content consists of three parts: comment context, aspect words, and sentiment polarity.

Experimental Parameter Settings. We use 300-dimensional GloVe word vectors to initialize word embedding for all baseline models. Furthermore, we apply the BERT pretrained word embedding model, to better our model's performance. The L2 regularization coefficient is set at 0.0001, and the dropout keep rate is set at 0.2. If the training loss does not drop after every three epochs, we decrease the learning rate by half.

3.2 Performance Comparison

Table 1. Model comparison results (%). Average accuracy and macro-F1 score with random initialization. The best two results with each dataset are in bold.

Model	Res14		Res15		Res16		Twitter	
	acc	f1	acc	f1	acc	f1	acc	f1
LSTM	77.05	64.39	75.09	52.11	85.88	55.96	68.93	66.77
TC-LSTM	72.43	62.92	73.99	52.09	83.44	53.45	70.81	68.90
TD-LSTM	78.48	66.94	77.68	**61.95**	85.88	61.04	69.94	67.93
ATAE-LSTM	77.77	64.91	78.04	57.15	83.60	52.05	68.35	66.71
AOA	79.97	70.42	78.17	57.02	87.5	66.21	72.3	70.2
MemNet	79.55	71.05	75.83	55.71	85.06	56.21	70.23	68.60
IAN	79.26	70.09	78.54	52.65	84.74	55.21	72.5	70.81
TSAN-Glove	**80.01**	**71.24**	78.78	52.98	**87.66**	66.91	**72.83**	**70.84**
TSAN-BERT	**81.96**	**71.48**	**81.55**	**64.11**	**89.61**	**72.02**	**73.84**	**72.11**

The table above shows the main experimental results of the TSAN model training. In the experiment, we took the average of three random seed results and recorded the experimental results, using Accuracy and Macro-Averaged F1 as evaluation criteria (Table 1).

The results showed that the TSAN model achieved better results on all datasets. As shown in the table, when using the same GloVe word vector for embedding, the TSAN model outperforms all other models across all datasets.

Compared to the model that relied solely on LSTM to capture text information, our performance improved by 2.51%, 3.69%, 1.78%, and 3.9% across four datasets, respectively. In addition, the IDA structure allows a better capture of semantic dependencies between text and aspect words, outperforming attention-oriented models such as ATAE-LSTM, MemNet, and IAN. When using BERT for word embedding instead of Glove, which captures more contextual semantic information, it further improves the performance of the TSAN model. Compared to all datasets, the test effect on the Twitter dataset is slightly worse than that on other datasets, probably because there are more syntax errors and non-standard expressions in the comment text on the social platform.

Based on the above analysis, it can be seen that the TSAN model can achieve better results in sentiment analysis tasks by adding text feature extraction methods and capturing the dependency relationship between text and aspect words.

3.3 Case Study

To gain a better understanding of how the TSAN model works, we use a heat map to visualize the influence of words on different aspects depending on attention scores (Table 2).

Table 2. Examples of final attention weights for sentences. The color indicates the importance degree of the weight in attention vector.

Aspect	Sentence	Predict / Answer
food	The food was extremely tasty.	+1/+1
food	Good food but the service was dreadful!	+1/+1
service	Good food but the service was dreadful!	-1/-1

For example, regarding the sentence "The food was extremely tasty." Our model successfully identified the words "extremely" and "tasty" as having the most significant impact on the emotional attitude of the aspect word "food". The second example "Good food but the service was dreadful!" contains two aspects "food" and "service" in one sentence. For the aspect "food", the weight of the word "good" in the text is relatively high, resulting in positive emotional polarity. For the aspect word "service", the word "dreadful" gets a higher weight, indicating that the attention mechanism has correctly identified the aspect and the description of the aspect.

4   Conclusion

In this paper, we propose the TSAN model, which captures both spatial and sequential features of text sequences, and includes an IDA attention module to extract semantic dependencies between aspect and context. Our experimental results on four ABSA datasets indicate the model's ability to extract aspect and context features. Furthermore, the case study demonstrates the model's ability to focus on critical words for predicting aspect sentiment polarity. In the future, we plan to improve feature extraction in information extraction by incorporating position information to further focus on the relative position relationship between aspect words and text. In addition, we will investigate the effectiveness of dependency information extraction between text and aspect words, with a particular focus on the contribution of text words to aspect words in polysemy cases.

References

1. Tang, D., Qin, B., Feng, X., Liu, T.: Effective LSTMS for target-dependent sentiment classification (2016)
2. Wang, Y., Huang, M., Zhu, X., Zhao, L.: Attention-based LSTM for aspect-level sentiment classification. In: Proceedings of the 2016 Conference on Empirical Methods in Natural Language Processing, pp. 606–615 (2016)
3. Ma, D., Li, S., Zhang, X., Wang, H.: Interactive attention networks for aspect-level sentiment classification. In: Proceedings of the 26th International Joint Conference on Artificial Intelligence (IJCAI 2017), pp. 4068–4074. AAAI Press (2017)
4. Cui, Y., Chen, Z., Wei, S., Wang, S., Liu, T., Hu, G.: Attention-over-attention neural networks for reading comprehension (2016)
5. Tang, D., Qin, B., Liu, T.: Aspect level sentiment classification with deep memory network. arXiv preprint arXiv:1605.08900 (2016)
6. Dong, L., Wei, F., Tan, C., Tang, D., Zhou, M., Xu, K.: Adaptive recursive neural network for target-dependent Twitter sentiment classification. In: Proceedings 52nd Annual Meeting Association Computational Linguistics, vol. 2, pp. 49–54 (2014)
7. Manandhar, S.: Semeval-2014 task 4: aspect based sentiment analysis. In: Proceedings of the 8th International Workshop on Semantic Evaluation (SemEval 2014) (2014)
8. Pontiki, M., Galanis, D., Papageorgiou, H., Manandhar, S., Androutsopoulos, I.: Semeval-2015 task 12: aspect based sentiment analysis. In: Proceedings of the 9th International Workshop on Semantic Evaluation (SemEval 2015), pp. 486–495 (2015)
9. Pontiki, M., et al.: Semeval-2016 task 5: aspect based sentiment analysis. In: International Workshop on Semantic Evaluation, pp. 19–30 (2016)

Celestial Machine Learning

From Data to Mars and Beyond with AI Feynman

Zi-Yu Khoo[1]([✉]), Abel Yang[1], Jonathan Sze Choong Low[2],
and Stéphane Bressan[1,3]

[1] National University of Singapore, 21 Lower Kent Ridge Rd,
Singapore 119077, Singapore
khoozy@comp.nus.edu.sg, {phyyja,steph}@nus.edu.sg
[2] Singapore Institute of Manufacturing Technology, Agency for Science,
Technology and Research (A*STAR), Singapore 138634, Singapore
sclow@simtech.a-star.edu.sg
[3] CNRS@CREATE LTD, 1 Create Way, Singapore 138602, Singapore

Abstract. Can a machine or algorithm discover or learn Kepler's first law from astronomical sightings alone? We emulate Johannes Kepler's discovery of the equation of the orbit of Mars with the Rudolphine tables using AI Feynman, a physics-inspired tool for symbolic regression.

1 Introduction

In 2020, Silviu-Marian Udrescu and Max Tegmark introduced *AI Feynman* [17], a symbolic regression algorithm that could rediscover one hundred equations from the *Feynman Lectures on Physics* [3]. Although the authors motivated their work with Johannes Kepler's discovery of the orbital equation of Mars, to our knowledge, they did not attempt to rediscover the same equation with AI Feynman. We show that AI Feynman can emulate Kepler's discovery of the orbital equation of Mars from the data published in the Rudolphine tables.

Around 1514, Nicolaus Copernicus described a heliocentric model of the orbit of terrestrial planets, with the orbit of Mars described by its deferent and two epicycles [13]. Copernicus may have used sightings such as those in the *Alfonsine Tables*. The model was later published in 1543 in *De revolutionibus orbium coelestium*. Less than a decade later, Erasmus Reinhold calculated and published a new set of astronomical tables, the *Prutenic Tables*, based on Copernicus' heliocentric model. It was Tycho Brahe, however, who recorded sufficiently frequent and rigorous sightings to allow the discovery of the *Laws of Planetary Motion*. Kepler, Brahe's assistant and successor as imperial mathematician to Rudolf II, transposed Brahe's sightings of Mars. In 1627, Kepler published the *Rudolphine Tables*, comprising a set of 180 heliocentric positions of the planet. Kepler could have described the motions of Mars as a generalised ovoid or added additional epicycles to the Copernican model, but instead described the orbit of the planet as elliptical in *Astronomia nova* in 1609 [9]. Kepler leveraged a combination of mathematics and physics knowledge and hypotheses to propose the most accurate and parsimonious model he could devise. We use AI Feynman to illustrate

C. Strauss et al. (Eds.): DEXA 2023, LNCS 14147, pp. 469–474, 2023.
https://doi.org/10.1007/978-3-031-39821-6_41

how an algorithm can combine optimisation and heuristics to emulate Kepler's discovery of the elliptical orbital equation of Mars from the Rudolphine tables.

2   Background and Related Work

Finding an equation from sample data, for instance, an equation describing the orbit of Mars from sightings of the planet and the Sun, is a combinatorial challenge [17]. To circumvent this, one may use universal function approximators such as multilayer perceptron neural networks [6].

Alternatively, symbolic regressions search for a parsimonious and elegant form of the unknown equation. There are three main classes of symbolic regression methods [11]: regression-based, expression tree-based and physics- or mathematics-inspired. We use AI Feynman, a physics-inspired algorithm [17].

Regression-based symbolic regression methods [11], given solutions to the unknown equation, find the coefficients of a fixed basis that minimise the prediction error. As the basis grows, the fit improves, but the functional form of the unknown equation becomes less sparse or parsimonious. Sparse regressions promote sparsity through regularisation, as proposed by Robert Tibshirani [16] who used the l_1 norm, thus inventing the Lasso regression. Steven Brunton et al.'s Sparse Identification of Nonlinear Dynamics [1] is a state-of-the-art sparse symbolic regression approach. It leverages regularisation and identifies a system's equations of motion using a sparse regression over a chosen basis. However, committing to a basis limits the applicability of regression-based methods.

Expression tree-based symbolic regression methods based on genetic programming [11] can instead discover the form and coefficients of the unknown equation. The seminal work by John Koza [10] represents each approximation of an unknown equation as a genetic programme with a tree-like data structure, with traits (or nodes in the tree) representing functions or operations and variables representing real numbers. The fitness of each genetic programme is its prediction error. Fitter genetic programmes undergo a set of transition rules comprising selection, crossover and mutation to find the optimal equation form iteratively. Genetic programmes may greedily mimic nuances of the unknown equation [15], limiting generalisability. David Goldberg [5] used Pareto optimisation to balance the objectives of fit and parsimony in symbolic regression. State-of-the-art symbolic regression using genetic programming includes Eureqa by Michael Schmidt and Hod Lipson [14] and PySR by Miles Cranmer [2]. However, if an expression tree-based method finds a reasonably accurate equation with the wrong functional form, it risks getting stuck at a local optimum [17].

Physics-inspired symbolic regression methods leverage properties of the unknown equation like symmetry and separability [17]. Udrescu and Tegmark [17], in AI Feynman, use a neural network to test for such properties and recursively break the search for the unknown equation into that of simpler equations [17]. Each equation is then regressed with a basis-set of nonlinear functions. This guarantees that more accurate approximations of an equation are symbolically closer to the truth [17]. AI Feynman outputs a sequence of increasingly complex equations with progressively better accuracy along a Pareto frontier, leveraging the work of Goldberg [5] and Guido Smits [15]. We use AI Feynman to rediscover the orbital equation of Mars.

3   Methodology

We use AI Feynman to rediscover the orbital equation of Mars from the Rudolphine Tables. We devise and compare four algorithms representing different combinations of observational and inductive biases that inform AI Feynman of the periodicity of Mars' orbit, and the trigonometric nature of the data.

Observational biases are introduced directly through data that embody the underlying physics. The predictions of a machine learning model trained on such data will reflect the physical structure of the data. Inductive biases correspond to prior assumptions incorporated by tailored interventions to a machine learning model architecture, so predictions will satisfy a set of given physical laws [7].

The Rudolphine Tables already embed the assumptions of heliocentricity and planarity of Mars' orbit. All algorithms, in this sense, are already observationally biased as the data in the Rudolphine Tables is a transposition of the terrestrial sightings of Brahe to an angle and a distance from the Sun. We highlight further biases embedded in the four algorithms.

The first algorithm directly applies AI Feynman with no further biases. The second algorithm is observationally biased. AI Feynman is informed of the periodicity of Mars' orbit and the trigonometric nature of the data by replacing angular values with their sine and cosine. The third algorithm is inductively biased. AI Feynman is informed of the periodicity of Mars' orbit and the trigonometric nature of the data by the restriction of the search space. This inference stems from the knowledge that exponential and logarithmic functions only transform dimensionless quantities, therefore cannot transform data representing physical quantities. The inductive bias limits the search space to trigonometric, polynomial and radical functions. The fourth algorithm combines both observational and inductive biases. AI Feynman is informed of the periodicity of Mars' orbit by replacing angular values and limiting the search space of functions.

4   Performance Evaluation

We conduct four experiments corresponding to the application of the four algorithms, respectively, to the data in the table *Tabula Aequationum MARTIS* of the Rudolphine Tables [8]. The *Tabula Aequationum MARTIS*, or Table of Corrections for Mars, contains two columns of primary data, *Anomalia coaequata* and *Intervallu*, and two columns of derived data *Anomalia eccentri* and *Intercolumnium*. The four columns represent the coequated or true anomaly, the distance between the Sun and Mars, the eccentric anomaly and an interpolating factor respectively. The table, see Fig. 1, was digitised for this experiment.

We apply AI Feynman to the primary data to recover the orbital equation of Mars. We convert *Anomalia coaequata*, an angle in degrees minutes seconds, to decimal degrees. We scale *Intervallu* by $E - 05$. The code and data are available at https://github.com/zykhoo/AI-Feynman.

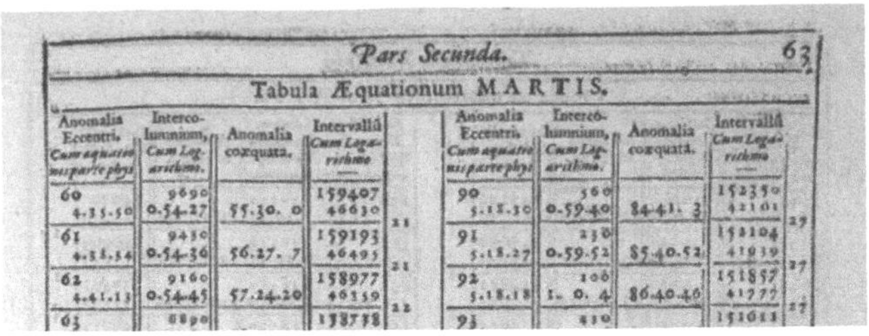

Pars Secunda. 63

Tabula Æquationum MARTIS.

Anomalia Eccentri. Cum aequatione parte phys	Intercolumnium, Cum Logarithmo.	Anomalia coaequata.	Intervallũ Cum Logarithmo		Anomalia Eccentri. Cum aequatione parte phys	Intercolumnium, Cum Logarithmo.	Anomalia coaequata.	Intervallũ Cum Logarithmo	
60 4.35.50	9690 0.54.27	55.30. 0	159407 46630	21	90 5.12.30	560 0.59.40	84.41. 3	152350 42101	27
61 4.38.54	9430 0.54.36	56.27. 7	159193 46495	21	91 5.15.27	230 0.59.52	85.40.52	152104 41939	27
62 4.41.13	9160 0.54.45	57.24.20	158977 46359	22	92 5.18.18	100 1. 0. 4	86.40.46	151857 41777	27
63	8890		158758		93	410		151611	

Fig. 1. Excerpt of the Table of Corrections for Mars from the Rudolphine Tables

We compare the equations discovered along the AI Feynman Pareto frontier with the orbital equation of Mars: $r = \frac{a}{1+\epsilon \times cos\theta}$ where r is the *Intervallu*, θ is the *Anomalia coaequata*, ϵ is the ellipse eccentricity, and a is the semi-major axis.

Results of the experiments are reported in Tables 1 and 2. We list the equations along the Pareto frontier in order of decreasing parsimony and increasing goodness of fit. We also indicate the mean description length loss [17,18] (DL) computed between each predicted and true *Intervallu* for each equation. The mean description length loss minimises the geometric mean instead of the arithmetic mean, which encourages improving already well-fit points [18].

Table 1. Results of Experiment 2, which took 1451 s seconds.

Eqn No.	Equation	MSE
(1a)	$r = \log\cos\theta + 5$	26.006
(1b)	$r = \frac{1}{7} \times \cos\theta + 1.5$	24.053
(1c)	$r = 1.5 \times \exp 0.1 \times \cos\theta$	23.512
(1d)	$\mathrm{r} = \frac{1}{\frac{2}{3} - 0.0556244812357114 \times \cos\theta}$	22.857
(1e)	$r = 1.5119670200057298 \times \exp(0.1 \times \cos\theta)$	22.457
(1f)	$r = 1.510965630582 + (\cos\theta/(\sin\theta + 6))$	21.070
(1g)	$r = 1.51366746425629 \times exp(0.0931480601429939 \times \cos\theta)$	20.762
(1h)	$\mathrm{r} = \frac{1}{0.662428796291351 - 0.0612906403839588 \times \cos\theta}$	19.781
(1i)	$\mathbf{r = (0.662428796291351 - 0.0612906403839588} \times \cos\theta)^{-1.00133872032166}$	12.211

In Experiments 1 and 3, none of the equations on the Pareto frontier match the orbital equation for Mars. We omit their detailed results in this paper. Experiments 1 and 3 had execution times of 748 and 621 s respectively. In Experiments 2 and 4, three out of nine equations on the Pareto frontier match the orbital equation for Mars. We omit results independent of θ. Relative to Experiment 1, the observational bias in Experiment 2 doubles the number of inputs to AI Feynman thereby doubling execution time. The inductive bias in Experiment 3 reduces the search space thereby reducing execution time. Experiment 4 has twice the execution time of Experiment 3 but is faster than Experiment 2.

Contrasting the Pareto fronts of Experiments 1 and 2 highlights the importance of an observational bias in guiding the search by AI Feynman. Contrasting the Pareto fronts of Experiments 2 and 4 highlight the importance of an

Table 2. Results of Experiment 4, which took 1184 s seconds.

Eqn No.	Equation	MSE
(2a)	$r = \frac{1}{7} \times \cos\theta + 1.5$	24.053
(2b)	$r = \cos\theta/(\sin\theta + 6) + 1.5$	23.617
(2c)	$r = \arccos(0.0420224035468255 - 0.142857142857143 \times \cos\theta)$	23.392
(2d)	$r = \dfrac{1}{\frac{2}{3} - 0.0566732120453772 \times \cos\theta}$	22.575
(2e)	$r = 1.511006320056 + (\cos\theta/(\sin\theta + 6))$	21.089
(2f)	$r = \tan(0.0425049090340329 \times \cos\theta + 0.986141372332807)$	20.057
(2g)	$r = \tan(0.0427569970488548 \times \cos\theta + 0.98658412694931)$	20.021
(2h)	$r = \dfrac{1}{0.662420213222504 - 0.0612917765974998 \times \cos\theta}$	19.747
(2i)	$\mathbf{r = (0.662420213222504 - 0.0612917765974998 \times \cos\theta)^{-1.00130701065063}}$	12.208

inductive bias in limiting the search space for AI Feynman. In Experiment 2, three equations utilise an exponential function applied to $\cos\theta$ (Eqs. 1c, 1e and 1g), and three utilise an inverse function applied to $\cos\theta$ (Eqs. 1d, 1h and 1i). Only the latter matches the orbital equation of Mars but both equation forms are equally prevalent. In Experiment 4, the equation form which matches the orbital equation of Mars (Eqs. 2d, 2h and 2i) is the most prevalent.

Therefore, AI Feynman, augmented with both an observational and inductive bias, can best rediscover Mars' orbital equation. The inductive bias also reduces the execution time for solving NP-hard in principle [17] symbolic regression problems. We observe Eqs. 1h, 1i, 2h, and 2i have forms that match the orbital equation of Mars and lowest mean description length loss of less than 20.

Lastly, while AI Feynman combines optimisation and heuristics in exploring the combinatorial space of candidate solutions, Kepler may instead have used thought experiments to hypothesise an elliptical orbit. Fitting the data from the Rudolphine tables to the orbital equation of Mars using non-linear least squares yields $\epsilon = 0.0926$ and $a = 1.5235$. For reference, the National Aeronautics and Space Administration reports values of 0.0934 and 1.5237, respectively [12]. We observe that Eqs. 1h and 2h suggest values of $\epsilon = 0.0925$ and $a = 1.5226$.

5   Conclusion

We have successfully shown that AI Feynman can rediscover the equation of the orbit of Mars from the Rudolphine Tables, provided it is further informed of the underlying physics of the periodicity of Mars' orbit and the trigonometric nature of the data. AI Feynman can be informed by the introduction of observational bias, inductive bias, and even more successfully, both.

We were not able to locate a copy of Brahe's original sightings. A contemporary version is, however, available from the National Aeronautics and Space Administration's Horizons system [4]. The challenge is to modify the symbolic regression algorithm to discover the change of referential from geocentric to heliocentric. We are also working on rediscovering the equations of the orbit of various celestial objects, such as Mercury, from heliocentric and geocentric data.

The discovery of laws of physics can be seen as an optimisation problem for the discovery of maximally accurate and parsimonious models, *Occam's razor*.

This discovery, by man or machine, whether directed by intuition and thought experiments or by combinatorial search and heuristics, can be effectively guided by the information of the underlying physics knowledge and hypotheses.

Acknowledgements. This research is supported by Singapore Ministry of Education, grant MOE-T2EP50120-0019, and by the National Research Foundation, Prime Minister's Office, Singapore, under its Campus for Research Excellence and Technological Enterprise (CREATE) programme as part of the programme Descartes.

References

1. Brunton, S.L., Proctor, J.L., Kutz, J.N.: Discovering governing equations from data by sparse identification of nonlinear dynamical systems. Proc. Natl. Acad. Sci. **113**(15), 3932–3937 (2016)
2. Cranmer, M.: PYSR: Fast & parallelized symbolic regression in python/Julia (2020). https://doi.org/10.5281/zenodo.4041459
3. Feynman, R.P.: The Feynman Lectures on Physics: Reading, Mass. Addison-Wesley Pub. Co., c1963–1965. (1965 c1963)
4. Giorgini, J.D., JPL Solar System Dynamics Group: NASA/JPL Horizons On-Line Ephemeris System. https://ssd.jpl.nasa.gov/horizons/
5. Goldberg, D.E.: Genetic Algorithms in Search, 1st edn. Optimization and Machine Learning. Addison-Wesley Longman Publishing Co., Inc, USA (1989)
6. Hornik, K., Stinchcombe, M., White, H.: Multilayer feedforward networks are universal approximators. Neural Netw. **2**(5), 359–366 (1989)
7. Karniadakis, G.E., Kevrekidis, I.G., Lu, L., Perdikaris, P., Wang, S., Yang, L.: Physics-informed machine learning. Nat. Rev. Phys. **3**(6), 422–440 (2021)
8. Kepler, J., Brahe, T., Eckebrecht, P.: Tabulæ Rudolphinæ, quibus astronomicæ scientiæ, temporum longinquitate collapsæ restauratio continetur (1627)
9. Kepler, J., Donahue, W.H.: New Astronomy. Cambridge University Press, Cambridge (1992)
10. Koza, J.R.: Genetic Programming: On the Programming of Computers by Means of Natural Selection. MIT Press, Cambridge, MA, USA (1992)
11. Makke, N., Chawla, S.: Interpretable scientific discovery with symbolic regression: a review (2022)
12. National Aeronautics and Space Administration: Mars fact sheet (2022). https://nssdc.gsfc.nasa.gov/planetary/factsheet/marsfact.html
13. Rosen, E.: The commentariolus of copernicus. Osiris **3**, 123–141 (1937). https://doi.org/10.1086/368473
14. Schmidt, M., Lipson, H.: Distilling free-form natural laws from experimental data. Science **324**(5923), 81–85 (2009)
15. Smits, G.F., Kotanchek, M.: Pareto-front exploitation in symbolic regression. In: O'Reilly, U.M., Yu, T., Riolo, R., Worzel, B. (eds.) Genetic Programming Theory and Practice II. Genetic Programming, vol. 8, pp. 283–299. Springer, Boston, MA (2005). https://doi.org/10.1007/0-387-23254-0_17
16. Tibshirani, R.: Regression shrinkage and selection via the lasso. J. Roy. Stat. Soc. B **58**, 267–288 (1994)
17. Udrescu, S.M., Tegmark, M.: AI Feynman: a physics-inspired method for symbolic regression. Sci. Adv. **6**(16), eaay2631 (2020)
18. Wu, T.: Intelligence, physics and information - the tradeoff between accuracy and simplicity in machine learning (2020)

Author Index

Printed in the United States
by Baker & Taylor Publisher Services